# Bionanotechnology, Microbiology and Genetic Engineering

# Bionanotechnology, Microbiology and Genetic Engineering

Edited by **David Rhodes**

**⬚SYRAWOOD**
PUBLISHING HOUSE

New York

Published by Syrawood Publishing House,
750 Third Avenue, 9th Floor,
New York, NY 10017, USA
www.syrawoodpublishinghouse.com

**Bionanotechnology, Microbiology and Genetic Engineering**
Edited by David Rhodes

© 2016 Syrawood Publishing House

International Standard Book Number: 978-1-68286-155-4 (Hardback)

# Contents

# Preface

This book brings forth some of the most innovative concepts and elucidates the unexplored aspects of bionanotechnology, microbiology, and genetic engineering. The biotechnological applications of microbiology and genetic engineering in the field of pharmaceuticals, bioenergy, food industry, agriculture, etc. have been discussed in this book. The interdisciplinary advances of these fields such as green chemistry, protein engineering, etc. have also been elucidated. It includes contributions of experts and scientists which will provide innovative insights into these fields. It is an essential guide for both research scholars and those who wish to pursue this discipline further.

The researches compiled throughout the book are authentic and of high quality, combining several disciplines and from very diverse regions from around the world. Drawing on the contributions of many researchers from diverse countries, the book's objective is to provide the readers with the latest achievements in the area of research. This book will surely be a source of knowledge to all interested and researching the field.

In the end, I would like to express my deep sense of gratitude to all the authors for meeting the set deadlines in completing and submitting their research chapters. I would also like to thank the publisher for the support offered to us throughout the course of the book. Finally, I extend my sincere thanks to my family for being a constant source of inspiration and encouragement.

**Editor**

# What else can we do to mitigate contamination of fresh produce by foodborne pathogens?

Shlomo Sela (Saldinger)[1]* and
Shulamit Manulis-Sasson[2]
[1]Department of Food Quality and Safety, Institute for Postharvest and Food Sciences, Bet Dagan, Israel.
[2]Department of Plant Pathology and Weed Research, Agricultural Research Organization (ARO), The Volcani Center, Bet-Dagan, Israel.

The beginning of the 21st century and the third millennium marks a revolution in science and technology with innovations in multiple areas, including space science, computer sciences, communication, biotechnology, nanotechnology and the human genome project, to name a few. Yet at the same time people are still getting ill from eating contaminated food, with a heavy toll on public health and economy. This is not merely the case in developing countries with poor sanitation but also in rich and developed countries. In fact, during the past few decades, increasing number of outbreaks of foodborne illnesses associated with raw or minimally processed fruits and vegetables in Europe and the USA was reported. This parallels the expansion in the consumption of raw and partially processed fresh produce, including ready-to-eat (RTE) salads, which are recommended by doctors and dieticians as nutritious and healthy food.

Unlike foods of animal origin, such as meat, poultry and dairy products, which can undergo a thermal process (e.g. cooking, pasteurization, etc.) in order to inactivate human pathogens, fresh produce cannot undergo such processes because they induce physiological damages and deteriorate the organoleptic properties of these products. In order to ensure the safety of partially processed RTE foods, chemical and physical decontamination treatments were developed. The most common one includes washing with tap water and dipping in disinfectant solutions, such as hypochlorite (Goodburn and Wallace, 2013). However, unlike the thermal processes which can achieve the 'gold standard' 5-log reduction in the numbers of

*For correspondence. E-mail shlomos@volcani.agri.gov.il

Funding Information S. Sela (Saldinger) is currently funded by grant No. 421-0265-14 from the Chief Scientist, Ministry of Agriculture, Israel.

the challenged pathogen of choice in food of animal origin, this goal cannot be met with the current industrial treatments in fresh produce (Goodburn and Wallace, 2013). This is quite surprising as foodborne pathogens, such as *Escherichia coli*, nontyphoidal *Salmonella*, *Campylobacter* and *Listeria monocytogenes*, are all sensitive to the disinfectants used by the industry, at the relevant concentrations and exposure time. The lack of an efficient kill-step forces the fresh produce industry to largely rely on preventive measures, such as good agricultural practice (GAP) and hazard analysis of critical control points (HACCP).

There are several plausible explanations for the limitation of disinfectants to completely eradicate foodborne pathogens on fresh produce. Two major explanations are (i) the localization of bacteria in protected niches, either in cracks and crevices on the plant/fruit surface, or within internal organs following internalization, which prevent direct contact with the disinfectant; and (ii) formation of biofilms, or integration of foodborne pathogens into existing biofilms on the plant's surfaces, which hampers the ability of the antimicrobial agent to kill the pathogen.

Contamination of fresh produce may occur during each step of the food chain from farm to fork, yet because postharvest treatments cannot be relied upon to eliminate pathogens, prevention of preharvest contamination on farm seems to be the most important step in reducing human health risk (Beuchat, 2006). Sources of human pathogens on the farm include low-quality irrigation water, application of contaminated organic fertilizers, such as untreated or partially treated animal manure, as well as close proximity to livestock operations. Preharvest contamination may result in the introduction of pathogens into the processing plant, establishment of biofilms on food-contact surfaces and subsequent cross-contamination of lots to be distributed at a national or international scale, leading to outbreak.

Current contamination routes of fresh produce in the field include surface contamination by splashes of irrigated water, rain or during floods, and internal contamination via root internalization, natural openings such as stomata or trichomes, as well as via physical or biological damage to the plant organ (Critzer and Doyle, 2010).

It is commonly accepted that bacteria residing on the surface of plant are exposed to environmental stresses, such as irradiation, desiccation and lack of nutrients, which limit bacterial survival. On the other hand, internal plant's tissues provide bacteria with a favourable environment, rich in nutrients and water, and protected from external stresses. Likewise, surface-attached bacteria will be more prone to bactericidal activity of external disinfectants used during industrial processing, while internally localized bacteria will be protected and might even proliferate under the appropriate conditions.

A common approach for hazard mitigation is the multiple hurdle approach. In this context, we would like to discuss here the introduction of additional hurdles, through agro-technology, which might be applied in future years to limit the contamination of plant in the field, thereby enhancing the safety of partially processed and RTE fresh produce. This can be achieved by adapting known agro-technologies previously utilized for plant protection against diseases also for minimizing plant contamination by foodborne pathogens. These include (i) development of breeding programmes towards cultivars/varieties that are less prone to colonization by foodborne pathogens, (ii) grafting as a mean to limit migration of foodborne pathogens from the rhizosphere to the phyllosphere, (iii) induced resistance against foodborne pathogens and (iv) application of antagonistic microorganisms to limit survival of foodborne pathogens in the plant environment.

## Breeding programmes for enhancing food safety

A vast amount of biodiversity exists in plants, which is utilized successfully by farmers, seed companies and researchers for decades as a mean to develop new cultivars/varieties with favourable traits, such as increased productivity, quality and resistance to diseases. Similarly, induction of mutation and the emergence of genetic engineering in plants have resulted in the advent of new cultivars with specific quality and disease resistance traits, which had a dramatic impact on the economy, globally. Although productivity, quality and plant protection remain the major incentive for screening and breeding programmes, it is envisaged that food safety will push the research also towards harnessing these tools for the generation of crops resistant to colonization by foodborne pathogens. This idea is supported by studies demonstrating dependency between colonization and plant's cultivar. For example, colonization of *Salmonella* in tomato (Barak *et al.*, 2011) and of *E. coli* O157:H7 in spinach leaves (Mitra *et al.*, 2009) was found to be affected by the nature of the cultivar. High variation in stomata internalization by *Salmonella* was also observed among different cultivars of iceberg lettuce grown in the field under identical agricultural setting (S. Sela, unpubl. data). Thus, a combination of screening for colonization-resistant plants' lines, accompanied by directed breeding programmes that

would take into account safety-related traits, should potentially contribute to the ongoing battle against contamination of fresh produce by human pathogens.

## Grafted plant as a novel approach to limit bacterial internalization through the root system

Internalization of plants through the root system is considered a possible contamination route by foodborne pathogens (Critzer and Doyle, 2010). *Salmonella* and *E. coli* strains were reported to enter through the root system of several plants, including lettuce, spinach and corn, yet root internalization seems to be cultivar-specific. Grafting susceptible scion onto internalization-resistant rootstock might be a promising alternative to limit entry of foodborne pathogens into vegetables. This idea was proved to be useful for inducing resistance against soilborne diseases (King *et al.*, 2008), as well as for minimizing the penetration of toxic contaminants from the soil into the plant (Otani and Seike, 2007). We envisage that such a technique, in vegetable crops, where grafting is possible, may prove a useful approach for generating additional hurdle against transport of foodborne pathogens from the rhizosphere to the phyllosphere.

## Stimulation of the plant's immune response

Accumulating evidence has led to the understanding that enteric bacterial pathogens are not merely contaminants of plants but can also adopt an endophytic lifestyle, using the plant as an alternative host (Holden *et al.*, 2009; Schikora *et al.*, 2012). Similar to plant's pathogens, foodborne pathogens are specifically sensed by the innate immune system of the host plant, which restricts their colonization and survival. The plant immune system is induced upon exposure to pathogen-associated molecular patterns (PAMPs) and initiate systemic defence response (PAMP-triggered immunity; PTI) in order to limit the dissemination of the pathogen in the plant. Recent data demonstrate the capability of some human enteric pathogens to induce plant's immune response (Holden *et al.*, 2009). Interestingly, colonization of *Salmonella* type 3 secretion system (T3SS) mutants on plant was found to be compromised, suggesting that human pathogens are utilizing T3SS effectors to suppress the plant immune response (Schikora *et al.*, 2012). Based on these findings and on similar studies, it can be speculated that stimulation of the plant's defence system may be utilized as a novel agro-technology to limit colonization of plants by foodborne pathogens. Indeed, exposure of *Nicotiana benthamiana* to the *Salmonella* flg22 peptide has induced PTI and limited the multiplication of the pathogen on the leaves (Meng *et al.*, 2013). These findings support the notion that elicitation of the plant immune response may act as an additional measure to limit plant contamination by foodborne pathogens. This approach certainly requires further studies to assess its effectiveness using different immune system elicitors, vegetable crops and human pathogens.

## Antagonistic microorganisms

The rhizosphere and the phyllosphere are rich with microorganisms that reside in close proximity to the plant tissues and have intimate interactions with the plant. It has been long known that microorganisms have multiple beneficial properties on plants, including increased crop productivity and resistance against multiple plant diseases by serving as biocontrol agents. Microorganisms secrete a multitude of metabolites, which might alter the physical and biochemical properties of the plant environment. For example, Bacteriocins, which are ecologically important family of metabolites, are produced by some bacteria to facilitate their colonization of specific niche by inhibiting or killing antagonistic microorganisms. Indeed, studies have demonstrated the presence of antagonistic microorganisms in the rhizosphere and phyllosphere of fresh produce and their potential utilization as safeguard against human pathogens (Goodburn and Wallace, 2013). The distribution of microorganisms depends on many variables, including the plant species, soil type, fertilizers, water, geographic region and climate, to name a few. The search for antagonistic microorganisms from the rhizosphere and phyllosphere of vegetable crops in combination with high-throughput screening programmes has the potential to identify antagonistic microorganisms with the ability to limit colonization of plant by foodborne pathogens.

In summary, we envisage that during the next years, modern agriculture will adopt an integrated food safety management system as part of a multiple hurdle approach to limit fresh produce contamination and protect public health. Safety of fresh produce should be strictly kept at all stages of the food chain from the farm to the consumer. In this essay, we have tried to point out on a number of potential, not necessarily new, agro-technologies that in combination with other approaches, such as GAP and HACCP, may be utilized to limit on-farm contamination of fresh produce. These include the use of natural biodiversity to screen for vegetables' cultivars resistant to foodborne pathogens, the generation of resistant plants via classical breeding or by genetic engineering, grafting technique to limit pathogens' migration from the rhizosphere to the phyllosphere, stimulation of the plant's immune response to restrict bacterial colonization, and the utilization of antagonistic microorganisms.

## Acknowledgement

S. Sela (Saldinger) is a member of the EU COST Action FA1202: A European Network for Mitigating Bacterial Colonisation and Persistence on Foods and Food Processing Environments (http://www.bacfoodnet.org/) and acknowledge this action for facilitating collaborative networking.

## Conflict of interest

The authors declare no conflict of interests.

## References

Barak, J.D., Kramer, L.C., and Hao, L. (2011) Colonization of tomato plants by *Salmonella enterica* is cultivar dependent, and type 1 trichomes are preferred colonization sites. *Appl Environ Microbiol* **77:** 498–504.

Beuchat, L.R. (2006) Vectors and conditions for preharvest contamination of fruits and vegetables with pathogens capable of causing enteric diseases. *Br Food J* **108:** 38–53.

Critzer, F.J., and Doyle, M.P. (2010) Microbial ecology of foodborne pathogens associated with produce. *Curr Opin Biotechnol* **21:** 125–130.

Goodburn, C., and Wallace, C.A. (2013) The microbiological efficacy of decontamination methodologies for fresh produce: a review. *Food Control* **32:** 418–427.

Holden, N., Pritchard, L., and Toth, I. (2009) Colonization outwith the colon: plants as an alternative environmental reservoir for human pathogenic enterobacteria. *FEMS Microbiol Rev* **33:** 689–703.

King, S.R., Davis, A.R., Liu, W., and Levi, A. (2008) Grafting for disease resistance. *HortScience* **43:** 1673–1676.

Meng, F., Altier, C., and Martin, G.B. (2013) *Salmonella* colonization activates the plant immune system and benefits from association with plant pathogenic bacteria. *Environ Microbiol* **15:** 2418–2430.

Mitra, R., Cuesta-Alonso, E., Wayadande, A., Talley, J., Gilliland, S., and Fletcher, J. (2009) Effect of route of introduction and host cultivar on the colonization, internalization, and movement of the human pathogen *Escherichia coli* O157:H7 in spinach. *J Food Prot* **72:** 1521–1530.

Otani, T., and Seike, N. (2007) Rootstock control of fruit dieldrin concentration in grafted cucumber (*Cucumis sativus*). *J Pestic Sci* **32:** 235–242.

Schikora, A., Garcia, A.V., and Hirt, H. (2012) Plants as alternative hosts for *Salmonella*. *Trends Plant Sci* **17:** 245–249.

# Methanol-based cadaverine production by genetically engineered *Bacillus methanolicus* strains

Ingemar Nærdal,[1†] Johannes Pfeifenschneider,[2†]
Trygve Brautaset[1,3] and Volker F. Wendisch[2*]
[1]*Sector for Biotechnology and Nanomedicine,
Department of Molecular Biology, SINTEF Materials and
Chemistry, Trondheim, Norway.*
[2]*Genetics of Prokaryotes, Faculty of Biology & CeBiTec,
Bielefeld University, Bielefeld, Germany.*
[3]*Department of Biotechnology, Norwegian University of
Science and Technology, Trondheim, Norway.*

## Summary

**Methanol is regarded as an attractive substrate for biotechnological production of value-added bulk products, such as amino acids and polyamines. In the present study, the methylotrophic and thermophilic bacterium *Bacillus methanolicus* was engineered into a microbial cell factory for the production of the platform chemical 1,5-diaminopentane (cadaverine) from methanol. This was achieved by the heterologous expression of the *Escherichia coli* genes *cadA* and *ldcC* encoding two different lysine decarboxylase enzymes, and by increasing the overall L-lysine production levels in this host. Both CadA and LdcC were functional in *B. methanolicus* cultivated at 50°C and expression of *cadA* resulted in cadaverine production levels up to 500 mg l$^{-1}$ during shake flask conditions. A volume-corrected concentration of 11.3 g l$^{-1}$ of cadaverine was obtained by high-cell density fed-batch methanol fermentation. Our results demonstrated that efficient conversion of L-lysine into cadaverine presumably has severe effects on feedback regulation of the L-lysine biosynthetic pathway in *B. methanolicus*. By also investigating the cadaverine tolerance level, *B. methanolicus* proved to be an exciting alternative host and comparable to the well-known bacterial hosts *E. coli* and *Corynebacterium glutamicum*. This study represents the first demonstration of microbial production of cadaverine from methanol.**

*For correspondence. E-mail volker.wendisch@uni bielefeld.de

**Funding Information** This work was supported by the research grants PROMYSE (EU, FP7 project 289540) and BioMet (RCN, GassMaks programme, project 224973). The authors acknowledge support of the publication fee by Deutsche Forschungsgemeinschaft and the Open Access Publication Funds of Bielefeld University.

## Introduction

There is a high societal demand for – and scientific interest in – more environmental-friendly and sustainable production processes for large quantity bulk products. As examples, amino acids and polyamines find applications as food/feed additives as well as in the pharmaceutical, plastics and polymer industry (Wendisch, 2014). The polyamine monomer 1,5-diaminopentane, commonly known as cadaverine, is a sought-after platform chemical used for production of various polyamides and is currently mainly fabricated by petroleum-based chemical synthesis. With the increasing focus on bio-economy and low-carbon footprints in the industry, efforts have been made to develop biotechnological production processes for several polyamines (Adkins *et al.*, 2012; Buschke *et al.*, 2013; Meiswinkel *et al.*, 2013a). Applying bacteria as microbial production hosts, certain polyamines can be obtained from amino acids including L-lysine, L-arginine and L-ornithine by thermodynamically favourable decarboxylation reactions (Schneider and Wendisch, 2011). These amino acids can be obtained by microbial fermentation processes and the worldwide production of the feed amino acid L-lysine amounts to almost 2 million tons per year (Wendisch, 2014). The common approach has been to establish L-lysine overproducing hosts for the concomitant engineering towards efficient production of cadaverine, as this compound is formed by a one-step conversion of L-lysine catalysed by lysine decarboxylase (Kind *et al.*, 2010; Kind and Wittmann, 2011; Qian *et al.*, 2011) (Fig. 1). In particular, the genes of the lysine decarboxylases found naturally in *Escherichia coli*, encoded by *cadA* and *ldcC*, have been applied and overexpressed. Also cadaverine secretion has been a target for optimization of production (Kind *et al.*, 2011; Li *et al.*, 2014). Typically, these production processes rely on *E. coli* and *Corynebacterium glutamicum* as microbial hosts using sugars from molasses or from starch hydrolysis as carbon and energy substrates leading to an unwanted competition with human food supply, and consequently nutrition prices are rising worldwide (Schrader *et al.*, 2009). As an alternative, e.g. recombinant *C. glutamicum* strains have been developed to accept alternative carbon sources such as glycerol from the biodiesel process (Meiswinkel *et al.*, 2013a), amino

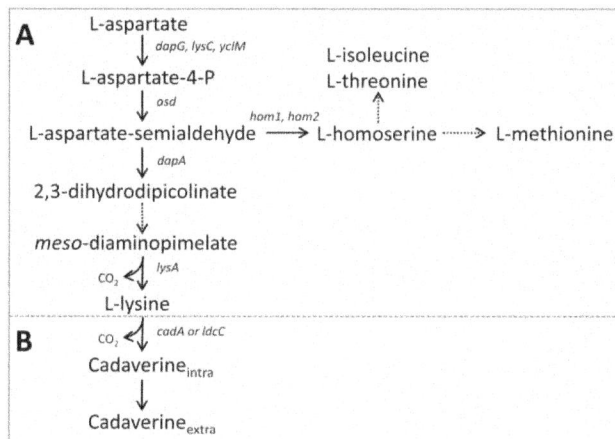

**Fig. 1.** Pathway for L-lysine and cadaverine biosynthesis in *B. methanolicus*. Gene names are indicated next to arrows representing reactions. Dotted arrows summarize several reactions. Reactions present in box A are endogenous in *B. methanolicus*, whereas reactions in box B involve the decarboxylation of L-lysine due to the heterologous expression of E. coli genes and export mechanisms of cadaverine to the extracellular medium.

sugars derived from chitin (Uhde *et al.*, 2013; Matano *et al.*, 2014) and pentoses present in lignocellulosic hydrolysates (Gopinath *et al.*, 2011; Meiswinkel *et al.*, 2013b). More generally, the possibility to produce polyamines, amino acids and other bulk products and biofuels from alternative non-food carbon sources has been in the research focus of biotechnology in recent years. The one-carbon substrate methanol has long been regarded as a convenient fuel and raw material for manmade hydrocarbon-based products (Olah, 2005). It occurs abundantly throughout nature, it is a pure raw material that can be completely utilized in microbial fermentation processes, and the price is expected to remain independent from and lower than sugar prices (Brautaset *et al.*, 2007; Schrader *et al.*, 2009). Based on all this, methanol is regarded as a highly attractive non-food substrate for microbial bioprocesses.

The Gram-positive and facultative methylotrophic bacterium *Bacillus methanolicus* is able to utilize methanol as sole carbon and energy source for growth (Müller *et al.*, 2014). As methanol growth is characterized by high oxygen demands leading to an increased heat output, it is an advantage that *B. methanolicus* has a growth optimum at 50–55°C, reducing the process cooling costs. The genome sequences of two wild-type *B. methanolicus* strains MGA3 and PB1 were recently published (Heggeset *et al.*, 2012; Irla *et al.*, 2014a) and its transcriptome characterized (Irla *et al.*, 2014b) serving as a solid basis for increased understanding of methylotrophy and product formation, e.g. L-glutamate and L-lysine, in this industrially relevant bacterium. It has been well documented that *B. methanolicus* has great potential for L-lysine overproduction through classical mutagenesis studies and selec-

tion of strains resistant to the L-lysine analog S-2-aminoethylcysteine (Hanson *et al.*, 1996; Brautaset *et al.*, 2010). Several key genes and enzymes of the aspartate pathway of *B. methanolicus* have been characterized, and insight into genetic repression and feedback inhibition has been established (Jakobsen *et al.*, 2009; Brautaset *et al.*, 2010). Furthermore, metabolic engineering of central metabolism and the aspartate pathway towards L-lysine in the MGA3 wild-type strain yielded significant L-lysine overproduction during shake flask experiments and fed-batch fermentations (Nærdal *et al.*, 2011). The theoretical maximum L-lysine yield from methanol has been calculated to $0.82 \, g \, g^{-1}$ in this bacterium (Brautaset *et al.*, 2007), which is comparable to the estimated maximum L-lysine yield from glucose in *C. glutamicum* (de Graaf, 2000; Wittmann and Becker, 2007). Thus, *B. methanolicus* was regarded as a potential promising host for production of cadaverine from methanol.

In the present study, we have investigated the potential of methanol-based biotechnological production of cadaverine at elevated temperature using wild-type and mutant *B. methanolicus* strains as hosts. Since inspection of the genome sequence did not reveal a gene putatively encoding a lysine decarboxylase (Fig. 1), synthetic cadaverine production modules based on the lysine decarboxylase isozymes LdcC and CadA from Gram-negative *E. coli* were constructed and heterologously expressed in *B. methanolicus* strains. Both enzymes proved functional and resulted in cadaverine production in *B. methanolicus*, and CadA overexpression provided the highest cadaverine production levels. This is to our knowledge the first demonstration of microbial cadaverine production from methanol.

## Results

### Bacillus methanolicus lacks cadaverine biosynthetic and degradation genes and tolerates up to 200 mM cadaverine before growth is severely affected

The genome sequencing of the wild-type *B. methanolicus* strains MGA3 and PB1 (Heggeset *et al.*, 2012; Irla *et al.*, 2014a) has identified all genes of the aspartate pathway leading to L-lysine, while genes putatively encoding L-lysine decarboxylases for conversion of L-lysine to cadaverine were not found. Furthermore, no putative cadaverine exporter genes were identified in the *B. methanolicus* genomes. Hence, heterologous expression of a lysine decarboxylase gene in *B. methanolicus* is a requirement for cadaverine production which has never been reported for this species.

To test the tolerance of *B. methanolicus* to cadaverine, this compound was added to exponentially growing cells and growth was monitored. For this purpose, the *B. methanolicus* strain M168-20 was used and cultivated

**Table 1.** Specific L-lysine decarboxylase activities, cadaverine and L-lysine production levels in recombinant *B. methanolicus* M168-20 strains.

| Plasmid | L-lysine decarboxylase specific activity nmol/min/mg protein | Cadaverine mg/L | L-lysine mg/L | Cadaverine + L-lysine mg/L |
|---|---|---|---|---|
| pHP13 | < 1 ± 0.2 | 0 | 140 ± 10 | 140 |
| pTH1mp-*ldcC* | 7.0 ± 1.0 | 130 ± 10 | 40 ± 5 | 170 |
| pTH1mp-*cadA* | 88.0 ± 11.0 | 420 ± 25 | 10 ± 2 | 430 |

The results shown are from triplicate (cadaverine and L-lysine) and duplicate (lysine decarboxylase activity) shake flask cultures. Activity was measured using crude extracts from exponentially growing cells, whereas the production levels were found from late stationary cultures, approximately 20 h after inoculation.

in shake flasks containing methanol ($MeOH_{200}$) medium. The cells were grown to an $OD_{600}$ of 0.4 before different concentrations of cadaverine dihydrochloride (0–200 mM, corresponding to 0–35 g $l^{-1}$) were added to triplicate cultures for each concentration. The control cultures without cadaverine supplementation grew with a specific growth rate ($\mu$) of 0.46 ± 0.01 $h^{-1}$ to an $OD_{600}$ of 8.7 ± 0.14. With the addition of 50 mM, 100 mM and 200 mM of cadaverine dihydrochloride, the maximum $OD_{600}$ values obtained were 7.5 ± 0.18, 6.2 ± 0.20 and 5.5 ± 0.22, respectively, and the accompanied specific growth rates were also reduced (0.40 ± 0.02 $h^{-1}$, 0.39 ± 0.01 $h^{-1}$ and 0.36 ± 0.01 $h^{-1}$). Thus, a minor growth inhibition by cadaverine was observed since addition of 200 mM (35 g $l^{-1}$) cadaverine dihydrochloride reduced the growth rate by about 20%.

Cadaverine may be degraded by certain bacteria and the involved genes have been identified (Schneider and Wendisch, 2011). However, inspection of the *B. methanolicus* MGA3 and PB1 genomes did not indicate that this bacterium is capable of catabolizing cadaverine. This was also experimentally confirmed in shake flask cultures by substituting methanol and ammonium sulphate, as carbon and nitrogen source, respectively, with cadaverine. Cadaverine did not support bacterial growth, and the cadaverine concentration did not decrease throughout the cultivation as analysed by reverse-phase high-performance liquid chromatography (data not shown).

*Heterologous expression of L-lysine decarboxylase genes enabled methanol-based cadaverine production by B. methanolicus classical mutant strain M168-20*

Since *B. methanolicus* lacks a lysine decarboxylase gene, the lysine decarboxylase genes *ldcC* and *cadA* from *E. coli* MG1655 were cloned into a pHP13 derivative expression vector carrying the strong *mdh* promoter for overexpression and used to transform the L-lysine overproducing classical mutant *B. methanolicus* strain M168-20. To assay for functional expression of *ldcC* and *cadA*, respectively, crude extracts of strains M168-20(pTH1mp-*ldcC*) and M168-20(pTH1mp-*cadA*) were prepared, and the speci-

fic L-lysine decarboxylase activities were determined (Table 1). The protein concentrations of the crude extracts of M168-20(pHP13), M168-20(pTH1mp-*ldcC*), and M168-20(pTH1mp-*cadA*) were 7.0 ± 0.3 mg, 7.8 ± 0.5 mg and 12.0 ± 1.5 mg respectively. L-lysine decarboxylase activity could not be detected in the empty vector control (< 1 nmol $min^{-1}$ $mg^{-1}$), whereas expression of *ldcC* and of *cadA* resulted in L-lysine decarboxylase activities of 7 ± 1 nmol $min^{-1}$ $mg^{-1}$ in M168-20(pTH1mp-*ldcC*) and of 88 ± 11 nmol $min^{-1}$ $mg^{-1}$ in M168-20(pTH1mp-*cadA*) (Table 1).

Subsequently, production experiments were carried out with *B. methanolicus* strains M168-20(pTH1mp-*ldcC*) and M168-20(pTH1mp-*cadA*) at 50°C in 500 ml shake flask cultures with $MeOH_{200}$ medium pH 7.2, and samples were harvested and analysed by HPLC, as described in *Experimental procedures*. As experimental control, the M168-20 strain transformed with the empty vector pHP13 was included. In accordance with previously reported data, the M168-20 (pHP13) strain produced 140 ± 10 mg $l^{-1}$ of L-lysine under these conditions (Nærdal *et al.*, 2011) and, as expected, no cadaverine production was detected. The heterologous expression of *ldcC* in *B. methanolicus* M168-20 resulted in production of 130 ± 10 mg $l^{-1}$ cadaverine and a L-lysine level of 40 ± 5 mg $l^{-1}$ (Table 1), confirming that the *ldcC* encoded lysine decarboxylase functions *in vivo* in *B. methanolicus* at 50°C. Similarly, heterologous expression of *cadA* entailed a surprisingly high cadaverine production level of 420 ± 25 mg $l^{-1}$ and only 10 ± 2 mg $l^{-1}$ L-lysine could be detected as side product (Table 1). Thus, methanol-based production of cadaverine by *B. methanolicus* was achieved. Notably, combined formation of cadaverine and L-lysine by the *cadA* and *ldcC* expressing strains was above threefold higher than L-lysine formation by the parent strain (Table 1), which might indicate feedback deregulation by L-lysine as consequence of a metabolic pull by lysine decarboxylase.

*Effect of the medium pH on cadaverine production by recombinant B. methanolicus*

Since LdcC and CadA function in pH homeostasis in *E. coli*, the effect of varying the pH of the production

**Table 2.** Production of cadaverine and L-lysine by recombinant *B. methanolicus* M168-20 strains cultivated at different medium pH.

| pH | M168-20(pHP13) | | M168-20(pTH1mp-*ldcC*) | | M168-20(pTH1mp-*cadA*) | |
|---|---|---|---|---|---|---|
| | Cadaverine | L-lysine | Cadaverine | L-lysine | Cadaverine | L-lysine |
| 6.5 | 0 | $50 \pm 10$ | $52 \pm 5$ | < 15 | $45 \pm 5$ | < 15 |
| 7.2 | 0 | $130 \pm 10$ | $135 \pm 10$ | $40 \pm 5$ | $430 \pm 20$ | < 30 |
| 7.6 | 0 | $140 \pm 10$ | $315 \pm 20$ | < 30 | $450 \pm 20$ | < 30 |
| 8.0 | 0 | $140 \pm 10$ | $305 \pm 30$ | < 15 | $500 \pm 30$ | < 30 |
| 8.5 | 0 | $140 \pm 10$ | $305 \pm 30$ | < 15 | $520 \pm 30$ | < 30 |

The mean values (mg/L) and standard deviation of triplicate shake flask cultures is presented. The production levels were found from late stationary cultures, from 20–30 h after inoculation.

media on cadaverine production was investigated. *B methanolicus* strains M168-20(pHP13), M168-20(pTH1mp-*ldcC*) and M168-20(pTH1mp-*cadA*) were cultivated in MeOH$_{200}$ medium adjusted to different pH values ranging from pH 6.5 to 8.5 prior to autoclaving. The standard MeOH$_{200}$ medium pH of 7.2 was included as control in these shake flask experiments for direct comparison. The control strain M168-20(pHP13) was included to test for any potential pH effects on L-lysine production. L-lysine production by M168-20(pHP13) was reduced to about $50 \pm 5$ mg l$^{-1}$ at slightly acidic pH (pH 6.5), but remained stable (130–140 mg l$^{-1}$) at slightly alkaline pH (pH 7.2 to 8.5). Cadaverine production by M168-20(pTH1mp-*ldcC*) was lower at pH 6.5 ($52 \pm 5$ mg l$^{-1}$) than at pH 7.2 ($135 \pm 10$ mg l$^{-1}$), but about twofold higher at pH values between 7.6 and 8.5 (about 300 mg l$^{-1}$; Table 2). However, the productivity was maximal at pH 7.6 since the growth rate decreased at higher pH values (data not shown). Strain M168-20(pTH1mp-*cadA*) accumulated similar concentrations of cadaverine (430 to 520 mg l$^{-1}$) at all tested pH values except at pH 6.5 ($45 \pm 5$ mg l$^{-1}$), a condition also characterized by reduced production of the immediate precursor L-lysine (Table 2).

As lysine decarboxylase activity is reported to depend on pyridoxal-5-phosphate (PLP) as cofactor, addition of pyridoxal-5-phosphate hydrate (1 mg l$^{-1}$) to MeOH$_{200}$ medium at pH 7.6 was tested. However, PLP supply in *B. methanolicus* was not limiting cadaverine production under the chosen conditions since production did not increase upon addition of pyridoxal phosphate (data not shown).

*Construction of cadaverine overproducing strains by using the wild-type B. methanolicus MGA3 as a host*

We have previously achieved L-lysine overproduction by engineering of the aspartate pathway and using wild-type *B. methanolicus* strain MGA3 as host. For example, overexpression of the genes *lysC* and *lysA*, encoding aspartokinase II and meso-diaminopimelate decarboxylase, respectively, resulted in L-lysine overproduction (Nærdal *et al.*, 2011). We hypothesized that coupled

overexpression of these two genes together with the *ldcC* and *cadA* genes in MGA3 could result in effective cadaverine production. The recombinant strains MGA3(pTH1mp-*ldcC*-*lysC*), MGA3(pTH1mp-*ldcC*-*lysA*) and MGA3(pTH1mp-*cadA*-*lysA*) were therefore constructed. To investigate if heterologous expression of *ldcC* and *cadA* alone entails cadaverine production in MGA3, strains MGA3(pTH1mp-*ldcC*) and MGA3(pTH1mp-*cadA*) were also established. Expression of *ldcC* alone resulted in only minor cadaverine production ($20 \pm 4$ mg l$^{-1}$), while coupled overexpression with endogenous *lysC* and *lysA* improved cadaverine production ($140 \pm 10$ and $190 \pm 10$ mg l$^{-1}$), and these strains produced 10 mg l$^{-1}$ of L-lysine (Table 3). Interestingly, L-lysine production was in each case lower (7, 55 and 150 mg l$^{-1}$, respectively; Table 3) for the three isogenic strains that do not express *ldcC*, i.e. MGA3(pHP13), MGA3(pTH1mp-*lysC*) and MGA3(pTH1mp-*lysA*), respectively, indicating that LdcC exerts a metabolic pull deregulating flux through the L-lysine biosynthesis pathway. This notion is supported by the finding that heterologous expression of *cadA* alone in MGA3 resulted in $450 \pm 30$ mg l$^{-1}$ cadaverine production (Table 3). The coupled overexpression of *cadA* with the endogenous *lysA* gene did not significantly increase cadaverine production further as $480 \pm 30$ mg l$^{-1}$ was measured.

**Table 3.** Cadaverine and L-lysine production by recombinant *B. methanolicus* MGA3 strains.

| Plasmid | Cadaverine mg/L | L-lysine mg/L |
|---|---|---|
| pHP13 | 0 | $7 \pm 1$[a] |
| pTH1mp-*lysC* | 0 | $55 \pm 5$[a] |
| pTH1mp-*lysA* | 0 | $150 \pm 10$[a] |
| pTH1mp-*ldcC* | $20 \pm 4$ | $7 \pm 1$ |
| pTH1mp-*ldcC*-*lysC* | $140 \pm 10$ | < 10 |
| pTH1mp-*ldcC*-*lysA* | $190 \pm 10$ | < 10 |
| pTH1mp-*cadA* | $450 \pm 30$ | < 10 |
| pTH1mp-*cadA*-*lysA* | $480 \pm 30$ | < 10 |

**a.** Data imported from (Nærdal *et al.*, 2011).
The production levels were found from late stationary shake flask cultures, approximately 20 h after inoculation.

**Table 4.** Fed-batch methanol fermentation production data of strains MGA3(pTH1mp-*cadA*) and MGA3(pHP13).

| Strain | CDW[b] g/L | μ[a] h⁻¹ | Asp[b] g/L | Glu[b] g/L | Ala[b] g/L | Lys[b] g/L | Cad[b] g/L |
|---|---|---|---|---|---|---|---|
| MGA3(pTH1mp-*cadA*) | 65.5 | 0.45 | 1.5 | 71.8 | 10.2 | 0.0 | 11.3 |
| MGA3(pHP13) | 45.0 | 0.49 | 1.1 | 59.0 | 12.0 | 0.4 | 0.0 |

**a.** Specific growth rates are maximum values calculated from the exponential growth period.
**b.** CDW, cadaverine and amino acid concentrations are maximum values and volume corrected (see 'Experimental Procedures' section).
The maximum mean values from early stationary (CDW) or late stationary growth phase are presented for the MGA3(pTH1mp-*cadA*) duplicate cultures and the deviation never exceed ten per cent. The MGA3(pHP13) data were imported from (Brautaset *et al.*, 2010). CDW, cell dry weight; μ, specific growth rate; Asp, L-aspartate; Glu, L-glutamate; Ala, L-alanine; Lys, L-lysine, Cad, cadaverine.

*Fed-batch methanol cultivation of strain MGA3(pTH1mp-cadA) lead to the substantial volumetric production level of 11.3 g l⁻¹*

We chose to investigate the promising cadaverine production strain MGA3(pTH1mp-*cadA*) during high-cell-density fed-batch methanol fermentation conditions. This strain was tested in duplicates and samples for cadaverine and amino acid analysis, cell dry weight and OD₆₀₀ were taken throughout the cultivation. Due to the significant increase in culture volume, all values were volume corrected by multiplying with the respective correction factor. We have previously cultivated strain MGA3(pHP13) at the same fed-batch conditions and reported volume corrected values as published in (Brautaset *et al.*, 2010). From these data we know that L-glutamate accumulate throughout the cultivation (59 g l⁻¹), whereas the L-lysine level remain low (0.4 g l⁻¹), and no cadaverine can be detected (Table 4). As also observed in shake flask studies, cadaverine accumulated during the fed-batch cultivation, but during fed-batch conditions MGA3(pTH1mp-*cadA*) reached a high volumetric yield, i.e. a volume-corrected concentration of 11.3 g l⁻¹ cadaverine (Table 4). At the same time, no L-lysine could be detected. Despite of the high cadaverine production, high levels of L-glutamate and biomass was still measured indicating that the cadaverine production did not negatively affect these parameters. However, a slight reduction of the specific growth rate was observed (Table 4). The MGA3(pTH1mp-*cadA*) production levels of L-aspartate and L-alanine were similar to previously reported values for MGA3(pHP13).

## Discussion

Methanol-based cadaverine production was shown here for the first time. The tolerance level of the thermophilic methylotroph *B. methanolicus* towards the end-product cadaverine was found to be similar to that of the natural cadaverine producer *E. coli*. 200 mM cadaverine added to the growth medium resulted in reduced growth rates by *B. methanolicus* and *E. coli* by 20% and 35% respectively (Qian *et al.*, 2011). Reports using agar plate assays suggested a slightly higher cadaverine tolerance of *C. glutamicum* (Mimitsuka *et al.*, 2007). Due to its tolerance to cadaverine and its proven inability to degrade this compound, *B. methanolicus* appears to be a suitable host for the production of cadaverine. Heterologous expression of both *ldcC* and *cadA* resulted in cadaverine production in *B. methanolicus*. Cadaverine production level was higher with *cadA* than with *ldcC* in both *B. methanolicus* host strains MGA3 and M168-20. Production of L-lysine as significant by-product was observed in an *ldcC* expressing strain (40 mg l⁻¹ by M168-20(pTH1mp-*ldcC*) at pH 7.2). The *in vitro* pH optima of LdcC and CadA are reported to be 7.6 (Yamamoto *et al.*, 1997; Lemonnier and Lane, 1998) and 5.7 (Moreau, 2007) respectively. The low pH optimum of CadA fits to its role in L-lysine dependent acid stress response of *E. coli* where *cadA* expression is induced at low pH and in the presence of L-lysine by the positive regulator CadC (Kuper and Jung, 2005). The intracellular pH of *B. methanolicus* has not yet been experimentally tested. A slightly acidic pH of the cultivation medium reduced L-lysine production, and as consequence lower cadaverine production was observed (Table 2). At slightly alkaline medium pH reduced L-lysine synthesis did not limit cadaverine production. Notably, in each isogenic strain pair analysed, cadaverine production due to heterologous L-lysine decarboxylase production was higher than L-lysine production by the respective parent strain. We propose that intracellular L-lysine concentrations are low as result of LdcC or CadA activity and that key aspartate pathway enzymes are relieved from feedback inhibition by L-lysine and/or their synthesis is relieved from repression by L-lysine. Indeed, AKII and DAP decarboxylase are known to be feedback inhibited by L-lysine (Mills and Flickinger, 1993; Jakobsen *et al.*, 2009).

Expression of *cadA* in *B. methanolicus* strains led to higher cadaverine production than expression of *ldcC* (Tables 1, 2 and 3). Two factors may explain this finding. First, *cadA* expression led to higher L-lysine decarboxylase activities in crude extracts as compared with *ldcC* expression (Table 1). Second, CadA is reported to display a higher affinity to L-lysine than LdcC with Km values for L-lysine of 0.84 mM and 0.27 mM respectively (Krithika *et al.*, 2010).

We could demonstrate high-level cadaverine production during high-cell-density fed-batch methanol fermentation of strain MGA3(pTH1mp-*cadA*). Whereas no

L-lysine accumulated during the fermentation, the volume corrected production level of cadaverine reached 11.3 g $l^{-1}$ after 30 h and remained stable throughout the cultivation time of 47 h. The volume corrected concentrations of biomass (65.5 g $l^{-1}$) and L-glutamate (71.8 g $l^{-1}$) obtained for MGA3(pTH1mp-cadA) were slightly higher than previously reported values for MGA3(pHP13) (Table 4). The finding that cadaverine could accumulate to higher concentrations in the fermenter than in shake flasks may in part be explained by the fact that the fermenter was pH-controlled and that the shake flask cultures acidified with time (data not shown). Moreover, higher cadaverine concentrations were tolerated by *B. methanolicus* since only minor negative effects on biomass and specific growth rate were observed upon addition of up to 35 g $l^{-1}$ (200 mM) pH-adjusted cadaverine. It was observed that the cadaverine concentration increased throughout the growth phase until the early stationary phase, as also reported previously for *E. coli* and *C. glutamicum* (Kind et al., 2011; Qian et al., 2011). Due to the significant accumulation of L-glutamate in strain MGA3(pTH1mp-cadA) during fed-batch fermentation, there should be a great potential to increase cadaverine production further, especially by coexpression of the 2-oxoglutarate dehydrogenase from *B. methanolicus* recently found to reduce L-glutamate production 5-fold and increase L-lysine production twofold in *B. methanolicus* M168-20 (Krog et al., 2013). An improved understanding of both L-lysine and cadaverine secretion in *B. methanolicus* and heterologous expression of relevant known exporter or permease genes like *cadB* from *E. coli* (Li et al., 2014) and *cg2893* from *C. glutamicum* (Kind et al., 2011) could certainly be valuable for future high-level methanol-based cadaverine production in *B. methanolicus*.

## Experimental Procedures

### Biological materials, deoxyribonucleic acid manipulations and growth conditions

Bacterial strains and plasmids used in this study are listed in Table 5. *E. coli* DH5α was used as a general cloning host. *E. coli* strains were cultivated in liquid and on solid lysogeny broth medium at 37°C and standard recombinant deoxyribonucleic acid (DNA) procedures were performed as described elsewhere (Sambrook et al., 2001). *B. methanolicus* strains were cultivated at 50°C and 200 r.p.m. in methanol (MeOH$_{200}$) medium (Jakobsen et al., 2006) containing salt buffer (4.1 g $l^{-1}$ K$_2$HPO4, 1.3 g $l^{-1}$ NaH$_2$PO4, 2.1 g $l^{-1}$ (NH$_4$)$_2$SO$_4$) and 0.025% yeast extract (Difco) adjusted to pH 7.2 unless stated otherwise. After autoclavation, the medium was supplemented with 1 mM MgSO$_4$, vitamins, trace metals and 200 mM methanol as described elsewhere (Schendel et al., 1990; Jakobsen et al., 2006). The transformation of *B. methanolicus* was performed by electroporation as described previously (Jakobsen et al., 2006). For classical *B. methanolicus* mutant strain M168-20 (Brautaset et al., 2010) the growth medium was supplemented with D,L-methionine (1.5 mM). Recombinant *E. coli* and *B. methanolicus* strains were cultivated in media supplemented with chloramphenicol (15 and 5 µg $ml^{-1}$ respectively). Bacterial growth was monitored by measuring optical density at 600 nm (OD$_{600}$). Tolerance of *B. methanolicus* to cadaverine was investigated by monitoring bacterial growth in the presence of different cadaverine concentrations. Cadaverine dihydrochloride (Sigma Aldrich Biochemie GmbH, Hamburg, Germany) was dissolved in MeOH$_{200}$ medium, and the solution was pH adjusted to 7.2 and pre-warmed before cadaverine was supplemented in different concentrations to the growing cell cultures. Control cultures without cadaverine were included.

### Construction of expression vectors

The *ldcC* gene of *E. coli* MG1655 was polymerase chain reaction (PCR) amplified from genomic DNA using primers

**Table 5.** Bacterial strains and plasmids used in this study.

| Strain or plasmid | Description | Reference |
|---|---|---|
| *E. coli* | | |
| DH5α | General cloning host | Stratagene |
| MG1655 | Wild type strain | ATCC 47076 |
| *B. methanolicus* | | |
| MGA3 | Wild-type strain | ATCC 53907 |
| M168-20 | AEC-resistant *hom1* MGA3 mutant | (Brautaset et al., 2010) |
| Plasmids | | |
| pHP13 | *E. coli-B. subtilis* shuttle vector, Clm$^r$ | (Haima et al., 1987) |
| pTH1mp-*lysC* | pHP13 derivate with *lysC* under control of *mdh* promoter | (Brautaset et al., 2010) |
| pTH1mp-*lysA* | pHP13 derivate with *lysA* under control of *mdh* promoter | (Nærdal et al., 2011) |
| pTH1mp-*ldcC* | pHP13 derivate with *ldcC* under control of *mdh* promoter | This study |
| pTH1mp-*cadA* | pHP13 derivate with *cadA* under control of *mdh* promoter | This study |
| pTH1mp-*ldcC-lysC* | pTH1mp-*ldcC* with *lysC* downstream of the *ldcC* gene | This study |
| pTH1mp-*ldcC-lysA* | pTH1mp-*ldcC* with *lysA* downstream of the *ldcC* gene | This study |
| pTH1mp-*cadA-lysA* | pTH1mp-*cadA* with *lysA* downstream of the *cadA* gene | This study |

Clm$^r$, chloramphenicol resistance.

ldcC-PciI-Fwd: 5'- GCTGCACATGTGAACATCATTGCCAT TATGG-3' and ldcC-XbaI-Rev 5'-GCTGCTCTAGATTATCC CGCCATTTTTAGGAC-3'. The resulting 2162 bp PCR product was digested with *PciI* and *XbaI* (restriction sites underlined) and ligated into corresponding sites of pTH1mp-*lysC* (replacing *lysC*) resulting in plasmid pTH1mp-*ldcC*. Vector pTH1mp-*lysC* was digested with *SpeI* and *NcoI* and the 2017 bp fragment containing *lysC* was ligated into the *XbaI* (compatible with *SpeI*) and *NcoI* sites of pTH1mp-*ldcC* resulting in plasmid pTH1mp-*ldcC-lysC*. Vector pTH1mp-*lysA* was digested with *SpeI* and *NcoI*, and the 1834 bp fragment containing *lysA* was ligated into the *XbaI* (compatible with *SpeI*) and *NcoI* sites of pTH1mp-*ldcC* resulting in plasmid pTH1mp-*ldcC-lysA*. The *cadA* gene (2148 bp) was PCR amplified from genomic DNA isolated from *E. coli* MG1655 using the following primer pair: cadA-fw: 5'-AGGAGGT AGTACATGTGAACGTTATTGCAATATTGAATC-3' and cadA-rv: 5'-CCTATGGCGGGTACCTTATTTTTTGCTTTCTTCTTT CAA-3'. The obtained PCR product was ligated into the vector pTH1mp-*lysC*, digested with *PciI/KpnI* (replacing the *lysC* gene), using the isothermal DNA assembly method (Gibson *et al.*, 2009) yielding expression vector pTH1mp-*cadA*. Vector pTH1mp-*lysA* was digested with *SpeI* and *NcoI*, and the 1834 bp fragment containing *lysA* was ligated into the *XbaI* (compatible with *SpeI*) and *NcoI* sites of pTH1mp-*cadA* resulting in plasmid pTH1mp-*cadA-lysA*.

*Lysine decarboxylase activity assays in B. methanolicus crude extracts*

The lysine decarboxylase activity was determined in *B. methanolicus* crude cell extracts. The preparation of crude cell extracts was performed as described elsewhere previously (Brautaset *et al.*, 2004). The cells were inoculated from a glycerol stock and grown in $MeOH_{200}$ medium overnight before they were transferred to fresh $MeOH_{200}$ medium and grown to an $OD_{600}$ of 1.5 to 2.0. Forty millilitre of the culture was harvested by centrifugation (4000 × g, 30 min, 4°C), washed in 100 mM sodium citrate buffer (pH 7.5) and stored at −20°C. The cells were disrupted by sonication (Brautaset *et al.*, 2003). The cell debris was removed by centrifugation (14.000 × g, 60 min, 4°C), and the supernatant was used as crude extract for measuring the lysine decarboxylase activity. Lysine decarboxylase activity was calculated by measuring the conversion of lysine to cadaverine over time using HPLC as described elsewhere (Kind *et al.*, 2010). The assays were carried out at 50°C, and one unit of lysine decarboxylase activity was defined as the amount of enzyme that formed 1 μmol of cadaverine per min at 50°C. Protein concentration was determined using the assay of Bradford (Bradford, 1976).

*Cadaverine and L-lysine shake flask production studies*

Production experiments were performed in 500 ml baffled shake flasks (Belco) containing 100 ml $MeOH_{200}$ medium (Jakobsen *et al.*, 2006). *B. methanolicus* strains were cultivated in triplicate cultures using inoculum made from exponentially growing cells (Brautaset *et al.*, 2010; Nærdal *et al.*, 2011). Samples for amino acid measurements were collected during the late exponential and stationary growth phases as described previously (Jakobsen *et al.*, 2009; Nærdal *et al.*, 2011), and measurements of cadaverine and amino acids were performed by using 9-fluorenylmethyl chloroformate (FMOC) or o-phthaldialdehyde derivatization and reverse-phase high-performance liquid chromatography (Jakobsen *et al.*, 2009; Brautaset *et al.*, 2010; Schneider and Wendisch, 2010). Concentrations for cadaverine are reported for the free base (MW of 102.18 g $mol^{-1}$).

*High-cell-density fed-batch methanol fermentation*

Fed-batch fermentation was performed at 50°C in UMN1 medium in Applikon 3-l fermentors with an initial volume of 0.75 litre essentially as described previously (Jakobsen *et al.*, 2009; Brautaset *et al.*, 2010). Chloramphenicol (5 μg $ml^{-1}$) was added to the initial batch growth medium, the pH was maintained at 6.5 by automatic addition of 12.5% (wt/vol) $NH_3$ solution, and the dissolved oxygen level was maintained at 30% saturation by increasing the agitation speed and using enriched air (up to 60% $O_2$). The methanol concentration in the fermentor was monitored by online analysis of the headspace gas with a mass spectrometer (Balzers Omnistar GSD 300 02). The headspace gas was transferred from the fermentors to the mass spectrometer in insulated heated (60°C) stainless steel tubing. The methanol concentration in the medium was maintained at a set point of 150 mM by automatic addition of methanol feed solution containing methanol, trace metals and antifoam 204 (Sigma), as described in (Brautaset *et al.*, 2010). The inoculum preparation protocol, the fermentation conditions and fermentation progress was as described previously (Brautaset *et al.*, 2010). All fermentations were run until the carbon dioxide content of the exhaust gas was close to zero (no cell respiration). Bacterial growth was monitored by measuring the optical density at 600 nm ($OD_{600}$). Dry cell weight was calculated using a conversion factor of one $OD_{600}$ unit corresponding to 0.24 g dry cell weight per litre (calculated based on multiple measurements of dry cell weight and $OD_{600}$ during the fermentation trial). Due to the significant the increase in culture volume throughout the fermentation, the biomass, cadaverine and amino acid concentrations were corrected for the increase in volume and subsequent dilution. The volume correction factor of 1.8 was used for values presented in Table 4. The actual concentrations measured in the bioreactors were therefore accordingly lower as described previously (Jakobsen *et al.*, 2009). Samples for determination of volumetric cadaverine and amino acid yields were collected from early exponential phase and throughout the cultivation (10–47 h) and analysed as described above.

## Acknowledgements

We thank Tone Haugen for help with production experiments and analyses.

## Conflict of interest

None declared.

## References

Adkins, J., Pugh, S., McKenna, R., and Nielsen, D.R. (2012) Engineering microbial chemical factories to produce renewable 'biomonomers'. *Front Microbiol* **3:** 313.

Bradford, M.M. (1976) A rapid and sensitive method for the quantitation of microgram quantities of protein utilizing the principle of protein-dye binding. *Anal Biochem* **72:** 248–254.

Brautaset, T., Williams, M.D., Dillingham, R.D., Kaufmann, C., Bennaars, A., Crabbe, E., and Flickinger, M.C. (2003) Role of the *Bacillus methanolicus* citrate synthase II gene, *citY*, in regulating the secretion of glutamate in L-lysine-secreting mutants. *Appl Environ Microbiol* **69:** 3986–3995.

Brautaset, T., Jakobsen, O.M., Flickinger, M.C., Valla, S., and Ellingsen, T.E. (2004) Plasmid-dependent methylotrophy in thermotolerant *Bacillus methanolicus*. *J Bacteriol* **186:** 1229–1238.

Brautaset, T., Jakobsen, O.M., Josefsen, K.D., Flickinger, M.C., and Ellingsen, T.E. (2007) Bacillus methanolicus: a candidate for industrial production of amino acids from methanol at 50 degrees C. *Appl Microbiol Biotechnol* **74:** 22–34.

Brautaset, T., Jakobsen, O.M., Degnes, K.F., Netzer, R., Naerdal, I., Krog, A., *et al.* (2010) *Bacillus methanolicus* pyruvate carboxylase and homoserine dehydrogenase I and II and their roles for L-lysine production from methanol at 50 degrees C. *Appl Microbiol Biotechnol* **87:** 951–964.

Buschke, N., Schafer, R., Becker, J., and Wittmann, C. (2013) Metabolic engineering of industrial platform micro-organisms for biorefinery applications – optimization of substrate spectrum and process robustness by rational and evolutive strategies. *Bioresour Technol* **135:** 544–554.

Gibson, D.G., Young, L., Chuang, R.Y., Venter, J.C., Hutchison, C.A., 3rd, and Smith, H.O. (2009) Enzymatic assembly of DNA molecules up to several hundred kilobases. *Nat Methods* **6:** 343–345.

Gopinath, V., Meiswinkel, T.M., Wendisch, V.F., and Nampoothiri, K.M. (2011) Amino acid production from rice straw and wheat bran hydrolysates by recombinant pentose-utilizing *Corynebacterium glutamicum*. *Appl Microbiol Biotechnol* **92:** 985–996.

de Graaf, A.A. (2000) Metabolic flux analysis of *Corynebacterium glutamicum*. In *Bioreaction Engineering*. Schügerl, K., and Bellgardt, K.H. (eds.). Berlin, Germany: Springer Verlag, pp. 506–555.

Haima, P., Bron, S., and Venema, G. (1987) The effect of restriction on shotgun cloning and plasmid stability in *Bacillus subtilis* Marburg. *Mol Gen Genet* **209:** 335–342.

Hanson, R.S., Dillingham, R., Olson, P., Lee, G.H., Cue, D., Schendel, F.J., *et al.* (1996) Production of L-Lysine and some other amino acids by mutants of *B. methanolicus*. In *Microbial Growth on C1 Compounds*. Lidstorm, M.E., and Tabita, F.R. (ed.). Berlin, Germany: Springer Verlag, pp. 227–236.

Heggeset, T.M., Krog, A., Balzer, S., Wentzel, A., Ellingsen, T.E., and Brautaset, T. (2012) Genome sequence of thermotolerant *Bacillus methanolicus*: features and regulation related to methylotrophy and production of L-lysine

and L-glutamate from methanol. *Appl Environ Microbiol* **78:** 5170–5181.

Irla, M., Neshat, A., Winkler, A., Albersmeier, A., Heggeset, T.M.B., Brautaset, T., *et al.* (2014a) Genome sequence of *Bacillus methanolicus* MGA3, a thermotolerant amino acid producing methylotroph. *J Biotechnol* **188:** 110–111.

Irla, M., Neshat, A., Brautaset, T., Rückert, C., Kalinowski, J., and Wendisch, V.F. (2014b) Transcriptome analysis of thermophilic methylotrophic *Bacillus methanolicus* MGA3 using RNA-sequencing provides detailed insights into its previously uncharted transcriptional landscape. *BMC Genomics* in press.

Jakobsen, O.M., Benichou, A., Flickinger, M.C., Valla, S., Ellingsen, T.E., and Brautaset, T. (2006) Upregulated transcription of plasmid and chromosomal ribulose monophosphate pathway genes is critical for methanol assimilation rate and methanol tolerance in the methylotrophic bacterium *Bacillus methanolicus*. *J Bacteriol* **188:** 3063–3072.

Jakobsen, O.M., Brautaset, T., Degnes, K.F., Heggeset, T.M., Balzer, S., Flickinger, M.C., *et al.* (2009) Overexpression of wild-type aspartokinase increases L-lysine production in the thermotolerant methylotrophic bacterium *Bacillus methanolicus*. *Appl Environ Microbiol* **75:** 652–661.

Kind, S., and Wittmann, C. (2011) Bio-based production of the platform chemical 1,5-diaminopentane. *Appl Microbiol Biotechnol* **91:** 1287–1296.

Kind, S., Jeong, W.K., Schroder, H., and Wittmann, C. (2010) Systems-wide metabolic pathway engineering in *Corynebacterium glutamicum* for bio-based production of diaminopentane. *Metab Eng* **12:** 341–351.

Kind, S., Kreye, S., and Wittmann, C. (2011) Metabolic engineering of cellular transport for overproduction of the platform chemical 1,5-diaminopentane in *Corynebacterium glutamicum*. *Metab Eng* **13:** 617–627.

Krithika, G., Arunachalam, J., Priyanka, H., and Indulekha, K. (2010) The two forms of Lysine decarboxylase; kinetics and effect of expression in relation to acid tolerance response in E. coli. *J Exp Sci* **1:** 10–21.

Krog, A., Heggeset, T.M., Ellingsen, T.E., and Brautaset, T. (2013) Functional characterization of key enzymes involved in L-glutamate synthesis and degradation in the thermotolerant and methylotrophic bacterium *Bacillus methanolicus*. *Appl Environ Microbiol* **79:** 5321–5328.

Kuper, C., and Jung, K. (2005) CadC-mediated activation of the *cadBA* promoter in *Escherichia coli*. *J Mol Microbiol Biotechnol* **10:** 26–39.

Lemonnier, M., and Lane, D. (1998) Expression of the second lysine decarboxylase gene of *Escherichia coli*. *Microbiology* **144:** 751–760.

Li, M., Li, D., Huang, Y., Liu, M., Wang, H., Tang, Q., and Lu, F. (2014) Improving the secretion of cadaverine in *Corynebacterium glutamicum* by cadaverine-lysine antiporter. *J Ind Microbiol Biotechnol* **41:** 701–709.

Matano, C., Uhde, A., Youn, J.W., Maeda, T., Clermont, L., Marin, K., *et al.* (2014) Engineering of *Corynebacterium glutamicum* for growth and L-lysine and lycopene production from N-acetyl-glucosamine. *Appl Microbiol Biotechnol* **98:** 5633–5643.

Meiswinkel, T.M., Rittmann, D., Lindner, S.N., and Wendisch, V.F. (2013a) Crude glycerol-based production of amino acids and putrescine by *Corynebacterium glutamicum*. *Bioresour Technol* **145:** 254–258.

Meiswinkel, T.M., Gopinath, V., Lindner, S.N., Nampoothiri, K.M., and Wendisch, V.F. (2013b) Accelerated pentose utilization by *Corynebacterium glutamicum* for accelerated production of lysine, glutamate, ornithine and putrescine. *Microb Biotechnol* **6:** 131–140.

Mills, D.A., and Flickinger, M.C. (1993) Cloning and sequence analysis of the meso-diaminopimelate decarboxylase gene from *Bacillus methanolicus* MGA3 and comparison to other decarboxylase genes. *Appl Environ Microbiol* **59:** 2927–2937.

Mimitsuka, T., Sawai, H., Hatsu, M., and Yamada, K. (2007) Metabolic engineering of *Corynebacterium glutamicum* for cadaverine fermentation. *Biosci Biotechnol Biochem* **71:** 2130–2135.

Moreau, P.L. (2007) The lysine decarboxylase CadA protects *Escherichia coli* starved of phosphate against fermentation acids. *J Bacteriol* **189:** 2249–2261.

Müller, J.E.N., Heggeset, T.M.B., Wendisch, V.F., Vorholt, J.A., and Brautaset, T. (2014) Methylotrophy in the thermophilic *Bacillus methanolicus*, basic insights and application for commodity productions from methanol. *Appl Microbiol Biotechnol*. DOI:10.1007/s00253-014-6224-3.

Nærdal, I., Netzer, R., Ellingsen, T.E., and Brautaset, T. (2011) Analysis and manipulation of aspartate pathway genes for L-lysine overproduction from methanol by *Bacillus methanolicus*. *Appl Environ Microbiol* **77:** 6020–6026.

Olah, G.A. (2005) Beyond oil and gas: the methanol economy. *Angew Chem Int Ed Engl* **44:** 2636–2639.

Qian, Z.G., Xia, X.X., and Lee, S.Y. (2011) Metabolic engineering of *Escherichia coli* for the production of cadaverine: a five carbon diamine. *Biotechnol Bioeng* **108:** 93–103.

Sambrook, J., Russell, D., and Russell, D. (2001) *Molecular Cloning: A Laboratory Manual*, Vol. **3** set. New York, NY, USA: Cold Spring Harbor Laboratory Press Cold Spring Harbor.

Schendel, F.J., Bremmon, C.E., Flickinger, M.C., Guettler, M., and Hanson, R.S. (1990) L-lysine production at 50 degrees C by mutants of a newly isolated and characterized methylotrophic *Bacillus* sp. *Appl Environ Microbiol* **56:** 963–970.

Schneider, J., and Wendisch, V.F. (2010) Putrescine production by engineered *Corynebacterium glutamicum*. *Appl Microbiol Biotechnol* **88:** 859–868.

Schneider, J., and Wendisch, V.F. (2011) Biotechnological production of polyamines by bacteria: recent achievements and future perspectives. *Appl Microbiol Biotechnol* **91:** 17–30.

Schrader, J., Schilling, M., Holtmann, D., Sell, D., Filho, M.V., Marx, A., and Vorholt, J.A. (2009) Methanol-based industrial biotechnology: current status and future perspectives of methylotrophic bacteria. *Trends Biotechnol* **27:** 107–115.

Uhde, A., Youn, J.W., Maeda, T., Clermont, L., Matano, C., Kramer, R., *et al.* (2013) Glucosamine as carbon source for amino acid-producing *Corynebacterium glutamicum*. *Appl Microbiol Biotechnol* **97:** 1679–1687.

Wendisch, V.F. (2014) Microbial production of amino acids and derived chemicals: synthetic biology approaches to strain development. *Curr Opin Biotechnol* **30C:** 51–58.

Wittmann, C., and Becker, J. (2007) The L-lysine story: from metabolic pathways to industrial production. In *Amino Acid Biosynthesis~ Pathways, Regulation and Metabolic Engineering*. Wendisch, V.F. (ed.). Berlin, Germany: Springer, pp. 39–70.

Yamamoto, Y., Miwa, Y., Miyoshi, K., Furuyama, J., and Ohmori, H. (1997) The *Escherichia coli ldcC* gene encodes another lysine decarboxylase, probably a constitutive enzyme. *Genes Genet Syst* **72:** 167–172.

# Beyond borders: investigating microbiome interactivity and diversity for advanced biocontrol technologies

*Gabriele Berg, Institute of Environmental Biotechnology, Graz University of Technology & ACIB Austrian Centre of Industrial Biotechnology, Graz, Austria.*

## Microbial biotechnology, crystal ball

Recent, primarily technology-based advances in microbial ecology have opened an immense treasure chest of microbial diversity that has been observed in the vast majority of all investigated habitats. Additionally, habitats which were assumed to be sterile for a long time are now known to be colonized by diverse microorganisms; e.g. within the placenta and stomach of humans or reproductive organs of plants. Through the implementation of new techniques, deeper insights into the structure of microbiomes mainly by amplicon sequencing and microscopy have been gained. Now the current focus of research is on analysis of their function implementing meta-(genomic/epigenomic/transcriptomic/proteomic) techniques. The management of the accumulated metadata and the gleaning of new information is the challenge we are facing, but that challenge will be solved in the near future (Jansson, 2013). By looking into the proverbial crystal ball, my intention is to highlight two scientific aspects in particular, which I believe have been overlooked in current approaches to ecological research: (i) the interactivity of microbiomes and (ii) the interplay between microorganisms derived from different kingdoms – in particular archaea, bacteria and fungi. Both aspects could inspire the future course of biotechnology.

In recent decades, deeper insight into many aquatic and terrestrial microbial communities was gained using omics approaches. Although this is now possible, considerable time must be invested in the future in order to assemble all pathways and understand the interactions within microbiomes (Jansson, 2013). In contrast to single microbiomes, the connection between microbiomes as well as mutual exchange between them is less understood. Although we live in a highly interconnected world, up to the present date, there are only a few examples of synergistic microbiomes, which have shown that there are important relationships between single microbiomes. The three presented here are the rhizosphere (root–soil connection), the gut microbiome (food–human connection) and the indoor microbiome (plant–inhabitant connection). First, the rhizosphere is one well-investigated example; the root–soil interface is influenced by the plant metabolism via root exudates (Philippot et al., 2013). After assessing many experimental data and a lengthy discussion, it was accepted that both the plant as well as the soil influence the composition of the rhizosphere community (Berg and Smalla, 2009). The extent of impact depends on the plant species/genotype, its metabolites and the soil quality. While the rhizosphere is an example of particular importance for plant health, another interesting example which is important for human health is the gut microbiome. David and colleagues (2014) recently provided evidence for the survival of food-borne microbes (both animal and plant-based diet) after transit through the digestive system, and that food-borne strains may have been metabolically active in the gut. Microbial diversity in our gut ecosystem has an enormous impact on the host and vice versa connected by gut–brain cross-talk, which was revealed as a complex, bidirectional communication system (Mayer, 2011). Interesting relationships were detected recently, such as those between the gut microbiome and the development of obesity, cardiovascular disease and metabolic syndromes (Blaser et al., 2013) and also on motivation and higher cognitive functions, including intuitive decision making (Mayer, 2011). However, less is known about the food microbiome, although in many countries food is monitored for the occurrence of pathogens while beneficials are often ignored. First studies show that the vegetable and fruit microbiome is highly diverse, and *Enterobacteriaceae* play a substantial role within the vegetable microbiome (Leff and Fierer, 2013; Berg et al., 2014). Hanski and colleagues (2012) found a correlation between environmental biodiversity, human microbiota, especially *Enterobacteriaceae*, and allergy, and showed an experimental correlation between bacterial diversity and atopy as shown through significant interactions with enteric bacteria. The current focus of research is placed on the impact of our diet on the composition of the gut microbiome (of particular importance); however, the microbiome in/on our diet opens many more potential insights into the complex interactions. A third example is the indoor microbiome, which has enjoyed enormous attention during the last years due to the fact that we spend most of our time in built environments. The indoor microbiome is also a mixture of variably fixed niches such as bathtubs, kitchen sinks, furniture etc., as well as more moving carriers of microbiomes including the inhabitants (humans and pets), outside air and also of plants (indoor plants as well as outside vegetation). Although we know who contributes to the composition of the indoor microbiome, we don't know how those influences affect the function of the microbial ecosystem. The indoor microbiome is also strongly shaped by cleaning procedures. Can indoor plants and their microbiome positively contribute to human health? Is there an optimal rate of microbial turnover and exchange? All these questions can be answered by studying the shared fractions of microbiome or 'microbiome connections'. In the three microbiomes presented above, organisms of all domains of life, archaea, bacteria and eukaryotic microorganisms interact to fulfill ecosystem functions. Most of the published studies so far focused only on one microbial group, and therefore it is necessary to make a short note on the interplay between microorganisms from different kingdoms. Recent research has shown that there is a lot of interaction between members of the different groups; e.g. bacteria and fungi interact in the broad range between symbiosis and antagonism (Frey-Klett *et al.*, 2011). Little is known about environmental archaea. Interestingly, they are everywhere – from human skin to the endosphere of plants – but often in a low proportion, and their function and interaction with other microorganisms is mainly unknown. Omics technologies can now assist in determining the types of interactions among these organismal groups, ideally by combination of different approaches.

The control of microbial growth is an important area of microbiology, which resulted in significant advances in agriculture, medicine and food science. Biological control is an environmentally sound and effective means of reducing pathogens and pests and their symptoms through the use of natural antagonists or enemies. While in the past mainly single organisms were used, often correlated with inconsistent effects, it is now possible to develop predictable microbiome-based biocontrol strategies (Berg *et al.*, 2013). In many cases, diseases are associated with microbiome imbalances (dysbiosis) or shifts, which make it promising to control the whole microbiome. I predict that analysing microbiome connections as well as the microbial interplay opens new doors for advanced biocontrol technologies (ABT). Moreover, ABTs can not only be used to suppress pathogens, they can also be effectively used to establish microbiomes in a desirable, beneficial composition. It should be possible to develop 'microbiome design' strategies for particular purposes in the future. Due to the impact of the microbiome on health, growth, size, height, weight, reproduction as well as development of their host, microbiome controls

are an attractive goal. Two principles should be considered for the development of ABTs: (i) microbial diversity is an important factor determining the invasion of pathogens (Van Elsas *et al.*, 2012), and (ii) synthetic ecology can support the selection of microbes (Dunham, 2007). In all three examples mentioned, biocontrol approaches are applied. Plant health has been the target of biological control for more than 100 years, and now safe and predictable control strategies are making the development of next generation biocontrol products possible (Berg *et al.*, 2013). Since we identified the origin and function of rhizosphere microorganisms, indigenous endophyte consortia are showing promise as effective biological control agents. Moreover, in metagenomic approaches to ancient plant-associated microbiomes such as *Sphagnum* mosses, stress protection is identified as the main function (Bragina *et al.*, 2014). This could be a valuable response against climate change. In comparison, biological control of the gut microbiome is a new but extremely promising development that has been supported by the enormous success of fecal transplantations (De Vrieze, 2013). Many more applications are, however, possible, e.g. beneficial food microbiomes promoting plant and human health. Further investigation into the impact of the vegetable microbiome on our health seems to be especially important and needs more attention in the future (Berg *et al.*, 2014). While functional food can sustain human health, targeted microbial treatment of liquid diets could be used as an additional therapy in hospitals. Last but not least, the indoor microbiome needs our attention and also biological solutions for control. Currently, especially in hospitals and clean rooms, the microbiome is chemically and UV treated, allowing only resistant microorganisms to survive. Additionally, hospital-acquired infections are permanently increasing and are especially caused by (multi)resistant pathogens. In 2014, the World Health Organization produced a global map of antimicrobial resistance and issued a warning that a 'post-antibiotic' world could soon become a reality. Woolhouse and Farrar (2014) realized that this phase has already started. They have also emphatically called for a dedicated, coordinated plan of action investigating the root causes of resistance, e.g. the misuse of antimicrobials especially in agriculture, and the development of new drugs and alternative therapies. In those areas, ABTs can make significant contributions to the development of new sanitary measurements and alternative therapies.

Looking even further ahead, I see the continued development and implementation of ABTs and advanced biocontrol products like pro-, pre- or synbiotics containing synergistic microbial consortia or those which induce specific beneficial microbiomes that will contribute to the maintenance and enhancement of microbial diversity for plant and human health and for our environment. The idea to restore the 'missing microbes' (Blaser, 2014) is not only important for humans and should be extended to our environment.

## Acknowledgements

I would like to thank Martin Grube, Christin Zachow and Timothy Mark (Graz) for the helpful discussions.

## References

Berg, B., Erlacher, A., Smalla, K., and Krause, R. (2014) Vegetable microbiomes: is there a connection between opportunistic infections, human health and our 'gut feeling'? *Microb Biotechnol* **7:** 487–495.

Berg, G., and Smalla, K. (2009) Plant species and soil type cooperatively shape the structure and function of microbial communities in the rhizosphere. *FEMS Microbiol Ecol* **68:** 1–13.

Berg, G., Zachow, C., Müller, H., Philipps, J., and Tilcher, R. (2013) Next-generation bio-products sowing the seeds of success for sustainable agriculture. *Agronomy* **3:** 648–656.

Blaser, M. (2014) *Missing Microbes*. London, UK: Oneworld Publications.

Blaser, M., Bork, P., Fraser, C., Knight, R., and Wang, J. (2013) The microbiome explored: recent insights and future challenges. *Nat Rev Microbiol* **11:** 213–217.

Bragina, A., Oberauner-Wappis, L., Zachow, C., Halwachs, B., Thallinger, G.G., Müller, H., and Berg, G. (2014) The *Sphagnum* microbiome supports bog ecosystem functioning under extreme conditions. *Mol Ecol* **23:** 4498–4510.

David, L.A., Maurice, C.F., Carmody, R.N., Gootenberg, D.B., Button, J.E., Wolfe, B.E., *et al.* (2014) Diet rapidly and reproducibly alters the human gut microbiome. *Nature* **505:** 559–563.

De Vrieze, J. (2013) Medical research. The promise of poop. *Science* **341:** 954–957.

Dunham, M.J. (2007) Synthetic ecology: a model system for cooperation. *Proc Natl Acad Sci USA* **104:** 1741–1742.

Frey-Klett, P., Burlinson, P., Deveau, A., Barret, M., Tarkka, M., and Sarniguet, A. (2011) Bacterial-fungal interactions: hyphens between agricultural, clinical, environmental, and food microbiologists. *Microbiol Mol Biol Rev* **75:** 583–609.

Hanski, I., von Hertzen, L., Fyhrquist, N., Koskinen, K., Torppa, K., Laatikainen, T., *et al.* (2012) Environmental bio-diversity, human microbiota, and allergy are interrelated. *Proc Natl Acad Sci USA* **109:** 8334–8339.

Jansson, J.K. (2013) Gleaning and assembling omics parts lists from soil. *Microb Biotechnol* **6:** 3–16.

Leff, J.W., and Fierer, N. (2013) Bacterial communities associated with the surfaces of fresh fruits and vegetables. *PLoS ONE* **8:** e59310.

Mayer, E.A. (2011) Gut feelings: the emerging biology of gut–brain communication. *Nat Rev Neurosci* **12:** 453–466.

Philippot, L., Raaijmakers, J.M., Lemanceau, P., and van der Putten, W.H. (2013) Going back to the roots: the microbial ecology of the rhizosphere. *Nat Rev Microbiol* **11:** 789–799.

Van Elsas, J.D., Chiurazzi, M., Mallon, C.A., Elhottova, D., Kristufek, V., and Salles, J.F. (2012) Microbial diversity determines the invasion of soil by a bacterial pathogen. *Proc Natl Acad Sci USA* **109:** 1159–1164.

Woolhouse, M., and Farrar, J. (2014) Policy: an inter-governmental panel on antimicrobial resistance. *Nature* **509:** 555–557.

# Apoptosis induced by *Pseudomonas aeruginosa*: a lonely killer?

**Alexis Broquet**[1][*] **and Karim Asehnoune**[1,2]

[1]*Faculté de Medicine, Laboratoire UPRES EA 3826, Université de Nantes, Nantes, France.*
[2]*Pôle Anesthésie Réanimations, Service d'Anesthésie Réanimation Chirurgicale, Hôtel Dieu, CHU Nantes, Nantes, France.*

Apoptosis is a fundamental biological process allowing tissue homeostasis through the regulation of cell populations by eliminating unnecessary elements. During infection, pathogens have evolved to take advantage of this process for their own and are able to induce the apoptosis of cells, i.e. immune cells by the host itself.

*Pseudomonas aeruginosa* is one of the most studied opportunistic bacteria due to its significant involvement worldwide in pneumonia, corneal infections and wound burns. Several research groups have pointed out the ability of these bacteria to interfere and/or evade host immune system by inducing apoptosis of the targeted cells. In May 2014, looking up '*Pseudomonas aeruginosa*' and 'apoptosis' keywords in PubMed search engine retrieve more than 300 hits. *Pseudomonas aeruginosa* seems to induce apoptosis through direct interaction with the host cells (the most studied system being the type-III secretion system: T3SS) or through secreting factors such as pyocyanin.

T3SS, the most well-studied virulence apparatus of *P. aeruginosa* is composed of a needle complex through which exoenzymes are injected into the host cells (Galle *et al.*, 2012). Recently, Beyaert's laboratory described an exotoxin-independent function of the T3SS in the killing of macrophages in an acute lung infection model (Galle *et al.*, 2012). Although, T3SS is a major virulence system, it is not fully required for the bacteria to display virulence as T3SS negative strains are shown to exihibit signficant virulence (example of Elsen's paper). In short, a wide variety of *P. aeruginosa* virulence factors are involved in inducing apoptosis by several distinct mechanisms, from the activation of the mitochondrial pathway,

*For correspondence. E-mail alexis.broquet@univ-nantes.fr

**Funding Information** No funding information provided.

the generation of reactive oxygen species to the activation of the caspase pathways (Table 1).

The ability of *P. aeruginosa* to induce apoptosis in various *in vitro* model of infection (macrophages, neutrophils, epithelial cells . . .) or *in vivo* models such as lungs, cornea and burn wounds infections is not mediated by a single bacterial cell but rather by a multicellular population of *P. aeruginosa*. Members of such population interact with each other through a number of chemical signals known globally on for quorum sensing (QS). Quoting Rutherford and Bassler (2012), 'Quorum sensing is a bacterial cell-cell communication process that involves the production, detection, and response to extracellular signaling molecules called autoinducers'. QS molecules were shown to regulate virulence factors such as toxins, exotoxin A, pyocyanin, . . . and so *in fine* apoptosis (Rutherford and Bassler, 2012). The best described QS signalling systems in *P. aeruginosa* are the N-acyl homoserine lactones systems Las and Rhl. The Las system produces and responds to N-oxododecanoyl homoserine lactone and the Rhl system to N-butanoyl homoserine lactone respectively. Las system is known to control the production of various virulence factors involved in host cell damages such as exotoxin A (Jones *et al.*, 1993). On the other hand, Rhl system was described to repress the expression of genes responsible for the assembly and function of the T3SS (Bleves *et al.*, 2005).

Last but not least, QS molecule such as 3-oxododecanoyl-L-homoserine lactone (3-oxo-C12-HSL) itself has been shown to induce apoptosis. Several studies have demonstrated that incubation of different cell lines with 3-oxo-C12-HSL molecule resulted in the induction of apoptosis involving calcium signalling, the mitochondrial pathway and caspase activations (Table 2). Interestingly, N-butanoyl-L-homoserine lactone (known as C4-HSL, the second major QS molecule in *P. aeruginosa*) harbouring a shorter fatty acid chain has not been shown to induce apoptosis compared with 3-oxo-C12-HSL (Tateda *et al.*, 2003; Holban *et al.*, 2014).

Knowing that *P. aeruginosa* is able to induce apoptosis through its QS systems molecule, studies focusing on apoptosis induction should be considered with the context of QS signalling. Particularly, QS considerations should be taken into account when comparing studies using

**Table 1.** Virulence factors inducing apoptosis in targeted host cells (non-exhaustive list).

| Virulence factors | Model used | Apoptosis pathway | Reference |
|---|---|---|---|
| Type-III secretion system (T3SS) | | | |
| Injector/needle complex | Macrophages/neutrophils | Caspase 3 | Galle et al., 2012 |
| Exoenzymes (ExoS, T, U, Y) | Macrophages | Caspase 3 | Galle et al., 2008 |
| | Epithelial cells/fibroblasts | Mitochondrial pathway, caspase 3 | Shafikhani et al., 2008 |
| Secreted virulence factors | | | |
| Pyocyanin | Neutrophils | Reactive oxygen intermediate, cAMP | Usher et al., 2002 |
| Exotoxin A | MEFs | Bak pathway | Du et al., 2010 |
| Protease | Macrophages | Caspase 3 | Zhang et al., 2003 |
| ExlA | HUVECs endothelial cells | Unknown | Elsen et al., 2014 |

cAMP, cyclic adenosine monophosphate; HUVECs, human umbilical veins; MEFs, mouse embryonic fibroblasts.

**Table 2.** Apoptosis pathways activated by 3-oxo-C12-HSL (non-exhaustive list).

| Quorum-sensing molecule | Abbreviation used in the study | Model used | Apoptosis pathway involved | Reference |
|---|---|---|---|---|
| N-3-(oxododecanoyl)-l-homoserine lactone | 3-oxo-C12-HSL | Macrophages/neutrophils | Caspase 3/8 | Tateda et al., 2003 |
| | OdDHL | Breast cancer cell lines | JAK/STAT pathway | Li et al., 2004 |
| | OdDHL | Jurkat cell line | Mitochondrial pathway | Jacobi et al., 2009 |
| | C12 | Airway epithelial cells | Cytochrome c, caspases 3/7, 8 and 9 | Schwarzer et al., 2012 |

different multiplicity of infection or bacteria preparation protocols, processes that influence QS molecules concentration and/or bacteria population numbers.

## Conflict of interest

None declared.

## References

Bleves, S., Soscia, C., Nogueira-Orlandi, P., Lazdunski, A., and Filloux, A. (2005) Quorum sensing negatively controls type III secretion regulon expression in Pseudomonas aeruginosa PAO1. J Bacteriol 187: 3898–3902.

Du, X., Youle, R.J., FitzGerald, D.J., and Pastan, I. (2010) Pseudomonas exotoxin A-mediated apoptosis is Bak dependent and preceded by the degradation of Mcl-1. Mol Cell Biol 30: 3444–3452.

Elsen, S., Huber, P., Bouillot, S., Couté, Y., Fournier, P., Dubois, Y., et al. (2014) A type III secretion negative clinical strain of Pseudomonas aeruginosa employs a two-partner secreted exolysin to induce hemorrhagic pneumonia. Cell Host Microbe 15: 164–176.

Galle, M., Schotte, P., Haegman, M., Wullaert, A., Yang, H.J., Jin, S., and Beyaert, R. (2008) The Pseudomonas aeruginosa type III secretion system plays a dual role in the regulation of caspase-1 mediated IL-1beta maturation. J Cell Mol Med 12 (5A): 1767–1776.

Galle, M., Carpentier, I., and Beyaert, R. (2012) Structure and function of the Type III secretion system of Pseudomonas aeruginosa. Curr Protein Pept Sci 13: 831–842.

Galle, M., Jin, S., Bogaert, P., Haegman, M., Vandenabeele, P., and Beyaert, R. (2012) The Pseudomonas aeruginosa type III secretion system has an exotoxin S/T/Y independ-

ent pathogenic role during acute lung infection. PLoS ONE 7: e41547.

Holban, A.M., Bleotu, C., Chifiriuc, M.C., Bezirtzoglou, E., and Lazar, V. (2014) Role of Pseudomonas aeruginosa quorum sensing (QS) molecules on the viability and cytokine profile of human mesenchymal stem cells. Virulence 5: 303–310.

Jacobi, C.A., Schiffner, F., Henkel, M., Waibel, M., Stork, B., Daubrawa, M., et al. (2009) Effects of bacterial N-acyl homoserine lactones on human Jurkat T lymphocytes-OdDHL induces apoptosis via the mitochondrial pathway. Int J Med Microbiol 299: 509–519.

Jones, S., Yu, B., Bainton, N.J., Birdsall, M., Bycroft, B.W., Chhabra, S.R., et al. (1993) The lux autoinducer regulates the production of exoenzyme virulence determinants in Erwinia carotovora and Pseudomonas aeruginosa. EMBO J 12: 2477–2482.

Li, L., Hooi, D., Chhabra, S.R., Pritchard, D., and Shaw, P.E. (2004) Bacterial N-acylhomoserine lactone-induced apoptosis in breast carcinoma cells correlated with down-modulation of STAT3. Oncogene 23: 4894–4902.

Rutherford, S.T., and Bassler, B.L. (2012) Bacterial quorum sensing: its role in virulence and possibilities for its control. Cold Spring Harb Perspect Med 2: a012427.

Schwarzer, C., Fu, Z., Patanwala, M., Hum, L., Lopez-Guzman, M., Illek, B., et al. (2012) Pseudomonas aeruginosa biofilm-associated homoserine lactone C12 rapidly activates apoptosis in airway epithelia. Cell Microbiol 14: 698–709.

Shafikhani, S.H., Morales, C., and Engel, J. (2008) The Pseudomonas aeruginosa type III secreted toxin ExoT is necessary and sufficient to induce apoptosis in epithelial cells. Cell Microbiol 10: 994–1007.

Tateda, K., Ishii, Y., Horikawa, M., Matsumoto, T., Miyairi, S., Pechere, J.C., et al. (2003) The Pseudomonas aeruginosa autoinducer N-3-oxododecanoyl homoserine lactone accelerates apoptosis in macrophages and neutrophils. *Infect Immun* **71:** 5785–5793.

Usher, L.R., Lawson, R.A., Geary, I., Taylor, C.J., Bingle, C.D., Taylor, G.W., and Whyte, M.K. (2002) Induction of neutrophil apoptosis by the *Pseudomonas aeruginosa*

exotoxin pyocyanin: a potential mechanism of persistent infection. *J Immunol* **168:** 1861–1868.

Zhang, J., Takayama, H., Matsuba, T., Jiang, R., and Tanaka, Y. (2003) Induction of apoptosis in macrophage cell line, J774, by the cell-free supernatant from *Pseudomonas aeruginosa*. *Microbiol Immunol* **47:** 199–206.

# A chromosomally encoded T7 RNA polymerase-dependent gene expression system for *Corynebacterium glutamicum*: construction and comparative evaluation at the single-cell level

Maike Kortmann, Vanessa Kuhl, Simon Klaffl and Michael Bott*

*Institute of Bio- and Geosciences, IBG-1: Biotechnology, Forschungszentrum Jülich, Jülich D-52425, Germany.*

## Summary

*Corynebacterium glutamicum* has become a favourite model organism in white biotechnology. Nevertheless, only few systems for the regulatable (over)expression of homologous and heterologous genes are currently available, all of which are based on the endogenous RNA polymerase. In this study, we developed an isopropyl-β-D-1-thiogalactopyranosid (IPTG)-inducible T7 expression system in the prophage-free strain *C. glutamicum* MB001. For this purpose, part of the DE3 region of *Escherichia coli* BL21(DE3) including the T7 RNA polymerase gene *1* under control of the *lac*UV5 promoter was integrated into the chromosome, resulting in strain MB001(DE3). Furthermore, the expression vector pMKEx2 was constructed allowing cloning of target genes under the control of the T7*lac* promoter. The properties of the system were evaluated using *eyfp* as heterologous target gene. Without induction, the system was tightly repressed, resulting in a very low specific eYFP fluorescence (= fluorescence per cell density). After maximal induction with IPTG, the specific fluorescence increased 450-fold compared with the uninduced state and was about 3.5 times higher than in control strains expressing *eyfp* under control of the IPTG-induced *tac* promoter with the endogenous RNA polymerase. Flow cytometry revealed that T7-based *eyfp* expression resulted in a highly uniform population, with 99% of all cells showing high fluorescence. Besides *eyfp*, the functionality of the corynebacterial T7 expression system was also successfully demonstrated by overexpression of the *C. glutamicum pyk* gene for pyruvate kinase, which led to an increase of the specific activity from 2.6 to 135 U mg$^{-1}$. It thus presents an efficient new tool for protein overproduction, metabolic engineering and synthetic biology approaches with *C. glutamicum*.

*For correspondence. E-mail m.bott@fz-juelich.de

**Funding Information** No funding information provided.

## Introduction

The recombinant production of proteins is a highly important issue in industrial biotechnology as well as in scientific research. Many different expression systems have been established in various eukaryotic and prokaryotic organisms (Demain and Vaishnav, 2009). Due to their easy handling and well-established genetic tools, bacteria are broadly used to express heterologous and homologous genes (Baneyx, 1999; Terpe, 2006; Chen, 2012). One of the most popular and commonly used systems for high-level protein production in *Escherichia coli* is the T7 expression system developed by Studier and Moffatt (1986). It is based on the RNA polymerase (RNAP) of bacteriophage T7, which shows a number of beneficial properties: (i) single-subunit enzyme in contrast to multi-subunit bacterial RNAP, (ii) high processivity, (iii) high specificity towards the T7 promoter, (iv) independence of auxiliary transcription factors, (v) production of very long transcripts, and (vi) termination only by class I and class II termination signals that differ significantly from bacterial transcription termination sites (Chamberlin and Ring, 1973; Macdonald *et al.*, 1994; Lyakhov *et al.*, 1998). Expression hosts like *E. coli* BL21(DE3) carry a single copy of gene *1* for T7 RNAP located chromosomally on a λDE3 lysogen (Studier and Moffatt, 1986). In strain *E. coli* BL21(DE3), transcription of gene *1* is controlled by a *lac*UV5 promoter, allowing repression by LacI and induction with isopropyl-β-D-1-thiogalactopyranoside (IPTG). The expression of desired target genes is controlled by the T7 promoter, which is usually present on a suitable expression vector. To minimize basal transcription, a LacI binding site can be introduced in front of the target gene, placing both gene *1* and the target gene under the control of the LacI repressor (Dubendorff and Studier, 1991). The characteristics of the T7 RNAP-dependent expression system permit a very efficient and exclusive expression of genes under control of the strong T7 promoter. Due to its favourable properties, the T7 RNAP-based expression system has also been established in a variety of other bacteria, such as *Pseudomonas aeruginosa* (Brunschwig and Darzins, 1992), *Pseudomonas putida* (Herrero *et al.*, 1993), *Ralstonia eutropha* (Barnard *et al.*, 2004), *Bacillus megaterium* (Gamer *et al.*, 2009), *Streptomyces lividans* (Lussier *et al.*, 2010), *Rhodobacter capsulatus* (Katzke *et al.*, 2010; Arvani *et al.*, 2012) and *Corynebacterium acetoacidophilum* (Equbal *et al.*, 2013).

*Corynebacterium glutamicum* is a Gram-positive soil bacterium of the order *Corynebacteriales* and serves in industry as the major host for production of amino acids, with L-glutamate and L-lysine being the most important ones. Efficient strains are available also for the synthesis of a variety of other amino acids, for example L-leucine (Vogt *et al.*, 2013), L-serine (Stolz *et al.*, 2007) or D-serine (Stäbler *et al.*, 2011). Furthermore, a variety of other commercially interesting metabolites can be produced with *C. glutamicum* (Becker and Wittmann, 2012), such as organic acids (Wendisch *et al.*, 2006; Okino *et al.*, 2008; Litsanov *et al.*, 2012a,b; Wieschalka *et al.*, 2013), diamines (Mimitsuka *et al.*, 2007; Kind and Wittmann, 2011; Schneider and Wendisch, 2011) or alcohols (Inui *et al.*, 2004; Smith *et al.*, 2010; Blombach *et al.*, 2011; Yamamoto *et al.*, 2013). Despite its complex cell envelope (Bansal-Mutalik and Nikaido, 2011; Marchand *et al.*, 2012; Laneelle *et al.*, 2013), *C. glutamicum* is also an efficient host for the secretory production of heterologous proteins (see Kikuchi *et al.*, 2008; Scheele *et al.*, 2013; Matsuda *et al.*, 2014; and references therein). Based on the broad spectrum of products and its robustness in large-scale production processes, *C. glutamicum* has become a platform and model organism in industrial biotechnology (Eggeling and Bott, 2005; Burkovski, 2008; Yukawa and Inui, 2013).

The development of production strains often requires the controlled expression of target genes or operons. All currently available systems for controlling gene expression in *C. glutamicum* are based on transcription by the endogenous RNA polymerase (Kirchner and Tauch, 2003; Eggeling and Reyes, 2005; Nesvera and Patek, 2011; Patek *et al.*, 2013). In this study, we constructed an IPTG-inducible expression system in *C. glutamicum* that is based on T7 RNAP. We characterized the properties of this system with the *eyfp* gene for enhanced yellow fluorescent protein (Perez-Jimenez *et al.*, 2006), which allows for analysing population heterogeneity by flow cytometry, and the homologous *pyk* gene for pyruvate kinase as a test case for overproduction of a cytosolic enzyme. The results obtained show that the T7 system allows very efficient and controllable protein overproduction in *C. glutamicum* to levels that outperform currently available systems.

## Results and discussion

### Construction of a T7 RNAP-dependent expression system for C. glutamicum

In this study, a T7 RNAP-dependent expression system was developed for *C. glutamicum* based on a chromosomally encoded T7 RNAP and a vector in which the target gene was placed under the control of a T7 promoter. For regulatable chromosomal expression of the T7 RNAP gene *1*, a 4.47 kb fragment (sequence is shown Fig. S1) was amplified by polymerase chain reaction (PCR) from the genome of *E. coli* BL21(DE3) (Table 1) with oligonucleotides DE3-for and DE3-rev (Table 2) that contains the repressor gene *lacI* under the control of its native promoter, *lacZα*, and T7 gene *1*, the latter two under the control of the *lac*UV5 promoter, including three LacI operator sites O1-O3 (Fig. 1A). The fragment was used to construct plasmid pK18*mobsacB*-DE3 (Table 1), in which the DE3 insert is flanked by adjacent 800 bp regions covering the genes cg1121 (encoding a permease of the major facilitator superfamily) and cg1122 (encoding a putative secreted protein) and their downstream regions (Fig. 1A). Via homologous recombination (Niebisch and Bott, 2001), the DE fragment was integrated into the intergenic region of cg1121-cg1122 within the genome of *C. glutamicum* MB001 (NC_022040.1), a prophage-free derivative of the type strain ATCC 13032, which showed a higher expression level for eYFP than the parent strain (Baumgart *et al.*, 2013b). The insertion site is located 340 bp downstream of the cg1121 stop codon. Kanamycin-sensitive and sucrose-resistant clones were checked by PCR and sequence analysis for the correct chromosomal insertion of the DE fragment and the generated strain was named *C. glutamicum* MB001(DE3).

For exclusive transcription by T7 RNAP, the target gene has to be under control of a T7 promoter. Therefore, a suitable expression vector was constructed based on the shuttle vector pJC1, which contains the pHM1519 replicon for *C. glutamicum* and the pACYC177 replicon for *E. coli* (Cremer *et al.*, 1990). In order to provide unique restriction sites for BamHI, XbaI and SalI in the multiple cloning site (MCS) to be inserted, the corresponding restriction sites in the backbone sequence of pJC1 were deleted in advance. For this purpose, pJC1 was digested with BamHI and SalI, the resulting 5′-overhangs filled in with Klenow polymerase and re-circularized via blunt end ligation, resulting in pJC1ΔBXS. A 1.97 kb fragment of plasmid pET52b(+) containing *lacI*, the T7 promoter, the *lac* operator and the downstream MCS was amplified with the oligonucleotides pETEx-for and pETEx-rev and inserted into the unique PstI restriction site of pJC1ΔBXS. After insertion of the expression cassette, an NcoI restriction site in the pJC1 backbone was removed by exchanging a single base (CCATTG → CCATAG). The resulting

**Table 1.** Bacterial strains and plasmids used in this study.

| Strain or plasmid | Relevant characteristics | Source |
|---|---|---|
| Strain | | |
| *E. coli* BL21(DE3) | F⁻ *ompT hsdSB*(rB⁻ mB⁻) *gal dcm* (*lcIts857 ind1 Sam7 nin5 lac*UV5-T7 gene *1*) | (Studier and Moffatt, 1986) |
| *E. coli* DH5α | F⁻ φ80*lacZ*ΔM15 Δ(*lacZYA-argF*)U169 *recA*1 *endA*1 *hsdR*17(rk⁻, mk⁺) *phoA supE*44 *thi*-1 *gyrA*96 *relA*1 λ⁻ | Invitrogen |
| *C. glutamicum* MB001 | Type strain ATCC 13032 with deletion of prophages CGP1 (cg1507-cg1524), CGP2 (cg1746-cg1752), and CGP3 (cg1890-cg2071) | (Baumgart *et al.*, 2013b) |
| *C. glutamicum* MB001(DE3) | MB001 derivative with chromosomally encoded T7 gene *1* (cg1122-P*lacI*-*lacI*-P*lacUV5*−*lacZα*-T7 gene *1*-cg1121) | This study |
| Plasmid | | |
| pEKEx2 | Kan^R; *C. glutamicum/E. coli* shuttle vector for regulated gene expression (P*tac*, *lacI*q, pBL1 *oriV_Cg*, pUC18 *oriV_Ec*) | (Eikmanns *et al.*, 1991) |
| pEKEx2-*eyfp* | Kan^R; expression plasmid carrying the *eyfp* gene under the control of the *tac* promoter | (Hentschel *et al.*, 2013) |
| pET-52b(+) | Amp^R; *E. coli* vector for expression of target genes under control of the T7 promoter (pBR322 *ori_Ec*, P_T7, *lacI*) | Novagen |
| pJC1 | Kan^R; *E. coli/C. glutamicum* shuttle vector (pHM1519 *ori_Cg*, pACYC177 *ori_Ec*); | (Cremer *et al.*, 1990) |
| pJC1-*venus*-term | Kan^R; pJC1 derivative containing *venus* gene and additional terminators | (Baumgart *et al.*, 2013a) |
| pJC1ΔBXS | Kan^R; pJC1 derivative lacking BamHI, XbaI, and SalI restriction sites | This study |
| pJC1-P*tac*-*eyfp* | Kan^R; pJC1 derivative containing the *eyfp* gene under the control of the P*tac* | This study |
| pK18*mobsacB* | Kan^R; vector for allelic exchange in *C. glutamicum* (oriT oriV_Ec sacB lacZα) | (Schäfer *et al.*, 1994) |
| pK18*mobsacB*-DE3 | Kan^R; pK18*mobsacB* derivative containing the 4.5 kb λDE3 region (P*lacI*, *lacI*, P*lacUV5*, gene *1*) from *E. coli* BL21(DE3) flanked by two 800-bp DNA regions for homologous recombination into the intergenic region of cg1121 and cg1122 | This study |
| pMKEx2 | Kan^R; *E. coli/C. glutamicum* shuttle vector based on pJC1 for expression of target genes under control of the T7 promoter (P*lacI*, *lacI*, P_T7, *lacO*1, N-term. Strep•tag II, MCS, C-term. His•tag, pHM1519 *ori_Cg*; pACYC177 *ori_Ec*) | This study |
| pMKEx2-*eyfp* | Kan^R; pMKEx2 derivative containing the *eyfp* gene under control of P_T7 | This study |
| pMKEx2-*pyk* | Kan^R; pMKEx2 derivative with the *C. glutamicum pyk* gene under control of P_T7 | This study |

expression vector was named pMKEx2 (GenBank accession number KM658503) and allows fusion of the target protein either with an N-terminal Streptag II or with a C-terminal decahistidine tag (Fig. 1B).

*Characterization of T7 RNAP-dependent expression system in* C. glutamicum *with the heterologous target protein eYFP and comparison with* P_tac-*based expression*

To analyse the functionality and efficiency of the newly constructed T7 expression system in *C. glutamicum*, the heterologous model protein eYFP (Perez-Jimenez et al., 2006) was used as target, as it allows easy detection also at the single-cell level by fluorescence microscopy and

fluorescence-activated cell sorting (FACS). The *eyfp* gene was amplified from the plasmid pEKEx2-*eyfp* (Hentschel et al., 2013) with oligonucleotides eYFP-for and eYFP-rev and cloned as NcoI-BamHI fragment into pMKEx2 under transcriptional control of the T7 promoter. The resulting plasmid pMKEx2-*eyfp* was transferred into *C. glutamicum* MB001(DE3), and for control purposes into *C. glutamicum* MB001. The synthesis of eYFP by strain MB001(DE3)/pMKEx2-*eyfp* was compared with strain MB001/pEKEx2-*eyfp*. The well-established expression vector pEKEx2 permits expression of target genes under control of the IPTG-inducible *tac* promoter by the endogenous RNA polymerase (Eikmanns et al., 1991).

The strains were cultivated in CGXII minimal medium with 4% (wt/vol) glucose using a BioLector system

**Table 2.** Oligonucleotides used in this study.

| Name | Sequence (5′→3′) | Restriction sites |
|---|---|---|
| DE3-for | CCGCTCGAGAACTGCGCAACTCGTGAAAGG | XhoI |
| DE3-rev | CGGAATTCGTTACGCGAACGCGAAGTC | EcoRI |
| pETEx-for | AACTGCAGGGAGCTGACTGGGTTGAAGG | PstI |
| pETEx-rev | AACTGCAGCTTAATGCGCCGCTACAGGG | PstI |
| pEKEx-for | GGAATTCCATATGTCGCTCAAGCCTTCGTCACTG | NdeI |
| pEKEx-rev | GGACTAGTTTATCTAGACTTGTACAGCTCGTCCATG | SpeI |
| eYFP-for | CATGCCATGGACAATAACGATCTCTTTCAGGCATCAC | NcoI |
| eYFP-rev | CGGGATCCTCAGCCCGCGAGCACC | BamHI |
| PykCg-for | CGGGATCCGGCGTGGATAGACGAACTAAG | BamHI |
| PykCg-rev | TGTACAGACACCACGTACAGTGTCAACGC | BsrGI |

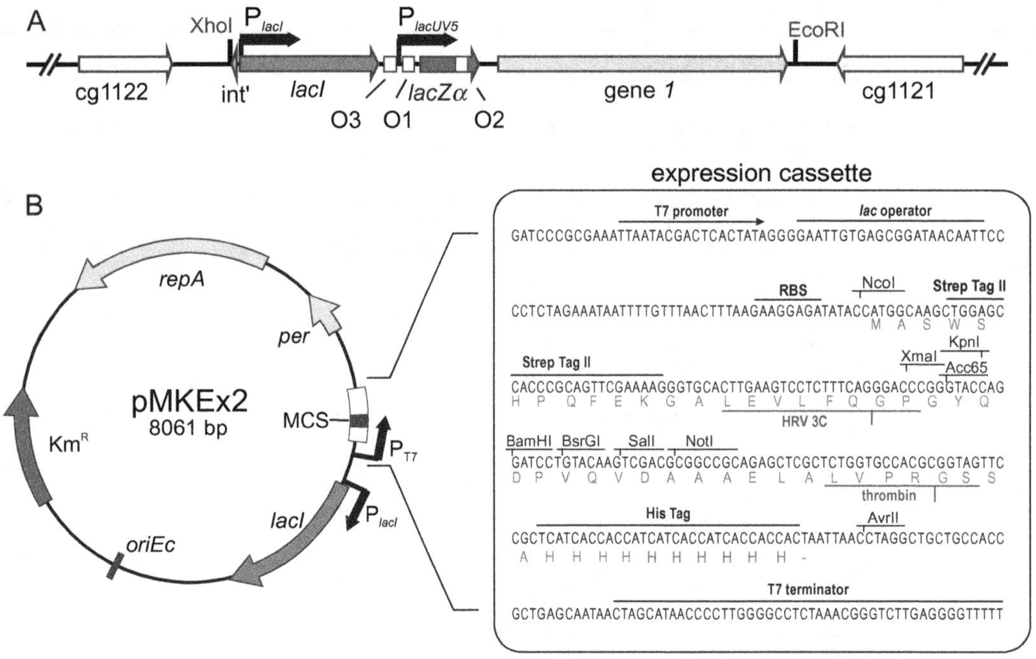

**Fig. 1.** A. genomic region of *C. glutamicum* MB001(DE3) carrying the DE3 insertion. A 4.5 kb DNA fragment was amplified from chromosomal DNA of *E. coli* BL21(DE3) and inserted into the intergenic region of cg1121-cg1122 of MB001(DE3). The fragment contains *lacI*, *lacZα* and T7 gene *1*, the latter two under the transcriptional control of the *lac*UV5 promoter and its three LacI operator sites O1-O3. B. Map of the expression plasmid pMKEx2, which is based on pJC1 and an expression cassette from pET52b. The region between the T7 promoter and the T7 terminator is shown in detail.

(Fig. S2) in the presence of 0, 5, 10, 15, 25, 50, 100 and 250 μM IPTG and expression of the target gene was determined by measuring the specific eYFP fluorescence (Fig. 2). For strain MB001/pMKEx2-*eyfp*, which lacks T7 RNAP, only background specific fluorescence was observed (< 0.001). In the case of strain MB001(DE3)/pMKEx2-*eyfp*, specific fluorescence was very low in the absence of IPTG (< 0.001), but increased up 0.30 when the medium was supplemented with 100 μM IPTG. These results demonstrate that the T7 promoter is not recognized by the corynebacterial RNAP, whereas specific and efficient expression occurs in the presence of T7 RNAP. In the case of strain MB001/pEKEx2-*eyfp*, the specific fluorescence in the absence of IPTG (0.001) was 1.6-fold higher than in the case of MB001(DE3)/pMKEx2-*eyfp* and increased up to 0.08 in the presence of 100 μM IPTG (Fig. 2). Thus, the T7 system shows a slightly lower expression level in the absence of IPTG and an up to fourfold higher maximal expression level in the presence of IPTG compared with pEKEx2-based expression of *eyfp*. Half-maximal specific fluorescence was obtained with 20 μM IPTG for strain MB001(DE3)/pMKEx2-*eyfp* and with 31 μM IPTG for strain MB001(DE3)/pEKEx2-*eypf*.

To exclude an influence of the different replication mechanisms of pMKEx2 and pEKEx2, the specific eYFP fluorescence was also determined in *C. glutamicum*

MB001(DE3) transformed with plasmid pJC1-P*tac*-*eyfp*. This plasmid contains the same replicon as pMKEx2 and the expression cassette of pEKEx2 (for construction see *Experimental procedures*). The results obtained with MB001/pJC1-P*tac*-*eyfp* were comparable to that of MB001/pEKEx2-*eyfp*, indicating that the difference to T7-based *eyfp* expression is not caused by the different replicons, but by the use of different RNAPs and promoters (Fig. 2).

To further characterize the T7 expression system in *C. glutamicum*, the four strains described above were analysed by SDS-PAGE, Western blotting and fluorescence microscopy. The Coomassie-stained SDS-polyacrylamide gel and more clearly the Western blot with anti-green-fluorescent protein (GFP) antiserum shown in Fig. 2B qualitatively confirmed the results of the specific fluorescence measurements described above. In the case of strains MB001/pEKEx2-*eyfp* and MB001(DE3)/pMKEx2-*eyfp*, no distinct band with a size of 27 kDa (calculated mass of eYFP) was visible in cells grown in the absence of IPTG, whereas a faint band was visible in the case of strain MB001/pJC1-P*tac*-*eyfp*. When cultivated in the presence of 250 μM IPTG, the 27 kDa eYFP protein band was clearly visible in strains MB001/pEKEx2-*eyfp*, MB001/pJC1-P*tac*-*eyfp* and pMB001(DE3)/pMKEx2-*eyfp*. For the latter strain, the intensity of the eYFP band continuously increased when the IPTG concentration was

A

**Fig. 2.** Comparison of eYFP synthesis in *C. glutamicum* with the newly constructed T7-based expression system and the pEKEx2 system using the *tac* promoter and the endogenous RNAP. The strains MB001/pEKEx2-*eyfp* (●/1), MB001/pJC1-P$_{tac}$-*eyfp* (△/2), MB001(DE3)/pMKEx2 (◊/3), MB001/pMKEx2-*eyfp* (□/4) and MB001(DE3)/pMKEx2-*eyfp* (■/5) were grown aerobically in CGXII minimal medium with 4% (wt/vol) glucose using a Biolector system at 30°C and 1200 r.p.m. Target gene expression was induced 2 h after inoculation by addition of 0–250 μM IPTG.
A. 25 h after starting the cultivation, the maximal specific eYFP fluorescence (ratio of fluorescence emission at 532 nm and backscatter value at 620 nm) was determined. Mean values of at least three independent experiments and standard deviations are shown.
B. For protein analysis, cells were disrupted by beat-beating and equivalent amounts of total protein (10 μg) of the cell-free extracts were subjected to SDS-PAGE and visualized by staining with Coomassie Brilliant Blue. In addition, eYFP was detected by Western blotting with an polyclonal anti-GFP antibody. The arrows indicate the predicted size of 27.2 kDa for eYFP.
C. Cells were analysed by fluorescence microscopy and images were taken with an exposure time of 40 ms. The red bar represents a length of 5 μm.

B

C

## Characterization of T7 RNAP-dependent expression system in C. glutamicum *with the heterologous target protein eYFP and comparison with* P$_{tac}$-*based expression at the single-cell level*

Flow cytometry was used to analyse *eyfp* expression at the single-cell level, allowing the detection of population heterogeneity (Figs 3 and S3). The gate used to define background fluorescence was set with *C. glutamicum* MB001(DE3)/pMKEx2, with 100% of the analysed cells falling into this gate (Fig. 3A). In the case of strain MB001/pEKEx2-*eyfp* (Fig. 3B), 7% of the cells cultivated in the absence of IPTG showed fluorescence above background, with an average intensity of $1.0 \times 10^2$, confirming the leakiness of the *tac* promoter used. When cultivated in the presence of 5, 10, 15, 25, 50, 100 and 250 μM IPTG, MB001/pEKEx2-*eyfp* cells formed subpopulations (Fig. 3E and Fig. S3A and C). At 250 μM IPTG, 19.5% of the cells showed a high average fluorescence signal of $2.63 \times 10^4$, whereas 78.5% of the cells had only a weak average fluorescence signal of $7.80 \times 10^3$, and 2% of the cells possessed only background fluorescence. The overall averaged fluorescence signal of MB001/pEKEx2-*eyfp* cells including all subpopulations was $1.13 \times 10^4$.

In the case of strain MB001/pJC1-P$_{tac}$-*eyfp* (Fig. 3C), which was only analysed after cultivation without and with 250 μM IPTG, more than 99% of the cells cultivated in the presence of 250 μM IPTG showed high fluorescence, with an average signal intensity of $7.38 \times 10^3$ and formed a homogenous population. But also in the absence of IPTG, about 90% of the cells showed a low fluorescence signal of $4.4 \times 10^2$ above background, confirming the results of the Western blot shown in Fig. 2B. As pEKEx2-*eyfp* and pJC1-P$_{tac}$-*eyfp* possess identical target gene expression determinants, the differences observed by flow cytometry presumably result from the different replicons. The vector

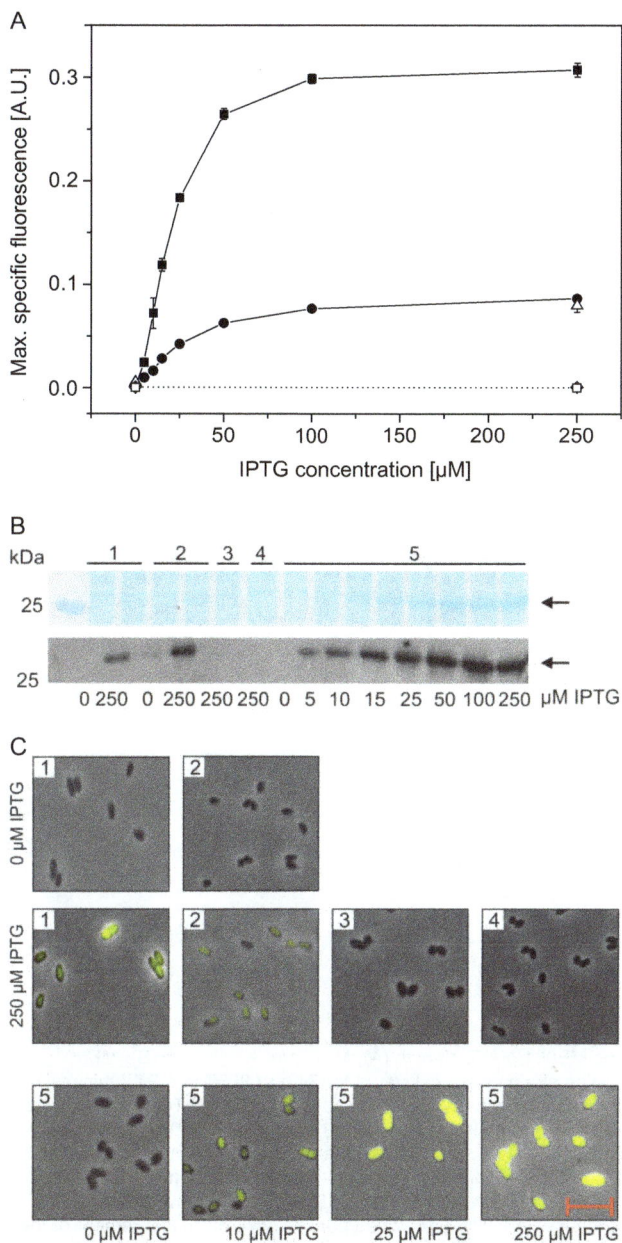

raised from 5 μM to 100 μM. The eYFP bands of MB001/pEKEx2-*eyfp* and MB001/pJC1-P$_{tac}$-*eyfp* in the presence of 250 μM IPTG were much fainter than those of strain MB001(DE3)/pMKEx2-*eyfp* in the presence of 50, 100 and 250 μM IPTG. As expected, no eYFP band was visible in the negative control strains MB001(DE3)/pMKEx2 and MB001/pMKEx2-*eyfp* in the presence of 250 μM IPTG. The fluorescence microscopy images shown in Fig. 2C were also in agreement with the results of the specific fluorescence measurements, the SDS polyacrylamide gels and the Western blots, with the most strongly fluorescent cells being those of strain MB001(DE3)/pMKEx2-*eyfp* cultivated in the presence of 250 μM IPTG.

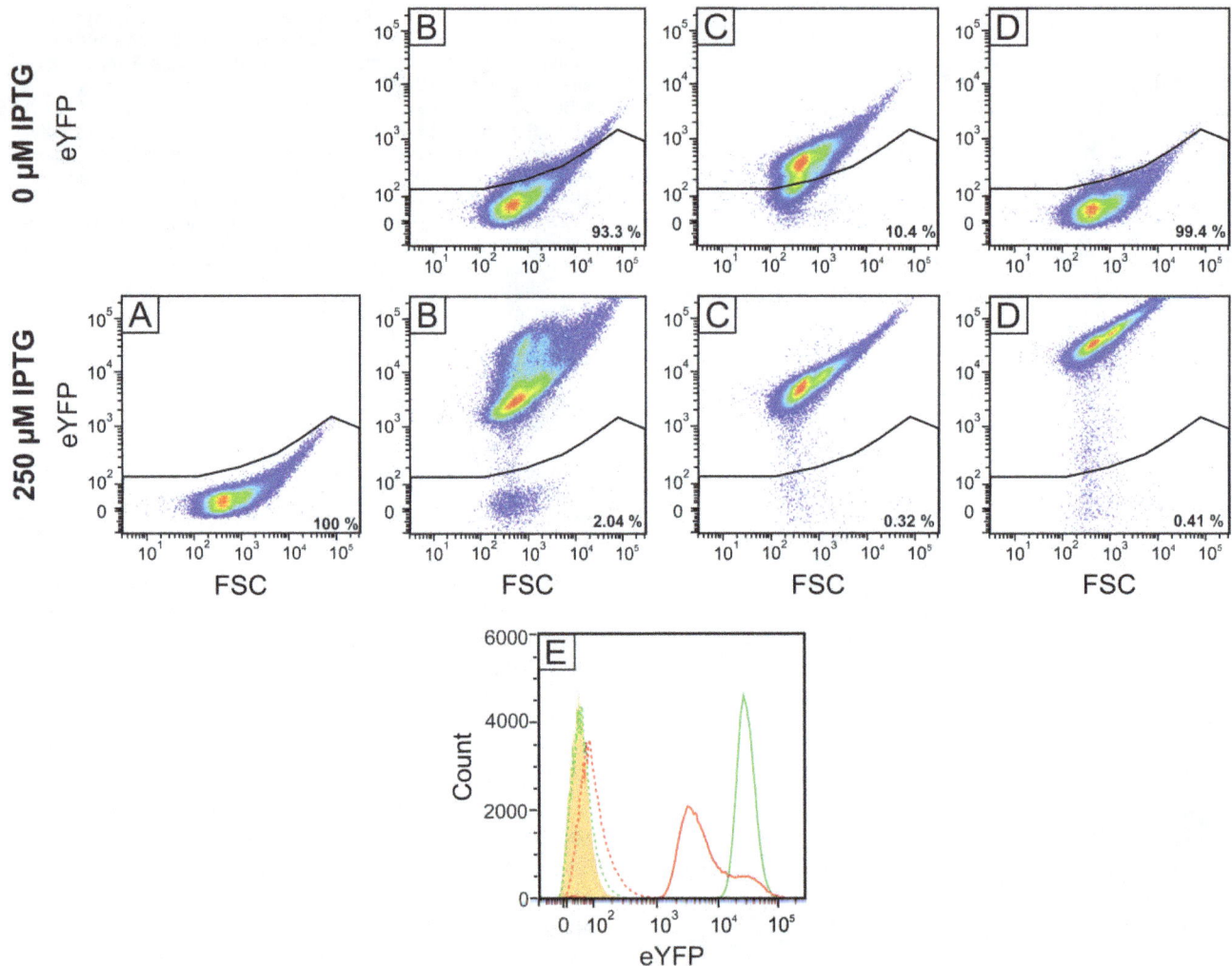

**Fig. 3.** Analysis of heterologous eYFP production in the *C. glutamicum* strains MB001/pMKEx2-*eyfp* (A), MB001/pEKEx2-*eyfp* (B), MB001/pJC1-P*tac*-*eyfp* (C) and MB001(DE3)/pMKEx2-*eyfp* (D) at the single-cell level. The strains were cultivated for 24 h at 30°C in CGXII minimal medium with 4% (wt/vol) glucose using a Biolector system. Induction of *eyfp* expression was triggered by adding 250 μM IPTG to the cultures after 2 h. Pseudo-coloured dot plots from flow cytometry analysis (excitation at 488 nm, emission at 533 nm) of at least 100 000 cells of each strain displaying the eYFP fluorescence signal against the forward scatter signal (FSC) are shown. The gate used to define non-fluorescent cells was set with *C. glutamicum* MB001(DE3)/pMKEx2 with 100% of the cells falling into this gate (data not shown). The number inside the blot indicates the percentage of cells inside this gate. Panel E shows a histogram of the strains MB001/pEKEx2-*eyfp* (red) and MB001(DE3)/pMKEx2-*eyfp* (green). The number of cells is plotted against the eYFP fluorescence intensity. The dotted peaks show the measurement of the uninduced culture, the continuous line the cultures grown in the presence of 250 μM IPTG. The orange peak represents the background set with strain MB001(DE3)/pMKEx2.

pEKEx2 (Eikmanns *et al.*, 1991) contains the replicon from plasmid pBL1 (Santamaria *et al.*, 1984), whereas pJC1 (Cremer *et al.*, 1990) contains the replicon from plasmid pHM1519 (Miwa *et al.*, 1984), which presumably is identical with the one from plasmids pCG1 (Ozaki *et al.*, 1984), pSR1 (Yoshihama *et al.*, 1985) and pCG100 (Trautwetter and Blanco, 1991), as described previously (Nešvera and Pátek, 2008). Both replicons mediate replication in the rolling circle mode, but the pBL1 replicons belong to pIJ101/pJV1 family, whereas the pHM1519 replicon belongs to pNG2 family (Pátek and Nešvera, 2013). The copy number of pBL1 and similar plasmids

was estimated to be between 10 and 30 copies per chromosome (Miwa *et al.*, 1984; Santamaria *et al.*, 1984), and that of pCG100 was also reported to be about 30 copies per chromosome (Trautwetter and Blanco, 1991). Apparently, the pNM1519 replicon is more stable than the pBL1 replicon, at least in the case of the expression vectors used in our study.

In the case of strain MB001(DE3)/pMKEx2-*eyfp* (Fig. 3D), less than 1% of the cells cultivated in the absence of IPTG showed fluorescence slightly above background. The fluorescence of cells cultivated in the presence of 5, 10, 15, 25, 50, 100 and 250 μM IPTG is shown in Fig. S3B and D. In the presence of 250 IPTG, 99.5% of the cells formed a very homogenous population, with an average fluorescence signal of about $5.22 \times 10^4$. Compared with *C. glutamicum* MB001/pEKEx2-*eyfp* and MB001/pJC1-P$_{tac}$-*eyfp*, the T7-based system showed a 4.6-fold and 7.1-fold higher eYFP signal after induction with 250 µM IPTG, respectively, whereas the signal in the uninduced state of the cells was at least 1.8-fold lower. These results confirm that the *C. glutamicum* T7 expression system allows tight repression of target gene expression in the absence of inducer, and a very uniform and strong expression level in the presence of inducer.

### Comparison of T7 RNAP-dependent expression of eyfp in C. glutamicum and E. coli

To compare the T7 RNAP-dependent expression system of *C. glutamicum* MB001(DE3) with the well-established *E. coli* BL21(DE3) system, the production of eYFP was analysed in both strains transformed with pMKEx2-*eyfp* and cultivated in 2xTY medium supplemented with different IPTG concentrations using the BioLector system (Fig. 4). The specific fluorescence of the culture in the absence of IPTG was lower for *C. glutamicum* MB001(DE3)/pMKEx2-*eyfp* (< 0.001) than for *E. coli* BL21(DE3)/pMKEx2-*eyfp* (0.003). The negative controls *C. glutamicum* MB001/pMKEx2-*eyfp* and *E. coli* BL21(DE3)/pMKEx2 showed only background fluorescence independent of the absence and presence of 250 µM IPTG (Fig. 4A). For *C. glutamicum* MB001(DE3)/ pMKEx2-*eyfp* and *E. coli* BL21(DE3)/pMKEx2-*eyfp*, comparable maximal values of $0.26 \pm 0.005$ and $0.25 \pm 0.002$ were recorded for the specific eYFP fluorescence, but at different IPTG concentrations of 250 µM and 50 µM, respectively (Fig. 4A). Half-maximal specific fluorescence was obtained with 31 µM IPTG for strain MB001(DE3)/ pMKEx2-*eyfp* and with 11 µM IPTG for strain BL21(DE3)/ pMKEx2-*eypf*. This difference is probably due to the presence of lactose permease in *E. coli*, which presumably is involved in IPTG uptake and allows *E. coli* to obtain higher intracellular IPTG concentrations than a strain lacking *lacY* (Fernandez-Castane *et al.*, 2012). In contrast to *E. coli*, *C. glutamicum* is unable to grow on lactose, but is able to do so when harbouring the *E. coli lac* operon, including *lacY*, which is essential for lactose uptake (Brabetz *et al.*, 1991; 1993).

Fluorescence microscopy (Fig. 4B) revealed that in the case of strain *E. coli* BL21(DE3)/pMKEx2-*eyfp* and *C. glutamicum* MB001(DE3)/pMKEx2-*eyfp*, almost all cells were fluorescent when cultivated in the presence of 250 µM IPTG. Whereas fluorescence was equally distributed over the entire cell in the case of the *C. glutamicum* strain, the majority of poles appeared non-fluorescent in the case of *E. coli* strain. The images taken at 10 and

25 µM IPTG confirm the results shown in Fig. 4A that *E. coli* requires lower IPTG concentrations for maximal expression.

Fluorescence-activated cell sorting analysis of cells of *E. coli* BL21(DE3)/pMKEx2-*eyfp* and *C. glutamicum* MB001(DE3)/pMKEx2-*eyfp* cultivated with 0, 10, 25 and 250 µM IPTG are depicted in Figs 4C and S4. Of the *E. coli* cells, 92% cultivated with 250 µM IPTG revealed an increased fluorescence, with an average signal intensity of $5.2 \times 10^4$. In the case of *C. glutamicum*, more than 99% of the cells cultivated with 250 IPTG showed an increased fluorescence, with an average signal intensity of $2.8 \times 10^4$. In contrast to the *E. coli* cells, the *C. glutamicum* cells formed a much more homogeneous population. When calculating the average fluorescence intensity of all analysed cells, the value for *C. glutamicum* MB001(DE3)/pMKEx2-*eyfp* ($2.8 \times 10^4$) was 1.8 times lower than the one for *E. coli* BL21(DE3)/pMKEx2-*eyfp* ($5.0 \times 10^4$). The FACS analysis of the strains cultivated in the absence of IPTG confirmed a lower basal *eyfp* expression in the *C. glutamicum* strain compared with the *E. coli* strain. Of the *E. coli* BL21(DE3)/pMKEx2-*eyfp* population, 1.4% showed an eYFP fluorescence above background, but none of the *C. glutamicum* MB001(DE3)/pMKEx2-*eyfp* cells.

### T7 RNAP-dependent overproduction of pyruvate kinase in C. glutamicum and E. coli

As an alternative target protein for analysing the properties of the newly established T7 expression system for *C. glutamicum*, we tested pyruvate kinase of *C. glutamicum*, which catalyses the conversion of phosphoenolpyruvate (PEP) and ADP to pyruvate and ATP. *Corynebacterium glutamicum* possesses a single *pyk* gene for pyruvate kinase (Gubler *et al.*, 1994), which was amplified by PCR from chromosomal DNA of *C. glutamicum* MB001 with the oligonucleotides Pyk-for and Pyk-rev and cloned into pMKEx2 using BamHI and BsrGI restriction sites. The resulting plasmid pMKEx2-*pyk* was transferred into *C. glutamicum* MB001(DE3) and *E. coli* BL21(DE3), and the overproduction of pyruvate kinase was analysed by measuring the specific activity in crude extract.

The results presented in Table 3 show that the expression of the *pyk* gene on plasmid pMKEx2-*pyk* in the absence of IPTG led to a small increase of the endogenous pyruvate kinase activity of 0.5 U mg$^{-1}$ in the case of *C. glutamicum* MB001(DE3) and of 0.9 U mg$^{-1}$ in the case of *E. coli* BL21(DE3). When cultivated in the presence of 250 µM IPTG, the pyruvate kinase activity increased more than 40-fold to 135 U mg$^{-1}$ in *C. glutamicum* carrying pMKEx2-*pyk* and 14-fold in *E. coli* carrying pMKEx2-*pyk*. In agreement with these data, SDS-PAGE of the cell

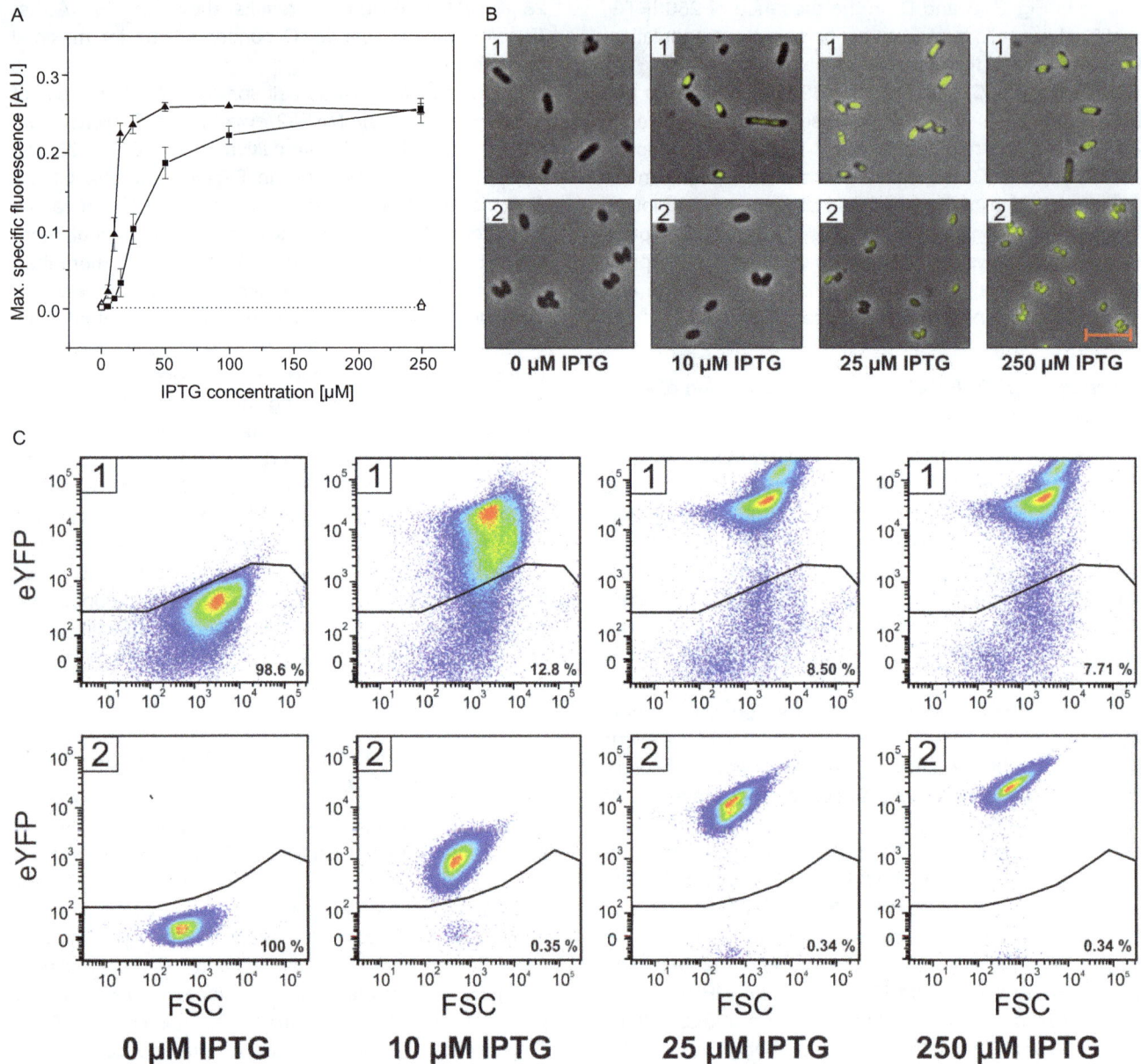

Fig. 4. T7 RNAP-dependent expression of *eyfp* in *C. glutamicum* and *E. coli*. The strains *C. glutamicum* MB001/pMKEx2-*eyfp* (□), *C. glutamicum* MB001(DE3)/pMKEx2-*eyfp* (■), *E. coli* BL21(DE3)/pMKEx2 (△) and *E. coli* BL21(DE3)/pMKEx2-*eyfp* (▲) were cultivated for 24 h aerobically in 2xTY medium using a BioLector system at 1200 r.p.m. and either 30°C (*C. glutamicum*) or 37°C (*E. coli*). Gene expression was induced 2 h after starting the cultivation by addition of 0–250 µM IPTG.
A. After 24 h, the maximal specific eYFP fluorescence was determined (ratio of fluorescence emission at 532 nm and backscatter value at 620 nm). Mean values and standard deviations of at least three independent replicates are shown.
B. Fluorescence microscopy images of *E. coli* BL21(DE3)/pMKEx2-*eyfp* (1) and *C. glutamicum* MB001(DE3)/pMKEx2-*eyfp* (2) cultivated with different IPTG concentrations. Images were taken with an exposure time of 40 ms. The red bar represents a length of 5 µm.
C. Flow cytometry analysis of *E. coli* BL21(DE3)/pMKEx2-*eyfp* (1) and *C. glutamicum* MB001(DE3)/pMKEx2-*eyfp* (2) cultivated with different IPTG concentrations. Pseudo-coloured dot plots of eYFP fluorescence versus forward scatter are shown.

extracts revealed a higher pyruvate kinase protein level in *C. glutamicum* compared with *E. coli* (Fig. 5). The more efficient overproduction of pyruvate kinase in the homologous host compared with *E. coli* might be due to a more efficient translation caused by differences in codon usage between the two species.

## Concluding remarks

In this study, a T7 RNAP-based expression system was developed for *C. glutamicum*. It is based on strain MB001(DE3), in which gene *1* encoding T7 RNAP is chromosomally encoded under control of the *lac*UV5

**Table 3.** Pyruvate kinase activity of different overexpression strains.

| Strain | Plasmid | Pyruvate kinase activity (U/mg) – IPTG | Increase (x-fold)[a] | Pyruvate kinase activity (U/mg) + 250 µM IPTG | Increase (x-fold)[b] |
|---|---|---|---|---|---|
| *C. glutamicum* MB001(DE3) | pMKEx2 | n.d. | n.a. | 2.6 ± 0.7 | n.a. |
| *C. glutamicum* MB001(DE3) | pMKEx2-*pyk* | 3.1 ± 0.2 | 1.2 | 135 ± 14 | 43.6 |
| *E. coli* BL21(DE3) | pMKEx2 | n.d. | n.a. | 0.6 ± 0.1 | n.a. |
| *E. coli* BL21(DE3) | pMKEx2-*pyk* | 1.5 ± 0.2 | 2.8 | 20.9 ± 7.1 | 13.9 |

Specific activity was measured in crude extracts of cells harvested 4 h after addition or non-addition of IPTG. Mean values and standard deviations of at least three independent measurements are shown. n.d., not determined; n.a., not applicable.
**a.** pMKEx2-*pyk* –IPTG versus pMKEx2 +IPTG.
**b.** pMKEx2-*pyk* –IPTG versus +IPTG.

promoter, and the expression vector pMKEx2 carrying the T7*lac* promoter. Thus, both gene *1* and the target gene are repressed by LacI. Using *eyfp* as target gene, the new system allowed tightly IPTG-regulatable gene expression to levels that were about four times higher than those obtained with expression vectors using the *tac* promoter and the endogenous RNA polymerase. It thus probably represents the strongest overexpression system currently available for *C. glutamicum*. A particular feature of the new system was revealed by flow cytometry: IPTG induction led to the formation of very homogeneous populations, in which about 99% of the cells showed high expression of the target protein. Half-maximal induction was obtained with IPTG concentrations between 20 and 30 µM, depending on the medium used. The T7 RNAP-based system can be useful for the overproduction of proteins for subsequent purification, but also for metabolic engineering studies in which strong overproduction of certain genes is required.

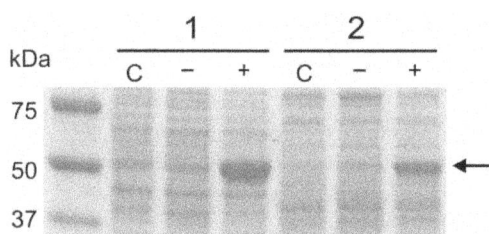

**Fig. 5.** Coomassie-stained SDS-polyacrylamide gel for analysing overproduction of pyruvate kinase in *C. glutamicum* MB001(DE3)/pMKEx2-*pyk* (1) and *E. coli* BL21(DE3)/pMKEx2-*pyk* (2). The strains were cultivated in M9 medium with 2% (wt/vol) glucose (*E. coli*) or in CGXII medium with 4% (wt/vol) glucose (*C. glutamicum*). Strains labelled with '–' were grown without IPTG, whereas strains labelled '+' were supplemented with 250 µM IPTG when the cultures had reached an OD$_{600}$ of 2. When the cultures had reached an OD$_{600}$ of 5, cells were harvested and used for preparation of cell extracts. Ten microgram total protein of these extracts were subjected to SDS-PAGE. The samples labelled with 'C' represent control strains, either *C. glutamicum* MB001(DE3)/pMKEx2 (1) or *E. coli* BL21(DE3)/pMKEx2 (2), that were cultivated in the presence of 250 µM IPTG. The arrow indicates the predicted size for *C. glutamicum* pyruvate kinase (54.4 kDa).

## Experimental procedures

### Bacterial strains, plasmids and growth conditions

All bacterial strains and plasmids used in this study are listed in Table 1. *Escherichia coli* was grown at 37°C in a complex tryptone-yeast extract medium (2xTY) or in M9 minimal medium with 4% (wt/vol) glucose (Sambrook and Russell, 2001). *Corynebacterium glutamicum* was routinely cultivated at 30°C in 2xTY medium or in defined CGXII minimal medium with 4% (wt/vol) glucose (Keilhauer *et al.*, 1993). If necessary, the media were supplemented with 50 mg l$^{-1}$ kanamycin for *E. coli* and 25 mg l$^{-1}$ kanamycin for *C. glutamicum*. *Escherichia coli* DH5α was used for plasmid construction, *E. coli* BL21(DE3) and *C. glutamicum* MB001(DE3) for overproduction of recombinant proteins. Cultures were inoculated to an optical density at 600 nm (OD$_{600}$) of 1, and target gene induction was triggered by adding 0–250 µM IPTG to the culture at an OD$_{600}$ of 2. To analyse *eyfp* expression, cells were grown as 800 µl cultures in 48-well microtitre plates (Flowerplates, m2p-labs, Baesweiler, Germany) at 80% humidity and 1200 r.p.m. using a BioLector system (m2p-labs, Baesweiler, Germany), which allows isochronal measurement of cell growth as backscattering light intensity at a wavelength of 620 nm and of eYFP fluorescence (ex/em 510/532 nm). For the overproduction of pyruvate kinase, all strains were cultivated at 120 r.p.m. and the required temperature in baffled 500 ml shake flasks with 50 ml CGXII or M9 medium, respectively.

### Recombinant DNA techniques

Standard DNA and cloning techniques were performed as described (Sambrook and Russell, 2001). Oligonucleotides were purchased from Eurofins MWG Operon (Ebersberg, Germany). Restriction enzymes (New England Biolabs, Frankfurt, Germany), shrimp alkaline phosphatase and Klenow fragment (both Thermo Scientific, Schwerte, Germany) were used according to the recommendations of the supplier. Introduction of a single point mutation in a pJC1 derivative was performed with the QuikChange Lightning Kit (Agilent Technologies, Waldbronn, Germany). Plasmid DNA of *E. coli* was isolated with QIAprep Spin Miniprep Kit (Qiagen, Hilden, Germany). Plasmid isolation from *C. glutamicum* was carried out with the same kit, but cells were pre-incubated in buffer P1 supplemented with 15 mg ml$^{-1}$ (wt/vol) lysozyme for 2 h at 30°C. Purification of

DNA fragments from agarose gels was done using the QIAex gel elution kit (Qiagen). RbCl-competent *E. coli* cells were transformed with plasmid DNA by the heat-shock method of Hanahan (1983), and *C. glutamicum* cells were transformed with plasmid DNA by electroporation as described previously (Tauch *et al.*, 2002).

For the construction of plasmid pJC1-P*tac*-*eyfp*, the plasmid pJC1-venus-term (Baumgart *et al.*, 2013a) was cut by NdeI and SalI to obtain a 6.58 kb fragment containing the same backbone (kanamycin resistance cassette, pCG1 replicon for *C. glutamicum*, pACYC177 replicon for *E. coli*) as pMKEx2. The expression cassette from plasmid pEKEx2-*eyfp* was amplified with oligonucleotides pEKEx-for and pEKEx-rev to obtain a 2.31 kb fragment containing the target gene *eyfp* under transcriptional control of the *tac* promoter and the *lacI* gene. After digestion with NdeI and SalI, this fragment was ligated with the 6.58 kb fragment from pJC1-venus-term to obtain pJC1-P*tac*-*eyfp*.

## Protein analysis

Recombinant protein production was analysed by SDS-PAGE. Cells were harvested by centrifugation and washed twice with lysis buffer (10 mM Tris-HCl, pH 8.0, 25 mM MgCl$_2$, 200 mM NaCl). Afterwards, *C. glutamicum* and *E. coli* cells were disintegrated by beat beating using a Precellys 24 device (Peqlab Biotechnologie, Erlangen, Germany). Intact cells and cell debris were sedimented by centrifugation (13 000 *g*, 20 min), and the supernatant was used further. The concentration of intracellular proteins was determined with the BCA assay (BC Assay Protein Quantitation Kit, Uptima, Interchim, Montlucon, France) as described (Smith *et al.*, 1985). Electrophoretic separation of proteins on SDS-polyacrylamide gels was performed by a standard procedure (Laemmli, 1970), and the gels were stained with Coomassie Brilliant Blue G-250 dye or used further for Western blot analysis. Immunological detection of eYFP was performed by using a polyclonal anti-GFP antibody (ab290, Abcam, Cambridge, UK) and a Cy5-conjugated goat-anti-rabbit antibody (GE Healthcare). Visualization and recording of fluorescent bands were performed using a Typhoon scanner (GE Healthcare) and the programme IMAGEQUANT TL 7.0.

## Fluorescence microscopy

For fluorescence microscopy, cells were fixed on soft-agarose coated glass slides. Images were taken on a Zeiss Axioplan 2 imaging microscope that was equipped with an AxioCam MRm camera and a Plan-Apochromat 100×, 1.40 Oil Ph3 immersion objective. Digital images were acquired and analysed with the AXIOVISION 4.6 software (Zeiss, Göttingen, Germany).

## Flow cytometry

Cells were grown under appropriate conditions in a BioLector system, harvested after 24 h and diluted to an OD$_{600}$ below 0.1 with sterile phosphate-buffered saline (37 mM NaCl, 2.7 mM KaCl, 10 mM Na$_2$HPO$_4$, 1.8 mM KH$_2$PO$_4$, pH 7.4). Expression of *eyfp* was analysed using a FACS ARIA II high-speed cell sorter (BD Biosciences, Franklin Lakes, NJ, USA) and the BD DIVA 6.1.3 software by measuring the eYFP fluorescence of single cells with an excitation wavelength of 488 nm and an emission wavelength of 533 ± 15 nm at a sample pressure of 70 psi. A threshold was set to exclude non-bacterial particles on the basis of forward versus side scatter area. There were 100 000 cells analysed for each measurement with a flow rate of 2000–4000 cells/s.

## Pyruvate kinase assay

Pyruvate kinase activity was determined spectrophotometrically using a coupled enzymatic assay with L-lactate dehydrogenase. The rate of NADH consumption was measured using an Infinite 200 PRO reader (Tecan, Männedorf, Switzerland) as the decrease of NADH absorbance at 340 nm ($\varepsilon_{NADH}$ = 6.22 mM$^{-1}$ cm$^{-1}$). The assay mixture contained 100 mM Tris-HCl buffer (pH 7.3), 15 mM MgCl$_2$, 1 mM ADP, 0.4 mM NADH, 5 U L-lactate dehydrogenase from pig heart, and 10 or 20 µl of cell extract (corresponding to 0.1–0.2 mg protein) in a total volume of 150 µl. The reaction was started by addition of 12 mM PEP. One unit of pyruvate kinase activity is defined as the amount of enzyme that converted 1 µmol of PEP to pyruvate per minute.

## Acknowledgements

We would like to thank Regina Mahr for help with the FACS analysis, Eva Hentschel for providing plasmid pEKEx2-*eyfp*, Meike Baumgart for providing plasmid pJC1-venus-term and strain *C. glutamicum* MB001, and Michael Vogt for making available a pK18mobsacB derivative.

## Conflict of interest

None declared.

## References

Arvani, S., Markert, A., Loeschcke, A., Jaeger, K.E., and Drepper, T. (2012) A T7 RNA polymerase-based toolkit for the concerted expression of clustered genes. *J Biotechnol* **159:** 162–171.

Baneyx, F. (1999) Recombinant protein expression in *Escherichia coli*. *Curr Opin Biotechnol* **10:** 411–421.

Bansal-Mutalik, R., and Nikaido, H. (2011) Quantitative lipid composition of cell envelopes of *Corynebacterium glutamicum* elucidated through reverse micelle extraction. *Proc Natl Acad Sci USA* **108:** 15360–15365.

Barnard, G.C., Henderson, G.E., Srinivasan, S., and Gerngross, T.U. (2004) High level recombinant protein expression in *Ralstonia eutropha* using T7 RNA polymerase based amplification. *Protein Expr Purif* **38:** 264–271.

Baumgart, M., Luder, K., Grover, S., Gätgens, C., Besra, G.S., and Frunzke, J. (2013a) IpsA, a novel LacI-type regulator, is required for inositol-derived lipid formation in Corynebacteria and Mycobacteria. *BMC Biol* **11:** 122.

Baumgart, M., Unthan, S., Rückert, C., Sivalingam, J., Grünberger, A., Kalinowski, J., *et al.* (2013b) Construction of a prophage-free variant of *Corynebacterium glutamicum* ATCC 13032 for use as a platform strain for basic research and industrial biotechnology. *Appl Environ Microbiol* **79:** 6006–6015.

Becker, J., and Wittmann, C. (2012) Bio-based production of chemicals, materials and fuels -*Corynebacterium glutamicum* as versatile cell factory. *Curr Opin Biotechnol* **23:** 631–640.

Blombach, B., Riester, T., Wieschalka, S., Ziert, C., Youn, J.W., Wendisch, V.F., and Eikmanns, B.J. (2011) *Corynebacterium glutamicum* tailored for efficient isobutanol production. *Appl Environ Microbiol* **77:** 3300–3310.

Brabetz, W., Liebl, W., and Schleifer, K.H. (1991) Studies on the utilization of lactose by *Corynebacterium glutamicum*, bearing the lactose operon of *Escherichia coli*. *Arch Microbiol* **155:** 607–612.

Brabetz, W., Liebl, W., and Schleifer, K.H. (1993) Lactose permease of *Escherichia coli* catalyzes active beta-galactoside transport in a gram-positive bacterium. *J Bacteriol* **175:** 7488–7491.

Brunschwig, E., and Darzins, A. (1992) A two-component T7 system for the overexpression of genes in *Pseudomonas aeruginosa*. *Gene* **111:** 35–41.

Burkovski, A. (ed.) (2008) *Corynebacteria: Genomics and Molecular Biology*. Norfolk, UK: Caister Academic Press.

Chamberlin, M., and Ring, J. (1973) Characterization of T7-specific ribonucleic acid polymerase: I. General properties of the enzymatic reaction and the template specificity of the enzyme. *J Biol Chem* **248:** 2235–2244.

Chen, R. (2012) Bacterial expression systems for recombinant protein production: *E. coli* and beyond. *Biotechnol Adv* **30:** 1102–1107.

Cremer, J., Eggeling, L., and Sahm, H. (1990) Cloning the *dapA dapB* cluster of the lysine-secreting bacterium *Corynebacterium glutamicum*. *Mol Gen Genet* **220:** 478–480.

Demain, A.L., and Vaishnav, P. (2009) Production of recombinant proteins by microbes and higher organisms. *Biotechnol Adv* **27:** 297–306.

Dubendorff, J.W., and Studier, F.W. (1991) Controlling basal expression in an inducible T7 expression system by blocking the target T7 promoter with *lac* repressor. *J Mol Biol* **219:** 45–59.

Eggeling, L., and Bott, M. (eds) (2005) *Handbook of Corynebacterium glutamicum*. Boca Raton, FL, USA: CRC Press, Taylor and Francis Group.

Eggeling, L., and Reyes, O. (2005) Experiments. In *Handbook of Corynebacterium Glutamicum*. Eggeling, L., and Bott, M. (eds). Boca Raton, FL, USA: CRC Press, Taylor and Francis Group, pp. 535–566.

Eikmanns, B.J., Kleinertz, E., Liebl, W., and Sahm, H. (1991) A family of *Corynebacterium glutamicum/Escherichia coli* shuttle vectors for cloning, controlled gene expression, and promoter probing. *Gene* **102:** 93–98.

Equbal, M.J., Srivastava, P., Agarwal, G.P., and Deb, J.K. (2013) Novel expression system for *Corynebacterium acetoacidophilum* and *Escherichia coli* based on the T7 RNA polymerase-dependent promoter. *Appl Microbiol Biotechnol* **97:** 7755–7766.

Fernandez-Castane, A., Vine, C.E., Caminal, G., and Lopez-Santin, J. (2012) Evidencing the role of lactose permease in IPTG uptake by *Escherichia coli* in fed-batch high cell density cultures. *J Biotechnol* **157:** 391–398.

Gamer, M., Frode, D., Biedendieck, R., Stammen, S., and Jahn, D. (2009) A T7 RNA polymerase-dependent gene expression system for *Bacillus megaterium*. *Appl Microbiol Biotechnol* **82:** 1195–1203.

Gubler, M., Jetten, M., Lee, S.H., and Sinskey, A.J. (1994) Cloning of the pyruvate kinase gene (*pyk*) of *Corynebacterium glutamicum* and site-specific inactivation of *pyk* in a lysine-producing *Corynebacterium lactofermentum* strain. *Appl Environ Microbiol* **60:** 2494–2500.

Hanahan, D. (1983) Studies on transformation of *Escherichia coli* with plasmids. *J Mol Biol* **166:** 557–580.

Hentschel, E., Will, C., Mustafi, N., Burkovski, A., Rehm, N., and Frunzke, J. (2013) Destabilized eYFP variants for dynamic gene expression studies in *Corynebacterium glutamicum*. *Microb Biotechnol* **6:** 196–201.

Herrero, M., de Lorenzo, V., Ensley, B., and Timmis, K.N. (1993) A T7 RNA polymerase-based system for the construction of *Pseudomonas* strains with phenotypes dependent on TOL-meta pathway effectors. *Gene* **134:** 103–106.

Inui, M., Kawaguchi, H., Murakami, S., Vertes, A.A., and Yukawa, H. (2004) Metabolic engineering of *Corynebacterium glutamicum* for fuel ethanol production under oxygen-deprivation conditions. *J Mol Microbiol Biotechnol* **8:** 243–254.

Katzke, N., Arvani, S., Bergmann, R., Circolone, F., Markert, A., Svensson, V., *et al.* (2010) A novel T7 RNA polymerase dependent expression system for high-level protein production in the phototrophic bacterium *Rhodobacter capsulatus*. *Protein Expr Purif* **69:** 137–146.

Keilhauer, C., Eggeling, L., and Sahm, H. (1993) Isoleucine synthesis in *Corynebacterium glutamicum*: molecular analysis of the *ilvB-ilvN-ilvC* operon. *J Bacteriol* **175:** 5595–5603.

Kikuchi, Y., Itaya, H., Date, M., Matsui, K., and Wu, L.F. (2008) Production of *Chryseobacterium proteolyticum* protein-glutaminase using the twin-arginine translocation pathway in *Corynebacterium glutamicum*. *Appl Microbiol Biotechnol* **78:** 67–74.

Kind, S., and Wittmann, C. (2011) Bio-based production of the platform chemical 1,5-diaminopentane. *Appl Microbiol Biotechnol* **91:** 1287–1296.

Kirchner, O., and Tauch, A. (2003) Tools for genetic engineering in the amino acid-producing bacterium *Corynebacterium glutamicum*. *J Biotechnol* **104:** 287–299.

Laemmli, U.K. (1970) Cleavage of structural proteins during the assembly of the head of bacteriophage T4. *Nature* **227:** 680–685.

Laneelle, M.A., Tropis, M., and Daffe, M. (2013) Current knowledge on mycolic acids in *Corynebacterium glutamicum* and their relevance for biotechnological processes. *Appl Microbiol Biotechnol* **97:** 9923–9930.

Litsanov, B., Brocker, M., and Bott, M. (2012a) Toward homosuccinate fermentation: metabolic engineering of *Corynebacterium glutamicum* for anaerobic production of succinate from glucose and formate. *Appl Environ Microbiol* **78**: 3325–3337.

Litsanov, B., Kabus, A., Brocker, M., and Bott, M. (2012b) Efficient aerobic succinate production from glucose in minimal medium with *Corynebacterium glutamicum*. *Microb Biotechnol* **5**: 116–128.

Lussier, F.X., Denis, F., and Shareck, F. (2010) Adaptation of the highly productive T7 expression system to *Streptomyces lividans*. *Appl Environ Microbiol* **76**: 967–970.

Lyakhov, D.L., He, B., Zhang, X., Studier, F.W., Dunn, J.J., and McAllister, W.T. (1998) Pausing and termination by bacteriophage T7 RNA polymerase. *J Mol Biol* **280**: 201–213.

Macdonald, L.E., Durbin, R.K., Dunn, J.J., and McAllister, W.T. (1994) Characterization of two types of termination signal for bacteriophage T7 RNA polymerase. *J Mol Biol* **238**: 145–158.

Marchand, C.H., Salmeron, C., Bou Raad, R., Meniche, X., Chami, M., Masi, M., *et al.* (2012) Biochemical disclosure of the mycolate outer membrane of *Corynebacterium glutamicum*. *J Bacteriol* **194**: 587–597.

Matsuda, Y., Itaya, H., Kitahara, Y., Theresia, N.M., Kutukova, E.A., Yomantas, Y.A., *et al.* (2014) Double mutation of cell wall proteins CspB and PBP1a increases secretion of the antibody Fab fragment from *Corynebacterium glutamicum*. *Microb Cell Fact* **13**: 56.

Mimitsuka, T., Sawai, H., Hatsu, M., and Yamada, K. (2007) Metabolic engineering of *Corynebacterium glutamicum* for cadaverine fermentation. *Biosci Biotechnol Biochem* **71**: 2130–2135.

Miwa, K., Matsui, H., Terabe, M., Nakamori, S., Sano, K., and Momose, H. (1984) Cryptic plasmids in glutamic acid-producing bacteria. *Agric Biol Chem* **48**: 2901–2903.

Nešvera, J., and Pátek, M. (2008) Plasmids and promoters in corynebacteria and their applications. In *Corynebacteria: Genomics and Molecular Biology*. Burkovski, A. (ed.). Norfolk, UK: Caister Academic Press, pp. 111–154.

Nesvera, J., and Patek, M. (2011) Tools for genetic manipulations in *Corynebacterium glutamicum* and their applications. *Appl Microbiol Biotechnol* **90**: 1641–1654.

Niebisch, A., and Bott, M. (2001) Molecular analysis of the cytochrome $bc_1$-$aa_3$ branch of the *Corynebacterium glutamicum* respiratory chain containing an unusual diheme cytochrome $c_1$. *Arch Microbiol* **175**: 282–294.

Okino, S., Noburyu, R., Suda, M., Jojima, T., Inui, M., and Yukawa, H. (2008) An efficient succinic acid production process in a metabolically engineered *Corynebacterium glutamicum* strain. *Appl Microbiol Biotechnol* **81**: 459–464.

Ozaki, A., Katsumata, R., Oka, T., and Furuya, A. (1984) Functional expression of the genes of *Escherichia coli* in gram-positive *Corynebacterium glutamicum*. *Mol Gen Genet* **196**: 175–178.

Pátek, M., and Nešvera, J. (2013) Promoters and plasmid vectors of *Corynebacterium glutamicum*. In *Corynebacterium Glutamicum: Biology and Biotechnology*. Yukawa, H., and Inui, M. (eds). Berlin, Heidelberg, Germany: Springer-Verlag, pp. 51–88.

Patek, M., Holatko, J., Busche, T., Kalinowski, J., and Nesvera, J. (2013) *Corynebacterium glutamicum* promoters: a practical approach. *Microb Biotechnol* **6**: 103–117.

Perez-Jimenez, R., Garcia-Manyes, S., Ainavarapu, S.R., and Fernandez, J.M. (2006) Mechanical unfolding pathways of the enhanced yellow fluorescent protein revealed by single molecule force spectroscopy. *J Biol Chem* **281**: 40010–40014.

Sambrook, J., and Russell, D. (2001) *Molecular Cloning: A Laboratory Manual*. Cold Spring Harbor, NY, USA: Cold Spring Harbor Laboratory Press.

Santamaria, R., Gil, J.A., Mesas, J.M., and Martin, J.F. (1984) Characterization of an endogenous plasmid and development of cloning vectors and a transformation system in *Brevibacterium lactofermentum*. *J Gen Microbiol* **130**: 2237–2246.

Schäfer, A., Tauch, A., Jäger, W., Kalinowski, J., Thierbach, G., and Pühler, A. (1994) Small mobilizable multipurpose cloning vectors derived from the *Escherichia coli* plasmids pK18 and pK19 – selection of defined deletions in the chromosome of *Corynebacterium glutamicum*. *Gene* **145**: 69–73.

Scheele, S., Oertel, D., Bongaerts, J., Evers, S., Hellmuth, H., Maurer, K.H., *et al.* (2013) Secretory production of an FAD cofactor-containing cytosolic enzyme (sorbitol-xylitol oxidase from *Streptomyces coelicolor*) using the twin-arginine translocation (Tat) pathway of *Corynebacterium glutamicum*. *Microb Biotechnol* **6**: 202–206.

Schneider, J., and Wendisch, V.F. (2011) Biotechnological production of polyamines by bacteria: recent achievements and future perspectives. *Appl Microbiol Biotechnol* **91**: 17–30.

Smith, K.M., Cho, K.M., and Liao, J.C. (2010) Engineering *Corynebacterium glutamicum* for isobutanol production. *Appl Microbiol Biotechnol* **87**: 1045–1055.

Smith, P.K., Krohn, R.I., Hormonson, G.T., Mallia, A.K., Gartner, F.H., Provenzano, M.D., *et al.* (1985) Measurement of protein using bicinchoninic acid. *Anal Biochem* **150**: 76–85.

Stäbler, N., Oikawa, T., Bott, M., and Eggeling, L. (2011) *Corynebacterium glutamicum* as a host for synthesis and export of D-amino acids. *J Bacteriol* **193**: 1702–1709.

Stolz, M., Peters-Wendisch, P., Etterich, H., Gerharz, T., Faurie, R., Sahm, H., *et al.* (2007) Reduced folate supply as a key to enhanced L-serine production by *Corynebacterium glutamicum*. *Appl Environ Microbiol* **73**: 750–755.

Studier, F.W., and Moffatt, B.A. (1986) Use of bacteriophage T7 RNA polymerase to direct selective high-level expression of cloned genes. *J Mol Biol* **189**: 113–130.

Tauch, A., Kirchner, O., Löffler, B., Gotker, S., Pühler, A., and Kalinowski, J. (2002) Efficient electrotransformation of *Corynebacterium diphtheriae* with a mini-replicon derived from the *Corynebacterium glutamicum* plasmid pGA1. *Curr Microbiol* **45**: 362–367.

Terpe, K. (2006) Overview of bacterial expression systems for heterologous protein production: from molecular and biochemical fundamentals to commercial systems. *Appl Microbiol Biotechnol* **72**: 211–222.

Trautwetter, A., and Blanco, C. (1991) Structural organization of the *Corynebacterium glutamicum* plasmid pCG100. *J Gen Microbiol* **137**: 2093–2101.

Vogt, M., Haas, S., Klaffl, S., Polen, T., Eggeling, L., van Ooyen, J., and Bott, M. (2013) Pushing product formation to its limit: metabolic engineering of *Corynebacterium glutamicum* for L-leucine overproduction. *Metab Eng* **22:** 40–52.

Wendisch, V.F., Bott, M., and Eikmanns, B.J. (2006) Metabolic engineering of *Escherichia coli* and *Corynebacterium glutamicum* for biotechnological production of organic acids and amino acids. *Curr Opin Microbiol* **9:** 268–274.

Wieschalka, S., Blombach, B., Bott, M., and Eikmanns, B.J. (2013) Bio-based production of organic acids with *Corynebacterium glutamicum*. *Microb Biotechnol* **6:** 87–102.

Yamamoto, S., Suda, M., Niimi, S., Inui, M., and Yukawa, H. (2013) Strain optimization for efficient isobutanol production using *Corynebacterium glutamicum* under oxygen deprivation. *Biotechnol Bioeng* **110:** 2938–2948.

Yoshihama, M., Higashiro, K., Rao, E.A., Akedo, M., Shanabruch, W.G., Follettie, M.T., *et al.* (1985) Cloning vector system for *Corynebacterium glutamicum*. *J Bacteriol* **162:** 591–597.

Yukawa, H., and Inui, M. (eds) (2013) *Corynebacterium glutamicum: Biology and Biotechnology*. Heidelberg, Germany: Springer.

## Supporting information

Additional Supporting Information may be found in the online version of this article at the publisher's web-site:

**Fig. S1.** Sequence of the 4,46 kb fragment amplified from the genome of *E. coli* BL21(DE3) and inserted into the intergenic region of cg1122-cg1121 of *C. glutamicum* MB001. 70 bp and 85 bp of the flanking *C. glutamicum* genome region are also shown with a grey background. The XhoI and EcoRI restriction sites used for cloning are shown in bold.

**Fig. S2.** Growth of *C. glutamicum* MB001/pEKEx2-*eyfp* (A) and *C. glutamicum* MB001(DE3)/pMKEx2-*eyfp* (B). The strains were inoculated to an OD600 of 1 and cultivated for 24 h at 30°C in CGXII minimal medium with 4% (wt/vol) glucose using a BioLector system. Induction of *eyfp* expression was triggered by adding 0 µM (■), 50 µM (), 100 µM ●(▲), or 250 µM (◊) IPTG to the cultures after 2 h.

**Fig. S3.** Analysis of heterologous eYFP production in the *C. glutamicum* strains MB001/pEKEx2-*eyfp* (A) and MB001(DE3)/pMKEx2-*eyfp* (B) at the single-cell level by flow cytometry. The strains were cultivated for 24 h at 30°C in CGXII minimal medium with 4% (wt/vol) glucose using a BioLector system. Induction of *eyfp* expression was triggered by adding the indicated concentrations of IPTG to the cultures after 2 h. Dot blots from FACS analysis (excitation at 488 nm, emission at 533 nm) of at least 100 000 cells of each strain displaying the eYFP fluorescence signal against the forward scatter signal (FSC). The gate used to define non-fluorescent cells was set with *C. glutamicum* MB001(DE3)/pMKEx2 with 100% of the cells falling into this gate (data not shown). The number inside the panels indicates the percentage of non-fluorescent cells inside this gate. In panels C and D, histograms of strains MB001/pEKEx2-*eyfp* (C) and MB001(DE3)/pMKEx2-*eyfp* (D) cultivated without IPTG or in the presence of 10, 25 and 250 µM IPTG are shown. The orange peaks indicate the background fluorescence set with strain MB001(DE3)/pMKEx2. The number of cells is plotted versus eYFP fluorescence.

**Fig. S4.** Analysis of heterologous eYFP production in *C. glutamicum* MB001(DE3)/pMKEx2-*eyfp* (A) and *E. coli* BL21(DE3)/pEKEx2-*eyfp* (B) at the single-cell level. The strains were cultivated for 24 h at 30°C in 2xTY medium using a BioLector system. Induction of *eyfp* expression was triggered by adding the indicated concentrations of IPTG to the cultures after 2 h. Dot plots from FACS analysis (excitation at 488 nm, emission at 533 nm) of at least 100 000 cells of each strain displaying the eYFP signal against the forward scatter signal (FSC) are shown. The gate used to define non-fluorescent cells was set with *C. glutamicum* MB001/pMKEx2 or *E. coli* BL21(DE3)/pMKEx2, respectively, with 100% of the cells falling into this gate (data not shown). The number inside the panels indicates the percentage of non-fluorescent cells. In panels C and D, histograms of strains *E. coli* BL21(DE3)/pMKEx2-*eyfp* (C) and *C. glutamicum* MB001(DE3)/pMKEx2-*eyfp* (D) cultivated without IPTG or in the presence of 10, 25 and 250 µM IPTG are shown. The orange peaks indicate the background fluorescence set with strain *E. coli* BL21(DE3)/pMKEx2 and *C. glutamicum* MB001(DE3)/pMKEx2. The number of cells is plotted versus eYFP fluorescence.

# 6

# Generalist hydrocarbon-degrading bacterial communities in the oil-polluted water column of the North Sea

Panagiota-Myrsini Chronopoulou,[1†§] Gbemisola O. Sanni,[1§] Daniel I. Silas-Olu,[1] Jan Roelof van der Meer,[2] Kenneth N. Timmis,[3] Corina P. D. Brussaard[4‡] and Terry J. McGenity[1*]

[1]School of Biological Sciences, University of Essex, Wivenhoe Park, Colchester CO4 3SQ, UK.
[2]Department of Fundamental Microbiology, University of Lausanne, Lausanne, Switzerland.
[3]Institute of Microbiology, Technical University Braunschweig, Braunschweig, Germany.
[4]Department of Biological Oceanography, Royal Netherlands Institute for Sea Research (NIOZ), Den Burg, The Netherlands.

## Summary

The aim of this work was to determine the effect of light crude oil on bacterial communities during an experimental oil spill in the North Sea and in mesocosms (simulating a heavy, enclosed oil spill), and to isolate and characterize hydrocarbon-degrading bacteria from the water column. No oil-induced changes in bacterial community (3 m below the sea surface) were observed 32 h after the experimental spill at sea. In contrast, there was a decrease in the dominant SAR11 phylotype and an increase in *Pseudoalteromonas* spp. in the oiled mesocosms (investigated by 16S rRNA gene analysis using denaturing gradient gel electrophoresis), as a consequence of the longer incubation, closer proximity of the samples to oil, and the lack of replenishment with seawater. A total of 216 strains were isolated from hydrocarbon enrichment cultures, predominantly belonging to the genus *Pseudoaltero-*

*For correspondence. E-mail tjmcgen@essex.ac.uk

**Funding Information** This work was carried out under the Sixth Framework project FACEiT, financed by the European Community (project no. 018391). PhD scholarship funds are gratefully acknowledged for P-MC (Greek State Scholarships Foundation), GOS and DS-O (Petroleum Technology Development Fund (PTDF), Nigeria).

*monas*; most strains grew on PAHs, branched and straight-chain alkanes, as well as many other carbon sources. No obligate hydrocarbonoclastic bacteria were isolated or detected, highlighting the potential importance of cosmopolitan marine generalists like *Pseudoalteromonas* spp. in degrading hydrocarbons in the water column beneath an oil slick, and revealing the susceptibility to oil pollution of SAR11, the most abundant bacterial clade in the surface ocean.

## Introduction

The global use of crude oil has plagued the marine environment with numerous major oil-pollution incidents, resulting in devastating environmental damage to marine habitats with serious socio-economic implications. A recent example is the largest offshore spill in the history of the USA, which occurred when the Deepwater Horizon rig exploded, releasing several hundred million litres of oil into the Gulf of Mexico (Crone and Tolstoy, 2010). Crude-oil components are toxic and stressful to marine organisms, including microorganisms (Sikkema *et al.*, 1995), and if they reach the shore they can persist for decades (Atlas and Bragg, 2009). However, many microbes have developed pathways for hydrocarbon metabolism, some to the extent that they thrive only in the presence of crude-oil components (Dyksterhouse *et al.*, 1995; Yakimov *et al.*, 1998; 2004a; Teramoto *et al.*, 2011; McGenity, 2014). Following oil contamination in marine systems, these obligate hydrocarbonoclastic bacteria (such as *Alcanivorax*, *Cycloclasticus*, *Thalassolituus*, and *Oleibacter* species) typically bloom and become dominant members of the prevailing microbial communities (Kasai *et al.*, 2002; Cappello *et al.*, 2007; McKew *et al.*, 2007b; Teramoto *et al.*, 2009; Vila *et al.*, 2010).

These specialized bacteria are often found in close contact with oil at the oil-water / oil-sediment interface (Schneiker *et al.*, 2006; Gertler *et al.*, 2009; Cappello and Yakimov, 2010). In experimental oil-contaminated tidal mudflat mesocosms, Coulon and colleagues (2012) showed that *Alcanivorax* sp. constituted almost 50% of the community in the tidal biofilms floating above oiled mudflat cores. *Thalassolituus* sp. dominated the microbial communities present in the oily phase of water samples obtained from production wells in Canada (Kryachko *et al.*, 2012). And, in beach-simulating mesocosms, *Alcanivorax* and *Cycloclasticus* dominated the surfaces of seawater-immersed oil-coated gravel after the addition of nutrients (Kasai *et al.*, 2002). Microbes like *Alcanivorax borkumensis* are exquisitely adapted to biofilm formation, oil solubilization and degradation, and oil-induced stress tolerance (Schneiker *et al.*, 2006; Sabirova *et al.*, 2008).

However, oil floating on the surface of marine waters is broken down into small droplets, dispersed and integrated into the water column to different degrees depending on many factors including wave action (ITOPF, 2002; Gros *et al.*, 2014). Known obligate hydrocarbonoclastic bacteria are not universally encountered in the water column below an oil slick, for example in the Gulf of Mexico (Redmond and Valentine, 2011). Therefore, we hypothesize that below the oil-water interface, specialized hydrocarbonoclastic bacteria are replaced by generalist marine bacteria, perhaps better adapted to grow on low levels or more soluble components of crude oil. Besides, there is paucity of information on the effect of crude oil spills on marine bacteria that play vital roles in global biogeochemical cycles. Therefore, we sought to investigate the impact of an oil spill on pelagic marine bacteria, both *in situ* and in mesocosms, that were not in direct contact with the oil slick; and to test the hypothesis that some common marine bacteria would be inhibited by the oil. Consequently, we have carried out comparative studies of bacterial community changes in the water column during a small experimental spill of light crude oil in the North Sea and in oil-enriched 1-m³ seawater mesocosms. Furthermore, we have isolated, identified and characterized generalist hydrocarbon-degrading microorganisms from both systems.

## Results

### Bacterial community changes during the experimental oil spill and in the oil-enriched on-board mesocosms

Temporal and treatment effects on the bacterial communities in the experimental oil spill at sea and the oil-enriched enclosed mesocosms on-board the ship were investigated by denaturing gradient gel electrophoresis (DGGE) analysis of the bacterial 16S rRNA gene. There was no obvious difference between water-column bacterial communities inside and outside of the experimental light crude oil spill at sea over 32 h (Fig. 1). In the on-board mesocosms, where samples could be taken closer to the surface (~ 15 cm), and the experiment was performed over a longer period, clear differences over time and between treatments were seen (Supplementary Fig. S1 and Fig. 2A). The community of the non-oiled mesocosm (BI) was relatively stable over the course of the experiment. In contrast, there were changes in the oiled mesocosms (BII and BIII) from day 2 until day 4, at

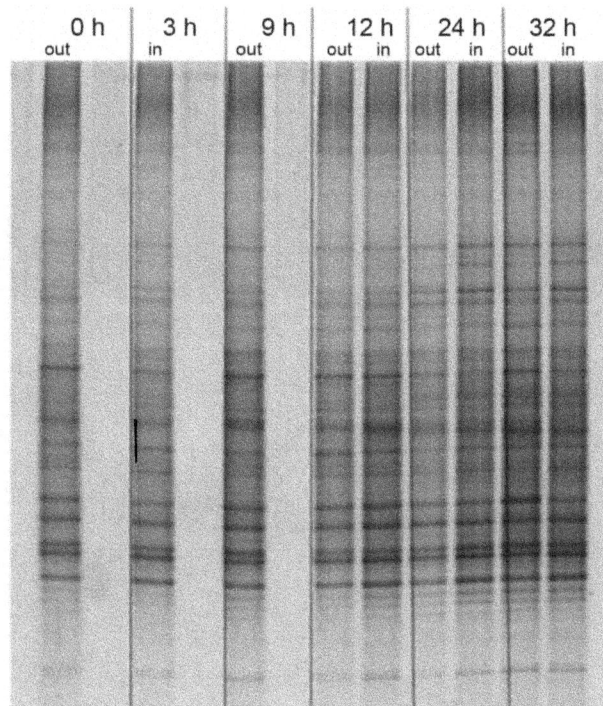

**Fig. 1.** DGGE profiles of amplified bacterial 16S rRNA genes from 1.5 m (out) and 3 m (in) below the sea surface during the experimental oil spill (outside and inside the spill respectively) at different times. Refer to the *Experimental Procedures* for an explanation of the experimental-spill and sampling strategy.

which time the communities became more stable (e.g. compare BI and BIII in Supplementary Fig. S1). The multidimensional scaling (MDS) plot (Fig. 2B) shows that addition of oil had a major impact on the bacterial community composition; the profiles of the two oiled mesocosms (BII, BIII) are distinct from the non-oiled mesocosm (BI), yet are relatively similar at corresponding time points, except that changes occur earlier in the heavily oiled mesocosm (BIII). The bacterial community surviving and then growing in the UV-treated mesocosm (BIV) was the most different, becoming more species rich over time (Fig. 2), showing that UV treatment did not kill all microbes, but selected for a distinct bacterial community.

Bacterial identification by sequencing DGGE bands was done for the on-board mesocosms, focusing on those bands that were most influenced by oil addition (Fig. 2A, Table 1). The main bacteria detected belong to the alphaproteobacterial SAR11 group and the gammaproteobacterial genus *Pseudoalteromonas*. Bands similar to SAR11 types (97–100% 16S rRNA gene identity) were present in the non-oiled mesocosm (BI), in the less oiled mesocosm (BII) and at early time points of the most oiled mesocosm (BIII). However, when oil was present at high concentration or lower concentration but for longer, many bands (D, M1, M2, N) representing SAR11 bacteria, were absent. Nevertheless, some SAR11-related bands

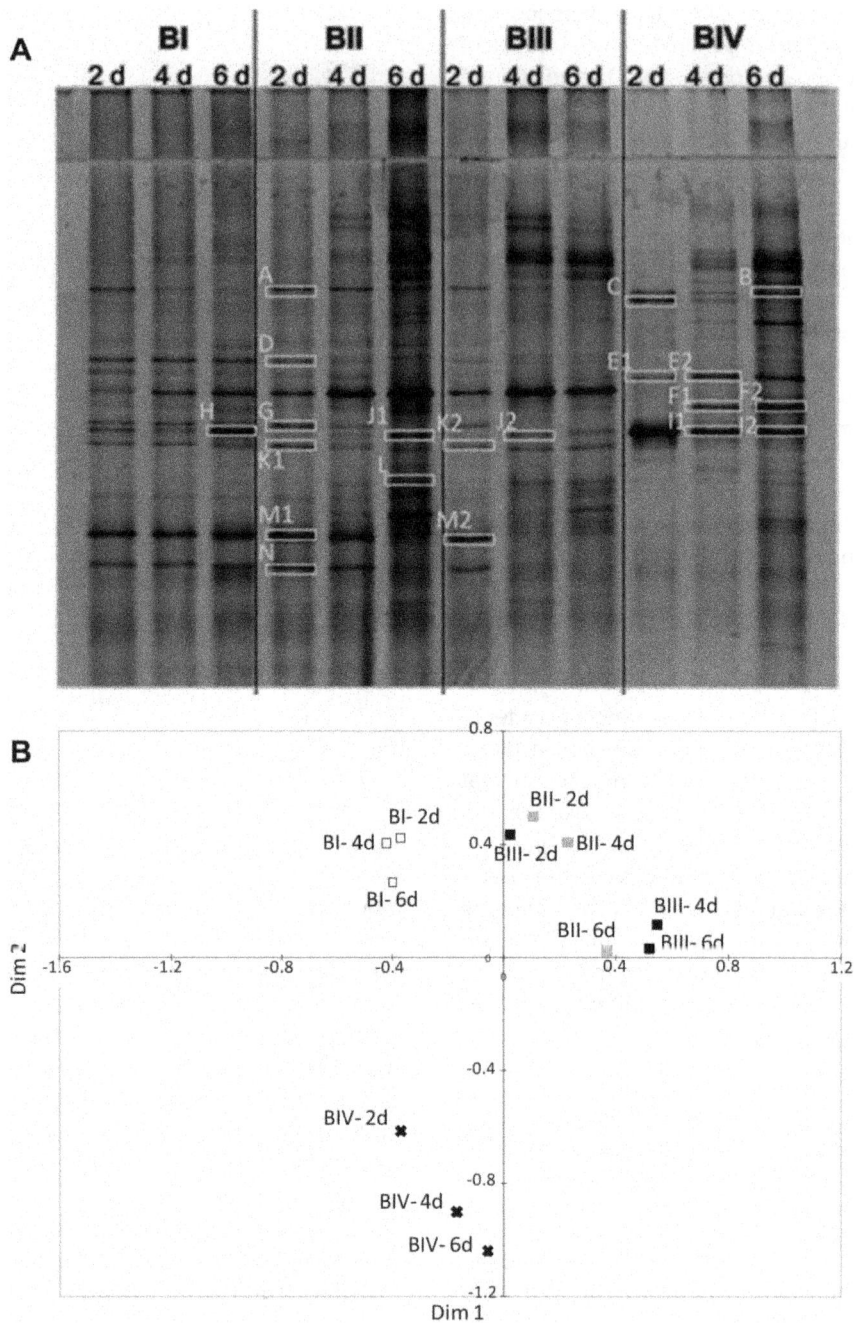

**Fig. 2.** DGGE profiles of amplified bacterial 16S rRNA genes from the on-board mesocosms (days 2, 4 and 6).
A. Profiles of days 2, 4 and 6 of all the four mesocosms are shown. The bands in white rectangles were excised and sequenced. BI: no oil (control); BII: 2 L of Arabian Light crude oil; BIII: 5.5 L of Arabian Light crude oil; BIV: 2 L of Arabian Light crude oil and UV treated (killed control).
B. MDS plot of the DGGE profiles of the on-board mesocosms. White squares indicate the non-oiled mesocosm (BI), grey squares the less oiled (BII), black squares the most oiled (BIII) and UV treated mesocosm is indicated by an 'x'.

persisted in oiled mesocosms, namely A, K1, K2. In contrast, the bands from *Pseudoalteromonas* spp. increased in intensity in oiled mesocosms, for example band J1, appearing at day 6 in the less-oiled mesocosm, and band J2 present at day 4 and 6 in the most oiled mesocosm (Fig. 2A). Band L, which was present at day 6 in the

less-oiled mesocosm, was also similar (98% 16S rRNA gene sequence identity) to *Pseudoalteromonas* sp. Although the UV-treated and oiled mesocosm (BIV) is a very artificial system, having selectively killed certain microbes, the dominant bands over time were most similar to *Pseudoalteromonas haloplanktis*. For example,

**Table 1.** Sequencing results of the 16S rRNA gene DGGE bands from the on-board mesocosms.

| Band | Oil Band int.[a] | Mesocosms -Time[b] | Closest Relative | Group[c] | Accession No | Id (%)[d] |
|---|---|---|---|---|---|---|
| A | — | BII-2 day | SAR11 bacterium | Alpha | JQ859613 | 99 |
| B | ↑ | BIV-6 day | *Alteromonas macleodii* | Gamma | CP003873 | 97 |
| C | ↓ | BIV-2 day | *Pseudoalteromonas* sp. M71_D76 | Gamma | FM992730 | 99 |
| D | ↓ | BII-2 day | SAR11 bacterium | Alpha | HM799000 | 99 |
| E1 | ↑ | BIV-2 day | *Pseudoalteromonas* sp. MSI02 | Gamma | EU420037 | 99 |
| E2 | ↑ | BIV-4 day | *Pseudomonas* sp. 28ox | Gamma | JQ411283 | 99 |
| F1 | ↑ | BIV-4 day | *Alteromonas* sp. B348 | Gamma | FN295800 | 99 |
| F2 | ↑ | BIV-6 day | *Alteromonas tagae* BCR 17571 | Gamma | NR_043977 | 100 |
| G | ↓ | BII-2 day | SAR11 bacterium | Alpha | JN710109 | 100 |
| H | ↓ | BI-6 day | Uncultured marine bacterium | Beta | AF419359 | 99 |
| I1 | ↑ | BIV-4 day | *Pseudoalteromonas* sp. BB68 | Gamma | FR693362 | 99 |
| I2 | ↑ | BIV-6 day | *Pseudoalteromonas* sp. BB68 | Gamma | FR693362 | 99 |
| J1 | ↑ | BII-6 day | *Pseudoalteromonas marina* ECSMB2 | Gamma | JX206469 | 100 |
| J2 | ↑ | BIII-4 day | *Pseudoalteromonas* sp. BSi20652 | Gamma | BADT01000269 | 98 |
| K1 | ↓ | BII-2 day | SAR11 bacterium | Alpha | HM799000 | 98 |
| K2 | — | BIII-2 day | SAR11 bacterium | Alpha | JQ859613 | 97 |
| L | ↑ | BII-6 day | *Pseudoalteromonas marina* ECSMB2 | Gamma | JX206469 | 98 |
| M1 | ↓ | BII-2 day | SAR11 bacterium | Alpha | JQ859613 | 100 |
| M2 | ↓ | BIII-2 day | SAR11 bacterium | Alpha | JQ859613 | 99 |
| N | ↓ | BII-2 day | SAR11 bacterium | Alpha | JQ859613 | 100 |

Sequences are based on approximately 130 bp of the 16S rRNA gene. A BLAST search was performed against the NCBI database.
**a.** Arrows indicate increase (↑) or decrease (↓) over time in oiled mesocosms; Dash symbol (—) indicates no change in band intensity.
**b.** BI: no oil (control); BII: 2 L of Arabian light crude oil; BIII: 5.5 L of Arabian light crude oil; BIV: 2 L of Arabian light crude oil and UV light-treated (killed control).
**c.** Alpha = Alphaproteobacteria; Beta = Betaproteobacteria; Gamma = Gammaproteobacteria.
**d.** Percentage partial 16S rRNA gene sequence identity.

bands I1 (99%) and I2 (99%), present at all time points were both similar to this species. Band E1 was also similar (99%) to *Pseudoalteromonas* sp. In this mesocosm there were also some dense bands, such as band B (97%) and bands F1 (99%) and F2 (100%), similar to *Alteromonas* spp.

### Isolation and characterization of oil-degrading strains

Although DGGE analysis suggests that *Pseudoalteromonas* spp. are probably the main microbes responsible for the degradation of oil in the on-board mesocosms, it is known that many species of this genus do not degrade hydrocarbons. Consequently, *Pseudoalteromonas*-related bacteria were cultivated to test the role they play in hydrocarbon degradation.

Samples from the experimental spills (Spill-in), accidental slicks (Acc-in) and their non-contaminated controls (Spill-out, Acc-out) and oil-enrichment mesocosms were inoculated into the hydrocarbon-enrichment media, and after subculturing, 216 strains were isolated (Table 2); 47 from the experimental spills [40 strains from inside the spill (Spill-in) and 7 from outside the spill (Spill-out)], 50 from the accidental slicks [25 each from inside the slicks (Acc-in) and outside (Acc-out)], 119 from the on-board mesocosms.

Most strains were isolated from the tetradecane (77 strains) and crude oil (76 strains) enrichment cultures,

while 19 strains were isolated from enrichment cultures with 1-methylnaphthalene as the sole carbon source. Sequencing of 16S rRNA genes from all 1-methylnaphthalene-grown isolates and selected isolates from other hydrocarbon-enrichments revealed a predominance of generalist hydrocarbon-degrading members

**Table 2.** Summary of strains isolated from hydrocarbon-enrichments of North Sea samples.

| | | Number of Isolates | | | | | | |
|---|---|---|---|---|---|---|---|---|
| | | | Carbon Source[d] | | | | Media | |
| Environment | | TOTAL | T | F | P | M | Solid | Liquid |
| ON-BOARD | BI | **25** | 6 | 8 | 9 | 2 | 12 | 13 |
| MESOCOSMS[a] | BII | **26** | 12 | 9 | 3 | 2 | 9 | 17 |
| | BIII | **36** | 15 | 12 | 6 | 3 | 16 | 20 |
| | BIV | **32** | 11 | 10 | 2 | 9 | 18 | 14 |
| EXPERIMENTAL | Spill-in | **40** | 13 | 17 | 10 | 0 | 19 | 21 |
| SPILL[b] | Spill-out | **7** | 3 | 1 | 3 | 0 | 6 | 1 |
| ACCIDENTAL | Acc-in | **25** | 9 | 10 | 5 | 1 | 14 | 11 |
| SLICKS[c] | Acc-out | **25** | 8 | 9 | 6 | 2 | 8 | 17 |
| | **TOTAL** | **216** | 77 | 76 | 44 | 19 | 102 | 114 |

**a.** BI: no oil (control); BII: 2 L of Arabian light crude oil; BIII: 5.5 L of Arabian light crude oil; BIV: 2 L of Arabian light crude oil and UV light-treated (killed control).
**b.** Spill-in: inside experimental oil spill; Spill-out: outside the experimental spill.
**c.** Acc-in: inside accidental slicks; Acc-out: outside accidental slicks.
**d.** T: tetradecane; F: weathered Forties crude oil; P: pristane; M: 1-methyl naphthalene.

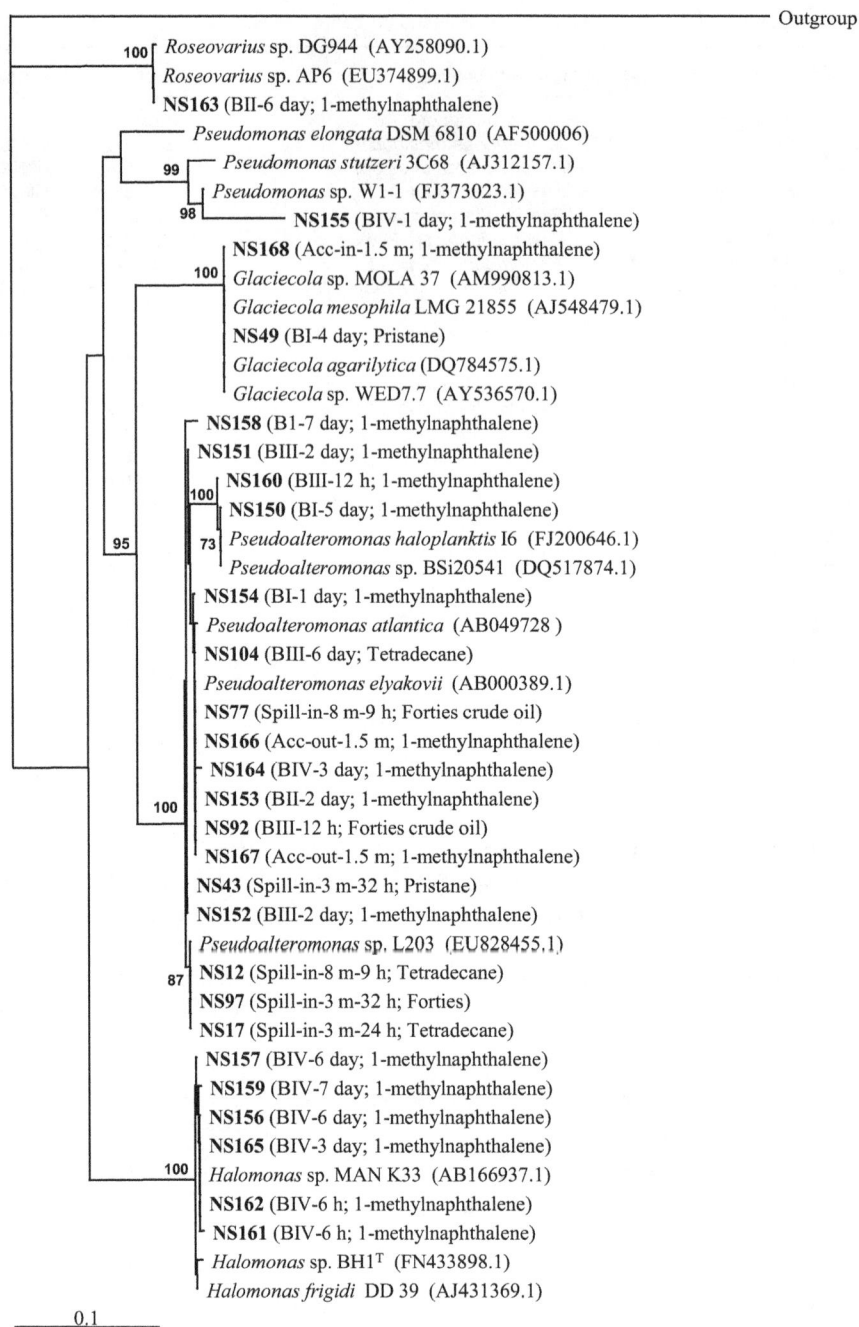

**Fig. 3.** Phylogenetic tree of North Sea isolates based on partial 16S rRNA (500 bp) sequences. Spill-in: inside experimental oil spill; Spill-out: outside experimental spill; Acc-in: inside accidental slicks; Acc-out: outside accidental slicks. On-board mesocosms – BI: no oil (control); BII: 2 L of Arabian Light crude oil; BIII: 5.5 L of Arabian Light crude oil; BIV: 2 L of Arabian Light crude oil and UV treated (killed control). The bar represents the average nucleotide substitution per base. The 16S rRNA gene of *Haloferax volcanii* NCIMB 2012 (AY425724.1) was used as outgroup.

belonging to Gammaproteobacteria but not obligate hydrocarbonoclastic bacteria (Fig. 3). With the exception of strain NS168 related to *Glaciecola* sp. (100% 16S rRNA gene sequence identity), all the isolates from samples obtained from both within and outside the spill and accidental slicks belonged to the genus *Pseudoalteromonas* (99.4–100% identity). Strain NS163

(99% identity to *Roseovarius* strain AP6), isolated from a 1-methylnaphthalene-enrichment, was the only strain belonging to Alphaproteobacteria.

A greater variety of microbes was isolated from the UV-treated control mesocosm (BIV) on-board the ship; in addition to *Pseudoalteromonas*, a member of the genus *Pseudomonas* (90% identity to *Pseudomonas stutzeri* 3C68, 93.2% identity to *Pseudomonas* sp. W1-1) and several isolates belonging to the genus *Halomonas* (99.4–99.8% identity) were isolated from 1-methylnaphthalene enrichments from mesocosm BIV.

## Utilization of different growth substrates

A growth experiment was set up in order to investigate the range of carbon sources that these isolates could utilize for growth, mainly focusing on hydrocarbons. Based on the phylogenetic tree (Fig. 3), eight phylogenetically distinct *Pseudoalteromonas* strains (two from each hydrocarbon enrichment), six phylogenetically distinct strains of *Halomonas*, and the *Glaciecola* and *Roseovarius* strains, were selected.

The results of these assays are summarized in Fig. 4. All the isolates were able to grow with Forties crude oil, pyruvate and the amino acids alanine and arginine as sole carbon sources. With the exception of strain *Glaciecola* NS168 all strains grew on glucose. Acetate and tetradecane supported growth of all *Pseudoalteromonas* strains, while only three strains of *Halomonas* (NS159, NS161, NS162) grew with tetradecane. However, no isolates grew on benzoate, cyclohexane, benzene, toluene, pyrene or biphenyl; while methanol supported growth of four strains of *Halomonas* (NS159, NS161, NS162 and NS165), *Glaciecola* (NS168) and *Roseovarius* (NS163). Some isolates grew on a wide range of hydrocarbons whereas others seemed to specialize in particular classes of hydrocarbons. For example, *Pseudoalteromonas* strain

NS50 grew on linear alkanes (decane, tetradecane and eicosane), branched alkanes (pristane and squalane) and PAHs (phenanthrene, anthracene and fluorene). *Pseudoalteromonas* strain NS151, *Halomonas* strains NS159, NS161, NS162 and *Pseudoalteromonas* NS164, isolated from 1-methylnaphthalene enrichments, did not grow on any other aromatic hydrocarbon but grew on *n*-alkanes and branched alkanes. Generally, linear alkanes supported the growth of most of the isolates.

## Crude-oil biodegradation by selected isolates

The components of crude oil consumed by isolates NS163 (99% 16S rRNA gene sequence identity to *Roseovarius* sp. AP6), NS168 (100% identity to *Glaciecola* sp. MOLA 37), NS159 (99.6% identity to *Halomonas* sp. MAN K33) and isolate NS164 (99% identity to *Pseudoalteromonas elyakovii*) were identified by growing them on 1% v/v weathered Forties crude oil for 12 days (except strain NS164 that was incubated for 28 days) and measuring the hydrocarbon components remaining. There was a significant reduction of Total Petroleum Hydrocarbon (TPH) (Tukey's HSD: $P < 0.05$) in all cultures, confirming that all the isolates were oil degraders, with isolates NS159 (*Halomonas* sp.) and NS168 (*Glaciecola* sp.) exhibiting similar degradative potentials (data not shown).

These four isolates exhibited a preference for alkanes to PAHs. Total alkanes significantly reduced by 36.9, 42.1, 44.7% for strains NS163, NS168 and NS159 respectively

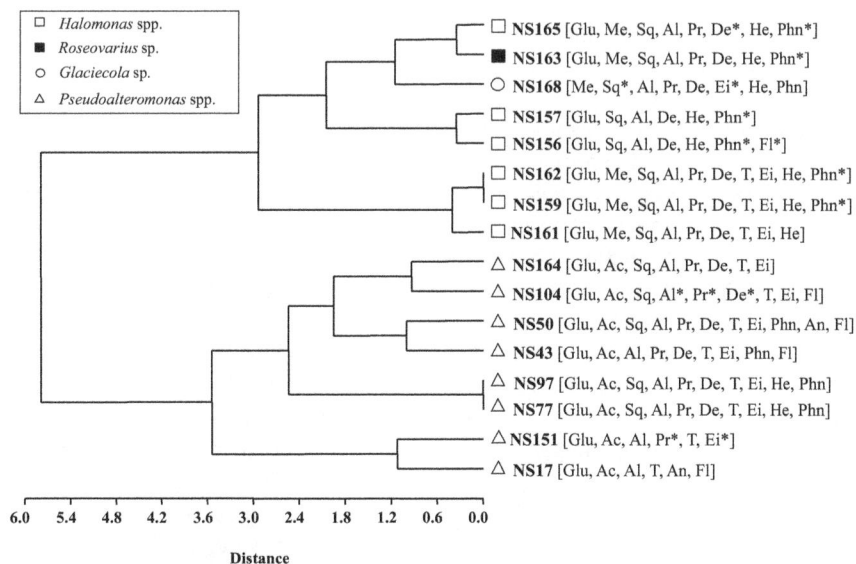

**Fig. 4.** Phenogram of sequenced isolates based on the pattern of growth substrate utilization. All negative for Biphenyl, Pyrene, Toluene, Benzene, Cyclohexane and Sodium Benzoate. All positive for Pyruvate, L-Alanine, L-Arginine, Forties crude oil. Square brackets contain substrates utilized by corresponding isolate. (Glu = Glucose; Al = L-Alanine; Ac = K-acetate; Me = Methanol; Sq = Squalane; T = Tetradecane; Pr = Pristane; De = Decane; He = Hexacosane; Phn = Phenanthrene; Ei = Eicosane; Fl = Fluorene; An = Anthracene). * = One out of two cultures grew on the substrate.

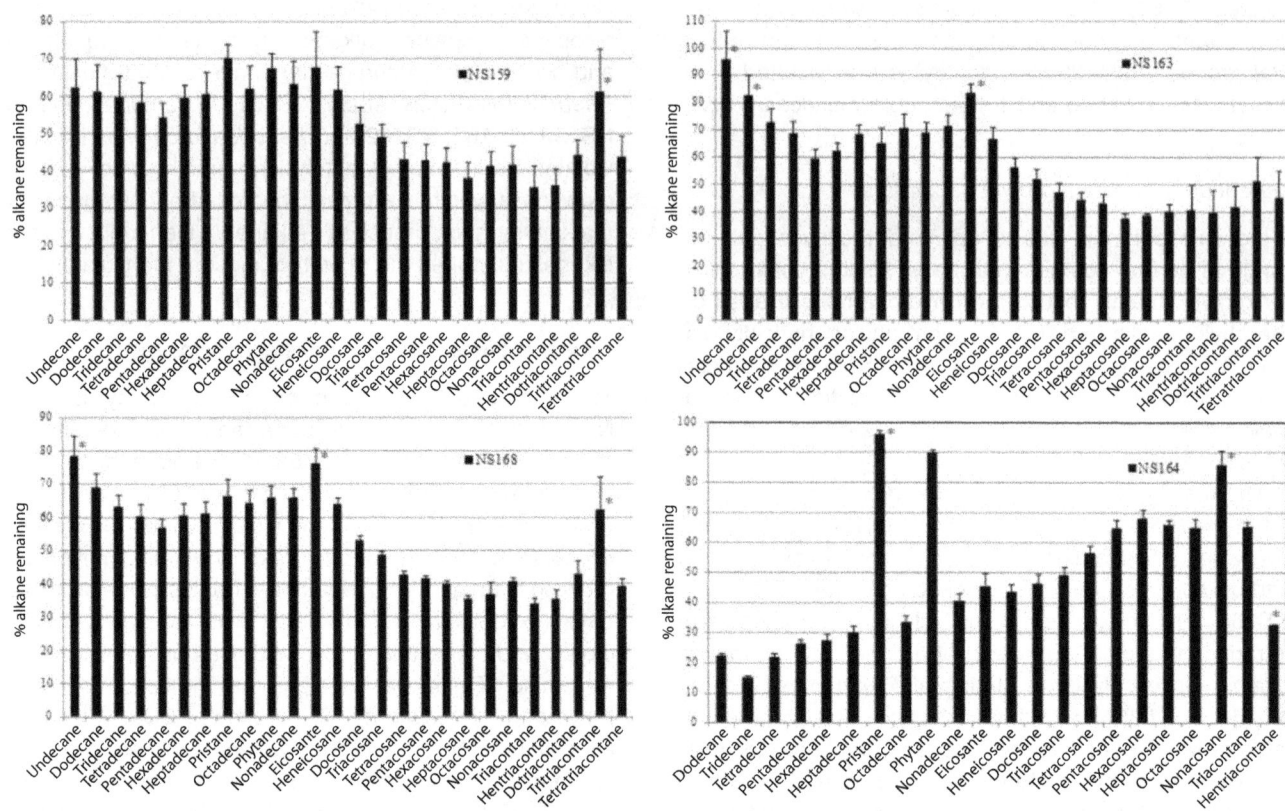

**Fig. 5.** Changes in alkane concentration in Forties crude oil after cultivation with four different isolates. The values represent percentage remaining of each alkane component relative to the negative control after 12 days of incubation (28 days for strain NS164). Vertical bars show the standard error; $n = 3$. The asterisk denotes no significant degradation (Tukey's HSD: $P > 0.05$; Dunnett's two sided: $P > 0.05$). The four strains are: NS159 (99.6% 16S rRNA sequence identity to *Halomonas* sp. MAN K33), NS163 (99% 16S rRNA sequence identity to *Roseovarius* sp. AP6), NS168 (100% 16S rRNA sequence identity to *Glaciecola* sp. MOLA 37) and NS164 (99% 16S rRNA sequence identity to *Pseudoalteromonas elyakovii*).

after 12 days of incubation, and by 54% for strain NS164 after 28 days (Tukey's HSD: $P < 0.05$). In contrast, total PAH values were unchanged.

The extent of biodegradation of individual alkane components ($C_{11}$–$C_{33}$) in Forties crude oil by the four isolates was assessed (Fig. 5). Strain NS159 significantly degraded all alkane components although those in the range of $C_{18}$–$C_{20}$ (including the branched-chain alkanes) were least degraded. Strain NS163 and NS168 did not significantly degrade $C_{11}$, $C_{12}$ and $C_{20}$ n-alkanes (Tukey's HSD: $P > 0.05$), but all other alkanes were degraded. *Pseudoalteromonas* strain NS164 had a different degradation profile, significantly degrading most of the alkane components (Tukey's HSD: $P < 0.05$) except pristane and $C_{29}$ and $C_{31}$ alkanes.

## Discussion

### Open-sea experimental spill versus the on-board mesocosms

We took this rare opportunity to investigate the effects of a light crude-oil spill on the bacterial community in the marine water column under natural conditions. When sampling from inside the experimental spill, it was essential to minimize disturbance of the oil so that other measurements would not be compromised and to avoid breakup of the slick, and so sampling was performed 3 m below the surface. The open-sea oil slick had dissipated by 32 h, and, for the reasons described by Brussaard and colleagues (2010), production of a larger, contained slick was not feasible. Therefore, in parallel, on-board mesocosms were established to simulate a more severe oil spill in a restricted area, e.g. in a harbour or bay. DGGE analysis revealed that bacterial communities in the on-board mesocosms changed rapidly upon oil addition, with a decline in most SAR11 phylotypes and a concomitant increase in *Pseudoalteromonas* spp. (see later discussion). In contrast, the bacterial community composition in the water column inside and outside the experimental oil spill did not change over the course of the 32-hour experiment. Three factors are likely to have contributed to this difference between open-sea and enclosed-mesocosm experiments. 1) The timescale of the on-board mesocosm was much longer (7 days), and indeed most differences became apparent after 2 days, and so insufficient time had elapsed to allow a large population of hydrocarbon degraders to develop in the open-sea experiment. 2) The availability of oil was lower in the open-sea experiment, due to: horizontal spreading of the Arabian light crude oil, lower oil loading per unit area, and greater depth of sampled water, and thus distance from the oil source. 3) In contrast to the simulation of an enclosed spill, the open sea would have received a flow of seawater from uncontaminated areas, containing a new supply of nutrients and microbes. These findings emphasize that complementary approaches, including experimental oil spills at sea, are required to investigate the microbial responses to crude oil; and that scaling-up data from mesocosms to the open sea requires careful consideration.

### Negative impact of crude oil on the SAR11 clade

Dominant bands in the uncontaminated mesocosm (BI) and in the less-oiled mesocosm (BII) or at early stages in the most oiled one (BIII) were similar (97–100%) to SAR11, a clade of Alphaproteobacteria and the most abundant phylotype in the surface ocean (Morris *et al.*, 2002). SAR11 varies widely in spatial and temporal abundance, but generally represents the dominant bacterioplankton taxon in the North Sea, constituting up to 37% (Giebel *et al.*, 2011) or 47% (Sperling *et al.*, 2012) of total bacterial 16S rRNA genes.

*Candidatus* Pelagibacter ubique strain HTCC1062 is the first cultured member of the clade and has one of the smallest genomes of autonomously replicating bacteria (Giovannoni *et al.*, 2005). Evidence from this strain and *in-situ* studies have shown that SAR11 bacteria are predominantly chemoorganoheterotrophs, utilizing glucose, amino acids, trimethylamine *N*-oxide and dimethyl-sulfoniopropionate (Malmstrom *et al.*, 2004; Laghdass *et al.*, 2012; Lidbury *et al.*, 2014), but when starved of organic carbon, growth is enhanced by light-energy capture via proteorhodopsin (Steindler *et al.*, 2011). Importantly, our study has provided experimental evidence that a prolonged dose of light crude oil has a negative impact on this bacterial clade that has such a pivotal role in marine ecology and biogeochemical processes: most SAR11-derived bands decreased in intensity with time in the oiled mesocosms compared with the non-oiled mesocosm, and were absent after day 4 in the mesocosm receiving most oil (BIII). The detrimental effects of oil on SAR11 phylotypes were also observed in diesel-polluted mesocosms containing surface water from a coastal area of southwest Mallorca (Lanfranconi *et al.*, 2010). Nogales and colleagues (2007) also noted the absence of SAR11 in polluted harbour waters. Moreover, Redmond and Valentine (2011) found SAR11 to be among

the dominant bacteria in surface water collected from three sites of the Deepwater Horizon spill that had a thin oil sheen, but they were undetected or represented by < 1% of 16S rRNA sequences in two sites with a thicker oil coating. Also in the Gulf of Mexico, but this time in the oil plume in deeper water, most of the abundant taxa showed an increase in gene expression; in contrast SAR11 transcript abundance was lower inside than outside the plume, but not significantly so (Rivers *et al.*, 2013). Little attention is given to the impact of crude oil on marine bacteria, yet we showed that there are clear dose- and time-dependent effects on SAR11. It will be important to determine what effect spills of different magnitude may have on such microbes, as well as the resulting impact on cycling of elements (e.g. N, P and Fe) that are required for efficient hydrocarbon degradation. The mechanisms by which crude oil inhibits SAR11 cells remain to be elucidated, but may include: direct interaction of hydrocarbons with cells, e.g. damaging membranes; reduced light availability, potentially affecting sensory processes or energy generation; and/or indirect processes, such as enhanced competition with hydrocarbon-degrading microbes for nutrients.

### The importance of nutritionally diverse bacteria, such as Pseudoalteromonas spp., in hydrocarbon degradation in the marine water column

*Pseudoalteromons* spp. are ubiquitous in marine environments, representing on average 2.6% of the surface bacterioplankton along a latitudinal gradient (Wietz *et al.*, 2010), but less than 1% in one location in the southern North Sea (Eilers *et al.*, 2000). *Pseudoalteromonas*-related bands became dominant in oiled mesocosms (BII, BIII, BIV), and *Pseudoalteromonas* spp. were the most commonly isolated bacteria, including from the open-sea spill. This, coupled with the demonstrated ability of these isolates to degrade a wide range of hydrocarbons, implies that they were the main microbes responsible for hydrocarbon degradation in the water column. No obligate hydrocarbonoclastic bacteria (OHCBs) were detected or isolated from the experimental spill or the mesocosms. This was not due to the choice of media, because in another study of mudflat sediments using the same procedures, OHCBs belonging to the genera *Alcanivorax, Cycloclasticus* and *Thalassolituus* were isolated in abundance (Sanni *et al.*, unpublished). OHCBs such as *Alcanivorax borkumensis* and *Marinobacter hydrocarbonoclasticus* form biofilms that aid their attachment to oil-water interfaces where they carry out degradation (Schneiker *et al.*, 2006; Duran, 2010), and so they may well have been present on the unsampled oil-water surface. However, an intriguing possibility is that the anti-biofilm activity of *Pseudoalteromonas* (Dheilly *et al.*,

2010; Klein *et al.*, 2011) may have inhibited oil-attached OHCBs.

The closest known relatives of the oil-degrading *Pseudoalteromonas* strains isolated in this study (e.g. *P. haloplanktis*, *P. elyakovii*, *P. distincta*, *P. atlantica*) are not known to degrade hydrocarbons. However, representatives of this genus have been detected in a variety of oil-contaminated environments and oiled microcosms, and shown to degrade hydrocarbons. Notably, in one of the two surface water samples from the Gulf of Mexico that had a thick layer of oil, and that was mentioned previously because of the absence of SAR11 (Redmond and Valentine, 2011), *Pseudoalteromonas* constituted 93% of the 16S rRNA sequences, and strains were isolated from plume water samples during the Deepwater Horizon spill (Gutierrez *et al.*, 2013). *Pseudoalteromonas* spp. increased in abundance in deeper waters of the Gulf-of-Mexico plume, after hydrocarbon exposure, especially after alkanes had been depleted and aromatics remained (Dubinsky *et al.*, 2013). Also, *Pseudoalteromonas* clones were detected in seawater microcosms amended with crude oil and incubated at 4°C and 20°C (Yakimov *et al.*, 2004b; Coulon *et al.*, 2007), and in PAH enrichment-microcosms of seawater amended with nutrients over a period of 9 weeks (McKew *et al.*, 2007a). *Pseudoalteromonas* spp. have been isolated from phenanthrene, chrysene and naphthalene-enrichment cultures of marine sediments (Melcher *et al.*, 2002; Hedlund and Staley, 2006; Cui *et al.*, 2008). The biodegradation potential of *Pseudoalteromonas* spp. for both aliphatic and aromatic compounds of crude oil has been confirmed in a consortium deriving from arctic ice and seawater (Deppe *et al.*, 2005) and in a consortium from the Korean Western coast (Cho and Oh, 2012), whereas a strain of *Pseudoalteromonas* has been found to preferentially metabolize short-chain alkanes in the range of $C_{12}$–$C_{20}$ in arctic sediments (Lin *et al.*, 2009). All of our *Pseudoalteromonas* strains grew on a range of carbon and energy sources, and varied in the breadth of utilization of hydrocarbons, with all but three strains able to grow on straight-chain and branched alkanes as well as PAHs, but none grew on toluene. In contrast, OHCBs, like *Alcanivorax* and *Cyloclasticus* spp., tend to be more specialized, degrading alkanes and PAHs, respectively (see Cappello and Yakimov, 2010; Staley, 2010).

Interactions between microbes are as important in oil degradation as they are in other biogeochemical processes (McGenity *et al.*, 2012), and the ability of *Pseudoalteromonas* spp. to remove potential competitors has been mentioned previously. *Pseudoalteromonas* spp. may also be beneficial to co-existing microbes by producing EPS that has the potential to serve as a biosurfactant (Nichols *et al.*, 2005; Saravanan and Jayachandran, 2008; Cho and Oh, 2012). It will be important to elucidate the relative contribution of generalists like *Pseudoalteromonas* and specialists like *Alcanivorax*, and to understand to what extent they interact. *Pseudoalteromonas* spp. will be at an advantage owing to their higher abundance in unpolluted seawater, allowing them to become established more rapidly after an oil spill; however, the burden of carrying genes that are not required for hydrocarbon biodegradation is likely to make them less competitive over time.

*Other hydrocarbon-degrading isolates*

Whereas *Pseudoalteromonas* was both shown to be abundant in oiled mesocosms and readily cultivable on hydrocarbon-enriched marine media, hydrocarbon-degrading strains from other genera were also isolated but not shown to have increased in abundance in the oiled mesocosms. Six strains were related to species of *Halomonas* (more than 99% percent identity), all of which derived from UV-treated and oiled mesocosm BIV and had been enriched with 1-methylnaphthalene as sole carbon source. Like *Pseudoalteromonas* strains, they grew on a wide range of hydrocarbons, including branched and straight-chain alkanes and PAHs. GC-MS analysis of one strain demonstrated metabolism of almost all the alkane components of crude oil but not the aromatic components within 12 days, implying a preference for alkanes when both alkane and PAHs are present as crude oil components. *Halomonas* spp. have been reported to degrade aromatic (Melcher *et al.*, 2002; Garcia *et al.*, 2004; Cui *et al.*, 2008) and aliphatic hydrocarbons (Pepi *et al.*, 2005; Mnif *et al.*, 2009) in marine and hypersaline environments.

One strain had 99% 16S rRNA sequence identity with two members of the Roseobacter clade: *Roseovarius* sp. AP6 which was isolated by Alonso-Gutierrez and colleagues (2009) from shoreline samples contaminated with oil from the *Prestige* spill, and *Roseovarius* sp. DG944 found in association with an alga, *Gymnodinium catenatum*, which causes paralytic shellfish poisoning in marine waters (Green *et al.*, 2004). *Roseobacter* spp. were found in abundance in both *n*-alkane- (McKew *et al.*, 2007a) and oil- (Brakstad and Løgeng, 2005) enrichments of seawater and also in oil-contaminated seawater (Prabagaran *et al.*, 2007). Similarly, species of *Roseovarius* have been implicated in the degradation of PAHs (Harwati *et al.*, 2007; Wang *et al.*, 2008; Vila *et al.*, 2010), *n*-alkanes and pristane (Harwati *et al.*, 2007).

Species of *Glaciecola* are usually found in polar seas and their coastal environments, Arctic and Antarctic sea-ice in association with marine invertebrates (Romanenko *et al.*, 2003; Yong *et al.*, 2007) and clones have been detected in oil-contaminated samples from such environments (Prabagaran *et al.*, 2007; Brakstad *et al.*, 2008; Røberg *et al.*, 2011). Yakimov and colleagues (2004b) also isolated a strain of *Glaciecola* from tetradecane-enrichments of Antarctic seawater. However, these studies did not present evidence to support the direct participation of these species in hydrocarbon degradation, whereas our *Glaciecola* strain NS168 grew on most of the *n*-alkanes tested and on pristane.

It is noteworthy that the four isolates (*Roseovarius* sp. strain NS163, *Glaciecola* sp. strain NS168, *Halomonas* sp. strain NS159 and *Pseudoalteromonas* sp. strain NS164) did not degrade any of the PAH components of crude oil during the period of incubation (as evidenced by GC-MS analysis), even though they were all isolated on 1-methylnaphthalene-enrichments and the first three strains grew on phenanthrene in the growth substrate assay. This implies that the strains have a preference for alkanes when both alkane and PAHs are present as crude oil components but could metabolize some PAHs in the absence of alkanes.

### Selection of Alteromonas *sp. by UV-light treatment*

Although the primary purpose of the UV-treated mesocosm (BIV) was to allow analysis of *in-situ* hydrocarbon biodegradation, it was a source of several strains of hydrocarbon-degrading microbes. Hydrocarbons were a selective force in this mesocosm, but also a dense band indicating the presence of *Alteromonas*, not detected in other oiled mesocosms, could be a consequence of the resistance of representatives of this genus to ultraviolet light (Agogué *et al.*, 2005) and their ability to grow rapidly on the organic matter released from cells killed by UV light in the same way that they are often associated with phytoplankton blooms (Tada *et al.*, 2012) and exudates (Nelson and Carlson, 2012).

### Environmental implications

Several factors have been proposed to explain the patchy horizontal distribution of important marine taxa, such as SAR11, in marine environments (Sperling *et al.*, 2012); however our study shows that consideration should be given to oil-pollution incidents. Also, a ten-fold enrichment of *Pseudoalteromonas* spp., and absence of SAR11, in the sea-surface microlayer compared with underlying North Sea water (Franklin *et al.*, 2005), may in part be explained by the accumulation of hydrocarbons in the sea-surface microlayer (Wurl and Obbard, 2004) and the demonstrated sensitivity of SAR11 to, and predilection of *Pseudoalteromonas* spp. for, hydrocarbons.

It has been proposed that ecological damage caused by oil spills might be reduced by the addition of microbes

like *Alcanivorax* spp. which, although naturally present in contaminated sites, require considerable time to reach population densities needed for maximum biodegradation rates. The work presented here shows that such bioaugmentations may only address biofilm-biodegradation of oil-slick or oil-droplet contamination and not dispersed hydrocarbons in the water column. Thus, bioaugmentation strategies may require multiple species with different specialities, including oil-attached specialists like *Alcanivorax* and planktonic generalists such as *Pseudoalteromonas*.

### Experimental procedures

#### Sample site and sample collection

*The experimental oil spill.* Details of the permitted, controlled, experimental spill and on-board mesocosms are provided by Brussaard and colleagues (2010). As part of an EC project FACEiT (Fast-Advanced Cellular and Ecosystem Information Technologies), an experimental oil spill was organized by the Royal Netherlands Institute for Sea Research (NIOZ), Texel, in collaboration with the Netherlands Ministry of Transport, Public Works and Water Management, RWS Noordzee (RWS-NZ); the spill was carried out on 8 May 2008 in Netherlands Exclusive Economic Zone waters. To aid the visualization and tracking of the oil slick, 5 kg of the fluorescent dye rhodamine WT (25 L 20% w/v) was mixed with 1 m³ of seawater and discharged together with 5000 L of Arabian Light crude oil into Dutch waters. In addition to the dye, two drifting buoys were used to locate the position of the oil slick especially when the rhodamine had diluted beyond the limit of detection and it thus became difficult to follow the slick. Time zero (0 h) corresponds with the end-point of oil and dye addition, which took approximately 30 min. In order to prevent RV Pelagia from disturbing the experimental spill, a rubber boat was dispatched to collect water samples from inside the spill (Spill-in) using a specially designed sampler that gently pumped seawater from 3 m depth into a metallic jerry can. The tube used was attached to a metallic inlet that was passed closed through the surface oil layer, preventing oil droplets from contaminating the sample. Samples were taken at 0, 3, 9, 12, 24 and 32 h after addition of oil. Within 30 min of these time points, corresponding control samples were collected from waters at least 500 m from the spill with no apparent oil on the surface, and ensuring no cross-contamination with the added oil by accounting for wind, tide and current. These samples, called 'Spill-out', were taken at a depth of 1.5 m using Niskin bottles (10 L each) mounted on a Rosette sampler equipped with Seabird conductivity-temperature-depth (CTD) sensors and a natural chlorophyll autofluorescence sensor.

*Accidental slicks.* Samples were collected from oil slicks encountered in shipping lanes within a section of the Dutch EEZ rich of oil platforms (southeast of Dogger bank) (Acc-in) at 1.5 m water depth. Corresponding control samples (Acc-out) were collected in clean waters outside the slicks at the same depth.

*Oil-enriched enclosed mesocosms.* In addition to the experimental spill, four 1 m³ mesocosms (called Cube vessels) were set up on board the RV Pelagia on the 6th May, 2008. The Cube vessels were filled with seawater (1000 L) from the Dogger Bank, a biologically productive location containing waters with no obvious oil contamination and high biomass. Cube 1 (BI) served as the non-oiled control, Cube 2 (BII) contained 2 L of Arabian Light Crude oil, Cube 3 (BIII) contained 5.5 L of the same crude oil, and Cube 4 (BIV) contained 2 L of oil, but was treated with UV light to sterilize the mesocosm ('killed control'). Each Cube vessel had three taps: at the top, middle and bottom. As the level of water in each mesocosm reduced, samples were taken from the tap closest to the surface of the water (approximately 15–20 cm from the surface).

The mesocosms were sampled on the 7th May, 2008 before the addition of oil at 08:15 (t = 0), then at 2 h, 6 h, 12 h after oil addition. Further samples were obtained daily at 13:00 for the next 7 days.

### Initial processing of samples

Aliquots of seawater sampled from all environments were immediately inoculated into hydrocarbon-enrichment media while the rest were stored on board the RV Pelagia and transported at 4°C back to the University of Essex lab. Microbial cells were also immediately collected on 0.2 μm pore size filters by filtering 1.2 L of seawater samples with Nalgene bottle-top filter units. Seawater samples (pristine, experimental spill and accidental spills) were filtered with Sartorius gridded cellulose nitrate filters (47 mm diameter, 0.2 μm pore size) while samples from the Cube vessels were filtered with Durapore PVDF filters (47 mm diameter, 0.22 μm pore size). The filters were initially stored at 4°C for 12 h in centrifuge tubes (50 ml) containing RNA*later*, and then frozen at − 80°C.

### Hydrocarbon enrichment cultures

Hydrocarbon enrichment cultures in liquid media were established by inoculating water samples (200 μl) into tubes containing 10 ml of ONR7a medium (Dyksterhouse *et al.*, 1995) using 1% v/v filter-sterilized tetradecane, weathered Forties crude oil, pristane or 1-methylnaphthalene as sole carbon sources. Tubes containing ONR7a medium with hydrocarbons had been prepared, capped and crimp-sealed with PTFE-lined silicon septa prior to inoculation. Subculturing was done by carrying out 10-fold serial dilutions of the cultures and spreading 50 μl from the $10^{-2}$, $10^{-5}$ and $10^{-8}$ dilutions on ONR7a agar plates containing sterile GF/C filter papers soaked with 125 μl of the hydrocarbons on the lids of the Petri dishes.

All hydrocarbon enrichment cultures using solid media were prepared with washed agar. These enrichments were prepared by spreading 100 μl of seawater samples (that had been preserved at 4°C for a maximum of three weeks) onto ONR7a agar with hydrocarbon-soaked GF/C filter papers placed on the lids of the Petri dishes. Each of the four hydrocarbons (125 μl) was used as sole carbon source. All cultures were incubated at 12°C, representing the *in-situ* temperature at the time of sampling.

### Nucleic acid extraction, polymerase chain reaction, denaturing gradient gel electrophoresis, and sequencing

DNA was extracted from cultures by boiling the bacterial cells in diethyl pyrocarbonate-treated water (DEPC water), followed by PCR amplification of bacterial 16S rRNA genes using primers described by Lane (1991). The 16S rRNA amplicons of bacterial isolates were sent to GATC Biotech (http://www.gatc-biotech.com/en/index.html) for sequencing.

DNA from filtered water was extracted following bacterial cell lysis with phenol : chloroform: isoamylalcohol (25:24:1 v/v) as described by McKew and colleagues (2007a). PCR of bacterial 16S rRNA gene from this DNA was performed using primers described by Muyzer and colleagues (1993), followed by denaturing gradient gel electrophoresis (DGGE) employing a denaturing gradient from 40 to 60% (McKew *et al.*, 2007a). Bands were excised and sequenced on an ABI Prism 3100 Genetic Analyser.

Full details of these methods and associated analyses are given in the supporting information.

### Accession numbers

The DNA sequences reported in this study were deposited in the EMBL database under the accession numbers HE961976 to HE962017.

### Growth substrate assays

The ability of selected isolates to utilize different carbon sources was examined by growing the isolates (100 μl) in duplicate tubes containing sterile ONR7a liquid medium (10 ml) and various carbon sources (0.1% v/v). The growth substrates included Forties crude oil, *n*-alkanes (decane, tetradecane, eicosane and hexacosane), branched alkanes (pristane and squalane), cyclic alkane (cyclohexane), aromatic hydrocarbons (benzene, toluene, phenanthrene, anthracene, pyrene, fluorene and biphenyl), amino acids (L-alanine and L-arginine), D-glucose, Na-benzoate, Na-pyruvate, K-acetate and methanol. Details are provided in the supporting information.

### Biodegradation of crude oil by selected isolates

Detailed biodegradation studies were carried out on four selected isolates. Washed, suspended cells (200 μl) were inoculated in triplicate into serum bottles (125 ml) containing sterilized ONR7a medium (20 ml) with weathered Forties crude oil (1% v/v) and incubated on a rotary shaker (Innova 2300; New Brunswick Scientific) in the dark at 100 rpm for 12 days (isolate NS164 was incubated for 28 days of 12 h light-dark cycles). After incubation, the cultures were vigorously vortexed for 10 s to dislodge bacterial cells from the crude oil and centrifuged in a Sorvall Biofuge Stratos centrifuge at 8,500 *g* for 15 min at 4°C to pellet cells. Total hydrocarbon was solvent extracted from the liquid contents of the serum bottles as described by Coulon and colleagues (2007), and hydrocarbons were quantified using a Thermo Trace GC gas chromatograph attached to a Thermo Trace DSQ® mass spectrometer. Details are given in the supporting information.

## Statistical analysis

Similarity between DGGE profiles was calculated based on a manually produced binary matrix of the presence/absence of DGGE bands. A proximity matrix based on the Pearson correlation coefficient was used to construct a Multidimensional Scaling (MDS) plot in XLSTAT Version 2008.1.02. A phenogram based on the growth substrate utilization pattern of sequenced isolates was generated with PAST (Paleontological statistics) 2.17 software package (Hammer et al., 2001) using the Ward's method and Euclidean distance to demonstrate the dissimilarities in substrate utilization by the isolates. Significant hydrocarbon degradation was determined by analysis of variance coupled with Tukey's HSD and Dunnett's two sided tests. These tests were performed with XLSTAT 2011 (Addinsoft™).

## Acknowledgements

We thank the captains and crews of the RV Pelagia and MV Arca for excellent support, and also partners of the FACEiT project for helpful information and discussion. Special thanks to Sjon Huisman, RWS, who made the access to the experimental oil spill possible. We appreciate the technical advice and assistance provided by Dr. Frédéric Coulon, John Green and Farid Benyahia.

## Conflict of interest

None declared.

## References

Agogué, H., Joux, F., Obernosterer, I., and Lebaron, P. (2005) Resistance of marine bacterioneuston to solar radiation. *Appl Environ Microbiol* **71:** 5282–5289.

Alonso-Gutierrez, J., Figueras, A., Albaiges, J., Jimenez, N., Vinas, M., Solanas, A.M., and Novoa, B. (2009) Bacterial communities from shoreline environments (Costa da Morte, northwestern Spain) affected by the Prestige oil spill. *Appl Environ Microbiol* **75:** 3407–3418.

Atlas, R., and Bragg, J. (2009) Bioremediation of marine oil spills: when and when not – the *Exxon Valdez* experience. *Microb Biotechnol* **2:** 213–221.

Brakstad, O.G., and Lø<br>deng, A.G. (2005) Microbial diversity during biodegradation of crude oil in seawater from the North Sea. *Microb Ecol* **49:** 94–103.

Brakstad, O.G., Nonstad, I., Faksness, L.G., and Brandvik, P.J. (2008) Responses of microbial communities in Arctic sea ice after contamination by crude petroleum oil. *Microb Ecol* **55:** 540–552.

Brussaard, C.P.D., Peperzak, L., Witte, Y., and Huisman, J. (2010) An experimental oil spill at sea. In *Handbook of Hydrocarbon and Lipid Microbiology*. Timmis, K.N., McGenity, T.J., van der Meer, J.R., and de Lorenzo, V. (eds). Berlin Heidelberg: Springer, pp. 3491–3502.

Cappello, S., and Yakimov, M.M. (2010) Alcanivorax. In *Handbook of Hydrocarbon and Lipid Microbiology*. Timmis, K.N., McGenity, T.J., van der Meer, J.R., and de Lorenzo, V. (eds). Berlin Heidelberg: Springer, pp. 1738–1748.

Cappello, S., Caruso, G., Zampino, D., Monticelli, L.S., Maimone, G., Denaro, R., et al. (2007) Microbial community dynamics during assays of harbour oil spill bioremediation: a microscale simulation study. *J Appl Microbiol* **102:** 184–194.

Cho, S.H., and Oh, K.H. (2012) Removal of Crude Oil by Microbial consortium isolated from oil-spilled area in the Korean western coast. *Bull Environ Contam Toxicol* **89:** 680–685.

Coulon, F., McKew, B.A., Osborn, M.A., McGenity, T.J., and Timmis, K.N. (2007) Effects of temperature and biostimulation on oil-degrading microbial communities in temperate estuarine waters. *Environ Microbiol* **9:** 177–186.

Coulon, F., Chronopoulou, P., Fahy, A., Païssé, S., Goñi-Urriza, M., Peperzak, L., et al. (2012) Central role of dynamic tidal biofilms dominated by aerobic hydrocarbonoclastic bacteria and diatoms in the biodegradation of hydrocarbons in coastal mudflats. *Appl Environ Microbiol* **78:** 3638–3648.

Crone, T.J., and Tolstoy, M. (2010) Magnitude of the 2010 Gulf of Mexico oil leak. *Science* **330:** 634.

Cui, Z., Lai, Q., Dong, C., and Shao, Z. (2008) Biodiversity of polycyclic aromatic hydrocarbon-degrading bacteria from deep sea sediments of the Middle Atlantic Ridge. *Environ Microbiol* **10:** 2138–2149.

Deppe, U., Richnow, H.-H., Michaelis, W., and Antranikian, G. (2005) Degradation of crude oil by an arctic microbial consortium. *Extremophiles* **9:** 461–470.

Dheilly, A., Soum-Soutera, E., Klein, G.L., Bazire, A., Compere, C., Haras, D., and Dufour, A. (2010) Antibiofilm activity of the marine bacterium *Pseudoalteromonas* sp. strain 3J6. *Appl Environ Microbiol* **76:** 3452–3461.

Dubinsky, E.A., Conrad, M.E., Chakraborty, R., Bill, M., Borglin, S.E., Hollibaugh, J.T., et al. (2013) Succession of hydrocarbon-degrading bacteria in the after-math of the Deepwater Horizon oil spill in the Gulf of Mexico. *Environ Sci Technol* **47:** 10860–10867.

Duran, R. (2010) Marinobacter. In *Handbook of Hydrocarbon and Lipid Microbiology*. Timmis, K.N., McGenity, T.J., van der Meer, J.R., and de Lorenzo, V. (eds). Berlin Heidelberg: Springer, pp. 1725–1735.

Dyksterhouse, S.E., Gray, J.P., Herwig, R.P., Lara, J.C., and Staley, J.T. (1995) *Cycloclasticus pugetii* gen. nov., sp. nov., an aromatic hydrocarbon-degrading bacterium from marine sediments. *Int J Syst Bacteriol* **45:** 116–123.

Eilers, H., Pernthaler, J., Glockner, F.O., and Amann, R. (2000) Culturability and in situ abundance of pelagic bacteria from the North Sea. *Appl Environ Microbiol* **66:** 3044–3051.

Franklin, M.P., McDonald, I.R., Bourne, D.G., Owens, N.J.P., Upstill-Goddard, R.C., and Murrell, J.C. (2005) Bacterial diversity in the bacterioneuston (sea surface microlayer): the bacterioneuston through the looking glass. *Environ Microbiol* **7:** 723–736.

Garcia, M.T., Mellado, E., Ostos, J.C., and Ventosa, A. (2004) *Halomonas organivorans* sp. nov., a moderate halophile able to degrade aromatic compounds. *Int J Syst Evol Microbiol* **54:** 1723–1728.

Gertler, C., Gerdts, G., Timmis, K.N., and Golyshin, P.N. (2009) Microbial consortia in mesocosm bioremediation trial using oil sorbents, slow-release fertilizer and bioaugmentation. *FEMS Microbiol Ecol* **69:** 288–300.

Giebel, H.-A., Kalhoefer, D., Lemke, A., Thole, S., Gahl-Janssen, R., Simon, M., and Brinkhoff, T. (2011) Distribution of *Roseobacter* RCA and SAR11 lineages in the North Sea and characteristics of an abundant RCA isolate. *ISME J* **5:** 8–19.

Giovannoni, S.J., Bibbs, L., Cho, J.C., Stapels, M.D., Desiderio, R., Vergin, K.L., *et al.* (2005) Proteorhodopsin in the ubiquitous marine bacterium SAR11. *Nature* **438:** 82–85.

Green, D.H., Llewellyn, L.E., Negri, A.P., Blackburn, S.I., and Bolch, C.J. (2004) Phylogenetic and functional diversity of the cultivable bacterial community associated with the paralytic shellfish poisoning dinoflagellate *Gymnodinium catenatum*. *FEMS Microbiol Ecol* **47:** 345–357.

Gros, J., Deedar, N., Würz, B., Wick, L.Y., Brussaard, C.P.D., Huisman, J., *et al.* (2014) First day of an oil spill on the open sea: early mass transfers of hydrocarbons to air and water. *Environ Sci Technol* **48:** 9400–9411.

Gutierrez, T., Singleton, D.R., Berry, D., Yang, T., Aitken, M.D., and Teske, A. (2013) Hydrocarbon-degrading bacteria enriched by the Deepwater Horizon oil spill identified by cultivation and DNA-SIP. *ISME J* **7:** 2091–2104.

Hammer, Ø., Harper, D.A.T., and Ryan, P.D. (2001) PAST: paleontological statistics software package for education and data analysis. *Palaeontol Electronica* **4:** 1–9.

Harwati, T.U., Kasai, Y., Kodama, Y., Susilaningsih, D., and Watanabe, K. (2007) Characterization of diverse hydrocarbon-degrading bacteria isolated from Indonesian seawater. *Microbes Environ* **22:** 412–415.

Hedlund, B., and Staley, J. (2006) Isolation and characterization of *Pseudoalteromonas* strains with divergent polycyclic aromatic hydrocarbon catabolic properties. *Environ Microbiol* **8:** 178–182.

ITOPF (2002) Fate of marine oil spills. The International Tanker Owners Pollution Federation Limited 2: 1–8.

Kasai, Y., Kishira, H., and Harayama, S. (2002) Bacteria belonging to the genus *Cycloclasticus* play a primary role in the degradation of aromatic hydrocarbons released in a marine environment. *Appl Environ Microbiol* **68:** 5625–5633.

Klein, G.L., Soum-Soutera, E., Guede, Z., Bazire, A., Compere, C., and Dufour, A. (2011) The anti-biofilm activity secreted by a marine *Pseudoalteromonas* strain. *Biofouling* **27:** 931–940.

Kryachko, Y., Dong, X., Sensen, C.W., and Voordouw, G. (2012) Compositions of microbial communities associated with oil and water in a mesothermic oil field. *Antonie Van Leeuwenhoek* **101:** 493–506.

Laghdass, M., Catala, P., Caparros, J., Oriol, L., Lebaron, P., and Obernosterer, I. (2012) High contribution of SAR11 to microbial activity in the North West Mediterranean Sea. *Microb Ecol* **63:** 324–333.

Lane, D.J. (1991) 16S/23S rRNA sequencing. In *Nucleic Acid Techniques in Bacterial Systematics*. Stackebrandt, E., and Goodfellow, M. (eds). Chichester: John Wiley & Sons, pp. 115–175.

Lanfranconi, M.P., Bosch, R., and Nogales, B. (2010) Short-term changes in the composition of active marine bacterial assemblages in response to diesel oil pollution. *Microb Biotechnol* **3:** 607–621.

Lidbury, I., Murrell, J.C., and Chen, Y. (2014) Trimethylamine *N*-oxide metabolism by abundant marine heterotrophic bacteria. *Proc Natl Acad Sci USA* **111:** 2710–2715.

Lin, X., Yang, B., Shen, J., and Du, N. (2009) Biodegradation of Crude Oil by an Arctic Psychrotrophic Bacterium *Pseudoalteromomas* sp. P29. *Curr Microbiol* **59:** 341–345.

McGenity, T.J. (2014) Hydrocarbon biodegradation in intertidal wetland sediments. *Curr Opin Biotechnol* **27:** 46–54.

McGenity, T.J., Folwell, B.D., McKew, B.A., and Sanni, G.O. (2012) Marine crude-oil biodegradation: a central role for interspecies interactions. *Aquat Biosyst* **8:** 10.

McKew, B.A., Coulon, F., Osborn, A.M., Timmis, K.N., and McGenity, T.J. (2007a) Determining the identity and roles of oil-metabolizing marine bacteria from the Thames estuary, UK. *Environ Microbiol* **9:** 165–176.

McKew, B.A., Coulon, F., Yakimov, M.M., Denaro, R., Genovese, M., Smith, C.J., *et al.* (2007b) Efficacy of intervention strategies for bioremediation of crude oil in marine systems and effects on indigenous hydrocarbonoclastic Bacteria. *Environ Microbiol* **9:** 1562–1571.

Malmstrom, R.R., Kiene, R.P., Cottrell, M.T., and Kirchman, D.L. (2004) Contribution of SAR11 bacteria to dissolved dimethylsulfoniopropionate and amino acid uptake in the North Atlantic ocean. *Appl Environ Microbiol* **70:** 4129–4135.

Melcher, R.J., Apitz, S.E., and Hemmingsen, B.B. (2002) Impact of irradiation and polycyclic aromatic hydrocarbon spiking on microbial populations in marine sediment for future aging and biodegradability studies. *Appl Environ Microbiol* **68:** 2858–2868.

Mnif, S., Chamkha, M., and Sayadi, S. (2009) Isolation and characterization of *Halomonas* sp. strain C2SS100, a hydrocarbon-degrading bacterium under hypersaline conditions. *J Appl Microbiol* **107:** 785–794.

Morris, R.M., Rappe, M.S., Connon, S.A., Vergin, K.L., Siebold, W.A., Carlson, C.A., and Giovannoni, S.J. (2002) SAR11 clade dominates ocean surface bacterioplankton communities. *Nature* **420:** 806–810.

Muyzer, G., De Waal, E.C., and Uitterlinden, A.G. (1993) Profiling of complex microbial populations by denaturing gradient gel electrophoresis analysis of polymerase chain reaction-amplified genes coding for 16S rRNA. *Appl Environ Microbiol* **59:** 695–700.

Nelson, C.E., and Carlson, C.A. (2012) Tracking differential incorporation of dissolved organic carbon types among diverse lineages of Sargasso Sea bacterioplankton. *Environ Microbiol* **14:** 1500–1516.

Nichols, C.M., Lardière, S.G., Bowman, J.P., Nichols, P.D., A.E. Gibson, J., and Guézennec, J. (2005) Chemical characterization of exopolysaccharides from Antarctic marine bacteria. *Microb Ecol* **49:** 578–589.

Nogales, B., Aguiló-Ferretjans, M.M., Martín-Cardona, C., Lalucat, J., and Bosch, R. (2007) Bacterial diversity, composition and dynamics in and around recreational coastal areas. *Environ Microbiol* **9:** 1913–1929.

Pepi, M., Cesaro, A., Liut, G., and Baldi, F. (2005) An antarctic psychrotrophic bacterium *Halomonas* sp. ANT-3b, growing on *n*-hexadecane, produces a new emulsyfying glycolipid. *FEMS Microbiol Ecol* **53:** 157–166.

Prabagaran, S.R., Manorama, R., Delille, D., and Shivaji, S. (2007) Predominance of *Roseobacter*, *Sulfitobacter*, *Glaciecola* and *Psychrobacter* in seawater collected off Ushuaia, Argentina, Sub-Antarctica. *FEMS Microbiol Ecol* **59:** 342–355.

Redmond, M.C., and Valentine, D.L. (2011) Natural gas and temperature structured a microbial community response to the Deepwater Horizon oil spill. *Proc Natl Acad Sci USA* **109:** 20292–20297.

Rivers, A.R., Sharma, S., Tringe, S.G., Martin, J., Joye, S.B., and Moran, M.A. (2013) Transcriptional response of bathypelagic marine bacterioplankton to the Deepwater Horizon oil spill. *ISME J* **7:** 1–15.

Romanenko, L.A., Zhukova, N.V., Rohde, M., Lysenko, A.M., Mikhailov, V.V., and Stackebrandt, E. (2003) *Glaciecola mesophila* sp. nov., a novel marine agar-digesting bacterium. *Int J Syst Evol Microbiol* **53:** 647–651.

Røberg, S., Østerhus, J.I., and Landfald, B. (2011) Dynamics of bacterial community exposed to hydrocarbons and oleophilic fertilizer in high-Arctic intertidal beach. *Polar Biol* **34:** 1455–1465.

Sabirova, J.S., Chernikova, T.N., Timmis, K.N., and Golyshin, P.N. (2008) Niche-specificity factors of a marine oil-degrading bacterium *Alcanivorax borkumensis* SK2. *FEMS Microbiol Lett* **285:** 89–96.

Saravanan, P., and Jayachandran, S. (2008) Preliminary characterization of exopolysaccharides produced by a marine biofilm-forming bacterium *Pseudoalteromonas ruthenica* (SBT 033). *Lett Appl Microbiol* **46:** 1–6.

Schneiker, S., Martins Dos Santos, V.A., Bartels, D., Bekel, T., Brecht, M., Buhrmester, J., *et al.* (2006) Genome sequence of the ubiquitous hydrocarbon-degrading marine bacterium *Alcanivorax borkumensis*. *Nat Biotechnol* **24:** 997–1004.

Sikkema, J., De Bont, J.A., and Poolman, B. (1995) Mechanisms of membrane toxicity of hydrocarbons. *Microbiol Rev* **59:** 201–222.

Sperling, M., Giebel, H.A., Rink, B., Grayek, S., Staneva, J., Stanev, E., and Simon, M. (2012) Differential effect of hydrographic and biogeochemical properties on SAR11 and *Roseobacter* RCA populations in the southern North Sea. *Aquat Microb Ecol* **67:** 25–34.

Staley, J.T. (2010) *Cycloclasticus*: A genus of marine polycyclic aromatic hydrocarbon degrading bacteria. In *Handbook of Hydrocarbon and Lipid Microbiology*. Timmis, K.N., McGenity, T.J., van der Meer, J.R., and de Lorenzo, V. (eds). Berlin Heidelberg: Springer, pp. 1782–1786.

Steindler, L., Schwalbach, M.S., Smith, D.P., Chan, F., and Giovannoni, S.J. (2011) Energy starved *Candidatus* Pelagibacter ubique substitutes light-mediated ATP production for endogenous carbon respiration. *PLoS ONE* **6:** e19725.

Tada, Y., Taniguchi, A., Sato-Takabe, Y., and Hamasaki, K. (2012) Growth and succession patterns of major phylogenetic groups of marine bacteria during a mesocosm diatom bloom. *J Oceanogr* **68:** 509–519.

Teramoto, M., Suzuki, M., Okazaki, F., Hatmanti, A., and Harayama, S. (2009) *Oceanobacter*-related bacteria are important for the degradation of petroleum aliphatic hydrocarbons in the tropical marine environment. *Microbiology* **155:** 3362–3370.

Teramoto, M., Ohuchi, M., Hatmanti, A., Darmayati, Y., Widyastuti, Y., Harayama, S., and Fukunaga, Y. (2011) *Oleibacter marinus* gen. nov., sp. nov., a bacterium that degrades petroleum aliphatic hydrocarbons in a tropical marine Environment. *Int J Syst Evol Microbiol* **61:** 375–380.

Vila, J., Maria Nieto, J., Mertens, J., Springael, D., and Grifoll, M. (2010) Microbial community structure of a heavy fuel oil-degrading marine consortium: linking microbial dynamics with polycyclic aromatic hydrocarbon utilization. *FEMS Microbiol Ecol* **73:** 349–362.

Wang, B., Lai, Q., Cui, Z., Tan, T., and Shao, Z. (2008) A pyrene-degrading consortium from deep-sea sediment of the West Pacific and its key member *Cycloclasticus* sp. P1. *Environ Microbiol* **10:** 1948–1963.

Wietz, M., Gram, L., Jørgensen, B., and Schramm, A. (2010) Latitudinal patterns in the abundance of major marine bacterioplankton groups. *Aquat Microb Ecol* **61:** 179–189.

Wurl, O., and Obbard, J.P. (2004) A review of pollutants in the sea-surface microlayer (SML): a unique habitat for marine organisms. *Mar Pollut Bull* **48:** 1016–1030.

Yakimov, M.M., Golyshin, P.N., Lang, S., Moore, E.R., Abraham, W.R., Lunsdorf, H., and Timmis, K.N. (1998) *Alcanivorax borkumensis* gen. nov., sp. nov., a new, hydrocarbon-degrading and surfactant-producing marine bacterium. *Int J Syst Bacteriol* **48:** 339–348.

Yakimov, M.M., Giuliano, L., Denaro, R., Crisafi, E., Chernikova, T.N., Abraham, W.R., *et al.* (2004a) *Thalassolituus oleivorans* gen. nov., sp. nov., a novel marine bacterium that obligately utilizes hydrocarbons. *Int J Syst Evol Microbiol* **54:** 141–148.

Yakimov, M.M., Gentile, G., Bruni, V., Cappello, S., D'Auria, G., Golyshin, P.N., and Giuliano, L. (2004b) Crude oil-induced structural shift of coastal bacterial communities of rod bay (Terra Nova Bay, Ross Sea, Antarctica) and characterization of cultured cold-adapted hydrocarbonoclastic bacteria. *FEMS Microbiol Ecol* **49:** 419–432.

Yong, J.J., Park, S.J., Kim, H.J., and Rhee, S.K. (2007) *Glaciecola agarilytica* sp. nov., an agar-digesting marine bacterium from the East Sea, Korea. *Int J Syst Evol Microbiol* **57:** 951–953.

## Supporting information

Additional Supporting Information may be found in the online version of this article at the publisher's web-site:

**Fig. S1.** DGGE profiles of bacterial communities in the non-oiled (BI) and the most oiled (BIII) on-board mesocosms over time, based on 16S rRNA gene.

**Appendix S1.** Supporting experimental procedures.

# 7

# Improved cytotoxic effects of *Salmonella*-producing cytosine deaminase in tumour cells

Beatriz Mesa-Pereira, Carlos Medina, Eva María Camacho, Amando Flores* and Eduardo Santero
*Centro Andaluz de Biología del Desarrollo, CSIC, Junta de Andalucía, Universidad Pablo de Olavide, Carretera de Utrera, Km. 1, Seville, 41013, Spain.*

## Summary

In order to increase the cytotoxic activity of a *Salmonella* strain carrying a salicylate-inducible expression system that controls cytosine deaminase production, we have modified both, the vector and the producer bacterium. First, the translation rates of the expression module containing the *Escherichia coli codA* gene cloned under the control of the Pm promoter have been improved by using the T7 phage gene 10 ribosome binding site sequence and replacing the original GUG start codon by AUG. Second, to increase the time span in which cytosine deaminase may be produced by the bacteria in the presence of 5-fluorocytosine, a 5-fluorouracyl resistant *Salmonella* strain has been constructed by deleting its *upp* gene sequence. This new *Salmonella* strain shows increased cytosine deaminase activity and, after infecting tumour cell cultures, increased cytotoxic and bystander effects under standard induction conditions. In addition, we have generated a *purD* mutation in the producer strain to control its intracellular proliferation by the presence of adenine and avoid the intrinsic *Salmonella* cell death induction. This strategy allows the analysis and comparison of the cytotoxic effects of cytosine deaminase produced by different *Salmonella* strains in tumour cell cultures.

*For correspondence. E-mail aflodia@upo.es

**Funding Information** This work was supported by the Grant 'Proyecto de Excelencia P07-CVI02518' from the Andalusian government and by a fellowship from the Andalusian government to B. M.-P. C. M. holds a JAE DOC contract from the Spanish National Research Council (CSIC).

## Introduction

Bacteria can be easily adapted to synthesize proteins with relevant biotechnological applications. Over the past decade, many genera of bacteria have been explored as cell factories for cancer therapy due to their ability to specifically target tumours (Pawelek *et al.*, 1997), reviewed in (Forbes, 2010). *Salmonella enterica* serovar Typhimurium (*S.* Typhimurium) is probably the intracellular pathogen that has been most extensively studied as an anti-tumour vector due to its intrinsic properties. These bacteria preferentially colonize and proliferate in solid tumours at ratios greater than 1000/1 compared with normal target organs, a behaviour that usually results in tumour growth inhibition (Pawelek *et al.*, 1997). In addition, as a facultative anaerobe, *Salmonella* can grow under aerobic and anaerobic conditions, which allows bacteria to accumulate in large solid tumours and invade metastases (Saltzman *et al.*, 1996; Yam *et al.*, 2010).

Administration of attenuated *Salmonella* strains expressing different anti-tumour agents has been attempted in recent years with promising results in tumour regression (Nemunaitis *et al.*, 2003; Barnett *et al.*, 2005; Zhao *et al.*, 2006; Royo *et al.*, 2007; Jeong *et al.*, 2014). One of the therapeutic genes successfully expressed in *S.* Typhimurium is the *Escherichia coli codA* gene, encoding cytosine deaminase (CD). This enzyme, present in fungi and bacteria but absent in mammalian cells (Nishiyama *et al.*, 1985), catalyses the conversion of cytosine to uracil and ammonia (Koechlin *et al.*, 1966). Cytosine deaminase can also deaminate the non-toxic cytosine analog, 5-fluorocytosine (5-FC) to the toxic metabolite, 5-fluorouracil (5-FU) that is widely used as a chemotherapeutic agent. This metabolite is then converted by cellular enzymes into 5-FdUMP, which inhibits DNA synthesis by blocking the activity of thymidylate synthase, 5-FUTP and 5-FdUTP, which are incorporated into RNA and DNA, respectively (Meyers *et al.*, 2003), thus leading to cell death (Polak and Scholer, 1975; Damon *et al.*, 1989). In addition, 5-FU can freely diffuse across the cell membrane and produce its cytotoxic effects in neighbouring cells, a phenomenon known as the bystander effect (Kuriyama *et al.*, 1998). Despite several co-administration studies that have demonstrated conversion of 5-FC to 5-FU and significant tumour growth reduction in animal models (King *et al.*, 2002; Nemunaitis *et al.*, 2003; Royo *et al.*, 2007), its application in cancer patients has been limited (Nemunaitis *et al.*, 2003). Clinical data suggest that the anti-tumour activity of 5-FU is directly related to both the duration of drug exposure and its concentration in the tumour (Nemunaitis *et al.*, 2003). However, in order to achieve a significant amount of active metabolites and cell killing, the required dose of the apparently harmless 5-FC may be high enough to cause adverse effects (reviewed in (Vermes *et al.*, 2000)). This 5-FC toxicity may be due, in part, to the conversion of 5-FC to 5-FU by human intestinal microflora (Harris *et al.*, 1986). Increasing the anti-tumour activity and minimizing the systemic toxicity would circumvent these problems, but to achieve this, it is necessary to improve the selective production of CD into the tumour. We have previously validated an *in vivo* salicylate-inducible cascade expression system that allows the controlled cytosine deaminase production. This system combines a set of salicylate-regulated elements from *Pseudomonas putida* that work in cascade, containing a regulatory module (NahR and XylS2 transcription regulators coding sequences) integrated in the chromosome of attenuated *S.* Typhimurium *aroA* (SL7207 strain) and an expression module, consisting in a *codA* gene cloned under the control of the Pm promoter either in a plasmid or integrated in the chromosome (Royo *et al.*, 2007). In the presence of salicylate, XylS2 promotes transcription from Pm. In order to increase the CD production rates, in this work we have improved the CD expression module by engineering *codA* to be translated from the T7 phage gene 10 ribosome binding site and changing the original CD GUG start codon to AUG in new salicylate induced expression vectors (Medina *et al.*, 2011). Since the microbial uracil phosphoribosyltransferase, encoded by *upp*, directly converts 5-FU to the metabolite 5-FUMP, from which the other toxic metabolites are produced, strains lacking this activity are more tolerant to 5-FU (Lundegaard and Jensen, 1999). To prevent killing of the producing bacteria during accumulation of toxic 5-FU, thus increasing the time span in which *Salmonella* produces CD, we have also constructed a 5-FU resistant *upp* mutant. Finally, in order to assess the effects of CD produced by improved strains and plasmids in tumour cell cycle distribution and bystander activity in long-term cell cultures, a *purD* mutation has been generated in the producer strains to avoid cell death induced by intracellular *Salmonella* proliferation (Leung and Finlay, 1991; Mesa-Pereira *et al.*, 2013).

## Results and discussion

### Construction of a Salmonella *strain with high salicylate-induced CD production rates*

In order to increase the amount of CD produced keeping standard induction conditions, we improved both the producing *Salmonella* strain and the CD expression plasmid. First, we transferred the new genome-integrated regulatory module previously developed in our laboratory (Medina *et al.*, 2011) to the SL7207 *Salmonella* strain, thus generating the MPO375 strain (bacterial strains and plasmid are listed in Supporting information Table S1,). This regulatory module contains a constitutively expressed *gfp* gene to track *Salmonella* during the infection process. Second, we modified this strain with the aim of avoiding host cell death induced by *Salmonella* intracellular proliferation (see below). To that end, we transduced a $\Delta purD$ mutation into MPO375 to get the strain MPO376. In this way, intracellular proliferation can be controlled by the amount of adenine in the culture medium (Leung and Finlay, 1991; Mesa-Pereira *et al.*, 2013). On the other hand, we constructed new plasmids with higher CD expression rates than pMPO16, the vector previously used in our laboratory to express CD (Royo *et al.*, 2007). The *E. coli codA* sequence cloned in this plasmid has the original GUG start codon and its own ribosome binding site (from now on, $CD_{GUG}$ sequence). To increase CD production, we changed the *codA* start codon to AUG and cloned the resulting sequence into the high copy number vector pMPO52 (Medina *et al.*, 2011) to produce pMPO88. In this vector, *codA* expression is under the control of the Pm promoter and the T7 gene10 ribosome binding site, a strong ribosome binding site that achieves high translation levels (from now on, $CD_{7AUG}$ sequence). We have previously reported that the salicylate-induced expression levels of vectors based in pWSK29, a low copy number vector that is stable through the whole *Salmonella* infection cycle without selection pressure, are comparable to that of their corresponding high copy number vectors (Medina *et al.*, 2011). To generate versions of the CD expression modules in low copy number plasmids, the engineered *codA* genes in pMPO88 and pMPO16 were subcloned in the pWSK29 derived vector pMPO20, thus generating plasmids pMPO90 and pMPO1088 respectively.

To compare the amount of CD produced by the different constructs, we analysed whole-cell protein extracts from cultures of the strain MPO376 carrying the low copy number vectors pMPO1088 ($CD_{GUG}$) or pMPO90 ($CD_{7AUG}$) by SDS-PAGE, in the presence or absence of salicylate. As shown in Fig. 1A (lanes 4 and 6), the pMPO90 vector produces more CD than pMPO1088 after salicylate induction. Afterwards, we determined the CD activity from these cell lysates by analysing the

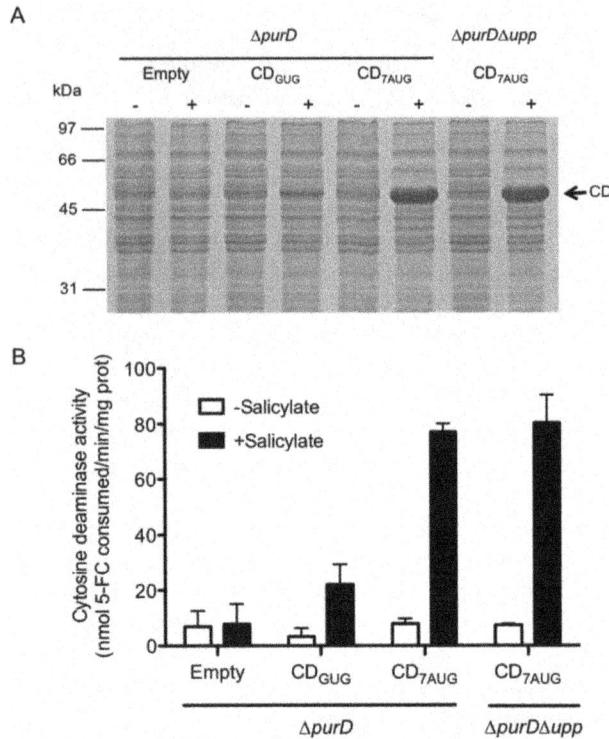

**Fig. 1.** Production of CD in low copy number expression vectors.
A. SDS-PAGE analysis of salicylate dependent overproduction of cytosine deaminase. Whole extracts of *Salmonella* MPO376 (Δ*purD*) bearing pMPO54 (empty vector), pMPO1088 (CD$_{GUG}$) or pMPO90 (CD$_{7AUG}$) plasmids and *Salmonella* strain MPO378 (Δ*purD*Δ*upp*) bearing pMPO90 plasmid, either uninduced (-) or induced by salicylate for 4 h (+). Three µl of supernatant was loaded in each track.
B. Analysis of conversion of 5-FC. Cytosine deaminase activity from cell extracts of *Salmonella* MPO376 bearing pMPO54, pMPO1088 or pMPO90 plasmids, and MPO378 bearing pMPO90 either induced by salicylate for 4 h, or not induced. Cytosine deaminase activity was assayed as previously described (Nishiyama *et al.*, 1985). Each bar represents the average of three independent experiments ± SD.

conversion of 5-FC to the cytotoxic agent 5-FU. The assays (Fig. 1B) revealed that, upon induction, the strain harbouring the pMPO90 (CD$_{7AUG}$) vector reached an activity about 3.5-fold higher than the same strain bearing pMPO1088 (CD$_{GUG}$). Thus, these results demonstrate that the new CD$_{7AUG}$ sequence produces higher amounts of CD and therefore 5-FU than CD$_{GUG}$ using the same *Salmonella* producer strain and concentrations of salicylate and 5-FC.

## Expression of CD with a 5-FU resistant *Salmonella* strain

As shown before, the new vector pMPO90 (CD$_{7AUG}$) allows production of a larger amount of CD than the former construct under salicylate induction. However, since *Salmonella* is sensitive to 5-FU, the maximum rate of synthesis of this cytotoxic metabolite could also be limited by the maximum tolerated concentration and not only by the amount of CD present in the bacterium. To test this prediction, we first determined the growth of *Salmonella* carrying different plasmids on plates containing 5-FC (Fig. 2). Cultures of strain MPO376 bearing either the empty vector or one of the two CD-expressing plasmids (pMPO1088 or pMPO90) were spotted on plates in the presence or absence of salicylate and supplemented with two different concentrations, 0.5 or 5 µM, of 5-FC. Consistent with the results mentioned above, the *Salmonella* strain carrying the plasmid pMPO90 presents a more severe growth defect even at the low 5-FC concentration than the strain bearing pMPO1088 at the high 5-FC concentration, which correlates with its higher expression of CD and, consequently, higher production

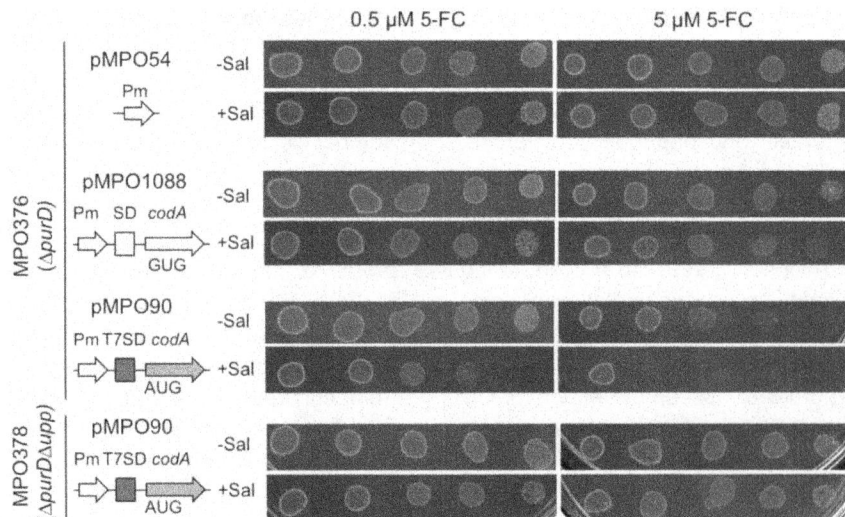

**Fig. 2.** Effect of 5-FU produced on the bacterial growth. Serial dilutions ($10^2$ to $10^6$) of cultures grown with or without salicylate were plated in supplemented minimal E medium in the presence of 0.5 µM or 5 µM of 5-FC and incubated for 24 h at 37°C.

rate of 5-FU. This clearly showed that 5-FU production could be limited by the bacterium sensitivity to it.

This observation prompted us to obtain a *Salmonella* mutant resistant to 5-FU, strain that could produce higher amounts of this drug and for a longer time than the isogenic sensitive strain using the same 5-FC dosage. The mutant was constructed by deleting the *upp* gene sequence, whose product is involved in 5-FU sensitivity (Glaab *et al.*, 2005), thus generating the strain MPO378 (Δ*purD*Δ*upp*). For the construction of *upp* mutant strain, the 'One Step Deletion' approach was used to replace target gene by the chloramphenicol resistance cassette (Datsenko and Wanner, 2000). We transformed this strain with the plasmid pMPO90 (CD$_{7AUG}$) and performed the same experiments to determine the amount of CD produced and the activity of whole-cell extracts of salicylate induced cultures (Nishiyama *et al.*, 1985). As shown in Fig. 1A and B, CD production and activity were independent of the *upp* mutation, since the strain behaved as its *upp*+ counterpart. Conversely, and as expected, the mutant was resistant to the 5-FU produced when grown on plates supplemented with 0.5 and 5 μM of 5-FC (Fig. 2) despite the high CD activity achieved. These results suggest that this strain and plasmid combination may represent an improvement in bacterial cancer therapy since it has the capacity of achieving a higher 5-FU concentration with a low 5-FC dosage, which, in turns, would reduce the deleterious effect of this compound in healthy eukaryotic cells.

### A novel strategy to analyse the cytotoxic effect of Salmonella-*producing 5-FU in tumour cell cultures*

Next, we decided to compare the consequences of 5-FU-controlled production by the different plasmids and strains obtained in this work in eukaryotic cell cultures. To determine the effects of 5-FU in eukaryotic cells, it is necessary to analyse the evolution of cell cultures for 6 days after the addition of this compound (Erbs *et al.*, 2000; Bourbeau *et al.*, 2004). However, once *Salmonella* has infected the eukaryotic cells, bacterial proliferation and expression of certain bacterial proteins during the first hours of infection induce host cell death within 18–24 h, hindering the study of the effect of the 5-FU produced by *Salmonella* in cell cultures (Kim *et al.*, 1998; Paesold *et al.*, 2002; Mesa-Pereira *et al.*, 2013). To analyse the effects of the 5-FU overproduced by *Salmonella* in cell cultures, we generated a mutation in the producer strain to prevent bacterial growth and protein production inside host cell. It has been previously reported that attenuated *purD* mutants are invasion proficient but unable to proliferate once inside the eukaryotic cell. Nevertheless, the addition of adenine to culture medium can temporally suppress this deficiency (Leung and Finlay, 1991); thus, in a *purD*-

background, intracellular proliferation and CD overproduction can be controlled by the presence of adenine and salicylate respectively. This strategy has been recently exploited in our laboratory to study the role of SpvB *Salmonella* effector protein in the infection process (Mesa-Pereira *et al.*, 2013). In the present work, we have used a similar experimental approach to study CD overproduction effects, and included a Δ*purD* mutation in all the strains used in this work. The strain MPO376 (Δ*purD*) bearing the empty vector, pMPO1088 (CD$_{GUG}$) or pMPO90 (CD$_{7AUG}$) and the strain MPO378 (Δ*purD*Δ*upp*) carrying pMPO90 were used to infect HeLa cells. After invasion, adenine was added to infected cell cultures. and 1 h later, once infection was established, *codA* expression was induced with salicylate. Five hours later, adenine concentration was reduced 40-fold to avoid bacterial proliferation, 50 μM of 5-FC was added and cells were incubated for 6 days in the presence of salicylate and 5-FC. As a control, uninfected cell cultures followed the same treatment but in the presence of 10 μM of 5-FU (Erbs *et al.*, 2000). The effects of *codA* overexpression and 5-FU production were analysed by flow cytometry and microscopy.

Figure 3A shows the cell cycle distribution of HeLa cell cultures. As expected, most of the cells (67%) of the control HeLa cell cultures in presence of 5-FC and absence of 5-FU were in G0/G1 phase of the cell cycle. Similarly, treatment with 5-FU produced the expected effects on the cell cycle distribution (Pizzorno *et al.*, 1995; Takeda *et al.*, 1999; Yoshikawa *et al.*, 2001; De Angelis *et al.*, 2006): cells in G0/G1 phase were reduced down to 30%, while the dead cell population, represented as the percentage of cells in sub-G1 phase of the cell cycle, and cells arrested in S and G2/M phase were increased. Conversely, cell cycle of cells infected with the Δ*purD* strain bearing the empty vector were not affected even 6 days after infection, which confirms that this mutant is unable to induce cell death in the absence of adenine. Interestingly, expression of CD by this proliferation deficient strain led to a substantial alteration of the cell cycle distribution in a way similar to 5-FU (increase in dead or arrested cells and decrease in normal cells). Consistent with the above experiments, infection with the Δ*purD* strain bearing pMPO1088 (CD$_{GUG}$) had a slight but observable effect on cell cycle distribution. However, the effect was much higher when the infecting bacteria harboured pMPO90 (CD$_{7AUG}$). In this case, there even was an increase in > 4N cells indicative of aberrant endoreduplication. In addition, the maximum effect on cell cycle distribution was achieved when the strain bearing pMPO90 (CD$_{7AUG}$) was the 5-FU resistant strain Δ*purD*Δ*upp* (more cells in sub-G1 than in G0/G1 phase), which was even more pronounced than that induced with 10 μM of 5-FU in the control culture. Accordingly, phase contrast microscopy

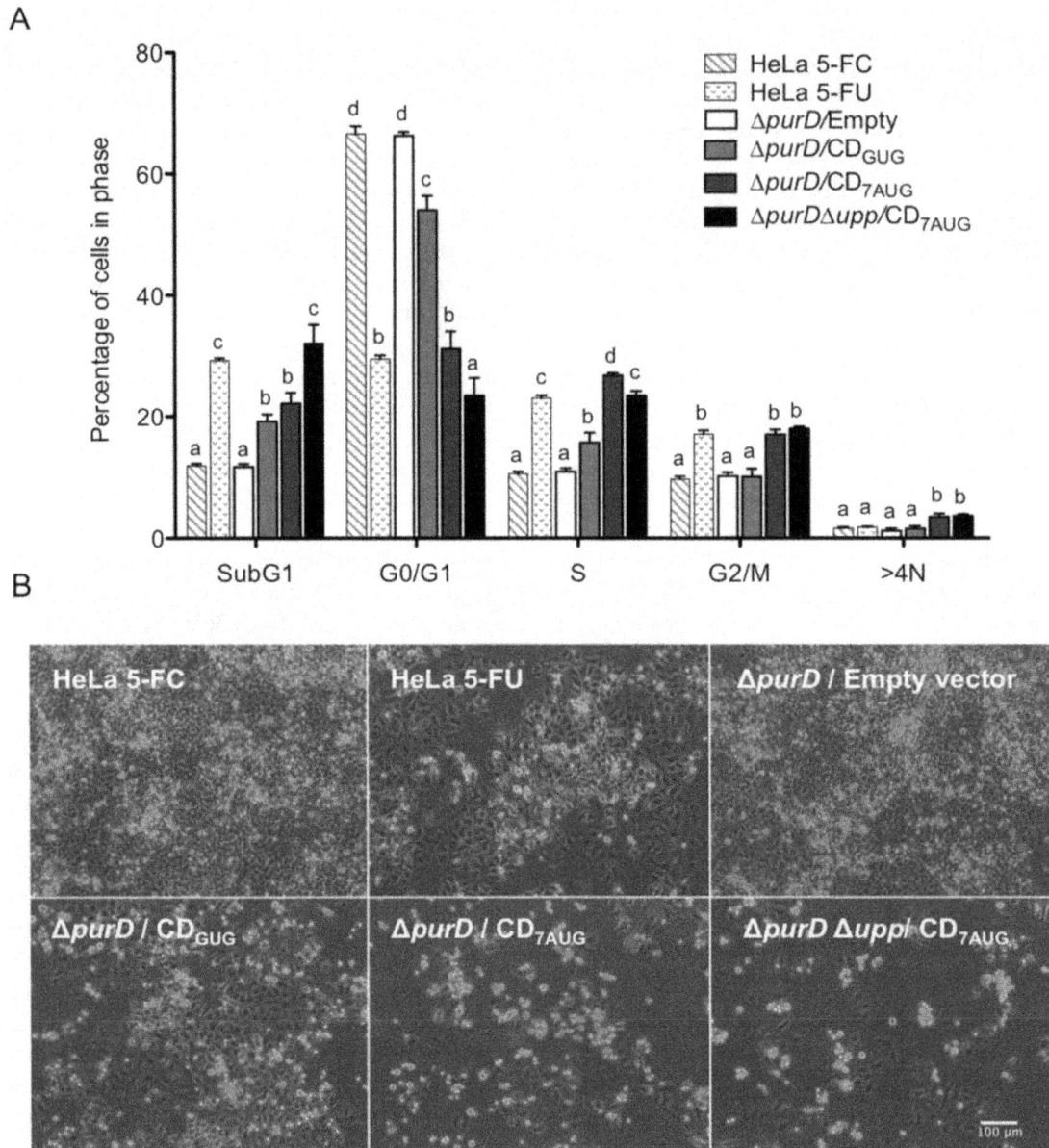

**Fig. 3.** *In vitro* sensitivity to 5-FU produced by *Salmonella* on infected HeLa cells.
A. Cell cycle distribution of HeLa cells infected with *Salmonella* MPO376 (*ΔpurD*) bearing pMPO54 (empty vector), pMPO1088 (CD$_{GUG}$) or pMPO90 (CD$_{7AUG}$) and MPO378 (*ΔpurDΔupp*) bearing pMPO90, at multiplicity of infection 50:1. The cells were cultured in the presence of 50 μM of 5-FC and harvested at 6 days post-induction. Ten thousand events were analysed by flow cytometry for each sample. Graphics represents the mean ± SD of three independent experiments. Non-infected HeLa cells treated with 50 μM of 5-FC or 10 μM of 5-FU were used as controls. One-way analysis of variance and Tukey HSD *post hoc* tests were applied to test for significant differences. Data from the same group marked with different alphabet are significantly different at *P* < 0.05.
B. Phase contrast microscopy of infected HeLa cells as well as uninfected control and HeLa cells treated with 5-FU are shown at day 6.

showed much less proliferating HeLa cells when *Salmonella* expressed CD from pMPO90 (CD$_{7AUG}$), an effect that was even more evident in these conditions than in 5-FU treated cultures (Fig. 3B). Additionally, to test whether the 5-FU produced from these vectors and strains has similar consequences on different tumour cell lines, we performed the same analysis to determine the effects on MCF-7 and HCT116 cell cycle distribution. As shown in

Fig. S1 (Supporting information), CD controlled expression produced similar effects on MCF-7 and HCT116 cell cycle, thus showing they are not restricted to HeLa cells. Taken together, these results demonstrate that upon salicylate induction, *Salmonella* bearing the new *codA* expressing construct produces more CD and converts more 5-FU than the same strain transformed with the former construct, which subsequently correlates with a

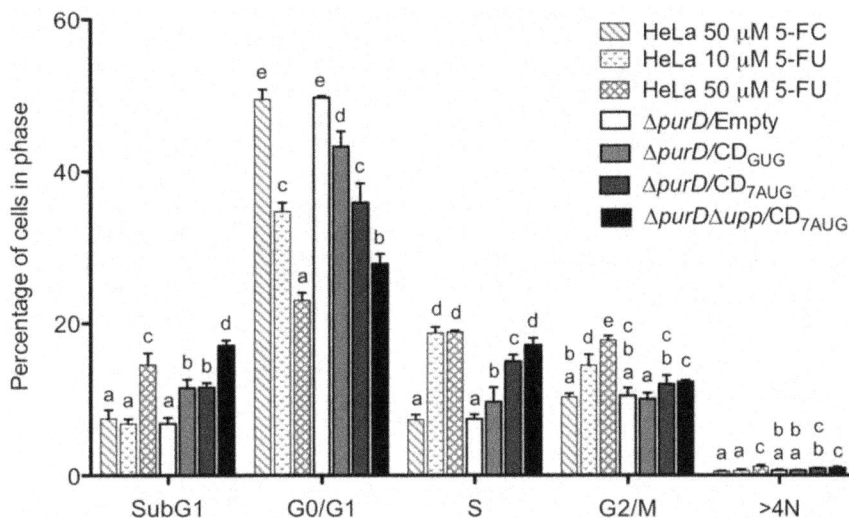

**Fig. 4.** Bystander effect: Cytotoxicity of supernatants from infected HeLa cells treated with 5-FC. HeLa cells were infected at multiplicity of infection 50:1 and were grown in presence of 5-FC 250 μM in medium supplemented with adenine. Media were collected 40 h later, diluted 1:5 and added to uninfected HeLa cells. Effect of 5-FU produced was determined at day 6 by cell cycle analysis. Graphics represents mean ± SD of three independent experiments. Non-infected HeLa cells treated with 50 μM of 5-FC, 10 μM and 50 μM of 5-FU were used as controls. One-way analysis of variance and Tukey HSD *post hoc* tests were applied to test for significant differences. Data from the same group marked with different alphabet are significantly different at $P < 0.05$.

higher cytotoxicity in HeLa, MCF-7 and HCT116 cells. These high levels of 5-FU reached are also toxic to the producer strain, although using an *upp* mutant to produce CD circumvents this limitation. Finally, this experimental approach combining the salicylate induced expression system, and the *purD* mutant has proven to be effective to analyse and compare the effect of different CD producing strains on eukaryotic cell cycle. Thus, it can be a useful tool to investigate the consequences on cell physiology of any other cytotoxic protein produced by *Salmonella*, evaluate its potential as anticancer therapy agent in cell cultures and select the most appropriate combination of strains and plasmids prior to their study in animal models.

### Bystander activity of the 5-FU produced in cell cultures

Although *Salmonella* is able to invade tumour cells *in vitro*, there is some controversy regarding bacterial localization *in vivo* and some data indicate that bacteria also proliferates extracellulary in the necrotic region of solid tumour (Westphal *et al.*, 2008; Crull *et al.*, 2011). Therefore, the effectiveness of CD expressed by *Salmonella* in cancer therapy depends on the bystander activity of the produced drug. 5-fluorouracil has such bystander activity since it passively diffuses from cell to cell. For that reason, we compared the bystander effect on HeLa cell cultures infected with the strain MPO376 (*ΔpurD*) bearing the empty vector, pMPO90 (CD7AUG) or pMPO1088 (CDGUG) and the strain MPO378 (*ΔpurDΔupp*) carrying pMPO90. Cytosine deaminase expression was induced with salicylate as described above but, in this case, adenine was

always present in the cultures at normal concentration. Forty hours after infection, supernatants of the different cultures were transferred to uninfected HeLa cells, and cell cycle distribution was analysed by flow cytometry 6 days later. Since supernatant transfer to fresh cultures resulted in a fivefold dilution, we used uninfected cultures treated with either 10 μM or 50 μM 5-FU as control, so they were also diluted to about 2 μM and 10 μM respectively.

The experiments, summarized in Fig. 4, revealed a bystander effect of the three CD-expressing *Salmonella* when compared with the strain carrying the empty vector. An increase in the sub-G1 population can be observed with these three strains. Although there is little difference between the effects detected with the *ΔpurD* strain carrying either pMPO90 (CD7AUG) or pMPO1088 (CDGUG), the higher effect was again achieved with the *ΔpurDΔupp* strain carrying pMPO90 plasmid. In fact, the consequences on cell cycle distribution caused by this combination of strain and plasmid were even greater than those generated in the control cultures.

Current gene delivery systems have low efficiency targeting tumour tissues and can transduce only a small percentage of cells within a tumour. In consequence, the clinical use of cancer gene therapy is limited. Gene therapy approaches to express CD by transformed tumour cells could surpass this limitation due to the bystander effect of the CD/5-FU system. Given that CD has a cytosolic location, a possible limitation of this approach is that CD-expressing cells are killed before cytotoxic concentrations of extracellular 5-FU are reached, limiting in this way the bystander effect and, therefore, the anti-tumour efficiency (Lawrence *et al.*, 1998). Better results would probably be obtained by expressing a secreted form of CD (Rehemtulla *et al.*, 2004) or, specially, using *Salmonella* as a delivery vector, because it selectively targets tumours and preferentially colonizes extracellular compartments (Pawelek *et al.*, 1997; Agorio *et al.*, 2007; Loessner *et al.*, 2007; Leschner and Weiss, 2010; Crull *et al.*, 2011). Bacterial production of CD also has limitations since *Salmonella* is also sensitive to the high concentrations of 5-FU achieved. However, the *upp* 5-FU resistance mutation in *Salmonella* would make the bacterial strain more competent in cancer therapy as it circumvents the suicide of the drug 'factory' and increases the bystander effect. The results of this work show that a *Salmonella* strain that combines high production levels of CD such as those achieved with pMPO90 with resistance to 5-FU due to the *upp* mutation is a better candidate to be used to intra-tumourally delivered 5-FU in cancer treatment.

## Acknowledgements

We are grateful to all members of the laboratory for their insights and helpful suggestions, and to Guadalupe Martín Cabello, Nuria Pérez Claros, Katherina García García and Corin Díaz Ramos for technical help. We also thank J. Casadesús, A. López-Rivas and D. Tuveson for the gift of HeLa, MCF-7 and HCT116 cell lines respectively.

## Conflict of interest

None declared.

## References

Agorio, C., Schreiber, F., Sheppard, M., Mastroeni, P., Fernandez, M., Martinez, M.A., and Chabalgoity, J.A. (2007) Live attenuated Salmonella as a vector for oral cytokine gene therapy in melanoma. *J Gene Med* **9**: 416–423.

Barnett, S.J., Soto, L.J., 3rd, Sorenson, B.S., Nelson, B.W., Leonard, A.S., and Saltzman, D.A. (2005) Attenuated *Salmonella typhimurium* invades and decreases tumor burden in neuroblastoma. *J Pediatr Surg* **40**: 993–997, discussion 997–998.

Bourbeau, D., Lavoie, G., Nalbantoglu, J., and Massie, B. (2004) Suicide gene therapy with an adenovirus expressing the fusion gene CD: UPRT in human glioblastomas: different sensitivities correlate with p53 status. *J Gene Med* **6**: 1320–1332.

Crull, K., Bumann, D., and Weiss, S. (2011) Influence of infection route and virulence factors on colonization of solid tumors by *Salmonella enterica serovar Typhimurium*. *FEMS Immunol Med Microbiol* **62**: 75–83.

Damon, L.E., Cadman, E., and Benz, C. (1989) Enhancement of 5-fluorouracil antitumor effects by the prior administration of methotrexate. *Pharmacol Ther* **43**: 155–185.

Datsenko, K.A., and Wanner, B.L. (2000) One-step inactivation of chromosomal genes in *Escherichia coli* K-12 using PCR products. *Proc Natl Acad Sci USA* **97**: 6640–6645.

De Angelis, P.M., Svendsrud, D.H., Kravik, K.L., and Stokke, T. (2006) Cellular response to 5-fluorouracil (5-FU) in 5-FU-resistant colon cancer cell lines during treatment and recovery. *Mol Cancer* **5**: 20.

Erbs, P., Regulier, E., Kintz, J., Leroy, P., Poitevin, Y., Exinger, F., *et al.* (2000) In vivo cancer gene therapy by adenovirus-mediated transfer of a bifunctional yeast cytosine deaminase/uracil phosphoribosyltransferase fusion gene. *Cancer Res* **60**: 3813–3822.

Forbes, N.S. (2010) Engineering the perfect (bacterial) cancer therapy. *Nat Rev Cancer* **10**: 785–794.

Glaab, W.E., Mitchell, L.S., Miller, J.E., Vlasakova, K., and Skopek, T.R. (2005) 5-fluorouracil forward mutation assay in *Salmonella*: determination of mutational target and spontaneous mutational spectra. *Mutat Res* **578**: 238–246.

Harris, B.E., Manning, B.W., Federle, T.W., and Diasio, R.B. (1986) Conversion of 5-fluorocytosine to 5-fluorouracil by human intestinal microflora. *Antimicrob Agents Chemother* **29**: 44–48.

Jeong, J.H., Kim, K., Lim, D., Jeong, K., Hong, Y., Nguyen, V.H., *et al.* (2014) Anti-tumoral effect of the mitochondrial target domain of Noxa delivered by an engineered *Salmonella typhimurium*. *PLoS ONE* **9**: e80050.

Kim, J.M., Eckmann, L., Savidge, T.C., Lowe, D.C., Witthoft, T., and Kagnoff, M.F. (1998) Apoptosis of human intestinal epithelial cells after bacterial invasion. *J Clin Invest* **102**: 1815–1823.

King, I., Bermudes, D., Lin, S., Belcourt, M., Pike, J., Troy, K., *et al.* (2002) Tumor-targeted *Salmonella* expressing cytosine deaminase as an anticancer agent. *Hum Gene Ther* **13**: 1225–1233.

Koechlin, B.A., Rubio, F., Palmer, S., Gabriel, T., and Duschinsky, R. (1966) The metabolism of 5-fluorocytosine-2-14-C and of cytosine-14-C in the rat and the disposition of 5-fluorocytosine-2-14-C in man. *Biochem Pharmacol* **15**: 435–446.

Kuriyama, S., Masui, K., Sakamoto, T., Nakatani, T., Kikukawa, M., Tsujinoue, H., *et al.* (1998) Bystander effect caused by cytosine deaminase gene and 5-fluorocytosine in vitro is substantially mediated by generated 5-fluorouracil. *Anticancer Res* **18**: 3399–3406.

Lawrence, T.S., Rehemtulla, A., Ng, E.Y., Wilson, M., Trosko, J.E., and Stetson, P.L. (1998) Preferential cytotoxicity of cells transduced with cytosine deaminase compared to bystander cells after treatment with 5-flucytosine. *Cancer Res* **58**: 2588–2593.

Leschner, S., and Weiss, S. (2010) Salmonella-allies in the fight against cancer. *J Mol Med (Berl)* **88**: 763–773.

Leung, K.Y., and Finlay, B.B. (1991) Intracellular replication is essential for the virulence of *Salmonella typhimurium*. *Proc Natl Acad Sci USA* **88**: 11470–11474.

Loessner, H., Endmann, A., Leschner, S., Westphal, K., Rohde, M., Miloud, T., *et al.* (2007) Remote control of tumour-targeted *Salmonella enterica serovar Typhimurium* by the use of L-arabinose as inducer of bacterial gene expression in vivo. *Cell Microbiol* **9:** 1529–1537.

Lundegaard, C., and Jensen, K.F. (1999) Kinetic mechanism of uracil phosphoribosyltransferase from *Escherichia coli* and catalytic importance of the conserved proline in the PRPP binding site. *Biochemistry* **38:** 3327–3334.

Medina, C., Camacho, E.M., Flores, A., Mesa-Pereira, B., and Santero, E. (2011) Improved expression systems for regulated expression in *Salmonella* infecting eukaryotic cells. *PLoS ONE* **6:** e23055.

Mesa-Pereira, B., Medina, C., Camacho, E.M., Flores, A., and Santero, E. (2013) Novel tools to analyze the function of *Salmonella* effectors show that SvpB ectopic expression induces cell cycle arrest in tumor cells. *PLoS ONE* **8:** e78458.

Meyers, M., Hwang, A., Wagner, M.W., Bruening, A.J., Veigl, M.L., Sedwick, W.D., and Boothman, D.A. (2003) A role for DNA mismatch repair in sensing and responding to fluoropyrimidine damage. *Oncogene* **22:** 7376–7388.

Nemunaitis, J., Cunningham, C., Senzer, N., Kuhn, J., Cramm, J., Litz, C., *et al.* (2003) Pilot trial of genetically modified, attenuated *Salmonella* expressing the *E. coli* cytosine deaminase gene in refractory cancer patients. *Cancer Gene Ther* **10:** 737–744.

Nishiyama, T., Kawamura, Y., Kawamoto, K., Matsumura, H., Yamamoto, N., Ito, T., *et al.* (1985) Antineoplastic effects in rats of 5-fluorocytosine in combination with cytosine deaminase capsules. *Cancer Res* **45:** 1753–1761.

Paesold, G., Guiney, D.G., Eckmann, L., and Kagnoff, M.F. (2002) Genes in the *Salmonella* pathogenicity island 2 and the *Salmonella* virulence plasmid are essential for *Salmonella*-induced apoptosis in intestinal epithelial cells. *Cell Microbiol* **4:** 771–781.

Pawelek, J.M., Low, K.B., and Bermudes, D. (1997) Tumor-targeted Salmonella as a novel anticancer vector. *Cancer Res* **57:** 4537–4544.

Pizzorno, G., Sun, Z., and Handschumacher, R.E. (1995) Aberrant cell cycle inhibition pattern in human colon carcinoma cell lines after exposure to 5-fluorouracil. *Biochem Pharmacol* **49:** 553–557.

Polak, A., and Scholer, H.J. (1975) Mode of action of 5-fluorocytosine and mechanisms of resistance. *Chemotherapy* **21:** 113–130.

Rehemtulla, A., Hamstra, D.A., Kievit, E., Davis, M.A., Ng, E.Y., Dornfeld, K., and Lawrence, T.S. (2004) Extracellular expression of cytosine deaminase results in increased 5-FU production for enhanced enzyme/prodrug therapy. *Anticancer Res* **24:** 1393–1399.

Royo, J.L., Becker, P.D., Camacho, E.M., Cebolla, A., Link, C., Santero, E., and Guzman, C.A. (2007) In vivo gene regulation in *Salmonella spp.* by a salicylate-dependent control circuit. *Nat Methods* **4:** 937–942.

Saltzman, D.A., Heise, C.P., Hasz, D.E., Zebede, M., Kelly, S.M., Curtiss, R., 3rd, *et al.* (1996) Attenuated *Salmonella*

*typhimurium* containing interleukin-2 decreases MC-38 hepatic metastases: a novel anti-tumor agent. *Cancer Biother Radiopharm* **11:** 145–153.

Takeda, H., Haisa, M., Naomoto, Y., Kawashima, R., Satomoto, K., Yamatuji, T., and Tanaka, N. (1999) Effect of 5-fluorouracil on cell cycle regulatory proteins in human colon cancer cell line. *Jpn J Cancer Res* **90:** 677–684.

Vermes, A., Guchelaar, H.J., and Dankert, J. (2000) Flucytosine: a review of its pharmacology, clinical indications, pharmacokinetics, toxicity and drug interactions. *J Antimicrob Chemother* **46:** 171–179.

Westphal, K., Leschner, S., Jablonska, J., Loessner, H., and Weiss, S. (2008) Containment of tumor-colonizing bacteria by host neutrophils. *Cancer Res* **68:** 2952–2960.

Yam, C., Zhao, M., Hayashi, K., Ma, H., Kishimoto, H., McElroy, M., *et al.* (2010) Monotherapy with a tumor-targeting mutant of *S. typhimurium* inhibits liver metastasis in a mouse model of pancreatic cancer. *J Surg Res* **164:** 248–255.

Yoshikawa, R., Kusunoki, M., Yanagi, H., Noda, M., Furuyama, J.I., Yamamura, T., and Hashimoto-Tamaoki, T. (2001) Dual antitumor effects of 5-fluorouracil on the cell cycle in colorectal carcinoma cells: a novel target mechanism concept for pharmacokinetic modulating chemotherapy. *Cancer Res* **61:** 1029–1037.

Zhao, M., Yang, M., Ma, H., Li, X., Tan, X., Li, S., *et al.* (2006) Targeted therapy with a *Salmonella typhimurium* leucine-arginine auxotroph cures orthotopic human breast tumors in nude mice. *Cancer Res* **66:** 7647–7652.

## Supporting information

Additional Supporting Information may be found in the online version of this article at the publisher's web-site:

**Fig. S1.** *In vitro* sensitivity to 5-FU produced by *Salmonella* on infected MCF-7 and HCT116 cells. A cell cycle distribution of MCF-7 (A) or HCT116 (B) cells infected with *Salmonella* MPO376 (Δ*purD*) bearing pMPO54 (empty vector), pMPO1088 (CD$_{GUG}$) or pMPO90 (CD$_{7AUG}$) and MPO378 (Δ*purD*Δ*upp*) bearing pMPO90, at multiplicity of infection 50:1. The cells were cultured in the presence of 50 μM of 5-FC and harvested at 6 days post-induction. 10 000 events were analysed by flow cytometry for each sample. Graphics represents the mean ± SD of three independent experiments. Non-infected cells treated with 50 μM of 5-FC or 10 μM of 5-FU were used as controls. One-way ANOVA and Tukey HSD *post hoc* tests were applied to test for significant differences. Data from the same group marked with different alphabet are significantly different at $P < 0.05$.

**Table S1.** Bacterial strains and plasmids used in or constructed for this study.

# 8

# Cytometric patterns reveal growth states of *Shewanella putrefaciens*

Susanne Melzer,[1,2] Gudrun Winter,[3] Kathrin Jäger,[4] Thomas Hübschmann,[5] Gerd Hause,[6] Frank Syrowatka,[7] Hauke Harms,[5] Attila Tárnok[2,8] and Susann Müller[5]*

[1]*LIFE – Leipzig Research Center for Civilization Diseases,*
[2]*Department of Pediatric Cardiology, Heart Center Leipzig,*
[4]*Interdisciplinary Center for Clinical Research, Core Unit Fluorescence-Technology,*
[8]*Translational Center for Regenerative Medicine, University of Leipzig,*
[5]*Department of Environmental Microbiology, Helmholtz Center for Environmental Research, Leipzig,*
[3]*Department of Biology: Physiology and Biochemistry of Plants, University of Konstanz, Konstanz,*
[6]*Biocenter and* [7]*Interdisciplinary Center for Material Sciences, Martin-Luther University Halle-Wittenberg, Halle-Wittenberg, Germany.*

*For correspondence. E-mail susann.mueller@ufz.de

Funding Information This work was integrated in the internal research and development programme of the UFZ and the IP controlling Chemicals Fate (CCF). The work was partially funded by the German Federal Ministry of Education and Research (PtJ-Bio, 1315883).

## Summary

Bacterial growth is often difficult to estimate beyond classical cultivation approaches. Low cell numbers, particles or coloured and dense media may disturb reliable growth assessment. Further difficulties appear when cells are attached to surfaces and detachment is incomplete.

Therefore, flow cytometry was tested and used for analysis of bacterial growth on the single-cell level. *Shewanella putrefaciens* was cultivated as a model organism in planktonic or biofilm culture. Materials of smooth and rough surfaces were used for biofilm cultivation. Both aerobic and anaerobic as well as feast and famine conditions were applied. Visualization of growth was also done using Environmental Scanning and Phase Contrast Microscopy. Bioinformatic tools were applied for data interpretation.

Cytometric proliferation patterns based on distributions of DNA contents per cell corresponded distinctly to the various lifestyles, electron acceptors and substrates tested. Therefore, cell cycling profiles of *S. putrefaciens* were found to mirror growth conditions.

The cytometric patterns were consistently detectable with exception of some biofilm types whose resolution remained challenging. Corresponding heat maps proved to be useful for clear visualization of growth behaviour under all tested conditions. Therefore, flow cytometry in combination with bioinformatic tools proved to be powerful means to determine various growth states of *S. putrefaciens*, even in constrained environments. The approach is universal and will also be applicable for other bacterial species.

## Introduction

Bacteria are unicellular organisms that live either planktonically or as biofilms in nearly all environments on earth, and they comprise essential functions in all respects (Houry *et al.*, 2012; Ishihama, 2012; Rizoulis *et al.*, 2013). Performances of bacteria in nature or in biotechnological processes correlate with abundances of cells and, thereby, their ability to grow. However, accurate analysis of bacterial growth can be challenging. Apart from using microscopic methods, quantification of bacterial growth is commonly done by biomass analysis. This classical approach has limitations, especially if media are opaque or biofilms are firmly attached to surfaces. Another problem is the enrichment of matrix particles by biomass harvest from porous surfaces resulting in error-prone biomass estimation by dry weight (DW) or optical density (OD) measurements (Uría *et al.*, 2011). The biomass analysis can also be strongly influenced by changes in the cell's morphology-like variations in granularity and cell size. Another typical problem is the accumulation of storage products that increase biomass values although no growth is ongoing. An alternative and promising method to study bacterial growth is the analysis of DNA pattern using flow cytometry (FCM), which is tested in this study.

Deoxyribonucleic acid patterns provide information on cell proliferation activity. Proliferation of bacterial cells can be followed by quantification of chromosome numbers per cell (Müller, 2007). Replication of chromosomes is the most prominent sign for growth of a cell. Flow cytometry uses the fact that the chromosome numbers per cell depend on proliferation activity. The bacterial cell cycle is subdivided into three periods: the time between the 'birth' and initiation of replication (usually one chromosome equivalent, $C_{1n}$), DNA replication (with two chromosome equivalents at the end of the phase, $C_{2n}$) and cell division ($C_{1n}$ again).

Flow cytometry measures thousands of single cells within few seconds. When measuring cellular DNA contents, the first sub-population of a histogram contains cells with one single chromosome equivalent, the $C_{1n}$ cells. The second sub-population [when having the double mean fluorescence intensity (MFI)] contains cells with the double chromosome equivalent, the $C_{2n}$ cells. In between are proliferating cells. But the bacterial cell cycle can be more complex (Müller and Nebe-von-Caron, 2010; Müller et al., 2010). The most widely studied bacterium in this respect is Escherichia coli. A special feature of the E. coli cell cycle is the uncoupled DNA synthesis: at optimal growth, new replication rounds are initiated in preceding generations (Cooper, 2006). In the DNA histogram, therefore, bacterial cells with multiple copies of the chromosome can be found for a fast-growing culture. Cell division usually occurs symmetrically and is defined as a dynamic event during which a cell divides into two equal cells. But this does not apply for all bacteria. The $\alpha$-Proteobacterium Caulobacter crescentus for instance divides asymmetrically producing two morphologically and functionally different cells (Tsokos and Laub, 2012). Asymmetric division also occurs in a way that chromosome numbers are distributed unequally. In this case, the distribution of chromosomes is not log2 but linearly scaled. An example is Desulfobacula toluolica grown on toluene under sulfate-reducing conditions (Vogt et al., 2005). Bacteria can also contain much more than one or two chromosomes. An overview is given by Müller (2007).

We used S. putrefaciens in this study to relate growth dynamics to particular micro-environmental conditions. The strain can grow in planktonic (Beliaev et al., 2001) or biofilm lifestyles (Carmona-Martínez et al., 2013) and under aerobic or anaerobic conditions (Martín-Gil et al., 2004). The organism belongs to the $\alpha$-Proteobacteria, is widely distributed in the environment in aquatic habitats and soils (Long and Hammer, 1941) and known to be a relevant human pathogen (Vogel et al., 1997). It is, besides using oxygen, capable of using various anaerobic electron acceptors like nitrate, nitrite, fumarate, thiosulfate and Mn(III) or Fe(III) oxides (Saffarini et al., 1994). Further-

more, the strain is known as an electricity-generating organism using electrodes as electron acceptors (Borole et al., 2011).

In this study, we aim for accurate analysis of cell growth in microbial samples that cannot be fully recovered from the environment without cell loss (e.g. biofilms), contain high amounts of particles or are cultivated in coloured media. To reach this aim, S. putrefaciens was used as model organism due to its versatile metabolism and lifestyle. The organism was cultivated under feast and famine conditions where we expected different proliferation activities. We compared classical methods for growth measurement with FCM and developed a workflow and a visualization procedure for population growth dynamics on the basis of single-cell analysis. Variation in growth behaviour under altering micro-environmental conditions was assessed by similarity analysis using bioinformatic tools.

## Results

### Classical growth analyses

Detection of growth in planktonic cultures: S. putrefaciens was grown both aerobically and anaerobically as planktonic cultures on peptone and lactate respectively (Fig. 1). Planktonic cultivation was chosen because of the relatively fast growth of cells and their simple harvest directly from the cultivation medium. Fastest growth of S. putrefaciens ($\mu max = 0.25$ h$^{-1}$; Table 1; Fig. 1A) was achieved by aerobic cultivation on peptone, measured at OD 700 nm. Growth on lactate under aerobic and anaerobic conditions was slower with $\mu max = 0.06$ h$^{-1}$ and $\mu max = 0.05$ h$^{-1}$ respectively (Table 1; Fig. 1B and C). Lactate was metabolized during 30–50 h of cultivation (Fig. S1). Ferric citrate was used as electron acceptor under anaerobic conditions and caused a colour change of the medium from yellow to green. In addition, aggregates of iron-complex-compounds precipitated, impeding an accurate analysis of growth activity by OD measurement.

Detection of growth in biofilms: S. putrefaciens was grown both aerobically and anaerobically on lactate on two different glass surfaces and graphite paper. The specific growth parameters are shown in Table 1. The glass slides had a relatively smooth surface, whereas the glass wool had a rougher surface and an artificial 3D structure of grooves and micro-niches (Fig. 2A). The biomass harvest was similar with 1.6 mg cm$^{-2}$ for glass wool and 1.3 mg cm$^{-2}$ for glass slides after 216 h. In addition to biofilm growth, the conspecific planktonic populations were also studied. In the glass wool experiment, growth rates were found of $\mu = 0.06$ h$^{-1}$ and $\mu = 0.02$ h$^{-1}$ for planktonic and biofilm grown cells respectively (Fig. 1D and E; Table 1). Small spots of the glass fibres were covered by

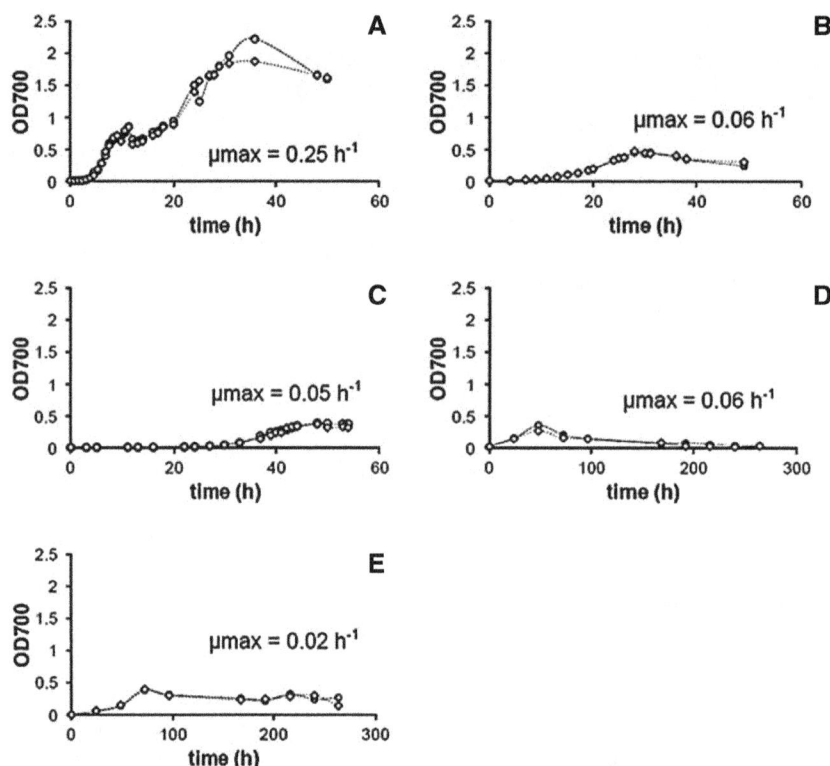

**Fig. 1.** Growth curves of *S. putrefaciens* cells in planktonic (A–D) and biofilm cultures (E). Bacterial growth was measured by increase in OD at 700 nm. Growth rates ($\mu$max) were calculated from the steepest rise of the growth curves. Planktonically grown cells were cultivated on peptone (feast/aerobic growth, A), lactate (famine/aerobic growth, B and D) and lactate/ferric–citrate (famine/anaerobic growth, C) in duplicates. Cultivation on lactate in the presence of glass wool was shown for planktonic cells in (D) and biofilm cells in (E).

biofilms already after 24 h incubation, as shown by environmental scanning electron microscopy (ESEM) (Fig. 2A) and phase contrast microscopy (PC; Fig. 2E). Despite strong agitation, biofilms developed into spot-like 3D structures over the next 216 h. A similar colonization behaviour was found for the static glass slide cultures (Fig. 2C).

Biomass of cells grown on graphite paper could not be estimated. Mechanical cell detachment was difficult due

to the cells' growth inside the porous structure of the rough surface. Therefore, biomass was low, and the sample preparations were rich in matrix particles. Cell attachment on graphite paper was imaged by ESEM after 240 h cultivation (Fig. 2B). The arrangement of the graphite paper fibres was comparable to that of the glass fibres, and dense biofilms grew until the end of the experiment (336 h). Nevertheless, less than $6 \times 10^6$ cells cm$^{-2}$ could be detached from the graphite paper at this time point.

**Table 1.** Overview of different cultivation experiments to analyse different lifestyles of *Shewanella putrefaciens*.

| Cultivation technique | Incubation method | C-source | Aerobic/Anaerobic | | Growth condition | Maximal growth rate | Microscopy | | | | FCM | Dalmatian plot |
|---|---|---|---|---|---|---|---|---|---|---|---|---|
| | | | | | | | PC | Fl | 3D | ESEM | | |
| Batch | Shaking flask | Peptone | X | – | feast | 0.25 h$^{-1}$ | X | X | n.d. | n.d. | X | X |
| Batch | Shaking flask | Lactate | X | – | famine | 0.06 h$^{-1}$ | X | X | n.d. | n.d. | X | X |
| Batch | Shaking flask | Lactate + iron(III)citrate | – | X | famine | 0.05 h$^{-1}$ | X | X | n.d. | n.d. | X | X |
| Flow through chamber | FTC with GS | Lactate | X | – | famine | n.d. | X | X | X | n.d. | X | X |
| Batch | Shaking flask with GW | Lactate | X | – | famine | 0.06 h$^{-1\,\text{(plankt)}}$ | X | n.d. | n.d. | n.d. | X | X |
| | | | X | – | famine | 0.02 h$^{-1\,\text{(biofilm)}}$ | X | n.d. | n.d. | X | X | X |
| Static chamber | Static chamber with GP | Lactate | – | X | famine | n.d. | n.d. | n.d. | n.d. | X | X | X |

C-source, carbon and energy source; PC, phase contrast microscopy; Fl, fluorescence microscopy; GS, glass slide; GW, glass wool, GP, graphite paper; n.d., not determined; X, was done in this study.

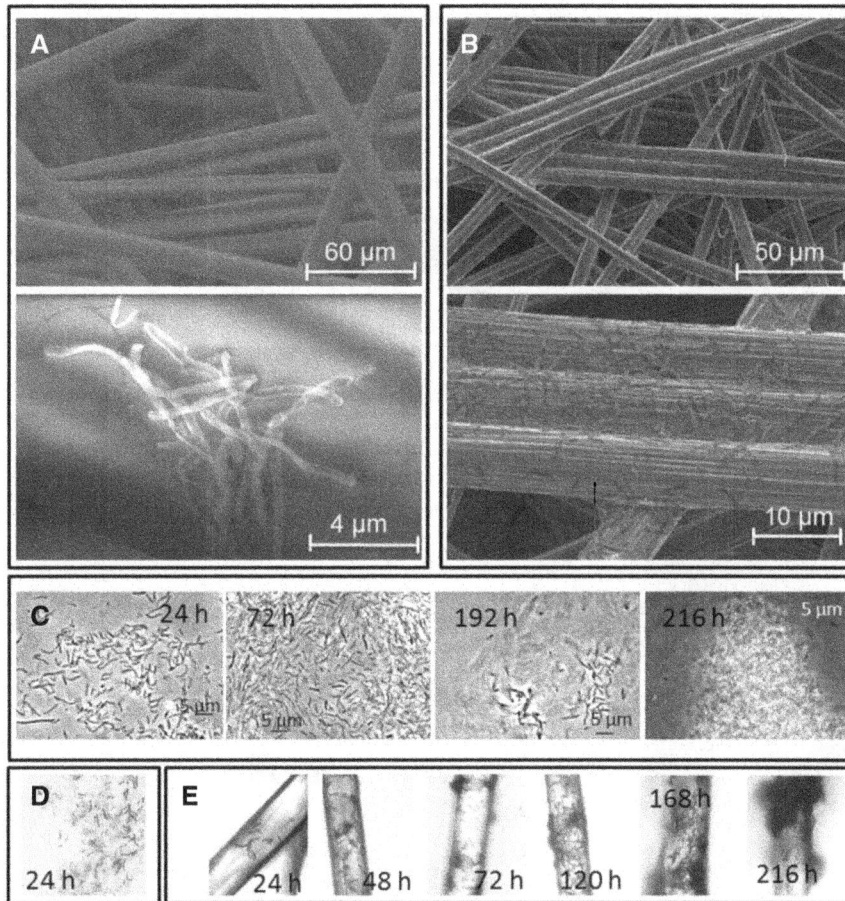

**Fig. 2.** Microscopical analysis of bacterial growth on different surfaces. Environmental scanning electron microscopy imaging of a 24 h old biofilm on glass wool (A) and a 240 h old biofilm on graphite paper (B) in different magnifications. Single rod-shaped cells cluster as initial structures on the glass wool surface. Cells were distributed with higher densities on the graphite fibre surface. Growth of biofilms on glass slides during 24 h to 216 h was analysed by phase contrast microscopy (C). During the first 24 h of growth cells attached in one layer on the glass surface, after 72 h multilayered stacks were found, resulting in mushroom-like 3D structures after 240 h cultivation. (D) Planktonically grown cells from the glass wool growth approach after 24 h incubation. (E) Biofilm growth on glass wool during 24 h to 216 h. Attachment of single cells (24 h) was observed, resulting in dense growth on area spots onto the total glass wool (72 h, 120 h, 168 h) and mushroom-like structures (216 h), as observed for growth on the glass slides.

## Detection of growth using FCM

The measurement of DNA patterns of single cells by FCM served to determine proliferation activities of cells. They are expected to mirror the growth rate of a population and be helpful in estimating growth activity if other techniques like DW or OD determination fail. In this study, we used FCM to analyse *S. putrefaciens* when grown in planktonic or biofilm lifestyles on different surfaces, various substrates and electron acceptors. Samples from all cultivation approaches were taken regularly, analysed by FCM, and the abundances of cells with respective chromosome equivalents were determined.

Four sub-populations with distinct DNA contents were clearly distinguished (Fig. 3). Geometric MFI values of 4'6-diamidino-2-phenylindole (DAPI) channel numbers were calculated for all samples to be 98.0 ($C_{1n}$), 183.0 ($C_{2n}$), 282.5 ($C_{3n}$) and 388.7 ($C_{4n}$), as shown in Fig. S2. Some samples showed more sub-populations with even higher DNA contents. Due to cell cycling events, cells virtually shifted between the DNA distributions over time, and their varying number per distribution was evaluated. The outcome is visualized in a coloured heat map (Fig. 3) by using a normalization approach (Fig. S3), enabling the comparison of all experiments (Fig. S4). The heat map in Fig. 3 shows cytometric DNA sub-population distribution over time containing cells with up to four chromosomes ($C_{1n}$–$C_{4n}$). The cell abundances at 0 h mirrored the proliferation state of the cells when inoculated. While, e.g. the inoculum for the complex peptone medium (Fig. 3A) originated from the exponential growth phase of a pre-culture, the inoculum for the anaerobic cultivation on minimal medium (Fig. 3C) was drawn from a respective stationary pre-culture.

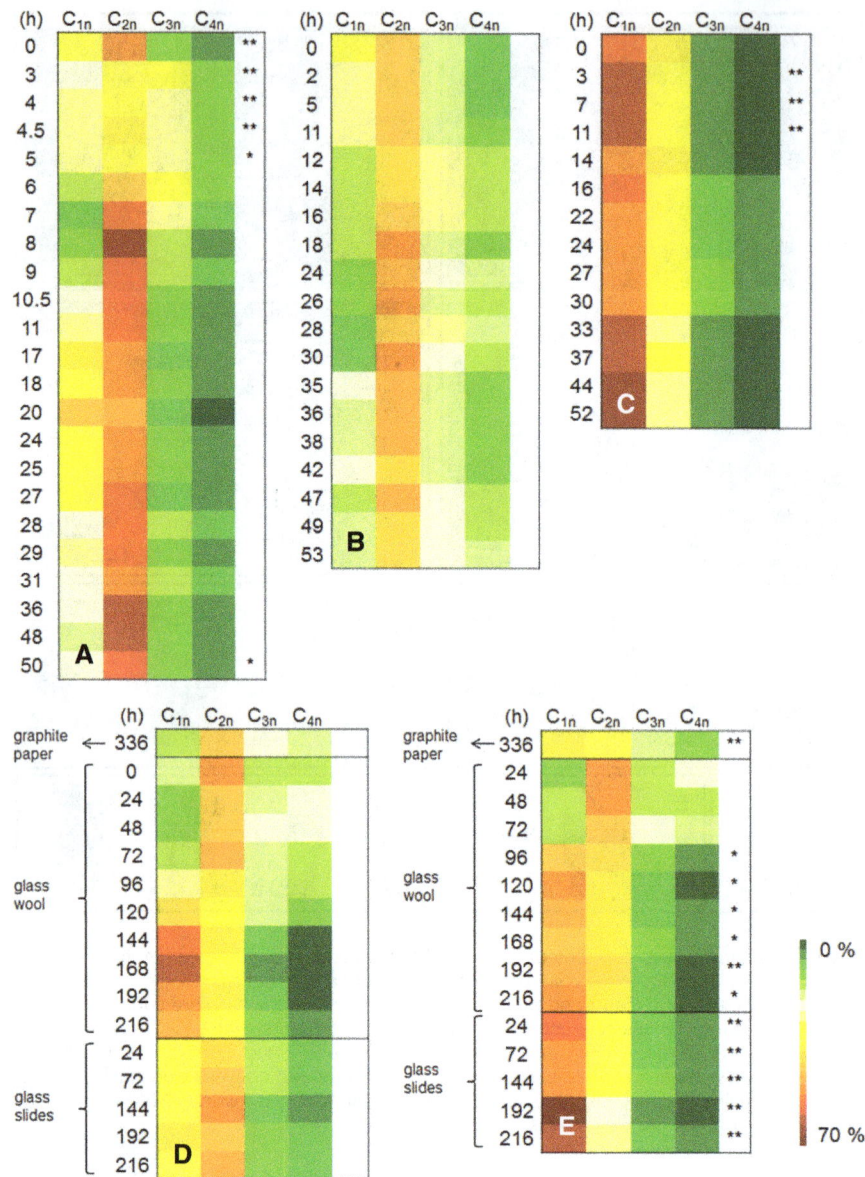

**Fig. 3.** Growth patterns of planktonic and biofilm cells of *S. putrefaciens*. Deoxyribonucleic acid distribution pattern of nine different growth conditions were compared in a heat map with each other. Distribution of DAPI stained cells with $C_{1n}$, $C_{2n}$, $C_{3n}$ and $C_{4n}$ chromosome contents were measured by FCM over cultivation time when planktonically grown under feast/ aerobic (A), famine/ aerobic (B), famine/ anaerobic (C) and in matrix containing approaches (D). Data of biofilm grown cells are shown in (E), corresponding to the planktonic cells of D. Time of cultivation is given in the first row of each heat map. The colour key marks abundance of cells in the $C_{1n}$-$C_{4n}$ sub-populations from 0–70 % (colour key) of DAPI-stained cells. Possibly emerging unstained cells are highlighted by one (frequency > 10%) and two stars (frequency > 25%) in the last column of each table. The results rely on one-parametric analysis of DAPI histograms. In depth analysis of two dimensions (scatter versus DAPI fluorescence) revealed dynamics in bacterial growth (Figs 4–6). Fig. S4 shows the abundances of all cells.

In this study, two types of pattern distributions were observed for *S. putrefaciens*.

The first type envisages major parts of the cells in the $C_{2n}$ distribution. Planktonic growth caused high abundances of cells in the $C_{2n}$ distribution (Fig. 3A and B). When *S. putrefaciens* was grown on a complex carbon and energy source like peptone (Fig. S5), or under aerobic conditions on lactate (Fig. S6), a major part of

cells maintained double chromosome contents during the whole cultivation time. Such behaviour was also observed for planktonic cells in the flow-through chamber (FTC; Fig. S7) and for the glass wool experiments (Fig. S8) during the first days of cultivation. Even in the graphite paper chamber, cells (Fig. S9) preferred the $C_{2n}$ instead of the $C_{1n}$ state, suggesting that growth was rather fast under such conditions (Fig. 3D). *Shewanella putrefaciens*

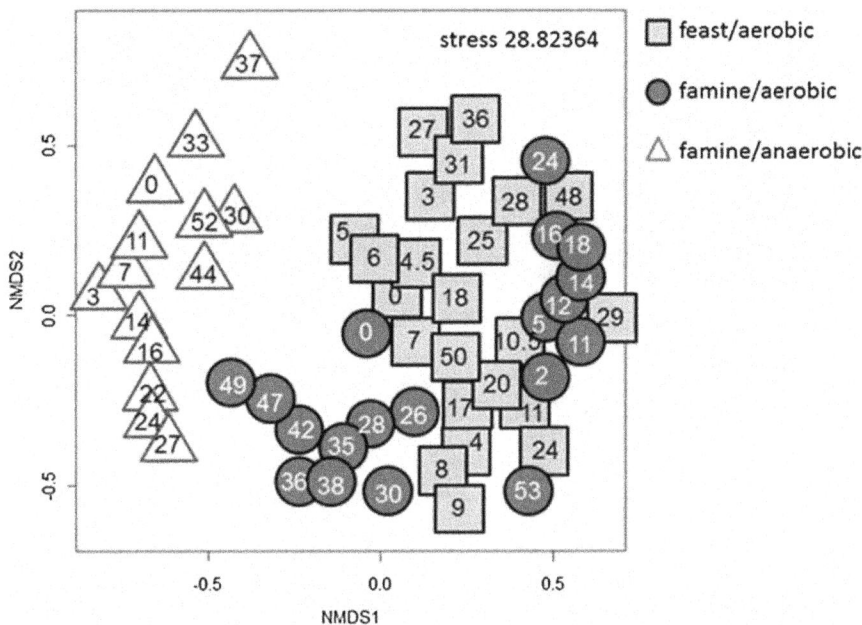

**Fig. 4.** Similarity analysis of SSC : DNA patterns of planktonic cells under feast/famine growth conditions. Dalmatian black and white SSC : DAPI plots of three different batch cultivation approaches are shown. In detail, planktonic cells on feast/aerobic (cube), famine/aerobic (circle) and famine/anaerobic (triangle) growth conditions were compared. Numbers within the geometric figure symbol the time points of the sample in hours.

showed even higher growth velocities for only 2 h on peptone in the early exponential growth phase. Here, the organism generated a uniform distribution of cells between sub-populations $C_{1n}$ to $C_{4n}$ (Fig. 3 and Fig. S4, 3–5 h), and is undergoing probably uncoupled DNA synthesis for a short-time interval.

The second type envisages major parts of the cells in the $C_{1n}$ distribution. Under anaerobic growth conditions (Fig. S10), the highest abundances of cells were found in the $C_{1n}$ distributions in all samples (Fig. 3C and D). The carbon source lactate caused obviously limiting proliferation activity for *S. putrefaciens* when oxygen was not

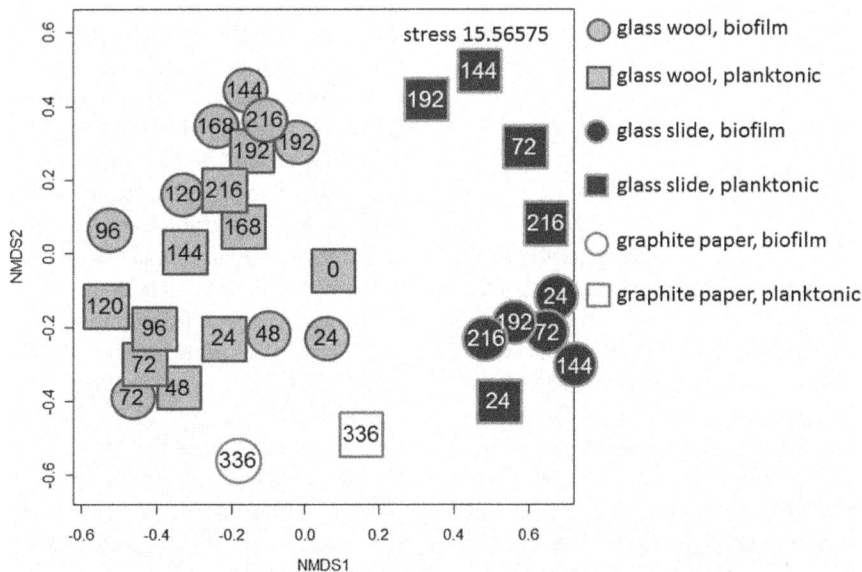

**Fig. 5.** Similarity analysis of SSC : DNA patterns of biofilm and planktonically grown cells. Dalmatian black and white SSC : DAPI plots are shown from biofilm grown cells on glass wool, glass slides and graphite paper as well as from corresponding planktonically grown cells. Aerobic cells from the glass wool approach in batch culture (biofilm: light-grey filled circle; planktonic cells: light-grey filled cube) were compared with aerobic biofilm (dark grey filled circle) and planktonic (dark grey filled cube) cells from the FTC experiment and with anaerobic biofilm cells from the graphite paper (unfilled circle plus the corresponding planktonic cells, unfilled cube).

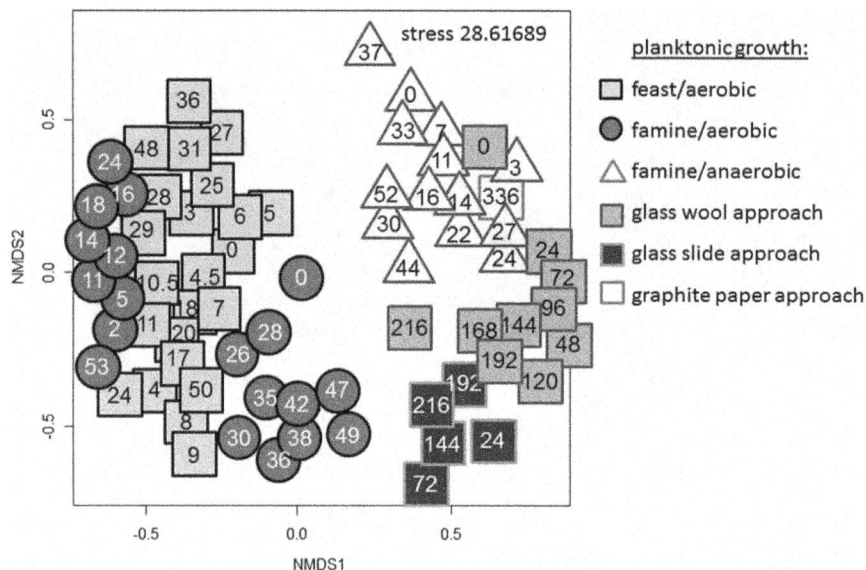

**Fig. 6.** Similarity analysis of SSC : DNA pattern of all planktonic cells. Dalmatian black and white SSC : DAPI plots are shown of all planktonic cells of batch cultures and the matrix containing growth approaches. The same symbols were used as in Figs 4 and 5.

available as electron acceptor. In addition, all biofilm samples showed accumulation of cells in $C_{1n}$ (Fig. 3E), with exception of early biofilms on glass wool (Fig. 3E, 24–72 h). However, also the planktonic cells in the aerobically grown culture that contained glass fibres as a compact attachment matrix accumulated in $C_{1n}$ after a longer cultivation time (Fig. 3D, 120–216 h). Besides the cells' distribution in $C_{1n}$ to $C_{4n}$ subsets, some of the cells remained unstained or showed cells with lower stainability (Fig. S8). High proportions of unstained cells were found in nearly all biofilm samples and, surprisingly, also in the first hours during cultivation under feast growth conditions (Fig. S5).

*Comparison of population dynamics using the Dalmatian tool*

To better describe and compare population behaviour of *S. putrefaciens* under diverse growth conditions, cell side-scatter characteristics (SSC) were followed in addition to chromosome numbers per cell. The two parameter FCM analysis created 2D plots and allowed a defined sub-population distinction. The workflow is the following: The 2D plots mirrored subsets of cells. Depending on growth conditions, theses subsets appeared or disappeared by changing the number and position of sub-populations in the 2D plot. These variations can be evaluated by using the Dalmatian tool (Bombach *et al.*, 2011). The tool compares 2D plots which each other and follows trends in population behaviour (Müller *et al.*, 2012). Side-scatter characteristics : deoxyribonucleic acid sub-populations were

defined for each 2D plot by gate settings, position corrected if required (Fig. S11), and the Dalmatian plot evaluation performed as described in *Experimental procedures*. Results allowed trend interpretations and similarity estimation in population dynamics.

The nMDS plot in Fig. 4 clearly showed similarities or dissimilarities in the proliferation activity of *S. putrefaciens* over time due to the different cultivation conditions. A clear dissimilarity was observed for the cells' proliferation behaviour under aerobic and anaerobic conditions. In addition, cells that had access to oxygen but were cultivated under famine or feast carbon conditions were distinguishable as well in the nMDS plot, although they clustered in near vicinity. Hence, the nMDS plot reflected the proliferation pattern shown in Fig. 3A and B where aerobically grown cells accumulated a major part of the cells in the $C_{2n}$ distribution, while anaerobically grown cells stayed longer in $C_{1n}$ (Fig. 3C).

Furthermore, the Dalmatian tool was used to show dissimilarities in the cells' proliferation behaviour when cells were grown in planktonic or biofilm lifestyles (Fig. 5). Glass slide, glass wool and graphite paper biofilms clustered distinctly, due to their different SSC : DNA characteristics (Fig. 5, circles). Plainly, the planktonic cells thriving in the identical environments were found in the vicinity of the respective biofilm clusters for all three approaches (Fig. 5, squares). The Dalmatian plot findings suggest that planktonic and biofilm cells show similar growth characteristics in identical environments. Such a behaviour is also reflected in Fig. 3D and E, although abundances of cells in sub-populations are better reflected in the heat map.

As the proliferation activity of the planktonic cells in the aerobic glass wool approach was surprisingly low in FCM (Fig. 3D) in contrast to OD analysis (Table 1), the nMDS plot was used to compare these findings with all other planktonically grown samples. As expected, these planktonic cells (light-grey cubes) clustered evidently dissimilar from the cells cultivated without the solid glass fibre matrix (Fig. 6) but were, nevertheless, similar to the anaerobically grown cells (unfilled triangles) and even to planktonic cells from the FTC (filled squares).

The evaluation of the various DNA : SSC dot plot patterns of *S. putrefaciens* using the Dalmatian tool verified the data obtained by evaluation of single parameter DNA patterns and showed the multiple behaviour of the organism under the various growth conditions. Lifestyle, in addition to carbon source and electron acceptor availability, influenced the proliferation velocity of *S. putrefaciens*. Flow cytometry in combination with bioinformatic tools was perfectly able to document these variations even when classical techniques as analyses of DW or OD are likely to fail.

## Discussion

Microbial growth is usually quantified by measurement of biomass over time. Besides DW analyses, OD is most often used for the purpose. Other techniques are analyses of increasing contents of macromolecules like total protein (e.g. Stickland, 1951), ribosomal RNA contents (e.g. Moller *et al.*, 1995) or other intrinsic cell markers (e.g. Unge *et al.*, 1999). Increase in cell counts can be determined by microscopy, coulter counter techniques or FCM. All of them can be error prone, especially, if detachment of cells from surfaces is difficult or particles are present.

We aimed to determine growth states of *S. putrefaciens* by measurement of chromosome numbers in individual cells (Müller, 2007; Müller and Nebe-von-Caron, 2010; Müller *et al.*, 2010). The advantage of this approach is that no further knowledge of biomass quantities or cell number is necessary. Already an amount of about 20 000 cells is sufficient to create a highly resolved cytometric dot plot. Complete recovery of cells from the growth matrix is not required. Neither coloured or opaque media (Neumeyer *et al.*, 2013) nor lower than equi-proportional particle numbers disturb such analyses because they do not specifically bind the used DNA specific fluorescent dye.

When applicable, classical analysis of biomass was used in this study to quantify growth of *S. putrefaciens* in comparison to FCM. Generally, when OD measurement was possible, the FCM data mirrored growth behaviour under the respective conditions. However, biofilm samples did not allow continuous sampling for biomass detection due partly to low cell density and tight attachment to surfaces. Further, following biofilm cell growth by microscopic techniques was elaborate. In addition, quantitative evaluation of biofilm mass would require specific software tools which are, to our knowledge, still not applicable. Therefore, only limited and flawed classical biomass information was available for the biofilm grown cells but also for some of the anaerobically grown samples.

In contrast, high-resolution FCM and the highly DNA-specific dye DAPI were useful for quantitative growth determination in all tested samples. The method gave cell cycle and growth behaviour information on the strain that was not available before.

In general, *S. putrefaciens* was surprisingly found to contain at least four chromosome equivalents and divide in an asymmetric way. Apart from symmetric cell division ($C_{2n}$) with sub-populations containing 1, 2 or 4 chromosome equivalents, a sub-population comprising a $C_{3n}$ chromosome amount was detected. Additionally, $C_{5n}$ states, also characteristic for asymmetric cell division, could often be distinguished in samples grown under famine conditions (Fig. S4). However, because cells in the $C_{5n}$ states were of low frequency, they were disregarded for the overview presentation of proliferation states in Fig. 3 (like all other additional sub-populations, see below). Nevertheless, extended heat maps contain abundance variation within all subsets of cells to enable growth velocity evaluation, as shown in Fig. S4. This figure provides even higher resolved information on proliferation states of the strain under the various conditions. An evaluation of growth states was possible by comparing the varying numbers of cells in the upcoming sub-populations using bioinformatics tools. Heat maps (Fig. 3; Fig. S4) and Dalmatian plots (Müller *et al.*, 2012) were created to facilitate the overview on the variations and found to mirror the various growth states of *S. putrefaciens* like fingerprints. Both tools showed similar outcomes although based on distinct bioinformatics evaluation principles and even independent cell-based information.

Fastest growth was found on aerobic peptone medium, a complex mixture of peptides and amino acids as carbon source. Here, even uncoupled DNA synthesis was observed for the time range of 3–5 h, a behaviour that was described in detail for *E. coli* under similar growth conditions (Cooper, 2006). When lactate was used as carbon source, the proliferation activity decreased. The slower growth was caused by the low carbon source quantity in contrast to the energy-rich peptide mixture. A further decrease in proliferation activity was found when cells were grown aerobically on lactate but in the presence of glass wool. Here, the population also preferred the $C_{1n}$ state after about 4 days of cultivation. It can only be speculated that the glass wool represent an interfering component during the aerobic cultivation approach. Shear stress on the glass fibres addressed the cells from all directions and might be the reason for the slow growth in these shake flasks experiments. Cells might need additional energy to repair damages occurring in the process.

Shewanella strains are known to oxidize lactate completely under aerobic conditions but can also use it under anaerobic conditions, here in the presence of iron(III)citrate. Due to this behaviour, the organism is often enriched at oxic/anoxic interfaces of aquatic habitats (Scott and Nealson, 1994). Under these conditions, very low proliferation activity was analysed. When substrate is limiting and electron acceptors have a low-reducing potential bacteria are known to undergo maintenance states (Russell and Cook, 1995). The reduced growth conditions obviously resulted in the accumulation of cells in the $C_{1n}$ state.

The study of *S. putrefaciens* is of importance, as the organism might be useful to convert chemical energy into electrical energy but also vice versa to produce bulk chemicals in an environmental friendly way (Wang and Ren, 2013). Some efforts to follow growth characteristics of *S. putrefaciens* have been made recently. Franks and colleagues (2008) developed a real-time three-dimensional imaging system to visualize live, active microbial biofilms in microbial fuel cells with confocal scanning laser microscopy. However, biofilms of *S. putrefaciens* are described to be often tenuously distributed on surfaces (Bagge *et al.*, 2001), which was also observed in this study. In contrast, growth of another electricity generating organism, an engineered mCherry-labelled *Geobacter sulfurreducens* strain, was described to grow more compact, but accurate quantification of growth remained still difficult (Franks *et al.*, 2009). A first attempt to use FCM for the purpose was already made (Harnisch *et al.*, 2011) and shows the potential of the method on *G. sulfurreducens*. In this study, *S. putrefaciens* was grown on graphite paper using the cultivation assembly described in the Supporting Information. In contrast to the former study, the detection of the proliferation pattern remained limited. A huge part of the cells could not be stained requiring further developmental steps of cell handling in future.

It should not remain undiscussed that high numbers of unstained cells were found in several experiments. It is known from other approaches that stressed cells with a dense cell wall withstand staining with fluorescent dyes (Müller and Nebe-von-Caron, 2010). Obviously, larger parts of cells in biofilms or in populations that were freshly inoculated into new media protect themselves against harsh or drastically changing conditions by increasing the density of the cell wall. This behaviour is often observed for medically relevant strains (Mah and O'Toole, 2001; Cui *et al.*, 2003). In addition to this technical phenomenon, we

also would like to point to the so called 'ghost cells' found in natural environments without nucleic acid contents (Zweifel and Hagstrom, 1995). Nevertheless, to decrease the proportion of unstained cells, the cell handling and staining procedures should be further improved. Nonetheless, the rest of the population gave a clear indication of sub-populations diverging towards different chromosome contents.

In general, the states and features of cells over time (proliferating, highly proliferating, non-proliferating, stained/less stained/unstained) and their variation in dependence of abiotic factors generate huge data sets that make up a so-called cytome. In its original definition, a cytome is defined as the entity of all cells in an organ that are interrelated and connected (Valet *et al.*, 2004), but also bacterial populations and communities are known to act in a concerted manner if required. Following behaviour and dynamics of whole bacterial cytomes may open another way to a better understanding of bacterial population performances under varying growth conditions (Koch *et al.*, 2014). The presented workflow for determining growth states of *S. putrefaciens* under varying conditions is universal. It simply requires sampling of at least 20 000 cells, fixation, staining of the DNA, FCM and bioinformatic approaches for cytometric data evaluation. But other bacteria will ultimately show a different growth behaviour, therefore, reference proliferation patterns will always be required.

## Conclusion

The study aimed to describe means for determination of cell growth and proliferation when classical approaches like DW, OD or cell count detections fail. Here, we tested biofilm versus planktonic growth, representing two main different lifestyles of bacteria. In addition, different growth velocities were investigated by variation in substrates (feast versus famine) and electron acceptors (e.g. oxygen versus iron(III)citrate). Flow cytometry and bioinformatics were used for sample analysis and evaluation tools.

The proliferation patterns of *S. putrefaciens* were directly influenced by the substrate source, the used electron acceptor and planktonic or sessile lifestyles. As a main trend, slow growth was accompanied by an accumulation of cells within the $C_{1n}$ state, while faster growth prompted cells into the $C_{2n}$ state. This knowledge may be useful for stimulation and optimization of various biotechnological processes, where microbial biofilms are often the main producers. Comparison of unknown cytometric patterns with well-known ones allows estimating the proliferation activity and, as a consequence, growth. It has thus the potential for trend interpretation and prediction of population behaviour and selective stimulation of bio-processes. In addition, correlation analyses of FCM data with other 'abiotic' parameters will further highlight potentials or shortcomings in bioprocesses in future.

## Experimental procedures

### Materials

All chemicals were, unless otherwise specified, obtained from Merck KGaA (Darmstadt, Germany), Sigma-Aldrich (St. Louis, MO, USA), Hoffmann-La Roche AG (Basel, Switzerland) and Carl Roth GmbH + Co (Karlsruhe, Germany), with at least technical grade.

The lactate minimal medium (pH 7.4) was used for cultivation of (I) aerobic planktonic cells in batch culture, (II) aerobic biofilms on glass wool, (III) aerobic biofilms on glass slides in a flow-through chamber and (IV) anaerobic biofilms on graphite paper. Its composition was as followed: lactate $2.67 \text{ g l}^{-1}$, $NH_4Cl$ $0.95 \text{ g l}^{-1}$, $K_2HPO_4 \times 3 H_2O$ $1.301 \text{ g l}^{-1}$, $KH_2PO_4$ $0.449 \text{ g l}^{-1}$, $NaHCO_3$ $0.168 \text{ g l}^{-1}$, L-arginine $0.02 \text{ g l}^{-1}$, L-glutamate $0.02 \text{ g l}^{-1}$, L-serine $0.02 \text{ g l}^{-1}$, $CaCl_2 \times 2 H_2O$ $0.057 \text{ g l}^{-1}$, $CoSO_4 \times 7 H_2O$ $1.41 \text{ mg l}^{-1}$, $CuSO_4 \times 5 H_2O$ $0.042 \text{ mg l}^{-1}$, ethylenediaminetetraacetic acid (EDTA) $0.02 \text{ g l}^{-1}$, $FeSO_4 \times 7 H_2O$ $1.7 \text{ mg l}^{-1}$, $H_3BO_3$ $2.8 \text{ mg l}^{-1}$, $MgSO_4 \times 7 H_2O$ $0.2 \text{ g l}^{-1}$, $MnSO_4 \times H_2O$ $0.17 \text{ mg l}^{-1}$, $(NH_4)_2SO_4$ $0.66 \text{ mg l}^{-1}$, $Na_2MoO_4 \times 2 H_2O$ $0.75 \text{ mg l}^{-1}$, $Na_2SeO_4$ $0.03 \text{ mg l}^{-1}$, $Na_2SO4$ $0.7 \text{ mg l}^{-1}$, $NiCl_2 \times 6 H_2O$ $1.365 \text{ mg l}^{-1}$, $ZnSO_4 \times 7 H_2O$ $0.24 \text{ mg l}^{-1}$. For cultivation of anaerobic planktonic cells in batch culture, lactate-Fe(III)citrate medium (pH 7.0) was used. The composition of the medium was equal to the lactate medium but was supplemented with Fe(III)citrate $2.5 \text{ g l}^{-1}$ and lacked $CaCl_2 \times 2 H_2O$ and EDTA. For growth on complex medium DSMZ No 948 (pH 7.0) was used.

*Shewanella putrefaciens* $6067^T$ was purchased from DSMZ (Leibniz Institute, German Collection of Microorganisms and Cell Cultures, Braunschweig, Germany).

### Analysis of growth

Optical density was measured at $OD_{700nm}$, $d = 5 \text{ mm}$, using the Spectrophotometer UV/VIS Ultrospec III (Amersham Pharmacia Biotech, Piscataway, NJ, USA).

Dry weight was measured using 500 µl cell solution that was centrifuged at 3200*g for 20 min. The supernatant was carefully removed. Duplicates of each sample were dried overnight in a hot drying oven (80°C) and finally weighed.

Phase contrast microscopy was used to observe planktonic and biofilm samples on glass wool and glass slides (Axioskop, Type B, Zeiss, Germany). Pictures were taken with Sony Progressive 30 CD camera and analysed with Openlab 3.1.4 (Improvision, Lexington, MA, USA). A Plan-Neofluar 100×/1.30 objective was used with oil. Fluorescence was analysed after excitation by a 100 W mercury arc lamp (HBO W/2, 260 VA, Zeiss). Three-dimensional imaging was done using an Axio ObserverZ.1 (Carl Zeiss, Jena, Germany) with an LD LC1 Plan-Achromat 25×/63× objective to focus more deeply into the sample. Images were reordered on a digital camera (AxioCam MRm, Carl Zeiss).

Environmental Scanning Electron Microscopy (ESEM XL 30 FEG, FEI) served to observe biofilm growth on glass wool. Samples were analysed natively under low pressure as well as fixated samples under vacuum for higher resolution. Before microscopic analysis samples were fixated as follows: The material was fixed with 3% glutaraldehyde in sodium cacodylate buffer (SCB) pH 7.2 for 1 h at room temperature (RT), washed with SCB, postfixed with 1% osmiumtetroxide in SCB and finally washed with pure water. ESEM parameters for the single images were as followed: 5/8 kV gaseous secondary electron (GSE) detector, wet mode, 1.3 mbar (for glass-wool samples) and 2/5 kV, SE detector, Hi-Vac mode (for graphite anodes).

### Cultivation of S. putrefaciens *as planktonic cells*

*Shewanella putrefaciens* was grown in 500 ml flasks filled with 150 ml medium at 150 r.p.m. (shaking incubator TH25, SM-30 control, Edmund Bühler GmbH, Tübingen, Germany) and RT. While the cells were cultivated on peptone (5 g $l^{-1}$ peptone) for feast growth, they were cultivated on aerobic lactate medium (lactate $2.67 \text{ g l}^{-1}$, pH 7.4) and anaerobic lactate-Fe(III)citrate medium (lactate $2.67 \text{ g l}^{-1}$, Fe(III)citrate $2.5 \text{ g l}^{-1}$, pH 7.0) for famine growth. Under anaerobic conditions, cells were grown in 200 ml injection vials at RT and sealed with gas tight septa. The medium was degassed for 3 h with nitrogen. Cells were pre-cultivated on 0.5% peptone-agar plates for 14–16 h at 30°C. Single colonies were used for starter cultures and pre-cultivated in 25 ml peptone medium at 150 r.p.m. and 30°C. For inoculation of the main culture, 5–10 ml of the pre-culture were used to start with an optical density of 0.05. Growth was followed by measuring respective OD values. Cells were harvested and centrifuged at 3.200*g at RT for 10 min. The supernatant was transferred to a new tube to determine lactate concentration by ion chromatography (DX 100, Dionex, Sunnyvale, CA, USA). About $10^9$ cells were fixed with 1 ml fixation buffer [10 % sodium azide, 5 mM $BaCl_2$ and 5 mM $NiCl_2$ (Günther *et al.*, 2008)]. Cells were stored at 4°C for up to 14 days. Each growth experiment was performed at least in biological duplicates.

### Cultivation of S. putrefaciens *as biofilms*

Glass slides, glass wool and graphite paper have different surfaces and were used as matrix for cultivation of *S. putrefaciens*. While the glass slide surface was smooth, those of glass wool was rougher. A real porous surface was provided by graphite paper. In this study, the attachment and growth of cells on these different surfaces was tested.

*Shewanella putrefaciens* was cultivated on glass wool in 150 ml lactate medium under aerobic conditions in 500 ml flasks filled with 2.5 g glass-wool (Merck chemicals, Cat. No. 104086, 15–25 µm in diameter with a total surface of approximately 3800 $\text{cm}^2$/flask). The medium was inoculated as described before and cultivated for up to 216 h at 250 r.p.m. at RT. Biofilm samples were taken after 24 h, 48 h, 72 h, 96 h, 120 h, 144 h, 168 h, 192 h and 216 h. Also, planktonic samples (5–10 ml from the supernatant) were taken at these time points plus a sample from the inoculum. To harvest biofilm cells, victim flasks were used. For every sample time, the supernatant from one flask was discarded, and the complete glass wool was washed twice with 30 ml sterile PBS solution with addition of fixative (8 g $l^{-1}$ NaCl, 0.2 g $l^{-1}$

KH$_2$PO$_4$, 1.15 g l$^{-1}$ Na$_2$HPO$_4$, 0.2 g l$^{-1}$ KCl, 0.2 g l$^{-1}$ NaN$_3$, pH 7.2) for 10 min at 150 r.p.m. and RT. The supernatant was discarded, the glass wool transferred to a plastic shaking bottle and 10–15 ml PBS, and 45 g sterile glass beads (Sigma-Aldrich, G8722, 6 mm in diameter) were added. The wool was shaken at 400–450 r.p.m. for 30 min at RT. In a pre-test, samples were taken every 5 min for microscopic analysis of the detached cells. We found that incubation with the glass beads for 30 min was successful to fully detach all cells and to ensure that the cells' morphology was undisturbed. Optical density was measured as described above. For FCM, cells were transferred into a glass tube, centrifuged at 3200*g and RT for 10 min and fixated as described elsewhere.

Shewanella putrefaciens was cultivated on glass slides in an FTC under aerobic conditions (Fig. S12). A Hellendahl staining compartment (Carl Roth, 6 × 5.5 × 10.5 cm, volume: 350 ml) was used for the purpose. The chamber contained five glass slides at a size of 2.54 × 7.62 × 0.1 cm each and with a total surface of nearly 40 cm$^2$ each (Carl Roth, Cat. No. 0656.1). The slides and the chamber were sterilized and inoculated with 1 ml starter culture and 150 ml lactate medium. With a flow rate of 4 ml h$^{-1}$ (pump: Watson-Marlow 205 U) fresh medium was added to the compartment. Waste was removed continuously. Glass slide samples were taken after 24 h, 72 h, 144 h, 192 h and 216 h of cultivation for microscopy and FCM. Planktonic samples were taken at the same time points. For FCM sampling, slides were transferred to a 3 × 8 × 1 cm$^3$ stainless steel chamber with 1.5 mm broad cavity. By sonification (ultrasonic bath USR-9, demand – P30/120 W, frequency f 35 kHz, Merck Eurolab NV) for 10 min, cells were detached from the glass surfaces. One hundred microlitres of the supernatant were used to determine DW as described before. The residual supernatant was transferred to a glass tube and centrifuged at 3200*g and RT for 10 min. Cells were fixated as described elsewhere. Growth conditions for cultivation of S. putrefaciens on graphite paper are described in detail in the Supporting Information section. All cultivation experiments are summarized in Table 1.

### DNA staining of cells and flow cytometry

To determine DNA distribution patterns, cells were stained with the AT-specific dye DAPI. Upon UV excitation, the DNA-bound dye fluoresces blue ($\lambda$max 461 nm) with a linear correlation between signal intensity and DNA content of the cell. With FCM, cells with different DNA contents were analysed, and distributions of cells with multiple DNA copies were represented in histograms. To stain DNA in S. putrefaciens fixated cells were centrifuged at 3200*g at RT for 10 min and re-suspended in PBS to an amount of 3 × 10$^8$ cells per ml. Cells were washed twice with 2 ml PBS. Afterwards. cells were treated with 1 ml solution A (0.1 M citric acid in 0.5% Tween 20; v/v) for 30 min at RT. After centrifugation at 3.200*g at RT for 10 min, the cell pellet was re-suspended in 2 ml solution B (0.68 µM DAPI in 0.4 M Na$_2$HPO$_4$/NaH$_2$PO$_4$, pH 7.0) and stained for 1 h at RT in the dark. Flow cytometry analysis was carried out using a MoFlo cytometer (DakoCytomatation, Fort Collins, CO, USA) equipped with two water-cooled argon-ion lasers Innova 90 C and Innova 70C (Coherent, Santa Clara, CO, USA). Excitation at 488 nm

(400 mW) was used for analysing the scatter signals. 4'6-Diamidino-2-phenylindole was excited by multi-line UV with 333–363 nm (100 mW). The orthogonal side scatter was first reflected by a beam splitter and then recorded after reflection by a 555 nm long-pass dichroic mirror; passage a 505 nm short-pass dichroic mirror and a band pass (BP) filter 488/10. DAPI fluorescence passed to a 450/65 BP filter prior to detection. Photomultiplier tubes were obtained from Hamamatsu Photonics (models R 928 and R 3896; Hamamatsu City). Amplification of signals was carried out at logarithmic scale, and the measurement of events was triggered by the SSC signal.

For system calibration and quality control cell counting, fluorescence beads (yellow-green fluorescent beads: 2 µm, FluoSpheres 505/515, F-8827, blue fluorescent beads: 1 µm, FluoSpheres 350/440, Molecular Probes Eugene, Oregon, USA) were used (Koch et al., 2013). An internal DAPI-stained bacterial cell standard was introduced for testing the sensitivity of the system by adjusting the coefficient of variation not higher than 5%.

### Analysis of flow cytometric data

Data were recorded and analysed using SUMMIT software V 4.3 (DakoCytomation, Fort Collins, CO) and FLOWJO V 7.6.4 (Tree Star, Ashland, OR) resulting in 2D plots of SSC versus DAPI fluorescence. Gating strategy and calculation of relative cell counts in per cent are described in Fig. S2. Quantitative information on DNA distributions were summed up in manually generated heat maps (Fig. 3, Fig. S4) for each growth experiment. Colours were defined by using the red-green-blue (RGB) colour space with values of 0–255: from dark green (30:100:0) to light green (200:250:50) in six steps with +20:+40:0 for step one to four and +50:+20:+75 for step five and six as well as light yellow (255:255:200) to dark red (140:50:0) in 14 steps. Following, the 2D plots were simplified into present/absent plots representing subpopulations as black coloured gates representing cells with certain SSC and DNA contents. With these so-called Dalmatian plots, similarity alignments were performed, resulting in Dalmatian nMDS plot. Within such plots, samples containing cells with similar features are grouped together, and samples with cells of different characteristics are found more distant (Bombach et al., 2011). For histogram comparison, pixel intensities and amounts were determined with ImageJ V 1.45. The resulting table with pixel values was used as a similarity matrix by an R-SCRIPT (R-2.11.0, Development Core Team, 2009, R PACKAGE VEGAN 1.17-1) to calculate nMDS similarity plots based on the modified Jaccard Index. To compare the various measurements over time, a normalization approach was performed. For this an intern bacterial standard was prepared, measured, and its changing relative MFI values were used to adjust the measurements on a daily base (Fig. S3). In addition, independence of plot position was tested for position-corrected versus non-position-corrected data comparison (Fig. S11).

### Conflict of interest

None declared.

## References

Bagge, D., Hjelm, M., Johansen, C., Huber, I., and Gram, L. (2001) Shewanella putrefaciens adhesion and biofilm formation on food processing surfaces. Appl Environ Microbiol 67: 2319–2325.

Beliaev, A.S., Saffarini, D.A., McLaughlin, J.L., and Hunnicutt, D. (2001) MtrC, an outer membrane decahaem c cytochrome required for metal reduction in Shewanella putrefaciens MR-1. Mol Microbiol 39: 722–730.

Bombach, P., Hübschmann, T., Fetzer, I., Kleinsteuber, S., Geyer, R., Harms, H., and Müller, S. (2011) Resolution of natural microbial community dynamics by community fingerprinting, flow cytometry, and trend interpretation analysis. Adv Biochem Eng Biotechnol 124: 151–181.

Borole, A.P., Reguera, G., Ringeisen, B., Wang, Z.-W., Feng, Y., and Kim, B.H. (2011) Electroactive biofilms: current status and future research needs. Energy Environ Sci 4: 4813–4834.

Carmona-Martínez, A.A., Harnisch, F., Kuhlicke, U., Neu, T.R., and Schröder, U. (2013) Electron transfer and biofilm formation of Shewanella putrefaciens as function of anode potential. Bioelectrochem 93: 23–29.

Cooper, S. (2006) Regulation of DNA synthesis in bacteria: analysis of the Bates/Kleckner licensing/initiation-mass model for cell cycle control. Mol Microbiol 62: 303–307.

Cui, L., Ma, X., Sato, K., Okuma, K., Tenover, F.C., Mamizuka, E.M., et al. (2003) Cell wall thickening is a common feature of vancomycin resistance in Staphylococcus aureus. J ClinMicrobiol 41: 5–14.

Franks, A.E., Nevin, K.P., Jia, H., Izallalen, M., Woodard, T.L., and Lovley, D.R. (2008) Novel strategy for three-dimensional real-time imaging of microbial fuel cell communities: monitoring the inhibitory effects of proton accumulation within the anode biofilm. Energy Environ Sci 2: 113–119.

Franks, A.E., Nevin, K.P., Glaven, R.H., and Lovley, D.R. (2009) Microtoming coupled to microarray analysis to evaluate the spatial metabolic status of Geobacter sulfurreducens biofilms. ISME J 4: 509–519.

Günther, S., Hübschmann, T., Rudolf, M., Eschenhagen, M., Röske, I., Harms, H., and Müller, S. (2008) Fixation procedures for flow cytometric analysis of environmental bacteria. J Microbiol Meth 75: 127–134.

Harnisch, F., Koch, C., Patil, S.A., Hübschmann, T., Müller, S., and Schröder, U. (2011) Revealing the electrochemically driven selection in natural community derived microbial biofilms using flow-cytometry. Energy Environ Sci 4: 1265–1267.

Houry, A., Gohar, M., Deschamps, J., Tischenko, E., Aymerich, S., Gruss, A., and Briandet, R. (2012) Bacterial swimmers that infiltrate and take over the biofilm matrix. PNAS 109: 13088–13093.

Ishihama, A. (2012) Detection of bacterial habits: single planktonic cells and assembled biofilm. In Nanomedicine in Diagnostics. Rozlosnik, N. (ed.). Enfield, NH, USA: Science Publishers, pp. 191–211.

Koch, C., Günther, S., Desta, A.F., Hübschmann, T., and Müller, S. (2013) Cytometric fingerprinting for analyzing microbial intracommunity structure variation and identifying subcommunity function. Nat Protoc 8: 190–202.

Koch, C., Müller, S., Harms, H., and Harnisch, F. (2014) Microbiomes in bioenergy production: from analysis to management. Curr Opin Biotechnol 27: 65–72.

Long, H.F., and Hammer, B. (1941) Classification of the organism important in dairy products: III. Pseudomonas putrefaciens. Res Bulletin 285: 176–195.

Mah, T.-F.C., and O'Toole, G.A. (2001) Mechanisms of biofilm resistance to antimicrobial agents. Trends in Microbiol 9: 34–39.

Martín-Gil, J., Ramos-Sánchez, M.C., and Martín-Gil, F.J. (2004) Shewanella putrefaciens in a fuel-in-water emulsion from the Prestige oil spill. Antonie Van Leeuwenhoek 86: 283–285.

Moller, S., Kristensen, C.S., Poulsen, L., Carstensen, J.M., and Molin, S. (1995) Bacterial growth on surfaces: automated image analysis for quantification of growth rate-related parameters. Appl Environ Microbiol 61: 741–748.

Müller, S. (2007) Modes of cytometric bacterial DNA pattern: a tool for pursuing growth. Cell Prolif 40: 621–639.

Müller, S., and Nebe-von-Caron, G. (2010) Functional single-cell analyses: flow cytometry and cell sorting of microbial populations and communities. FEMS Microbiol Rev 34: 554–587.

Müller, S., Harms, H., and Bley, T. (2010) Origin and analysis of microbial population heterogeneity in bioprocesses. Current Opinion in Biotechnol 21: 100–113.

Müller, S., Hübschmann, T., Kleinsteuber, S., and Vogt, C. (2012) High resolution single cell analytics to follow microbial community dynamics in anaerobic ecosystems. Methods 57: 338–349.

Neumeyer, A., Hübschmann, T., Müller, S., and Frunzke, J. (2013) Monitoring of population dynamics of Corynebacterium glutamicum by multiparameter flow cytometry. Microbial Biotechnol 6: 157–167.

Rizoulis, A., Elliott, D.R., Rolfe, S.A., Thornton, S.F., Banwart, S.A., Pickup, R.W., and Scholes, J.D. (2013) Diversity of planktonic and attached bacterial communities in a phenol-contaminated sandstone aquifer. Microb Ecol 66: 84–95.

Russell, J.B., and Cook, G.M. (1995) Energetics of bacterial growth: balance of anabolic and catabolic reactions. Microbiol Rev 59: 48–62.

Saffarini, D.A., DiChristina, T.J., Bermudes, D., and Nealson, K.H. (1994) Anaerobic respiration of Shewanella putrefaciens requires both chromosomal and plasmid-borne genes. FEMS Microbiol Lett 119: 271–277.

Scott, J.H., and Nealson, K.H. (1994) A biochemical study of the intermediary carbon metabolism of Shewanella putrefaciens. J Bacteriol 176: 3408–3411.

Stickland, L.H. (1951) The determination of small quantities of bacteria by means of the biuret reaction. Journal of General Microbiol 5: 698–703.

Tsokos, C.G., and Laub, M.T. (2012) Polarity and cell fate asymmetry in Caulobacter crescentus. Current Opinion in Microbiol 15: 744–750.

Unge, A., Tombolini, R., Molbak, L., and Jansson, J.K. (1999) Simultaneous monitoring of cell number and metabolic activity of specific bacterial populations with a dual gfp-luxAB marker system. *Appl Environ Microbiol* **65:** 813–821.

Uría, N., Muñoz Berbel, X., Sánchez, O., Muñoz, F.X., and Mas, J. (2011) Transient storage of electrical charge in biofilms of *Shewanella oneidensis* MR-1 growing in a microbial fuel cell. *Environ Sci Technol* **45:** 10250–10256.

Valet, G., Leary, J.F., and Tárnok, A. (2004) Cytomics – new technologies: towards a human cytome project. *Cytometry A* **59:** 167–171.

Vogel, B.F., Jørgensen, K., Christensen, H., Olsen, J.E., and Gram, L. (1997) Differentiation of *Shewanella putrefaciens* and *Shewanella alga* on the basis of whole-cell protein profiles, ribotyping, phenotypic characterization, and 16 S rRNA gene sequence analysis. *App Environ Microbiol* **63:** 2189–2199.

Vogt, C., Lösche, A., Kleinsteuber, S., and Müller, S. (2005) Population profiles of a stable, commensalistic bacterial culture grown with toluene under sulphate-reducing conditions. *Cytometry* **66A:** 91–102.

Wang, H., and Ren, Z.J. (2013) A comprehensive review of microbial electrochemical systems as a platform technology. *Biotechnol Adv* **31:** 1796–1807.

Zweifel, U.L., and Hagstrom, A. (1995) Total counts of marine bacteria include a large fraction of non-nucleoid-containing bacteria (Ghosts). *Appl Environ Microbiol* **61:** 2180–2185.

## Supporting information

Additional Supporting Information may be found in the online version of this article at the publisher's web-site:

**Fig. S1.** Substrate consumption of *S. putrefaciens*.
**Fig. S2.** Gating strategy for differentiation of sub-populations with distinct DNA contents.
**Fig. S3.** Instrument stability over time.
**Fig. S4.** Dynamics in bacterial growth pattern of planktonic and biofilm cells.
**Fig. S5.** DNA distribution patterns for feast growth under aerobic conditions.
**Fig. S6.** DNA distribution patterns for famine growth under aerobic conditions.
**Fig. S7.** DNA distribution pattern for planktonic and biofilm cells for the growth on glass slides.
**Fig. S8.** DNA distribution pattern for planktonic and biofilm cells for the growth on glass wool.
**Fig. S9.** DNA distribution pattern for planktonic and biofilm cells for growth on graphite paper.
**Fig. S10.** DNA distribution pattern for famine growth under anaerobic conditions.
**Fig. S11.** Dalmatian plot position tendency of comparative analysis via nMDS.
**Fig. S12.** Schematic illustration of the flow-through-chamber system (FTC).

# 9

# Exploring ComQXPA quorum-sensing diversity and biocontrol potential of *Bacillus* spp. isolates from tomato rhizoplane

A. Oslizlo,[1] P. Stefanic,[1] S. Vatovec,[1] S. Beigot Glaser,[2] M. Rupnik[2,3,4] and I. Mandic-Mulec[1]*

[1]*Department of Food Science and Technology, Biotechnical Faculty, University of Ljubljana, Ljubljana, Slovenia.*

[2]*National Laboratory for Health, Environment and Food, Maribor, Slovenia.*

[3]*Faculty of Medicine, University of Maribor, Maribor, Slovenia.*

[4]*Centre of Excellence for Integrated Approaches in Chemistry and Biology of Proteins, Ljubljana, Slovenia.*

## Summary

*Bacillus subtilis* **is a widespread and diverse bacterium t exhibits a remarkable intraspecific diversity of the ComQXPA quorum-sensing (QS) system. This manifests in the existence of distinct communication groups (pherotypes) that can efficiently communicate within a group, but not between groups. Similar QS diversity was also found in other bacterial species, and its ecological and evolutionary meaning is still being explored. Here we further address the ComQXPA QS diversity among isolates from the tomato rhizoplane, a natural habitat of** *B. subtilis,* **where these bacteria likely exist in their vegetative form. Because this QS system regulates production of anti-pathogenic and biofilm-inducing substances such as surfactins, knowledge on cell–cell communication of this bacterium within rhizoplane is also important from the biocontrol perspective. We confirm the presence of pherotype diversity within** *B. subtilis* **strains isolated from a rhizoplane of a single plant. We also show that** *B. subtilis* **rhizoplane isolates show a remarkable diversity of surfactin production and potential plant growth promoting traits. Finally, we discover that effects of surfactin deletion on biofilm formation can be strain specific and unexpected in the light of current knowledge on its role it this process.**

*For correspondence. E-mail ines.mandic@bf.uni-lj.si

**Funding Information** This work was supported by Slovenian Research Agency Grant J4-3631 (to I.M.-M.), ARRS Program Grant JP4-116 and an ARRS Young Investigator Grant.

## Introduction

It was already suggested by Darwin (1859) that intraspecific diversity increases the species adaptive potential to changing conditions, simply by making the population better prepared for the unexpected. Recently, this assumption was confirmed experimentally for various species, such as eelgrass where diversity within species contributes to its survival in fluctuating environments (Hughes and Stachowicz, 2004) and for *Pseudomonas aeruginosa* where strain diversity was shown to increase its stress resistance in biofilms (Boles *et al.*, 2005). Moreover, experiments reveal that genetically uniform populations of bacteria can readily diverge, especially in structured environments that offer various niche opportunities (Rainey and Travisano, 1998; Poltak and Cooper, 2011).

*Bacillus subtilis*, ubiquitous and highly diverse Gram-positive bacterium, has been most often isolated in the form of heat resistant spores from soil and plant rhizosphere, as well as from other habitats such as aquatic systems, animal guts and various foods (Earl *et al.*, 2008; Mandic-Mulec and Prosser, 2011). Its intraspecies diversity is reflected by high number of ecotypes (Koeppel *et al.*, 2008; Stefanic *et al.*, 2012; Kopac *et al.*, 2014) and mirrored in diversification of distinct 'communication' groups or pherotypes (Tran *et al.*, 2000; Tortosa *et al.*, 2001; Ansaldi *et al.*, 2002; Stefanic and Mandic-Mulec, 2009), which are defined as groups of bacteria that are able to communicate through signalling molecules (peptides) that elicit a response in strains sharing the same pherotype but not (or to significantly lesser extent) in those of a different pherotype (Ansaldi *et al.*, 2002). Stefanic and colleagues (2012) proposed that pherotype diversity could be an adaptation to ecological diversity and showed that one pherotype dominates an ecotype with other pherotypes being present with lower frequency within an ecotype. Therefore, additional studies are needed to better understand the pherotype puzzle and its ecological meaning.

Diversification into pherotypes is coupled to striking polymorphisms of the ComQXPA quorum-sensing (QS) system (Tran *et al.*, 2000; Ansaldi *et al.*, 2002; Stefanic and Mandic-Mulec, 2009). It was shown that pherotypes can coexist in soil at a millimetre scale (Stefanic and Mandic-Mulec, 2009) and that the communication diversification is under Darwinian selection (Ansaldi and Dubnau, 2004) and is present also in other *Bacillus* species that encode the *comQXPA* homologues loci (Dogsa *et al.*, 2014). Still, it is not clear how this diversification is manifested in other parts of the genome and how it is adaptive for the species. The ComQXPA lingual system operates the QS (Fuqua *et al.*, 1994), a process where secreted signalling molecules (ComX) accumulate to critical concentration and by activation of cognate receptors (ComP), trigger the expression of target genes. More precisely, ComX signal is initially synthesized as 55 amino acid-long prepeptide that is processed by ComQ and secreted from the cells (Magnuson *et al.*, 1994; Ansaldi *et al.*, 2002; Schneider *et al.*, 2002). Signal production by *B. subtilis* serves as a negative feedback mechanism which modulates QS response of producing cells (Oslizlo *et al.*, 2014). When ComX accumulates, it activates ComP receptor, which then phosphorylates response regulator ComA, and this one in turn modulates the transcription of many genes (Weinrauch *et al.*, 1990; Comella and Grossman, 2005). Microarray studies revealed that *srfAA-D* operon encoding the surfactin synthetase (Nakano *et al.*, 1988) accounts for the most affected target by the ComQXPA regulon (Comella and Grossman, 2005).

Surfactin is a powerful lipopeptide biosurfactant and an antibiotic that acts against many bacteria and fungi, including plant pathogens like *Pseudomonas syringae* (Bais *et al.*, 2004). In addition, surfactin by inducing a potassium leakage in the cells (Lopez *et al.*, 2009) indirectly serves as a signal that triggers biofilm formation, which is essential for colonization of roots (Beauregard *et al.*, 2013; Zeriouh *et al.*, 2013) and protection of plants against pathogens (Bais *et al.*, 2004; Chen *et al.*, 2013; Zeriouh *et al.*, 2013). In fact, *B. subtilis* is regarded to be a PGPR (plant growth-promoting rhizobacterium) and has been well known for its biocontrol potential (Barea *et al.*, 2005; Berg, 2009; van Elsas and Mandic-Mulec, 2013). It was even proposed that the vegetative form of this species is normally associated with plant roots and that soil is predominantly inhabited by its dormant spores (Norris and Wolf, 1961). If this holds true, it is of major importance to complement our current knowledge on the communication diversity of *B. subtilis* isolates that live on plant root surfaces (rhizoplane), especially because the diverse ComQXPA system controls the expression of biocontrol agents, like surfactin (Nakano *et al.*, 1988; Zeriouh *et al.*, 2013). PGPR are widely accepted as ecofriendly alternatives to chemical pesticides, and they have been in commercial use for several years (Nakkeeran *et al.*, 2005). Knowledge on QS diversity in rhizoplane and on how this diversity manifests in QS-regulated traits could then contribute to optimal design of PGPR-based formulations.

Most work on *B. subtilis* ecology has been performed by studying spores isolated from soil (Mandic-Mulec and Prosser, 2011). Here, we use a set of *B. subtilis* isolates and close relatives isolated from tomato rhizoplane to address the genetic and functional diversity of spore formers. Our aim was to examine whether different QS pherotypes of *B. subtilis* vegetative cells can coexist within the rhizoplane of a single plant. We further addressed whether being a member of a certain pherotype in the rhizoplane manifests in similar expression of known ComQXPA-regulated biocontrol properties like production of surfactin or biofilm formation. In addition, we compare direct plant growth promotion and potential PGP traits between isolates derived from a single plant or within a pherotype. We find that *B. subtilis* strains living on roots of a single plant carry diverse QS pherotypes and are highly diverse with respect to their biocontrol potential. The strains show differences in biofilm formation, surfactin production and other PGPR traits and most interestingly, behave differently after silencing of the *srfA* operon. We discuss what such intraspecies diversity could mean for the bacteria, for the plant and for biocontrol by rhizoplane communities.

## Results

### *Isolation and characterization of* Bacillus *spp. isolates*

Rhizoplanes of 21 different tomato plants grown at two different locations (A and B; see *Experimental procedures*) were screened for *Bacillus subtilis*-like colonies. Strains ($n = 20$), phenotypically resembling colonies of *B. subtilis*, were obtained from 10 out of 21 tomato plants. Sequencing of 16S rRNA confirmed that they all belong to the *Bacillus* genus; however, due to high percentage of sequence similarity, we could not determine the species affiliation for all isolates based on 16S rRNA sequences. We therefore further identified the strains on the basis of the *gyrA* gene, which can be used as an alternative phylogenetic marker because of higher rates of molecular evolution as compared with 16S rRNA (Chun and Bae, 2000). This approach allowed us to identify *B. subtilis* (13 strains), *B. licheniformis* (4), *B. amyloliquefaciens* (1), *B. pumilus* (1) and *B. megaterium* (1) (Fig. 1A).

This approach was preferred over MIDI (Microbial Indentification System) Similarity index, which provided taxonomic identification only for 17 out of 20 strains (Table S1). *Bacillus subtilis* strains were isolated from 5

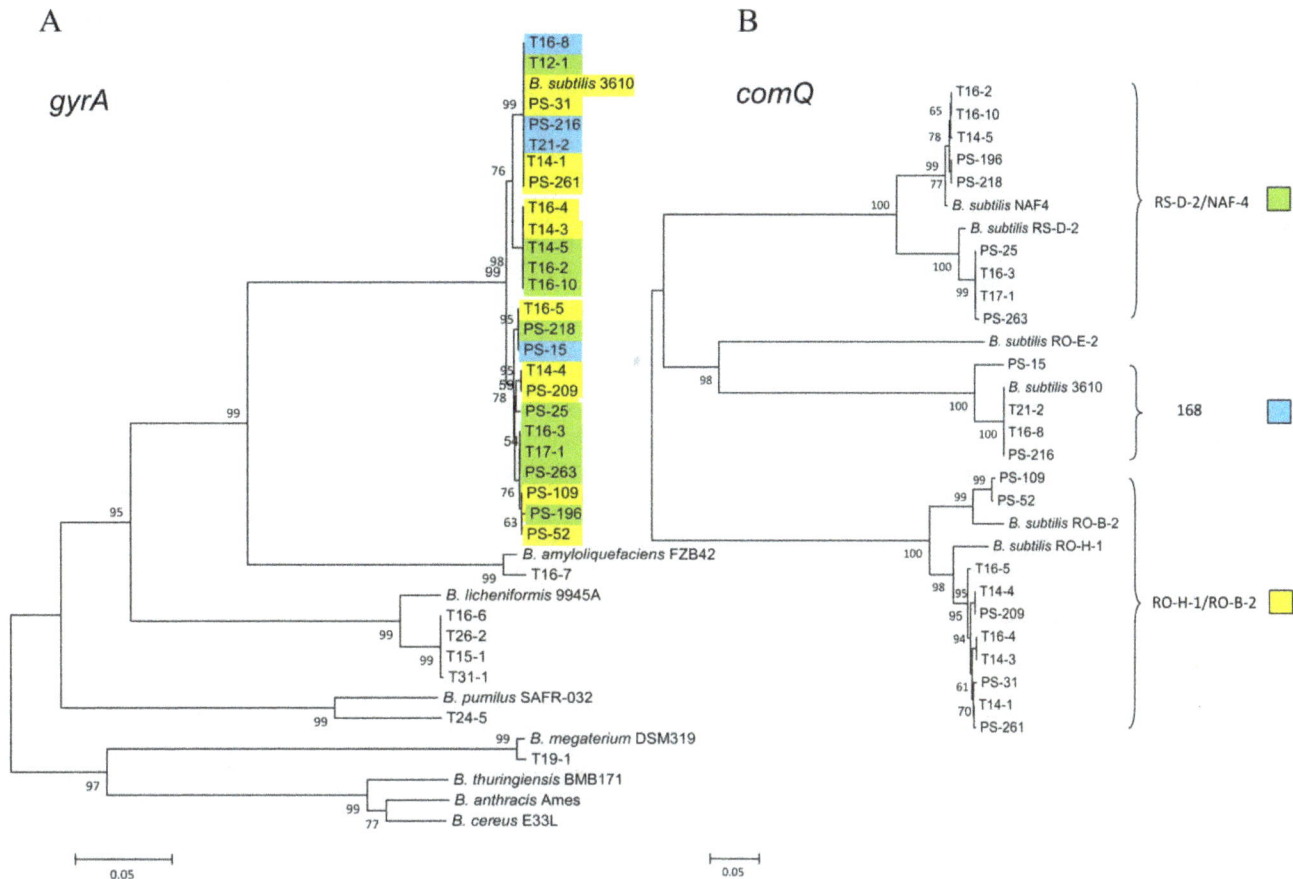

**Fig. 1.** Minimum evolution trees based on (A) partial gyrA nucleotide sequences (610 bp) and (B) partial comQ sequences (701 bp).
A. Strains representing pherotype 168, RO-H-1/RO-B-2 and RS-D-2/NAF-4 are higlighted in blue, yellow and green respectively.
B. Pherotypes with this colour code are indicated on the right side of the corresponding clades.

out of 21 plants suggesting heterogeneity among viable rhizoplane *Bacilli* and all *B. subtilis* strains listed in Table S1 were used for further experiments. The *B. subtilis* gyrA sequences were highly conserved (~ 99% sequence identity), but still some sequence subclusters could be identified on the gyrA ME (minimum evolution) similarity tree (Fig. 1A). These subclusters included *B. subtilis* strains isolated from one plant, different plants and even from plants at different locations. Moreover, gyrA sequences of rhizoplane isolates were at minimal genetic distances (~ 100% sequence identity) with gyrA of *B. subtilis* isolated from the Sava riverbank soil, Slovenia (Stefanic and Mandic-Mulec, 2009), and no habitat-dependent clustering was observed (Fig. 1A).

### Polymorphism of comQXP *locus*

High polymorphism within *comQXP* genes was previously confirmed for soil *B. subtilis* isolates (Ansaldi *et al.*, 2002; Stefanic and Mandic-Mulec, 2009). In this study, we find a

similar pattern for rhizoplane isolates (Fig. 1B). The *comQ* sequences were highly polymorphic and fall into three clusters: 168 (blue), RS-D-2/NAF4 (green) and RO-H-1/RO-B-2 (yellow), with only 65–70% identity at the nucleotide level between clusters (Fig. 1B). Each of the three clusters, depicted by the ME similarity tree of 12 rhizoplanes and few representative *comQ* sequences of the soil isolates (Fig. 1B), contained sequences from rhizoplane and soil bacteria. Also isolates derived from one plant (plant 16: strains T16) were randomly distributed along the tree, and each cluster contained at least one T16 isolate.

Diversity of *comQ* was also found within sequence similarity clusters as noticed in our previous study (Stefanic and Mandic-Mulec, 2009). Rhizoplane *comQ* sequences within the RS-D-2/NAF4 cluster split into two distinct subclusters, with ~ 85% *comQ* sequence identity between and 100% within the subclusters. However, within RO-H-1/RO-B-2 and 168 clusters, the *comQ* sequence diversity of the rhizoplane isolates was low, with 99% or 100% similarity respectively. This could be due to significantly lower number of isolates analysed here as compared with the previous studies involving soil isolates. As expected, ME trees based on *gyrA* and *comQ* were not congruent and over 30% *comQ* divergence was found among strains carrying clonal *gyrA* sequences (Fig. 1).

### Specificity of the comQXP *QS loci*

All *Bacillus spp.* isolates were tested for their activation of QS response of six tester strains representing different pherotypes. Tester strains carrying a P*srfA*-*lacZ* reporter fusion were grown in conditioned media of the rhizoplane isolates and then tested for β-galactosidase activity representing transcriptional response of *srfA* to ComX present in the condition medium. Specific producer strains were used as positive controls.

On the basis of strong and moderate activation responses, 15 out of 20 strains could be classified to three distinct pherotypes (Fig. 2). These pherotypes were consistent with clustering of *comQ* sequences (Fig. 1B), and we concluded that two strains (T16-8 and T21-2) belong to the pherotype 168; five strains (T14-1, T14-3, T14-3, T16-4 and T16-5) to the pherotype RO-H-1/ RO-B-2; and six strains (T12-1, T14.5, T16-2, T16-3, T16-10 and T17-1) were confirmed as the pherotype RSD-2/NAF-4. Interestingly, only three isolates (T12-1, T16-2 and T16-10) out of six induced a strong response of the NAF4 tester strain, suggesting a possible split of the pherotype and continuous evolution of the QS genes as previously indicated for soil microscale isolates (Stefanic and Mandic-Mulec, 2009). All strains that induced strong response of the tester (marked as '++') belonged to

| Pherotype group and isolate from which conditioned media was obtained | B-galactosidase activity of tester strain | | | | | |
|---|---|---|---|---|---|---|
| **168** | 168 | RS-D-2 | NAF 4 | RO-B-2 | RO-H-1 | RO-E-2 |
| *B. subtilis* T16-8 | ++ | – | – | +/– | +/– | +/– |
| *B. subtilis* T21-2 | ++ | +/– | +/– | +/– | +/– | + |
| **RS-D-2/NAF4** | | | | | | |
| *B. subtilis* T12-1 | – | ++ | ++ | +/– | – | + |
| *B. subtilis* T14-5 | – | ++ | – | +/– | – | +/– |
| *B. subtilis* T16-2 | – | ++ | ++ | + | +/– | + |
| *B. subtilis* T16-3 | – | + | – | +/– | – | +/– |
| *B. subtilis* T16-10 | – | ++ | ++ | +/– | +/– | +/– |
| *B. subtilis* T17-1 | – | ++ | – | +/– | +/– | + |
| **RO-B-2/RO-H-1** | | | | | | |
| *B. subtilis* T14-1 | – | – | – | ++ | ++ | +/– |
| *B. subtilis* T14-3 | – | – | – | ++ | ++ | + |
| *B. subtilis* T14-4 | – | – | – | ++ | ++ | + |
| *B. subtilis* T16-4 | – | – | – | ++ | ++ | +/– |
| *B. subtilis* T16-5 | – | – | – | ++ | ++ | + |
| **none** | | | | | | |
| *B. licheniformis* T15-1 | – | – | – | – | – | + |
| *B. licheniformis* T16-6 | – | – | – | – | – | – |
| *B. amyloliquefaciens* T16-7 | – | – | – | – | – | – |
| *B. megaterium* T19-1 | – | – | – | – | – | – |
| *B. pumilus* T24-5 | – | – | – | – | – | – |
| *B. licheniformis* T26-2 | – | – | – | – | – | +/– |
| *B. licheniformis* T31-1 | +/– | – | – | – | – | +/– |

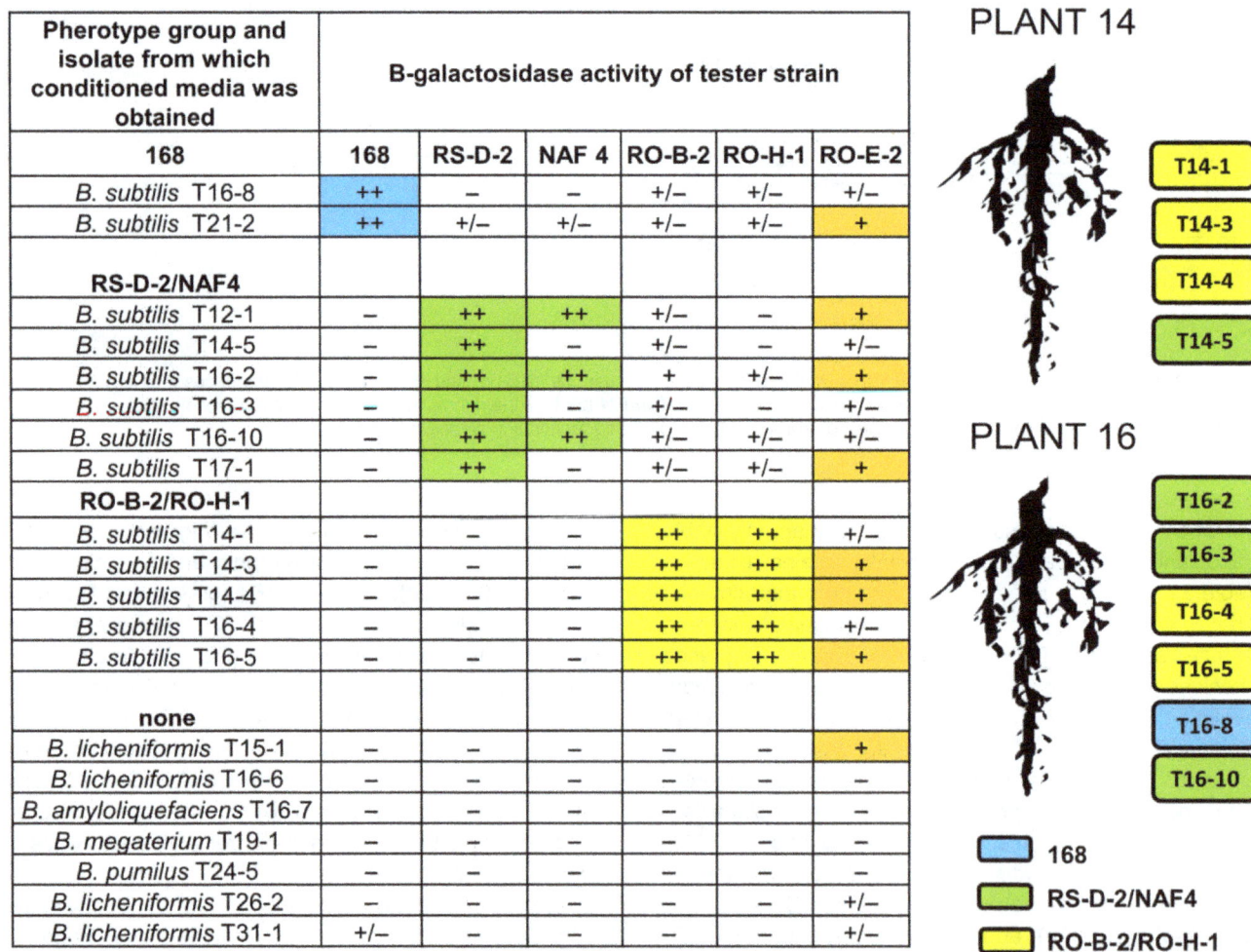

PLANT 14

T14-1
T14-3
T14-4
T14-5

PLANT 16

T16-2
T16-3
T16-4
T16-5
T16-8
T16-10

168
RS-D-2/NAF4
RO-B-2/RO-H-1

**Fig. 2.** Specific activation of the QS response was measured using tester strains able to detect one of the four previously determined pherotypes through activation of the srfA-lacZ reporter. The testers were specific for the pherotype 168 (blue), the pherotype RS-D-2 /NAF4 (green); the pherotype RO-B-2/RO-H-1 (yellow); and the pherotype RO-E-2 (orange). Rhizosphere isolates were grown in CM and conditioned media were sampled 1 h after entry into the stationary phase (T1). Tester strains were then inoculated (1:50) into conditioned medium mixed with an equal volume of fresh CM medium, grown for 16 h and assayed for β-galactosidase activity as indicated in *Experimental procedures*. Symbols: ++, strong response, similar to positive control; +, moderate response, approximately 50% of the positive-control response; +/–, weak but reproducible response; –, no activation. On the right: schematic drawing representing pherotype diversity of rhizoplane isolates from plant 14 and plant 16.

*B. subtilis* species. Some cross-talk of *B. subtilis* resulting in a moderate or very low response ('+' or '+/−') with testers outside the primary pherotype could be observed; however, the cross-talk never induced response as strong as the pherotype specific communication (Fig. 2). Finally, the *B. licheniformis* T15-1 conditioned medium induced moderate response of *B. subtilis* RO-E-2 tester, suggesting the presence of cross species communication. The results together with phylogenetic analysis confirmed that *B. subtilis* strains of distinct pherotypes can be isolated from the rhizoplane of a single plant supporting a ubiquity of the pherotype diversity (Fig. 2).

### Diversity of potential biocontrol properties within *B. subtilis* pherotypes and plants

As both surfactin production (Comella and Grossman, 2005) and indirectly also biofilm formation (Lopez *et al.*, 2009) are under ComQXPA QS control, we examined the variability of these traits within the rhizoplane collection of *B. subtilis* isolates. We were particularly interested whether this diversity is evident also within a pherotype. We used the *B. subtilis* BD2833 strain (Tortosa *et al.*, 2001), which is deficient in surfactin production and biofilm formation (BD 2883 derives from 168 described in McLoon *et al.*, 2011) as a negative control. In addition, the *B. subtilis* GB03 strain was used as a positive control for biofilm formation (Beauregard *et al.*, 2013).

We observed that all but one *B. subtilis* isolates produced pellicle biofilms with higher biomass as compared with negative control and that three isolates (T16-8,

T14-4 and T16-5) produced larger biofilm biomass as compared with the positive control (GB03 strain that exhibited 1.19. ± 0.3 value) (Fig. 3, Fig. S1). Within one pherotype, strains differed in both tested traits: biofilm formation and surfactin production (Fig. 3, Fig. S1). For example, within the pherotype RO-H-1/RO-B-2, one strain formed a copious biofilm and another was comparable with a negative control (Fig. 3). The diversity of surfactin production was also very pronounced, and within each pherotype we could find very strong biosurfactant producers (up to eight times more as compared with the control) but also strains that showed very weak surfactin activity (measured by haemolytic activity) or no such activity at all (Fig. 3). Similar diversity could be observed when strains were grouped in respect to the plant they were isolated from (Fig. 4A). For example, each plant (plant 14 and plant 16) contained a very strong surfactant producer (T14-3 and T16-8 respectively), a moderate producer or even a non-producer (T16-4) (Fig. 3). In order to confirm that the haemolytic assay measured surfactin-specific activity of the conditioned media produced by rhizoplane isolates, we managed to inactivate the *srfA* gene in five isolates: T16-8, T16-10, T16-4, T16-5 and T16-2 by incorporation of the *ΔsrfA* mutation (see *Experimental procedures*) using a standard transformation protocol applied for naturally competent *B. subtilis*. We successfully transformed five out of eight strains presumably due to differences in transformation efficiency of natural isolates (data not shown). We compared the haemolytic activity of the mutants with their ancestors. All 5 *ΔsrfA* mutants

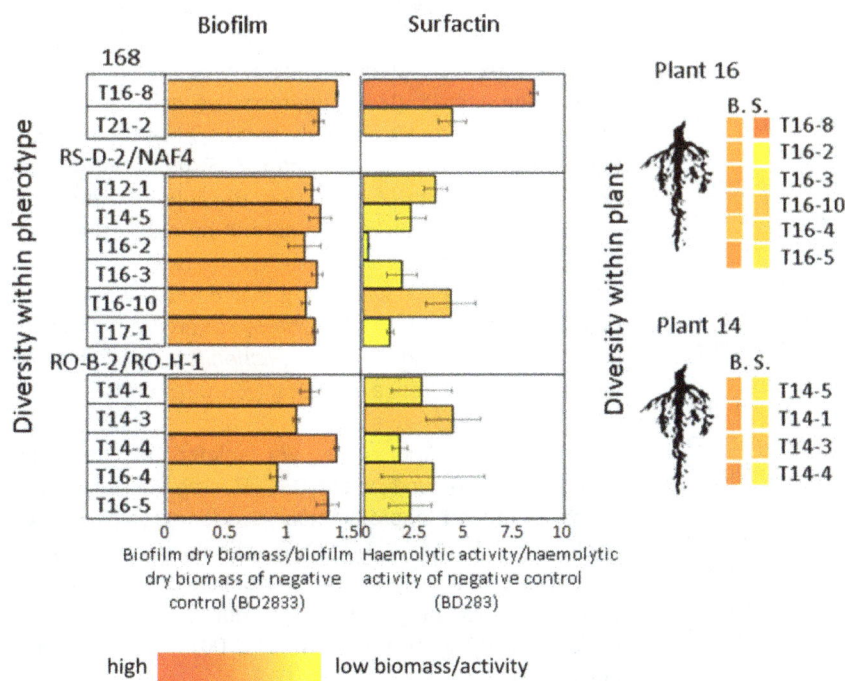

**Fig. 3.** Biofilm biomass and surfactin activity of *B. subtilis* isolates from tomato rhizoplane. Biofilms were harvested and their dry mass was determined. The dry mass of each biofilm was then divided by the dry mass of biofilm formed by the negative control BD2833. Conditioned media produced during biofilms growth were filter-sterilized, and the presence of biosurfactants was determined by heamolytic test. Strain BD2833 which is deficient in surfactin production was used as negative control. Percent of haemolysis obtained for each conditioned medium was divided by the value obtained for negative control. Columns on the graph indicating the diversity of the response are marked with RGB intensities that directly correspond to quantitative values measured for each trait (as shown below the table). Data represent average of three independent replicates with SE (standard error) indicated. The RGB colours also indicate the diversity of biofilm dry biomass (B.) and haemolytic activity (S.) at the plant level (on the right).

A

**Droplet collapse test**

|          | T16-8 | T16-10 | T16-2 | T16-4 | T16-5 |
| --- | --- | --- | --- | --- | --- |
| wild types: | | | | | |
| ΔsrfA mutants: | | | | | |

hem. act. of the mutant/
hem. act. of the wild type:  1.8 ±1   1.6 ±1   3.2 ±2   2.5 ±1   3.4 ±1

B

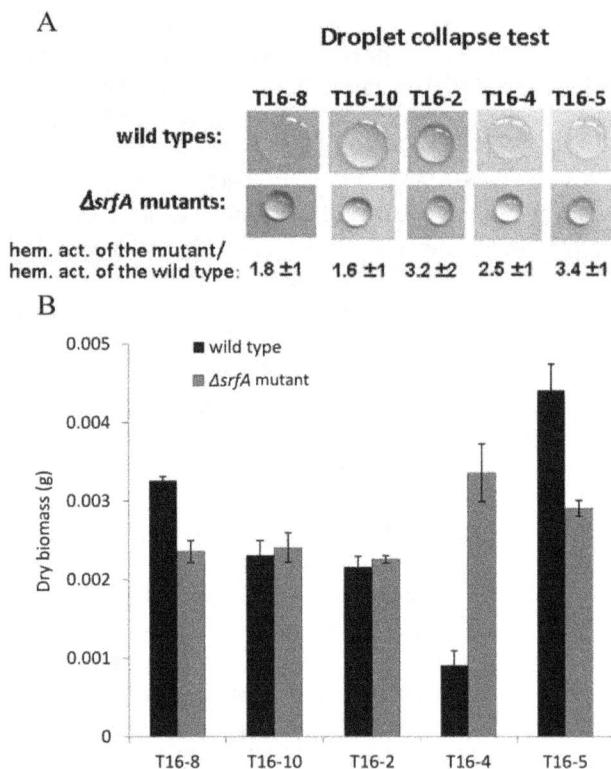

**Fig. 4.** (A) Conditioned media (30 μ droplets) of wild isolates T16-8, T16-2, T16-10, T16-4 and T16-5 (first raw) and of isogenic ΔsrfA mutants (second raw) from pellicle cultures grown in MSN medium at 28°C were sampled after 48 h. The numbers below the droplet pictures represent percent of haemolytic activity measured in conditioned medium of isogenic ΔsrfA mutants as compared with their ancestor wild type strains. (B) Biofilms were harvested after 48 h of incubation, and their dry mass was determined. Data represent average of three independent replicates (independent experiments).

showed over 95% decrease of haemolytic activity (Fig. 4A). Therefore, haemolytic activity of wild rhizoplane isolates was surfactin dependent. Moreover, the drop collapse test (Jain *et al.*, 1991), which is an alternative, qualitative method of surfactin detection, correlated with the haemolytic assay (Fig. 4A).

Next we tested the effects of surfactin deletion on biofilm biomass production. Interestingly, not all strains were affected equally. In the T16-8 and T16-5 strains, surfactin deficiency decreased biofilm biomass by 28% ($P < 0.02$) and 34% ($P < 0.02$), respectively, while the T16-2 and T16-10 mutants formed biofilms in biomass comparable with the wild-type ancestors. Surprisingly, the T16-4-ΔsrfA mutant had 3.5-fold ($P < 0.02$) larger biomass than the parental strain (Fig. 4B). Therefore, the regulatory role of surfactin in biofilm formation might be strikingly different among natural isolates of *B. subtilis*, and it will be interesting to identify genetic differences behind this observation in the future.

*PGPR properties within* B. subtilis *pherotypes*

It is known that bacteria can directly promote the growth of plant by various secretions (Kloepper *et al.*, 1980; López-Bucio *et al.*, 2007). We therefore tested whether secreted molecules of the rhizoplane isolates can influence the growth of plant roots and leaves of the model plant *Arabidopsis thaliana*. Bacteria were inoculated on 0.2× MS (Murashige and Skoog) solid medium (see *Experimental procedures*) app. 5 cm away from the *A. thaliana* seedlings, and the plant biomass versus control (no bacteria inoculated) was measured after 10 days (Fig. 5). Isolates sharing a pherotype or being derived from the rhizoplane of a single plant (Fig. 5) had different effects on roots and leaves biomass. For example, two strains of the same pherotype RS-D-2/NAF4, namely T16-2 and T16-4, increased the biomass of roots ($P < 0.06$ and $P < 0.002$, respectively) and leaves ($P < 0.03$ and $P < 0.006$, respectively) twofold, while the strain T16-3 of the same pherotype had no effect on the plant biomass (Fig. 5).

Similar diversity was found among isolates from one plant, with plant 16 isolates giving very strong PGP effect or no effect; and isolates from plant 14 showing weak positive effects, no effects, or even negative effects on plant biomass (Fig. 5).

In addition, we analysed three phenotypic traits that can influence plant growth: production of indole-3-acetic acid (IAA) (López-Bucio *et al.*, 2007) or siderophores (Kloepper *et al.*, 1980) and the ability to solubilize phosphate (Molla *et al.*, 1984). These properties were also variable within a pherotype, and there was no correlation between plant growth promotion measured directly and these traits. For example, strain T16-4, which had strong positive effect on plant biomass, was negative in all PGP traits tested (IAA, siderophores and phosphate solubilization) (Fig. 6). Similarly, strains T16-10 and T16-7 were positive in two out of three PGP traits but did not promote the growth of roots (Fig. 6). This indicates that direct effects on plant growth cannot be easily predicted only by testing the established PGP traits.

In addition to PGP bacteria, we also found inhibitors of plant growth: T14-5 (*B. subtilis*, phenotype RS-D-2/NAF-4) (Fig. 5) and T31-1 (*B. licheniformis*) (Table S2) induced 30% decrease in root's ($P < 0.08$ and $P < 0.05$, respectively) and leave's ($P < 0.09$ and $P < 0.06$ respectively) biomass. Additionally, we isolated an interesting PGP candidate from plant 19, *B. megaterium* T19-1, which produced approximately five times higher concentration of IAA (25 μg ml⁻¹) compared with other strains, showed very strong production of siderophores and also strongly (twofold) promoted the growth of roots ($P < .004$) and leaves ($P < .002$) of *A. thaliana* (Table S2).

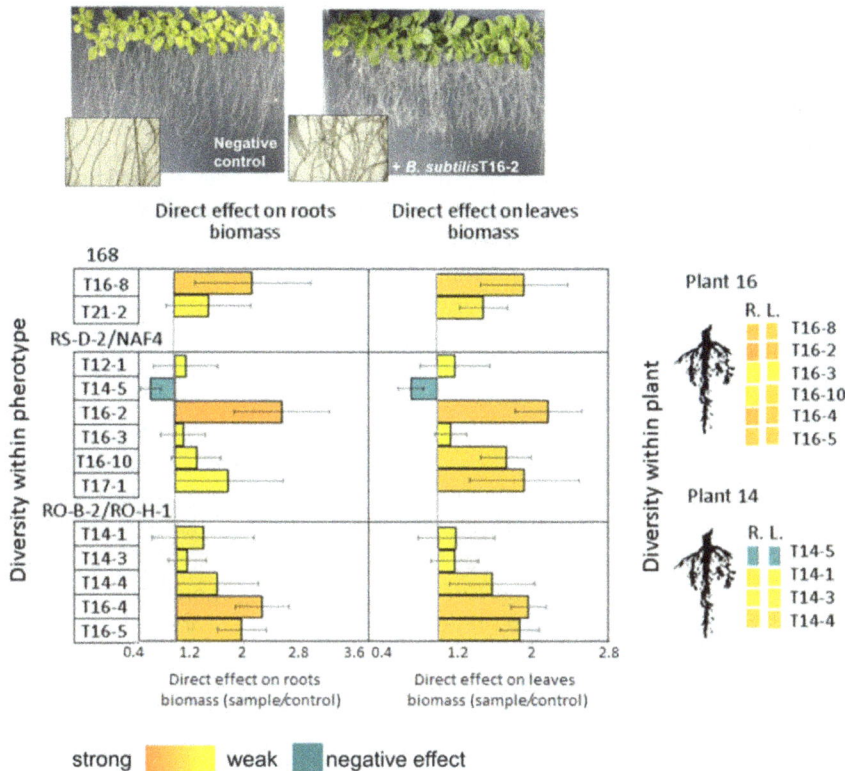

**Fig. 5.** Influence of *B. subtilis* isolates from tomato rhizoplane on *A. thaliana* roots and leaves biomass. Plants were grown for 14 days in app. 5 cm distance from the bacterial inoculums. Positive effect of *B. subtilis* isolate presence on growth of the model plant is shown on the pictures above. Final biomass of roots and leaves was divided by biomass of control plants (roots and leaves) that were grown on sterile medium. To better show the diversity in the response within each indicated phenotype (168, RS-D-2 /NAF4, RO-B-2/RO-H-1), columns on the graph were marked with RGB intensities that directly correspond to quantitative values measured for each trait (as shown below the graphs). Data represent average of the three independent replicates with SE indicated. The colours representing effect on roots biomass (R.) and leaves biomass (L.) were also used to demonstrate the diversity at the plant level (on the right).

## Discussion

*Bacillus subtilis* strains that persist in soil are classified to three to four distinct QS groups – pherotypes (Ansaldi *et al.*, 2002; Stefanic and Mandic-Mulec, 2009). As plant rhizosphere was proposed the main habitat of *B. subtilis* vegetative existence (Norris and Wolf, 1961), we asked here whether the pherotype diversity is also found among rhizoplane isolates.

We confirmed that different QS groups (pherotypes) can coexist on roots of a single plant. This was shown by *comQ* sequence analysis and by specific induction of QS response. Despite low number of *B. subtilis* isolates per plant (6 isolates from plant 16 and 4 isolates from plant 14), we found a comparable diversity, manifested in three pherotypes, which was previously observed for soil milli-metre scale (Stefanic and Mandic-Mulec, 2009), confirming that *comQXP* diversity is widespread and easy to find.

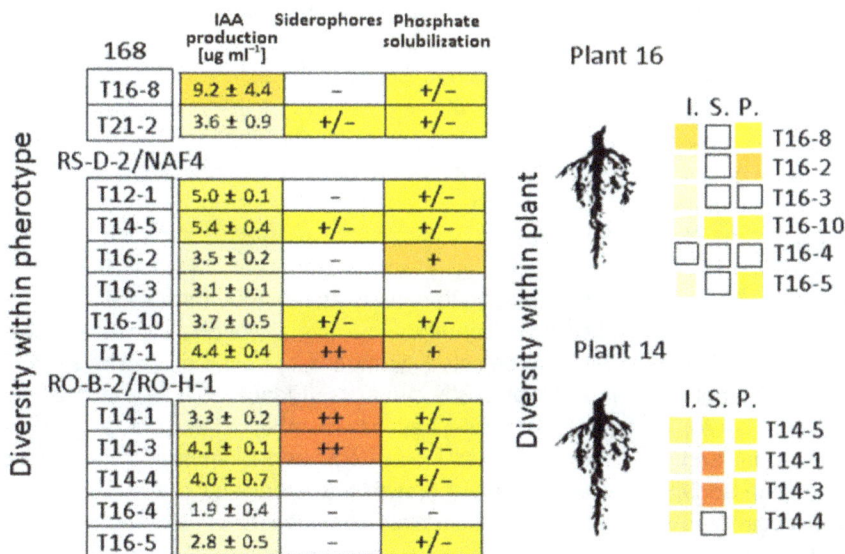

**Fig. 6.** Comparison of potential PGP traits: IAA production, siderophores production and phosphate solubilization. Symbols: ++, strong effect, +, moderate effect (significantly lower as compared to maximal effect observed); +/−, weak but reproducible effect, −, no effect. The RGB intensity was used to better show the diversity of the traits. The RGB colours are used to indicate the diversity of PGP traits: IAA concentration (I.), siderophores (S.) and phosphate solubilization (P.) at the plant level (on the right).

This again brings up the fundamental question on adaptive role of this diversity, which is also found in other gram-positive bacteria, including *Staphylococcus aureus*, *Streptococcus pneumoniae* (Pozzi *et al.*, 1996; Ji *et al.*, 1997; Whatmore *et al.*, 1999; Carrolo *et al.*, 2014) and *B. cereus* (Slamti and Lerecius, 2005). Stefanic and colleagues (2012) found that most but not all *B. subtilis* strains of the same ecotype belong to the same pherotype and proposed that pherotypes may at least in part be under ecological selection. Rhizoplane strains were not analysed for ecotype association, but their high phenotypic diversity (e.g. variability in surfactin production and other biocontrol properties) suggests that many of them may be ecologically distinct. Interestingly, isogenic strains of *S. pneumoniae* of distinct pherotypes differed in their ability to form biofilms, possibly because of pherotype-dependent strength of QS signalling (Carrolo *et al.*, 2014). Here we found no correlation between QS type (pherotype) and the expression of a QS-regulated trait – surfactin production. This is in contrast to observation by Carrolo and colleagues (2014), who noticed such interrelationship, but it should be stressed that they used isogenic strains, which only differed in pherotypes. We used wild isolates of different genetic backgrounds that may additionally influence the surfactin synthesis and secretion and dominate over pherotype association.

Strong variability in surfactin production among rhizoplane isolates, even among isolates of a single plant, may also influence social life of this species. Because surfactin is secreted and probably shared between neighbouring *B. subtilis* strains, it may, under certain conditions, serve as a public good (West *et al.*, 2007). Differences in surfactin production could also result in disproportions in metabolic investment and fitness among strains (Oslizlo *et al.*, 2014), allowing weak surfactin producers to benefit from strong producers when plant pathogen invades and needs to be opposed. Coexistence of social and less social phenotypes was previously found in neighbouring strains of *Myxococcus xanthus* (Kraemer and Velicer, 2014) or *P. aeruginosa* (Wilder *et al.*, 2011). In case of *M. xanthus* it was shown that 'social' strains can promote the persistence of less social isolates without negative effects on the social's fitness, but sometimes when highly abundant, the less social strains can decrease fitness of the whole community (Kraemer and Velicer, 2014). It was recently suggested that spatial segregation facilitates the evolution of cooperation in *B. subtilis* biofilms (van Gestel *et al.*, 2014). Therefore, high spatial segregation of different pherotypes might help to stabilize cooperative traits and public goods production, like surfactin. Consequently, selection for cooperation might lead to assortment and diversity of pherotypes even at small distances. In addition, it was recently discovered that *B. subtilis* colonizes hyphae of *Aspergilus niger* where surfactin expression is

downregulated by these fungi (Benoit *et al.*, 2014). Could then the decreased level of surfactin secretion in certain *B. subtilis* strains (like T16-2 orT14-4) be an adaptation to peaceful coexistence with fungi, which are also common rhizoplane inhabitants?

It is not known how presence of poor surfactin producers influences the performance of the whole *B. subtilis* community in fighting plant pathogens, and it is an interesting problem of sociomicrobiology that could be addressed in the future. It is also worth noting that in certain isolates (like T16-10 or T16-2), the presence of surfactin showed no influence on biofilm biomass and in others (like T16-8 and T16-5) even exerted a negative influence on biofilm biomass. It is known that surfactin serves as a paracrine signal for biofilm formation (Lopez *et al.*, 2009). Our results, however, indicate that surfactin role in biofilm formation may be more complex and strain specific. This result also further supports previous assumptions about high genetic and phenotypic diversity of strains sharing a pherotype. Moreover, because surfactin is required for root colonization (Bais *et al.*, 2004; Chen *et al.*, 2013), it remains to be tested how the observed differences in surfactin production translate into root-colonization abilities of the *B. subtilis* isolates. Also, it will be interesting to address whether the $\Delta srfA$ mutants (T16-2; T16-5: T16-5: T16-8 and T16-10) lose the ability to colonize the plant roots, despite different effects of *srfA* deletion on pellicle biofilms.

We observed only moderate diversity at the level of biofilm formation. Although we did not look into biofilm-related gene expression, our results are in line with previous data showing that genetic diversity within *B. subtilis* is especially high with respect to antibiotic-related genes (like surfactin) and low with respect to biofilm-related genes (Earl *et al.*, 2008). The reason for the later may be associated with attached growth being essential for rhizocompetence and probably represents a competitive advantage in rhizoplane.

While our strain isolation strategy does not allow us to speculate on the original spatial distribution among the *Bacillus* spp. isolates on roots, the results confirm that different pherotypes reside on roots of a single plant.

Stefanic and colleagues (2012) proposed that ratios of pherotypes continuously cycle in nature because of induction of costly products released by high-frequency pherotype and temporary advantage of the low-frequency pherotype. Therefore, the diversity of pherotypes could be naturally selected by means of social conflict based on release of costly products (Eldar, 2011; Stefanic *et al.*, 2012). However, because members of one pherotype dramatically differ in QS-regulated biocontrol traits, direct effects on plant growth and other potential PGP behaviours (IAA secretion, production of siderophores or phosphate solubilization), pherotype diversity would not threaten the biocontrol function of the *Bacillus* community – eventually each pherotype contains a strong surfactin producers, or/and direct plant growth promoters. Therefore, the diversity of pherotypes on plant roots may promote coexistence of different strains and thus positively influence the biocontol potential of this species.

It was previously shown that plant growth promotion by *B. subtilis* depends on production of volatile 2,3-butanediol (Ryu *et al.*, 2003). Because we found no correlation between PGP effects and analysed secretions (IAA, siderophores and phosphatases), it is possible that volatile molecules could influence the growth of a model plant in our experiment; however, this awaits further studies.

The idea to use microbes as biocontrol agents emerged many years ago, and many studies screening for PGP properties of rhizosphere and rhizoplane isolates were performed. Some strains were patented and its commercial use is steadily increasing (Maheshwari, 2011). Although it is known that diverse community of microbes determines the plant health, PGPR-based preparations are still based on monocultures (Maheshwari, 2011). There were several successful attempts of applying multispecies formulations (Raupach & Kloepper 1998; Singh *et al.*, 1999), but sometimes, simply because of interspecies competition, the effects can be just opposite to the expectations (Chiarini *et al.*, 1998; de Boer *et al.*, 1999). In fact, in terms of sharing a niche, one should rather expect a competition instead of a synergy between microbial species (Foster and Bell, 2012). It is therefore important to bring more attention to intraspecies genetic and phenotypic diversity in the rhizosphere and rhizoplane, where next to strong indirect competition, more cooperation, especially within a pherotype, would be predicted. Testing biocontrol properties or plant growth-promotion effects of mixed *B. subtilis* communities should be the next step to answer this question.

We show here that *B. subtilis* residing on roots differ in QS pherotypes, potential PGPR traits and the ability to influence the growth of *A. thaliana*. We believe this study opens new interesting questions about the role of strain diversity in arms race between *B. subtilis* and plant pathogens, or in interactions with host plant. We also believe that applying diverse strains of one genus or species carrying diverse biocontrol properties should be considered as an alternative to non-monoculture-based biocontrol agents design.

## Experimental procedures

### Bacterial strains

Strains used in the study are listed in Table 1 and Table 2. In Table 1, we listed isolates from the tomato rhizoplane and also three control strains: 6051, GB03 and FZB42, which

**Table 1.** *Bacillus spp.* isolates used in this study.

| Strain | Location | Plant | *Bacillus* species | Reference |
|---|---|---|---|---|
| T12-1 | A | 12 | *B. subtilis* | This work |
| T14-1 | A | 14 | *B. subtilis* | This work |
| T14-3 | A | 14 | *B. subtilis* | This work |
| T14-4 | A | 14 | *B. subtilis* | This work |
| T14-5 | A | 14 | *B. subtilis* | This work |
| T16-2 | A | 16 | *B. subtilis* | This work |
| T16-3 | A | 16 | *B. subtilis* | This work |
| T16-4 | A | 16 | *B. subtilis* | This work |
| T16-5 | A | 16 | *B. subtilis* | This work |
| T16-8 | A | 16 | *B. subtilis* | This work |
| T16-10 | A | 16 | *B. subtilis* | This work |
| T15-1 | A | 15 | *B. licheniformis* | This work |
| T16-6 | A | 16 | *B. licheniformis* | This work |
| T16-7 | A | 16 | *B. amyloliquefaciens* | This work |
| T17-1 | B | 17 | *B. subtilis* | This work |
| T21-2 | B | 21 | *B. subtilis* | This work |
| T19-1 | B | 19 | *B. megaterium* | This work |
| T24-5 | B | 24 | *B. pumilus* | This work |
| T26-2 | B | 26 | *B. licheniformis* | This work |
| T31-1 | B | 31 | *B. licheniformis* | This work |
| GB03 | na | na | *B. subtilis* | BGSC 3A37 |
| FZB42 | na | na | *B. amyloliquefaciens* | BGSC 10A6 |
| 6051 | na | na | *B. subtilis* | ATCC DSM 10 |

na, does not apply; strains were obtained from the Bacillu Genetic Stock Center (BGSC) and from The Leibniz Institute DSMZ.

were previously reported to show PGP properties (Bais *et al.*, 2004; Idris *et al.*, 2004; Xie *et al.*, 2009). Tomato rhizoplane isolates are marked with T and the following number corresponds with the plant so that, for example, strains T14-1, T14-3 and T14-4 were isolated from the rhizoplane of the same plant (T14). Table 2 includes engineered *B. subtilis* strains that were used to test the specificity of rhizoplane and riverbank isolates in activating the QS response in six tester strains, representing four currently recognized pherotypes. In addition Table 2 includes five Δ*srfA* mutants that were obtained by transforming eight randomly chosen *B. subtilis* rhizoplane T-isolates with chromosomal DNA isolated from the OKB120 strain (Nakano *et al.*, 1988; Vollenbroich *et al.*, 1996). To induce natural competence, *B. subtilis* T-isolates were grown in competence medium (CM) (Albano *et al.*, 1987) for 6 h. Briefly, 0.5 ml of early stationary phase cultures were mixed with 1 μg of OKB120 chromosomal DNA suspension and incubated for 30 min at 37°C with vigorous shaking. Next, the cultures were supplemented with 0.5 ml of Luria–Bertani (LB) medium and incubated for additional 60 min. Transformants were selected on agar plates containing lincomycin (12.5 μg ml⁻¹) and erythromycin (0.5 μg ml⁻¹).

### Isolation and characterization of T-strains from tomato rhizoplane

In October 2011, 21 tomato plants from two home vegetable gardens (A and B) located 60 km apart in north-west Slovenia were sampled. Bulk soil was carefully removed from the roots, and samples were transferred to sterile plastic bags. Roots were then washed with sterile saline solution. Six pieces of 1 cm long roots were collected in 1 ml sterile saline

**Table 2.** Engineered *B. subtilis* strains used in this study.

| Producer strains | | |
|---|---|---|
| BD2833 | *his leu met srfA-lacZ* (*tet*) | Tortosa and colleagues (2001) |
| BD2913 | *his met srfA-lacZ* (*tet*) *amyE::xylR Pxyl-comK* (*ery*) (*comQ comX comP* replaced by genes from *B. mojavensis* RO-H-1) | Tortosa and colleagues (2001) |
| BD2915 | *his met srfA-lacZ* (*tet*) *amyE::xylR Pxyl-comK* (*ery*) (*comQ comX comP* replaced by genes from *B. subtilis* natto NAF4) | Tortosa and colleagues (2001) |
| BD2936 | *his met srfA-lacZ* (*tet*) *amyE::xylR Pxyl-comK* (*cat*) (*comQ comX comP* replaced by genes from *B. mojavensis* RO-B-2) | Tortosa and colleagues (2001) |
| BD2940 | *his leu met srfA-lacZ* (*tet*) *amyE::xylR Pxyl-comK* (*cat*) (*comQ comX comP* replaced by genes from *B. subtilis* RO-E-2) | Ansaldi and colleagues (2002) |
| BD2949 | *his leu met srfA-lacZ* (*tet*) *amyE::xylR Pxyl-comK* (*cat*) (*comQ comX comP* replaced by genes from *B. subtilis* RS-D-2) | Ansaldi and colleagues (2002) |
| **Tester strains** | | |
| BD2876 | *his leu met srfA-lacZ* (*tet*) *comQ*::Km | Tortosa and colleagues (2001) |
| BD2877 | *his leu met srfA-lacZ* (*tet*) (*comQ::phl comX comP* replaced by genes from *B. subtilis* natto NAF4) | Tortosa and colleagues (2001) |
| BD2962 | *his met srfA-lacZ* (*tet*) *amyE::xylR Pxyl-comK* (*ery*) (*comQ::pED345 comX comP* replaced by genes from *B. mojavensis* RO-H-1) | Tortosa and colleagues (2001) |
| BD2983 | *his leu met srfA-lacZ* (*tet*) *amyE::xylR Pxyl-comK* (*cat*) (*comQ::pED345 comX comP* replaced by genes from *B. mojavensis* RO-B-2) | Ansaldi and colleagues (2002) |
| BD3019 | *his leu met srfA-lacZ* (*tet*) *amyE::xylR Pxyl-comK* (*cat*) (*comQ::pED375 comX comP* replaced by genes from *B. subtilis* RS-D-2) | Ansaldi and colleagues (2002) |
| BD3020 | *his leu met srfA-lacZ* (*tet*) *amyE::xylR Pxyl-comK* (*cat*) (*comQ::pED375 comX comP* replaced by genes from *B. subtilis* RO-E-2) | Ansaldi and colleagues (2002) |
| *ΔsrfA* mutants | | |
| T16-8ΔsrfA | *srfA::Tn917* (*mls*)[a] | This work |
| T16-10ΔsrfA | *srfA::Tn917* (*mls*)[a] | This work |
| T16-2ΔsrfA | *srfA::Tn917* (*mls*)[a] | This work |
| T16-4ΔsrfA | *srfA::Tn917* (*mls*)[a] | This work |
| T16-5ΔsrfA | *srfA::Tn917* (*mls*)[a] | This work |

a. srfA::Tn917 (mls) insertion was derived from genomic DNA of OKB120 strain (Nakano *et al.*, 1988; Vollenbroich *et al.*, 1996).

solution and homogenized with pestles for 3 min. The suspension was heat shocked (Brain Heart Infusion) at 70°C for 10 min, and serial dilutions were plated on BHI (Brain Heart Infusion) agar. After 48 or 72 h of incubation at room temperature, *Bacillus*-like colonies were subcultured. Crude DNA was isolated from pure culture by boiling one loop of culture in 100 μl of 5% Chelex® (BIO-RAD). To confirm, *B. subtilis*-specific polymerase chain reaction (PCR) was carried out using the primers gyrAF (5′-CAGTCAGGAAATGCGTACG TCCTT-3′) and gyrAR1 (5′-CAATGAGAGTATCCGTTGTG CGTC-3′). For additional analysis, the Sherlock Microbial Identification System (Agilent Technologies) was used.

## DNA amplification and sequencing

DNA from strains was extracted using classical phenol-chloroform method. The 16S rRNA genes were amplified by PCR with primers 27F (5′-AGAGTTTGATCMTGGCTCAG-3′) and 1406R (5′-ACGGGCGGTGTGTRCAA-3′) (Nubel *et al.*, 1996). The 16S rRNA genes were sequenced using the reverse primer 1406R (5′ACGGGCGGTGTGTRCAA-3′). The *gyrA* genes were amplified by PCR with primers gyrAR6 (5′-3′) and gyrAF5 (5′-3′), and *gyrA* gene was sequenced using the reverse primer gyrAR6 (5′-3′). The *comQ* genes were amplified by PCR with primers Uni-comQ1 (5′-GGG AGGGGGGAAGTCGTTATTG-3′) and P1 (5′-AAGAACCG

AATCGTGGAGATCGCG-3′) (Tortosa *et al.*, 2001). The *comQXP* locus was sequenced with the forward primer Uni-comQ1 (5′-GGGAGGGGGGAAGTCGTTATTG-3′); PCR products were sequenced by Macrogen (Seoul, Korea). Standard BLAST algorithm (standard nucleotide BLAST available at http://blast.ncbi.nlm.nih.gov/Blast.cgi) using default settings was used for sequence analysis/provisional identification of *Bacillus* spp.

## Phylogenetic analyses

Phylogenetic analyses were conducted using MEGA version 4 (Tamura *et al.*, 2007) for neighbour-joining and ME analyses using Tamura–Kumar model of evolution with heterogeneous patterns among lineages and gamma-distributed rates among sites. The evolutionary history was inferred using the ME method (Rzhetsky and Nei, 1992). The bootstrap consensus tree inferred from 1000 replicates (Falsenstein, 1985) was taken to represent the evolutionary history of the taxa analysed. Branches corresponding to partitions reproduced in less than 50% bootstrap replicates were collapsed. The percentage of replicate trees in which the associated taxa clustered together in the bootstrap test (1000 replicates) were shown next to the branches (Felsenstein, 1985). The tree was drawn to scale, with branch lengths in the same units as those of the evolutionary distances used to infer the phylogenetic tree. For *gyrA* sequences and *comQ* sequences, the evolutionary distances were computed using the Tamura 3-parameter method (Tamura, 1992) and maximum composite likelihood method (Tamura *et al.*, 2004), respectively, and are in the units of the number of base substitutions per site. The ME tree was searched using the close-neighbour-interchange algorithm (Nei and Kumar, 2000) at a search level of 1. The neighbour-joining algorithm (Saitou and Nei, 1987) was used to generate the initial tree. Codon positions included were 1st+2nd+3rd+ non-coding. All positions containing gaps and missing data were eliminated from the dataset (complete deletion option). For *gyrA* and *comQ*-based analysis, there were a total of 822 and 827 positions in the final dataset respectively.

## β-Galactosidase assay

β-Galactosidase was assayed using a Multiscan Spectrum Microplate Reader (Thermo Scientific). The absorbance at 420 nm was measured at 30°C immediately after the addition of ortho-nitrophenyl-β-galactoside substrate. Tester strains were incubated with shaking for 16 h at 28°C in 100 µl fresh CM medium and 100 µl of conditioned media produced by rhizoplane and riverbank isolates and harvested in T1. After the incubation, cells were centrifuged (4 °C; 1,800 × g) and re-suspended in 200 µl Z-buffer with 5.6% (vol/vol) β-mercaptoethanol before adding 10 µl toluene and incubating the cultures on ice for 30 min. The plate was then warmed to 30°C, 50 µl ortho-nitro-phenyl-β-galactoside substrate was added and the absorbance (420 nm) was immediately determined at 30°C.

## Biofilm formation assay

Assay was performed using MSN (minimal salts nitrogen) medium (5 mM potassium phosphate buffer pH 7, 0.1 M Mops pH 7, 2 mM MgCl2, 0.05 mM MnCl2, 1 µM ZnCl2, 2 µM thiamine, 700 µM CaCl2, 0.2% NH4Cl) supplemented with 0.5% pectin as performed by Beauregard and colleagues (20132). Strains were grown overnight in LB, and suspensions were inoculated (2%) into MSN pectin media distributed in 10 ml Petri dishes and grown for 48 h at 28°C. Pellicles were harvested, dried and weighted. Conditioned media were sampled for biosurfactant assay.

## Biosurfactant antibiotics production

Cells were grown as described above (biofilm formation assay), and conditioned media were collected. Biosurfactants activity was measured using haemolytic assay (Moran *et al.*, 2002). Bovine red blood cells (RBC) were washed two times with isotonic buffer (140 mM NaCl and 20 mM Tris pH 7.4) and once with 0.9% NaCl. The RBC were then resuspended in 0.9% NaCl to optical density 0.7. Dilution series of conditioned media were prepared and 100 µl of supernatant fraction, 30 µl of 96% ethanol and 100 µl of RBC were mixed on microtitre plate. Optical density ($\lambda$ = 650 nm) was measured immediately after addition of RBC and after 15 min of incubation at room temperature. The percentage of decrease in optical density that was in linear correlation to sample dilution

value was transformed into percentage of haemolysis and divided by corresponding biofilm dry biomass.

In addition, drop collapse test was performed (Jain *et al.*, 1991). The drops of 30 µl of conditioned media were transferred to smooth parafilm surface, and image was taken after 5 min of incubation in room temperature. Flatten droplets indicated presence of biosurfactant.

## Plant growth promotion assay

Arabidopsis ecotype Col-0 seeds were surface sterilized using 2% sodium hypochlorite solution. Briefly, seeds were incubated in 2% sodium hypochlorite with mixing on an orbital mixer for 20 min and then washed five times with sterile distilled water.

Seeds were germinated and grown on agar plates containing MS medium (Murashige and Skoog basal salts mixture; Sigma) (2.2 g l$^{-1}$) supplemented with 1% sucrose. After 3 days of incubation at 4°C, plates were transferred to plant growth chamber (photoperiod of 16 h of light, 8 h of darkness, light intensity, constant temperature of 24°C) and placed vertically at an angle of 65 degrees. After 10 days, homogenous 1 cm long seedlings were selected for growth promoting experiments.

Bacterial strains were grown in LB medium until late exponential phase, cells were washed twice with 0.9% NaCl and 50 µl suspensions were inoculated on MS agar plates in a line, approximately 2 cm from the bottom of the plate. Plates were incubated overnight in 37°C. Next, 16 1 cm-long Col-0 seedlings were transplanted to the MS plates, approximately 5 cm away from the bacterial line and arranged similarly as described by López-Bucio and colleagues (2007). After 10 days of incubation in plant growth chamber, the seedlings were removed from the agar, the roots were washed with distilled water, separated from the leaves and measured with a ruler. Roots and leafs from each MS plate were collected and weighted before and after drying. Roots and leaves mass obtained from bacteria-inoculated plates was compared with control, where seedlings were grown on sterile plates.

## IAA production

*Bacillus* spp. isolates were grown in LB medium supplemented with L-tryptophan (Sigma, T0254) (1 mg ml$^{-1}$ final concentration) for 48 h at 28°C with shaking 200 r.p.m. Supernatants were collected and IAA production was determined with the use of iron and perchloric acid according to modified method of Solon and Weber (1950). Briefly FeCl3-HClO$_4$ reagent (1.0 ml of 0.5 M FeCl3, 50 ml 35% HClO$_4$) was mixed with the culture supernatant in 1:2 ratio and incubated for 15 min. Next 5 µl of orthophosphortic acid was added for reaction enhancement and absorbance at $\lambda$ = 510 nm was determined. IAA concentration was calculated from a standard curve prepared using commercial IAA (Sigma, I3750).

## Siderophore production

Bacterial strains were grown overnight on solid media (Schwyn and Neilands, 1987) prepared as follows: 100 ml of Minimal Media 9 (MM9) stock solution (15 g KH$_2$PO$_4$, 25 g

NaCl, 50 g NH$_4$Cl dissolved in 500 ml of ddH$_2$O) was mixed with 750 ml of MiliQ and 15 g of agar. After autoclaving and cooling to 50°C, the medium was supplemented with 30 ml of 10% sterile casamino acid solution (BD, 223050) and 10 ml of sterile 20% glucose solution. MM9-based media with *Bacillus* growth were each overlaid by 10 ml of the following medium: 60.5 mg chrome azurol S, 72.9 mg hexadecyltrimetyl ammonium bromide, 30.24 g piperazine-1,4-bis(2-ethanesulfonic acid) and1 mMFeCl3·6H2O in 10 mMHCl 10 ml, 9 g of agar per 1 l of MiliQ. After a maximum period of 15 min, a change in colour from blue to purple around the colonies indicated the siderophore producers (Pérez-Miranda *et al.*, 2007). Qualitative estimation of siderophore production was performed as follows: – negative (no orange halo), from + /– to ++, positive.

*Phosphate solubilization*

The ability of the strains to solubilize inaccessible phosphate was determined using Pikovskaya agar (Pikovskaya, 1948). Cells were grown overnight in LB medium, next they were washed twice with 0.9% NaCl and re-suspended in 0.9% NaCl to produce equal cell densities among all the isolates. Solutions were inoculated on the agar plates and incubated in 37°C for 7 days. The size of halo (zone of solubilization) around the bacterial colony indicated phosphate solubilizing abilities of each strain. Qualitative estimation of phosphate solubilization was performed as follows: – negative (no halo), from + /– to ++, positive.

## Acknowledgements

We would like to thank J.I Prosser, A.T. Kovacs and N. Lyons for discussions and valuable comments and for proofreading the manuscript. We would also like to thank Simona Leskovec for technical assistance and to Bernarda Kovac for help with PGPR assays.

## References

Albano, M., Hahn, J., and Dubnau, D. (1987) Expression of competence genes in Bacillus subtilis. *J Bacteriol* **169:** 3110–3117.

Ansaldi, M., and Dubnau, D. (2004) Diversifying selection at the *Bacillus* quorum-sensing locus and determinants of modification specificity during synthesis of the ComX pheromone. *J Bacteriol* **186:** 15–21.

Ansaldi, M., Marolt, D., Stebe, T., Mandic-Mulec, I., and Dubnau, D. (2002) Specific activation of the *Bacillus* quorum-sensing systems by isoprenylated pheromone variants. *Mol Microbiol* **44:** 1561–1573.

Bais, H.P., Fall, R., and Vivanco, J.M. (2004) Biocontrol of *Bacillus subtilis* against infection of *Arabidopsis* roots by *Pseudomonas syringae* is facilitated by biofilm formation and surfactin production. *Plant Physiol* **134:** 307–319.

Barea, J.M., Pozo, M.J., Azcón, R., and Azcón-Aguilar, C. (2005) Microbial co-operation in the rhizosphere. *J Exp Bot* **56:** 1761–1778.

Beauregard, P.B., Chai, Y., Vlamakis, H., Losick, R., and Kolter, R. (2013) *Bacillus subtilis* biofilm induction by plant polysaccharides. *Proc Natl Acad Sci USA* **110:** 1621–1630.

Benoit, I., van den Esker, M.H., Patyshakuliyeva, A., Mattern, D.J., Blei, F., Zhou, M., *et al.* (2014) *Bacillus subtilis* attachment to *Aspergillus niger* hyphae results in mutually altered metabolism. *Environ Microbiol.* doi:10.1111/1462-2920.12564.

Berg, G. (2009) Plant–microbe interactions promoting plant growth and health: perspectives for controlled use of microorganisms in agriculture. *Appl Microbiol Biotechnol* **84:** 11–18.

de Boer, M., Van der Sluis, I., Van Loon, L.C., and Bakker, P.A.H.M. (1999) Combining fluorescent *Pseudomonas spp.* strains to enhance suppression of fusarium wilt of radish. *Eur J Plant Pathol* **105:** 201–210.

Boles, B.R., Thoendel, M., and Singh, P.K. (2005) Genetic variation in biofilms and the insurance effects of diversity. *Microbiology* **151:** 2816–2818.

Carrolo, M., Rodrigues Pinto, F., Melo-Cristino, J., and Ramirez, M. (2014) Pherotype influences biofilm growth and recombination in *Streptococcus pneumoniae*. *PLoS ONE* **9:** e92138. doi:10.1371/journal.pone.0092138.

Chen, Y., Yan, F., Chai, Y., Liu, H., Kolter, R., Losick, R., and Guo, J.H. (2013) Biocontrol of tomato wilt disease by *Bacillus subtilis* isolates from natural environments depends on conserved genes mediating biofilm formation. *Environ Microbiol* **15:** 848–864.

Chiarini, L., Bevivino, A., Dalmastri, C., Nacamulli, C., and Tabacchioni, S. (1998) Influence of plant development, cultivar and soil type on microbial colonization of maize root. *Appl Soil Ecol* **8:** 11–18.

Chun, J., and Bae, K.S. (2000) Phylogenetic analysis of *Bacillus subtilis* and related taxa based on partial *gyrA* gene sequences. *Antonie Van Leeuwenhoek* **78:** 123–127.

Comella, N., and Grossman, A.D. (2005) Conservation of genes and processes controlled by the quorum response in bacteria: characterization of genes controlled by the quorum-sensing transcription factor ComA in *Bacillus subtilis*. *Mol Microbiol* **57:** 1159–1174.

Darwin, C. (1859) *On the Origin of Species by Means of Natural Selection, or, the Preservation of Favoured Races in the Struggle for Life.* London: John Murray.

Dogsa, I., Choudhary, K.S., Marsetic, Z., Hudaiberdiev, S., Vera, R., Pongor, S., and Mandic-Mulec, I. (2014) ComQXPA quorum sensing systems may not be unique to Bacillus subtilis: A census in prokaryotic genomes. *PloS One,* **9:** e96122, doi: 10.1371/journal.pone.0096122.

Earl, A.M., Losick, R., and Kolter, R. (2008) Ecology and genomics of *Bacillus subtilis*. *Trends Microbiol* **16:** 269–275.

Eldar, A. (2011) Social conflict drives the evolutionary divergence of quorum sensing. *Proc Natl Acad Sci USA* **108:** 13635–13640.

van Elsas, J.D., and Mandic-Mulec, I. (2013) Advanced molecular tools for analysis of bacterial communities and their interactions in the rhizosphere. In *Molecular Microbial Ecology of the Rhizosphere.* Bruijn, F.J. (ed.). Hoboken, NJ, USA: Wiley-Blackwell, pp. 115–124.

Falsenstein, J. (1985) Confidence limits of phylogenies: an approach using the bootstrap. *Evolution* **39:** 783–791.

Foster, K.R., and Bell, T. (2012) Competition, not cooperation, dominates interactions among culturable microbial species. *Curr Biol* **22:** 1845–1850.

Fuqua, W.C., Winans, S.C., and Greenberg, P.E. (1994) Quorum sensing in bacteria: the LuxR-LuxI family of cell density-responsive transcriptional regulators. *J Bacteriol* **176:** 269–275.

van Gestel, J., Weissing, F.J., Kuipers, O.P., and Kovacs, A.T. (2014) Density of founder cells affects spatial pattern formation and cooperation in *Bacillus subtilis* biofilms. *ISME J* **8:** 2069–2079.

Hughes, A.R., and Stachowicz, J.J. (2004) Genetic diversity enhances the resistance of a seagrass ecosystem to disturbance. *Proc Natl Acad Sci USA* **101:** 8998–9002.

Idris, E.E., Bochow, H., and Borriss, R.R. (2004) Use of *Bacillus subtilis* as biocontrol agent VI. Phytohormone-like action of culture filtrates prepared from plant growth-promoting *Bacillus amyloliquefaciens* FZB24, FZB42, FZB45 and Bacillus subtilis FZB37. *J Plant Dis Prot* **111:** 2583–2597.

Jain, D.K., Collins-Thompson, D.L., Lee, H., and Trevors, J.T. (1991) A drop-collapsing test for screening biosurfactant-producing microorganisms. *J Microbiol Methods* **13:** 271–279.

Ji, G., Beavis, R., and Novick, R.P. (1997) Bacterial interference caused by autoinducing peptide variants. *Science* **276:** 2027–2030.

Kloepper, J.W., Leong, J., Teintze, M., and Schroth, M.N. (1980) Enhanced plant growth by siderophores produced by plant growth-promoting rhizobacteria. *Nature* **286:** 885–886.

Koeppel, A., Perry, E.B., Sikorski, J., Krizanc, D., Warner, A., Ward, D.M., Rooney, A.P., Brambilla, E., Connor, N., Ratcliff, R.M., Nevo, E., and Cohan, F.M. (2008) Identifying the fundamental units of bacterial diversity: A paradigm shift to incorporate ecology into bacterial systematic. *Proc Natl Acad Sci USA* **105:** 2504–2509.

Kopac, S., Wang, Z., Wiedenbeck, J., Sherry, J., Wu, M., and Cohan, F.M. (2014) Genomic heterogeneity and ecological speciation within one subspecies of *Bacillus subtilis*. *Appl Environ Microbiol* **80:** 4842–4853.

Kraemer, S.A., and Velicer, G.J. (2014) Social complementation and growth advantages promote socially defective bacterial isolates. *Proc Biol Sci* **281:** 1781. doi:.org/10.1098/rspb.2014.0036.

López-Bucio, J., Campos-Cuevas, J.C., Hernández-Calderón, E., Velásquez-Becerra, C., Farías-Rodríguez, R., Macías-Rodríguez, L.I., and Valencia-Cantero, E. (2007) *Bacillus megaterium* rhizobacteria promote growth and alter root-system architecture through an auxin- and ethylene-independent signaling mechanism in *Arabidopsis thaliana*. *Mol Plant Microbe Interact* **20:** 207–217.

Lopez, D., Vlamakis, H., Losick, R., and Kolter, R. (2009) Paracrine signaling in a bacterium. *Genes Dev* **23:** 1631–1638.

McLoon, A.L., Guttenplan, S.B., Kearns, D.B., Kolter, R., and Losick, R. (2011) Tracing the domestication of a biofilm-forming bacterium. *J Bacteriol* **193:** 2027–2034.

Magnuson, R., Solomon, J., and Grossman, A.D. (1994) Biochemical and genetic characterization of a competence pheromone from *Bacillus subtilis*. *Cell* **77:** 207–216.

Maheshwari, D.K. (2011) *Bacteria in Agrobiology: Plant Growth Responses*. Berlin, Germany: Springer-Verlag.

Mandic-Mulec, I., and Prosser, J.I. (2011) Diversity of endospore-forming bacteria in soil: characterization and driving mechanisms. In *Endospore: Forming Soil Bacteria*. Logan, N.A., and De Vos, P. (eds). Berlin, Germany: Springer, pp. 31–59.

Molla, M.A.Z., Chowdhury, A.A., Islam, A., and Hoque, S. (1984) Microbial mineralization of organic phosphate in soil. *Plant Soil* **78:** 393–399.

Moran, A.C., Martinez, M.A., and Sineriz, F. (2002) Quantification of surfactin in culture supernatants by hemolytic activity. *Biotechnol Lett* **24:** 177–180.

Nakano, M.M., Marahiel, M.A., and Zuber, P. (1988) Identification of a genetic locus required for biosynthesis of the lipopeptide antibiotic surfactin in *Bacillus subtilis*. *J Bacteriol* **170:** 5662–5668.

Nakkeeran, S., Dilantha Fernando, W.G., and Siddiqui, Z.A. (2005) Plant growth promoting rhizobacteria formulation and it scope in commercializationfor the management of pest and diseases. In *PGPR: Biocontrol and Biofertilization*. Siddiqui, Z.A. (ed.). Dordrecht, the Netherlands: Springer, pp. 257–296.

Nei, M., and Kumar, S. (2000) *Molecular Evolution and Phylogenetics*. New York, USA: Oxford University Press.

Norris, J.R., and Wolf, J. (1961) A study of antigens of the aerobic spore-forming bacteria. *J Appl Bacteriol* **24:** 42–56.

Nubel, U., Engelen, B., Felske, A., Snaidr, J., Wieshuber, A., Amann, R.I., Ludwig, W. and Backhaus, H. (1996) Sequence heterogeneities of genes encoding 16S rRNAs in Paenibacillus polymyxa detected by temperature gradient gel electrophoresis. *J Bacteriol* **178:** 5636–5643.

Oslizlo, A., Stefanic, P., Dogsa, I., and Mandic-Mulec, I. (2014) The private link between signal and response in *Bacillus subtilis* quorum sensing. *Proc Natl Acad Sci USA* **111:** 1586–1591.

Pérez-Miranda, S., Cabirol, N., George-Téllez, R., Zamudio-Rivera, L.S., and Fernández, F.J. (2007) O-CAS, a fast and universal method for siderophore detection. *J Microbiol Methods* **70:** 127–131.

Pikovskaya, R.I. (1948) Mobilization of phosphorus in soil in connection with the vital activity of some microbial species. *Mikrobiologiya* **17:** 362–370.

Poltak, S.R., and Cooper, V.S. (2011) Ecological succession in long-term experimentally evolved biofilms produces synergistic communities. *ISME J* **5:** 369–378.

Pozzi, G., Masala, L., Iannelli, F., Manganelli, R., Havarstein, L.S., and Piccoli, L. (1996) Competence for genetic transformation in encapsulated strains of *Streptococcus pneumoniae*: two allelic variants of the peptide pheromone. *J Bacteriol* **178:** 6087–6090.

Rainey, P.B., and Travisano, M. (1998) Adaptive radiation in a heterogeneous environment. *Nature* **394:** 69–72.

Raupach, G.S., and Kloepper, J.W. (1998) Mixtures of plant growth-promoting rhizobacteria enhance biological control of multiple cucumber pathogens. *Phytopathology* **88:** 1158–1164.

Ryu, C., Farag, M.A., Hu, C., Reddy, M.S., Wei, H., Pare, P.W., and Kloepper, J.W. (2003) Bacterial volatiles promote growth in *Arabidopsis*. *Proc Natl Acad Sci USA* **8:** 4927–4932.

Rzhetsky, A., and Nei, M. (1992) A simple method for estimating and testing minimum evolution trees. *Mol Biol Evol* **9**: 945–967.

Saitou, N., and Nei, M. (1987) The neighbor-joining method: a new method for reconstructing phylogenetic trees. *Mol Biol Evol* **4**: 406–425.

Schneider, K.B., Palmer, T.M., and Grossman, A.D. (2002) Characterization of comQ and comX, two genes required for production of ComX pheromone in *Bacillus subtilis*. *J Bacteriol* **184**: 410–419.

Schwyn, B., and Neilands, J.B. (1987) Universal chemical assay for the detection and determination of siderophores. *Anal Biochem* **160**: 47–56.

Singh, P.P., Shin, Y.C., Park, C.S., and Chung, Y.R. (1999) Biological control of *Fusarium* wilt of cucumber by chitinolytic bacteria. *Phytopatology* **89**: 92–99.

Slamti, L., and Lerecius, D. (2005) Specificity and polymorphism of the PlcR-PapR quorum-sensing system in the *Bacillus cereus* group. *J Bacteriol* **187**: 1182–1187.

Solon, A., and Weber, G. (1950) Colorimetric estimation of indoleactetic acid. *Plant Physiol* **26**: 192–195.

Stefanic, P., and Mandic-Mulec, I. (2009) Social interactions and distribution of *Bacillus subtilis* pherotypes at microscale. *J Bacteriol* **191**: 1756–1764.

Stefanic, P., Decorosi, F., Viti, C., Petito, J., Cohan, F.M., and Mandic-Mulec, I. (2012) The quorum sensing diversity within and between ecotypes of *Bacillus subtilis*. *Environ Microbiol* **14**: 1378–1389.

Tamura, K. (1992) Estimation of the number of nucleotide substitutions when there are strong transition-transversion and G + C-content biases. *Mol Biol Evol* **9**: 678–687.

Tamura, K., Dudley, J., Nei, M., and Kumar, S. (2007) MEGA4: Molecular Evolutionary Genetics Analysis (MEGA) software version 4.0. *Mol Biol Evol* **24**: 1596–1599.

Tamura, K., Nei, M., and Kumar, S. (2004) Prospects for inferring very large phylogenies by using the neighbor-joining method. *Proc Natil Acad Sci USA* **101**: 11030–11035.

Tortosa, P., Logsdon, L., Kraigher, B., Itoh, Y., Mandic-Mulec, I., and Dubnau, D. (2001) Specificity and genetic polymorphism of the *Bacillus* competence quorum-sensing system. *J Bacteriol* **183**: 451–460.

Tran, L.S., Nagai, T., and Itoh, Y. (2000) Divergent structure of the ComQXPA quorum-sensing components: molecular basis of strain-specific communication mechanism in *Bacillus subtilis*. *Mol Microbiol* **37**: 1159–1171.

Vollenbroich, D., Mehta, N., Zuber, P., Vater, J., and Kamp, R.M. (1996) Analysis of surfactin synthetase subunits in srfA mutants of *Bacillus subtilis* OKB105. *J Bacteriol* **176**: 395–400.

Weinrauch, Y., Penchev, R., Dubnau, E., Smith, I., and Dubnau, D. (1990) A *Bacillus subtilis* regulatory gene product for genetic competence and sporulation resembles sensor protein members of the bacterial two-component signal-transduction systems. *Genes Dev* **4**: 860–872.

West, S.A., Griffin, A.S., Gardner, A., and Diggle, S.P. (2007) Social evolution theory for microorganisms. *Nat Rev Microbiol* **4**: 597–608.

Whatmore, A.M., Barcus, V.A., and Dowson, C.G. (1999) Genetic diversity of the streptococcal competence (com) gene locus. *J Bacteriol* **181**: 3144–3154.

Wilder, C.N., Diggle, S.P., and Schuster, M. (2011) Cooperation and cheating in *Pseudomonas aeruginosa*: the roles of the las, rhl and pqs quorum-sensing systems. *ISME J* **5**: 1332–1343.

Xie, X., Zhang, H., and Pare, P.W. (2009) Sustained growth promotion in *Arabidopsis* with long-term exposure to the beneficial soil bacterium *Bacillus subtilis* (GB03). *Plant Signal Behav* **4**: 948–953.

Zeriouh, H., de Vicente, A., Perez-Garcia, A., and Romero, D. (2013) Surfactin triggers biofilm formation of *Bacillus subtilis* in melon phylloplane and contributes to the biocontrol activity. *Environ Microbiol* **16**: 2196–2211.

## Supporting information

Additional Supporting Information may be found in the online version of this article at the publisher's web-site:

**Fig. S1.** Images of pellicle biofilms were taken after 48 h of pellicles growth in MSN medium (see *Experimental procedures*) at 28°C. Strain BD2833 which is a derivative of laboratory strain IS75 was used as negative control, and strain GB03 which is known biopesticide was used as positive control.

**Table S1.** The MIDI Sherlock microbial identification system was used to identify *Bacillus* spp. isolates based on the fatty acid methyl ester profile of the bacteria. The program 'Sherlock microbial identification system' was used to compare the fatty acid methyl ester profiles of the bacteria and strains were identified based on similarity index. Last column indicates the identification based on gyrA sequence identity.

**Table S2.** Data on other *Bacillus* spp. rhizoplane isolates: effects on roots and leaves biomass and IAA production – determined quantitatively; and qualitative estimation of siderophores production and phosphate solubilization. Symbols: ++, strong effect, +, moderate effect (significantly lower as compared to maximal effect observed); +/–, weak but reproducible effect, –, no effect. RGB (Red, Green, Blue) intensity was used to better show the diversity.

# 10

# Integrated omics for the identification of key functionalities in biological wastewater treatment microbial communities

Shaman Narayanasamy, Emilie E. L. Muller,
Abdul R. Sheik and Paul Wilmes*

*Luxembourg Centre for Systems Biomedicine, University of Luxembourg, 7 avenue des Hauts-Fourneaux, Esch-Sur-Alzette L-4362, Luxembourg.*

## Summary

Biological wastewater treatment plants harbour diverse and complex microbial communities which prominently serve as models for microbial ecology and mixed culture biotechnological processes. Integrated omic analyses (combined metagenomics, metatranscriptomics, metaproteomics and metabolomics) are currently gaining momentum towards providing enhanced understanding of community structure, function and dynamics *in situ* as well as offering the potential to discover novel biological functionalities within the framework of Eco-Systems Biology. The integration of information from genome to metabolome allows the establishment of associations between genetic potential and final phenotype, a feature not realizable by only considering single 'omes'. Therefore, in our opinion, integrated omics will become the future standard for large-scale characterization of microbial consortia including those underpinning biological wastewater treatment processes. Systematically obtained time and space-resolved omic datasets will allow deconvolution of structure–function relationships by identifying key members and functions. Such knowledge will form the foundation for discovering novel genes on a

*For correspondence. E-mail paul.wilmes@uni.lu

**Funding Information** This work was supported by an ATTRACT programme grant (A09/03) and a European Union Joint Programming in Neurodegenerative Diseases grant (INTER/JPND/12/01) to PW and Aide à la Formation Recherche (AFR) grants to SN (PHD-2014-1/7934898), EELM (PDR-2011-1/SR) and ARS (PDR-2013-1/5748561) all funded by the Luxembourg National Research Fund (FNR).

much larger scale compared with previous efforts. In general, these insights will allow us to optimize microbial biotechnological processes either through better control of mixed culture processes or by use of more efficient enzymes in bioengineering applications.

## Biological wastewater treatment as a model system for Eco-Systems Biology

Biological wastewater treatment (BWWT), including the standard activated sludge process and other ancillary processes, relies on microbial community-driven remediation of municipal and industrial wastewater. Biological wastewater treatment plants host diverse and dynamic microbial communities possessing varied metabolic capabilities over changing environmental conditions, e.g. microorganisms accumulating various storage compounds of biotechnological importance. Given their structural and functional diversity, BWWT processes hold great potential for future sustainable production of various commodities from wastewater as well as from other mixed substrates (Muller *et al.*, 2014; Sheik *et al.*, 2014). Eco-Systems Biology is an integrative framework that includes systematic measurements, data integration, analysis, modelling, prediction, experimental validation (e.g. through targeted perturbations) and ultimately control of microbial ecosystems (Muller *et al.*, 2013). This framework will aid in the understanding of BWWT processes by dissecting interactions among its constituent populations, their genes and the biotope, with the ultimate aim of maximizing biotechnological outcomes through various control strategies (Muller, Pinel *et al.*, 2014; Sheik *et al.*, 2014).

Biological wastewater treatment plants typically possess a relatively homogeneous environment (compared with most natural ecosystems) with well-defined physico-chemical boundaries and are widespread in developed and developing countries (Daims *et al.*, 2006; Muller, Pinel *et al.*, 2014; Sheik *et al.*, 2014). Furthermore, contrary to other microbial habitats, e.g. the marine environment, acid mine drainage biofilms, the human gastrointestinal tract, etc., BWWT plants represent a

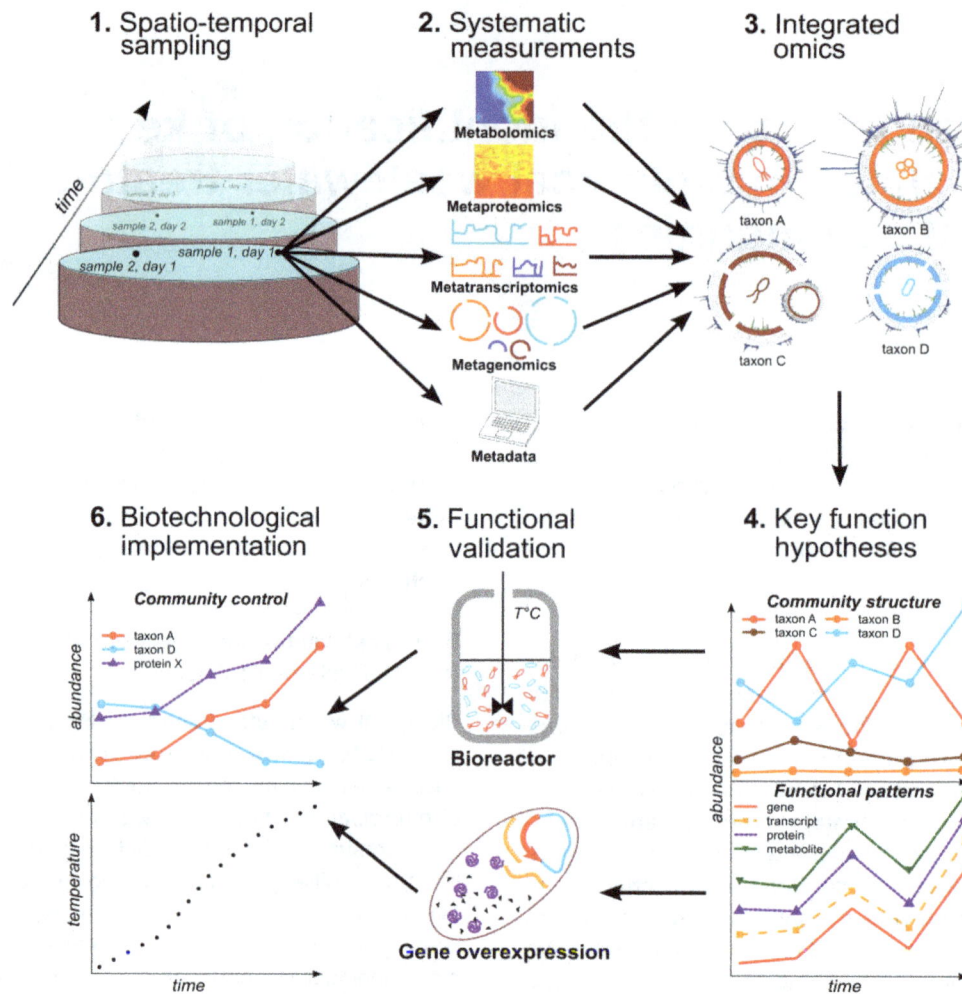

**Fig. 1.** The path from large-scale integrated omics to hypothesis testing and biotechnological application in the context of biological wastewater treatment.

convenient and virtually unlimited source of spatially and temporally resolved samples (Fig. 1; step 1). Physico-chemical parameters such as temperature, pH, oxygen and nutrient concentrations are routinely monitored and recorded, thereby facilitating hypothesis formulation and verification in rapid succession. This allows for example, the establishment of causal links between the influence of certain environmental parameters on microbial community structure and/or function derived from temporal sampling. Importantly, microbial consortia from BWWT plants are very amenable to experimental validation at differing scales, ranging from laboratory-scale bioreactors to full-scale plants (see section "From Eco-Systems Biology to biotechnology" below).

While being highly dynamic, microbial communities within BWWT plants maintain a medium to high range of diversity/complexity, thereby exhibiting a baseline stability over time such that there is temporal succession of repeatedly few quantitatively dominant populations

(Albertsen *et al.*, 2012; Zhang *et al.*, 2012; Muller, Pinel *et al.*, 2014; N. Pinel, pers. comm.). These characteristics reduce the complexity of downstream omic data analyses. In particular, given sufficient sequencing depth, current *de novo* metagenomic assemblers are highly effective for medium complexity communities, such as BWWT plant microbial communities (Segata *et al.*, 2013; Muller, Pinel *et al.*, 2014). Representative population-level genomic reconstructions can now be obtained for abundant community members (Albertsen *et al.*, 2013; Muller, Pinel *et al.*, 2014), and such genomic information is vital for the meaningful interpretation of additional functional omic data. Overall, BWWT plant microbial communities represent an important intermediary step/model between communities of lower diversity, e.g. acid mine drainage biofilms (Denef *et al.*, 2010), and complex communities such as those from soil environments (Mocali and Benedetti, 2010), while retaining important hallmarks of both extremes including, for example, quantitative dominance of specific taxa (a characteristic of acid mine drainage biofilm communities), rapid stochastic environmental fluctuations (a characteristic of soil environments). Therefore, BWWT plant microbial communities exhibit important properties rendering them an ideal model for microbial ecology (Daims *et al.*, 2006), and more specifically eco-systematic omic studies in line with a discovery-driven planning approach (Muller *et al.*, 2013).

## Laboratory protocols, systematic measurements and *in silico* analyses

Mixed microbial communities, such as those present in BWWT plants, exhibit varying degrees of inter- and intra-sample heterogeneity, rendering standard (i.e. originally designed for pure isolate culture systems) biomolecular extractions protocols and computational analyses ineffective (Muller *et al.*, 2013; Roume *et al.*, 2013a). In our opinion, it is therefore absolutely essential to apply tailored and systematic approaches such as the biomolecular isolation protocol designed by Roume and colleagues (Roume *et al.*, 2013a) to microbial communities. The protocol allows the sequential isolation of high-quality genomic deoxyribonucleic acid (DNA), ribonucleic acid (RNA), small RNA, proteins and metabolites from a single, undivided sample for subsequent systematic multi-omic measurements (Fig. 1, step 2). Importantly, this eliminates the need for subsampling the heterogeneous biomass and, therefore, reduces the noise arising from incongruous omics data in the subsequent downstream integration and analysis steps (Fig. 1, step 3; Muller *et al.*, 2013; Roume *et al.*, 2013a,b).

Following standardized and systematized biomolecular isolations, multi-omic datasets are generated in addition to the physico-chemical parameters recorded at the time of sampling (Fig. 1; step 2). The multi-omic data are then subjected to bioinformatic pre-processing and analyses. Preliminary characterization of microbial communities can be facilitated either by high-throughput ribosomal RNA gene amplicon sequencing to determine broad community composition from shotgun metagenomic analyses to resolve the overall structure as well as the functional potential of the communities (Vanwonterghem *et al.*, 2014). More importantly, hybrid *de novo* assemblies of metagenomic and metatranscriptomic reads promises higher quality compared with conventional *de novo* metagenomic assemblies due to the ability to reconstruct and resolve genomic complements of low abundance (i.e. low metagenomic coverage) yet highly active populations (i.e. high metatranscriptomic coverage for expressed genes; Muller, Pinel *et al.*, 2014). Hybrid assemblies allow high-quality population-level genomic reconstructions after the application of binning/classification methods, such as those developed for a single sample (Laczny

*et al.*, 2014) or for spatio-temporally resolved samples (Albertsen *et al.*, 2013; Alneberg *et al.*, 2014; Nielsen *et al.*, 2014). Furthermore, hybrid metagenomic and metatranscriptomic data assemblies allow the resolution of genetic variations with higher confidence through replication and highlights their potential relative importance, thereby allowing more detailed short-term evolutionary inferences regarding specific populations and while increasing sensitivity for downstream metaproteomic analysis (Muller, Pinel *et al.*, 2014). Thus, the generation of metatranscriptomic and metaproteomic data is crucial to fully understand the functional capacity of microbial communities. Therefore, we believe that the integrated omic approach as elucidated by Muller and colleagues (Muller, Pinel *et al.*, 2014), from systematic measurements to *in silico* analysis, is highly effective in: (i) minimizing errors by cancelling out noise and biases stemming from single omic analyses and (ii) optimizing/maximizing overall data usage.

Although high-throughput metagenomics and metatranscriptomics allow deep profiling of microbial communities at relatively low cost, existing sequence-based approaches do have some important limitations. Given the availability of omic technologies and their non-prohibitive costs (in particular for metagenomics and metatranscriptomics), fully integrated omic analyses should be applied routinely in the study of microbial consortia for greater effectiveness. For instance, despite this wealth of information, current metagenomic assemblies and analysis schemes, metagenomic (and metatranscriptomic) data resulting from the use of current short-read sequencing and assembly approaches do not allow the comprehensive resolution of microdiversity, e.g. genetic heterogeneity of microbial populations (Wilmes *et al.*, 2009). Furthermore, RNAseq technologies are subject to biases stemming from the extensive, yet compulsory pre-processing steps (Lahens *et al.*, 2014), thereby affecting the resulting metatranscriptomic data. On the other hand, chromatography and mass spectrometry-based metaproteomics and metabolomics currently remain limited in their profiling depth. While the situation for metaproteomics is rapidly improving (Hettich *et al.*, 2012), community-wide metabolomic studies are still limited in their scope due to the poor detection/sensitivity of high-throughput metabolomic instruments and high dependency on a limited knowledgebase reflected in current metabolite databases. Overall, we anticipate significant technological advancements in all high-throughput measurement techniques particularly in the area of long-read sequencing, chromatography as well as mass spectrometry. Naturally, these technological improvements will be complemented by equally sophisticated *in silico* data processing and analysis methods, which in turn will allow integrated omics to provide comprehensive multi-level snapshots of microbial population structures and functions *in situ* (Fig. 1; step 3).

In our opinion, the real power of the integrated omics approach within the Eco-Systems Biology framework will stem from applying the approach to temporally and spatially resolved samples (Fig. 1, steps 1 to 4; Muller *et al.*, 2013; Zarraonaindia *et al.*, 2013). In combination with appropriate statistical and mathematical modelling methods, the deconvolution of the data will unveil unprecedented insights into the structure and function of microbial communities (Fig. 1; step 4; Muller *et al.*, 2013; Segata *et al.*, 2013; Zarraonaindia *et al.*, 2013). Data mining, machine learning and/or modelling approaches will be useful for extracting features of interest, e.g. known and unknown populations/genes, and also to derive associations (or links) between desired features utilizing measures such as correlation, co-occurrence, mutual information and hyper-geometric overlap (Muller *et al.*, 2013; Segata *et al.*, 2013). Such associations may allow the prediction of gene functions using the concept of 'guilt by association' and interactions/dependencies between community members (Wolfe *et al.*, 2005; Segata *et al.*, 2013; Solomon *et al.*, 2014). Biological wastewater treatment plants offer particularly exciting opportunities to link responses in community structure and function to fluctuating environmental conditions because of the relative ease of sampling and routine recording of metadata (Muller *et al.*, 2013; Segata *et al.*, 2013; Vanwonterghem *et al.*, 2014). Systematic omic analyses of BWWT microbial communities may therefore uncover (i) the effect of physico-chemical parameters on the expression of specific genes or phenotypes and (ii) the linkage of unknown genes to specific metabolites as well as to both known and unknown community members. However, the derived associations will always be 'mere' hypotheses, which will require rigorous testing through targeted laboratory experiments (Fig. 1; step 5) and/or *in situ* perturbation experiments followed by additional omic measurements (Muller *et al.*, 2013; Segata *et al.*, 2013).

### Moving beyond associations and hypotheses

Although integrated omics-based approaches are highly effective for large-scale analysis and formulation of hypotheses (including within the context of BWWT plant communities), these efforts are limited due to current high-throughput measurement methods (see previous section) and the reliance on *a priori* knowledge for both taxonomical and functional inferences (Röling *et al.*, 2010). Hence, there is a need to validate newly generated hypotheses using full-scale plants, customized laboratory-based experiments, such as batch cultures, bioreactors or pilot plants (Fig. 1; step 5) and/or single-cell methods. Hypotheses may be tested using additional inte-

grated omic datasets generated from ancillary samples (e.g. Muller, Pinel *et al*, 2014) by using molecular biology techniques such as heterologous gene expression (e.g. Wexler *et al.*, 2005; Maixner *et al.*, 2008) or single-cell approaches using microautoradiography-fluorescent in situ hybridisation (MAR-FISH), nano-scale secondary-ion mass spectrometry (nanoSIMS) and/or Raman spectroscopy (e.g. Huang *et al.*, 2007; Lechene *et al.*, 2007; Musat *et al.*, 2012). Such a combination of technologies can be used to test hypotheses regarding (i) community dynamics, (ii) gene expression patterns/interactions, (iii) metabolite abundances, (iv) effect of physico-chemical factors on distinct microbial species and functionalities, (v) gene function associations between any of these. Identified patterns may be subsequently formulated as cues and can be used as input to facilitate knowledge-driven control of different microbial community structures and/or functions (Fig. 1; step 6). Thus, large-scale integrated omic analyses of *in situ* biological samples (section "Laboratory protocols, systematic measurements and *in silico* analyses"), coupled to carefully controlled laboratory experiments, will allow the effective elucidation of novel functions within BWWT plant microbial communities with potential biotechnological applications.

### From Eco-Systems Biology to biotechnology

Knowledge of gene function, regulation and physiological potential derived from integrated omic data over different spatial and temporal scales holds great promise in harnessing the biotechnological potential of microbial consortia. In particular, advancements in integrated omics followed by hypothesis testing may generate new knowledge (Muller *et al.*, 2013), which may for example be exploited in new approaches for the optimized production of biotechnologically relevant compounds under varying environmental conditions (Chen and Nielsen, 2013). The derived knowledge-base may further be used to fine-tune metabolic pathways at the transcriptional, translational and post-translational levels using the ever-expanding synthetic biology toolbox (Peralta-Yahya *et al.*, 2012). Examples of possible future applications may include, for instance the bioengineering of fatty acid utilization and production for the production of biodiesel from 'dirty' mixed substrates, the engineering of different gene combinations for the production of various alcohols from mixed substrates (Lee *et al.*, 2008) and the generation of hybrid processes by combining biological and chemical production steps resulting in new compounds that could serve as biofuels (Román-Leshkov *et al.*, 2007). Through exploration of BWWT plant microbial consortia using integrated omics, we are therefore poised to unravel key functionalities, which will find applications in a whole range of different biotechnologies. In this context, integrated omics through facilitating direct linkages between genetic potential and final phenotype may become an essential tool in future bioprospecting. Therefore, in our opinion, integrated omics will become the standard means of analysing microbial consortia in the near future and will allow meta-omics to fulfil their promise for the comprehensive discovery of biotechnology-relevant microbial traits in natural consortia.

## Conflict of interest

None declared.

## References

Albertsen, M., Hansen, L.B.S., Saunders, A.M., Nielsen, P.H., and Nielsen, K.L. (2012) A metagenome of a full-scale microbial community carrying out enhanced biological phosphorus removal. *ISME J* **6**: 1094–1106.

Albertsen, M., Hugenholtz, P., Skarshewski, A., Nielsen, K.L., Tyson, G.W., and Nielsen, P.H. (2013) Genome sequences of rare, uncultured bacteria obtained by differential coverage binning of multiple metagenomes. *Nat Biotechnol* **31**: 533–538.

Alneberg, J., Bjarnason, B.S., de Bruijn, I., Schirmer, M., Quick, J., Ijaz, U.Z., *et al.* (2014) Binning metagenomic contigs by coverage and composition. *Nat Methods* **11**: 1–7.

Chen, Y., and Nielsen, J. (2013) Advances in metabolic pathway and strain engineering paving the way for sustainable production of chemical building blocks. *Curr Opin Biotechnol* **24**: 965–972.

Daims, H., Taylor, M.W., and Wagner, M. (2006) Wastewater treatment: a model system for microbial ecology. *Trends Biotechnol* **24**: 483–489.

Denef, V.J., Mueller, R.S., and Banfield, J.F. (2010) AMD biofilms: using model communities to study microbial evolution and ecological complexity in nature. *ISME J* **4**: 599–610.

Hettich, R.L., Sharma, R., Chourey, K., and Giannone, R.J. (2012) Microbial metaproteomics: identifying the repertoire of proteins that microorganisms use to compete and cooperate in complex environmental communities. *Curr Opin Microbiol* **15**: 373–380.

Huang, W.E., Stoecker, K., Griffiths, R., Newbold, L., Daims, H., Whiteley, A.S., and Wagner, M. (2007) Raman-FISH: combining stable-isotope Raman spectroscopy and fluorescence in situ hybridization for the single cell analysis of identity and function. *Environ Microbiol* **9**: 1878–1889.

Laczny, C.C., Pinel, N., Vlassis, N., and Wilmes, P. (2014) Alignment-free visualization of metagenomic data by non-linear dimension reduction. *Sci Rep* **4**: 4516; 1–12.

Lahens, N.F., Kavakli, I.H., Zhang, R., Hayer, K., Black, M.B., Dueck, H., *et al.* (2014) IVT-seq reveals extreme bias in RNA-sequencing. *Genome Biol* **15**: R86.

Lechene, C.P., Luyten, Y., McMahon, G., and Distel, D.L. (2007) Quantitative imaging of nitrogen fixation by individual bacteria within animal cells. *Science* **317**: 1563–1566.

Lee, S.K., Chou, H., Ham, T.S., Lee, T.S., and Keasling, J.D. (2008) Metabolic engineering of microorganisms for biofuels production: from bugs to synthetic biology to fuels. *Curr Opin Biotechnol* **19**: 556–563.

Maixner, F., Wagner, M., Lücker, S., Pelletier, E., Schmitz-esser, S., Hace, K., *et al.* (2008) Environmental genomics reveals a functional chlorite dismutase in the nitrite-oxidizing bacterium 'Candidatus Nitrospira defluvii'. *Environ Microbiol* **10**: 3043–3056.

Mocali, S., and Benedetti, A. (2010) Exploring research frontiers in microbiology: the challenge of metagenomics in soil microbiology. *Res Microbiol* **161**: 497–505.

Muller, E.E.L., Glaab, E., May, P., Vlassis, N., and Wilmes, P. (2013) Condensing the omics fog of microbial communities. *Trends Microbiol* **21**: 325–333.

Muller, E.E.L., Sheik, A.R., and Wilmes, P. (2014) Lipid-based biofuel production from wastewater. *Curr Opin Biotechnol* **30C**: 9–16.

Muller, E.E.L., Pinel, N., Laczny, C.C., Hoopmann, M.R., Narayanasamy, S., Lebrun, L.A., *et al.* (2014) Community integrated omics links the dominance of a microbial generalist to fine-tuned resource usage. *Nat Commun.* **5**: 5603; 1–10.

Musat, N., Foster, R., Vagner, T., Adam, B., and Kuypers, M.M.M. (2012) Detecting metabolic activities in single cells, with emphasis on nanoSIMS. *FEMS Microbiol Rev* **36**: 486–511.

Nielsen, H.B., Almeida, M., Juncker, A.S., Rasmussen, S., Li, J., Sunagawa, S., *et al.* (2014) Identification and assembly of genomes and genetic elements in complex metagenomic samples without using reference genomes. *Nat Biotechnol* **32**: 1–11.

Peralta-Yahya, P.P., Zhang, F., del Cardayre, S.B., and Keasling, J.D. (2012) Microbial engineering for the production of advanced biofuels. *Nature* **488**: 320–328.

Román-Leshkov, Y., Barrett, C.J., Liu, Z.Y., and Dumesic, J.A. (2007) Production of dimethylfuran for liquid fuels from biomass-derived carbohydrates. *Nature* **447**: 982–985.

Röling, W.F.M., Ferrer, M., and Golyshin, P.N. (2010) Systems approaches to microbial communities and their functioning. *Curr Opin Biotechnol* **21**: 532–538.

Roume, H., Heintz-Buschart, A., Muller, E.E.L., and Wilmes, P. (2013a) Sequential isolation of metabolites, RNA, DNA, and proteins from the same unique sample. *Methods Enzymol* **531**: 219–236.

Roume, H., Muller, E.E.L., Cordes, T., Renaut, J., Hiller, K., and Wilmes, P. (2013b) A biomolecular isolation framework for eco-systems biology. *ISME J* **7**: 110–121.

Segata, N., Boernigen, D., Tickle, T.L., Morgan, X.C., Garrett, W.S., and Huttenhower, C. (2013) Computational meta'omics for microbial community studies. *Mol Syst Biol* **9**: 666; 1–15.

Sheik, A.R., Muller, E.E.L., and Wilmes, P. (2014) A hundred years of activated sludge: time for a rethink. *Front Microbiol* **5**: 47; 1–7.

Solomon, K.V., Haitjema, C.H., Thompson, D.A., and O'Malley, M.A. (2014) Extracting data from the muck: deriving biological insight from complex microbial communities and non-model organisms with next generation sequencing. *Curr Opin Biotechnol* **28C**: 103–110.

Vanwonterghem, I., Jensen, P.D., Ho, D.P., Batstone, D.J., and Tyson, G.W. (2014) Linking microbial community structure, interactions and function in anaerobic digesters using new molecular techniques. *Curr Opin Biotechnol* **27:** 55–64.

Wexler, M., Bond, P.L., Richardson, D.J., and Johnston, A.W.B. (2005) A wide host-range metagenomic library from a waste water treatment plant yields a novel alcohol/aldehyde dehydrogenase. *Environ Microbiol* **7:** 1917–1926.

Wilmes, P., Simmons, S.L., Denef, V.J., and Banfield, J.F. (2009) The dynamic genetic repertoire of microbial communities. *FEMS Microbiol Rev* **33:** 109–132.

Wolfe, C.J., Kohane, I.S., and Butte, A.J. (2005) Systematic survey reveals general applicability of 'guilt-by-association' within gene coexpression networks. *BMC Bioinformatics* **6:** 227; 1–10.

Zarraonaindia, I., Smith, D.P., and Gilbert, J.A. (2013) Beyond the genome: community-level analysis of the microbial world. *Biol Philos* **28:** 261–282.

Zhang, T., Shao, M.-F., and Ye, L. (2012) 454 pyrosequencing reveals bacterial diversity of activated sludge from 14 sewage treatment plants. *ISME J* **6:** 1137–1147.

# 11

# Whole genome and transcriptome analyses of environmental antibiotic sensitive and multi-resistant *Pseudomonas aeruginosa* isolates exposed to waste water and tap water

Thomas Schwartz,[1]* Olivier Armant,[2] Nancy Bretschneider,[3] Alexander Hahn,[3] Silke Kirchen,[1] Martin Seifert[3] and Andreas Dötsch[1]

[1]*Institute of Functional Interfaces (IFG)* and
[2]*Institute of Toxicology and Genetics (ITG), Campus North, Karlsruhe Institute of Technology (KIT), Hermann von Helmholtz Platz 1, Eggenstein-Leopoldshafen D-76344, Germany.*
[3]*Genomatix GmbH, Bayerstr. 85a, Munich D-80335, Germany.*

## Summary

The fitness of sensitive and resistant *Pseudomonas aeruginosa* in different aquatic environments depends on genetic capacities and transcriptional regulation. Therefore, an antibiotic-sensitive isolate PA30 and a multi-resistant isolate PA49 originating from waste waters were compared via whole genome and transcriptome Illumina sequencing after exposure to municipal waste water and tap water. A number of different genomic islands (e.g. PAGIs, PAPIs) were identified in the two environmental isolates beside the highly conserved core genome. Exposure to tap water and waste water exhibited similar transcriptional impacts on several gene clusters (antibiotic and metal resistance, genetic mobile elements, efflux pumps) in both environmental *P. aeruginosa* isolates. The MexCD-OprJ efflux pump was overexpressed in PA49 in response to waste water. The expression of resistance genes, genetic mobile elements in PA49 was independent from the water matrix. Consistently, the antibiotic sensitive strain PA30 did not show any difference in expression of the intrinsic resistance determinants and genetic mobile elements. Thus, the exposure of both isolates to polluted waste water and oligotrophic tap water resulted in similar expression profiles of mentioned genes. However, changes in environmental milieus resulted in rather unspecific transcriptional responses than selected and stimuli-specific gene regulation.

*For correspondence. E-mail thomas.schwartz@kit.edu

**Funding Information** We gratefully acknowledge the financial support by the BioInterfaces (BIF) program of the Karlsruhe Institute of Technology (KIT) in the Helmholtz Association.

## Introduction

The increasing numbers of infections by multi-resistant bacteria turn out to be a great threat to our daily life. Bacteria develop resistance against antibiotics used in human health care, agriculture and animal husbandry or against pollutants from industry by accumulating genetic adaptations or acquisition of mobile genetic elements via horizontal gene transfer. To overcome multi-drug resistance, we have to undergo a thorough study to unravel how bacteria adapt to different habitats, to finally discover novel strategies to handle such infections.

One of the most prominent bacterial pathogens that is infamous for its high potential to develop multi-drug resistance is *Pseudomonas aeruginosa*, a Gram-negative, ubiquitous opportunistic bacterium that can cause acute and chronic infections especially in patients in intensive care or suffering from predisposing conditions like cystic fibrosis. The rate of infections in human body differs according to the site of infection as 2% on skins, 3.3% on nasal mucosa, 6.6% for the throat, 24% for fecal samples (Morrison and Wenzel, 1984). *Pseudomonas aeruginosa* is found in hospital waste water, respiratory equipment, solutions, medicines, disinfectants, sinks, mops, food mixtures and vegetables (Trautmann *et al.*, 2005). An important characteristic of *P. aeruginosa* is its ability to form biofilms as an adaptation to adverse environmental conditions. The microbes attach to the surface and embed themselves in extracellular polymeric substances such as proteins (e.g. extracellular enzymes), lipids and nucleic acids (Flemming and Wingender, 2001), usually leading to increased resistance towards harsh conditions such as temperature changes, pH fluctuations, presence of antibiotics (Kwon and Lu, 2006) and immune cells of humans (Donlan and Costerton, 2002).

Sequencing of several *P. aeruginosa* strains genomes revealed that a large fraction (around 10%) of the genome is dedicated to gene regulation, which is consistent with its high versatility (Stover *et al.*, 2000; Mathee *et al.*, 2008). This high versatility enables evolutionary adaptations and facilitates the bacterium to colonize vigorous and diverse ecological niches. The core genome is usually highly conserved between different *Pseudomonas* strains (Mathee *et al.*, 2008; Klockgether *et al.*, 2011).

It has a disparate variety of metabolism; it can degrade very distinct compounds such as alcohols, fatty acids, sugars, di- and tri-carboxylic acids, aromatics, amines and amino acids, which can be used up as sources of carbon. *Pseudomonas aeruginosa* has both aerobic and anaerobic metabolism. It is capable of anaerobic metabolism by converting nitrate to nitrite (Schreiber *et al.*, 2007). Additionally, the genome harbours a huge repertoire of enzymes and efflux pumps that contribute to a high intrinsic resistance towards different classes of antibiotics. Additional resistance can easily develop by mutation or horizontal gene transfer, rendering *P. aeruginosa* a common cause of multi-drug resistant infections (Breidenstein *et al.*, 2011). Regular use of high amounts of antibiotics in hospitals and other practices were assumed to be the sources of origin of antibiotics in the waste water systems and responsible for supporting emergence of multi-resistant bacteria (Rizzo *et al.*, 2013). These resistances may not only develop from chromosomally encoded genes but also from mobile genetic elements like plasmids or integrons (Merlin *et al.*, 2011). Not only waste water systems contribute to the development of resistance in bacteria, but also pollutants from industries and agricultural activities where the antibiotics and pollutants are directly released into the environmental water like rivers and lakes, creating selective pressure on these bacteria and making them evolve as resistance strains. The sensitive strains accept the resistant genes from these resistant donors and propagate as resistant strains. The concentrations of antibiotics in waste water might not be high enough to stimulate inhibitory effects but stimulate stress response mechanisms, which contribute to horizontal gene transfer and relevant transcriptional activities. It has also been proven by mutant investigations that sub-inhibitory concentrations of antibiotics can drive the evolution of antimicrobial resistance (Pedró *et al.*, 2011). It all depends on the substance, concentration and strain present in the waste water systems. There is no final suggestion about long terms effects of sub-inhibitory concentration antibiotics and other micro-pollutants. It is commonly accepted that beside the linkage between antibiotics and antibiotic resistance, co-selection and the presence of heavy metal ions in the environments contributes to increasing resistance mechanisms due to the localization of resistance genes in close neighborhood on genetic mobile elements (Seiler and Berendonk, 2012).

Beside antibiotic and heavy metal stress, starvation is another widespread adverse stimulus present in many aquatic environments where *P. aeruginosa* is found in nature (Bernier *et al.*, 2013). Tap water represents an oligotrophic matrix with very low organic matter, and *P. aeruginosa* has recently been shown to persist and proliferate as biofilms in municipal drinking water distribution systems (Wang *et al.*, 2012). The molecular responses of *P. aeruginosa* strains to starvation stress in tap water are so far unknown. In this study, we compared the transcriptional response of an antibiotic sensitive and a multi-resistant *P. aeruginosa* waste water isolate cultivated in municipal waste water and tap water focusing on regulatory mechanisms that could promote the development of antibiotic resistance.

## Results and discussion

Bacteria have developed highly orchestrated processes to respond to environmental stresses, which when elicited alter the cellular physiology in a manner that enhances the organism's survival and its ability to cause disease. This study focused on the behaviour of two natural isolates of *P. aeruginosa* as a Gram-negative bacterium exposed to municipal waste water containing complex mixtures of xenobiotics and, as a second scenario, exposed to tap water simulating nutrient limitation (starvation). Since bacteria have to deal with unfavourable growth conditions in addition to diverse stresses in nature, bacteria that reached the stationary growth phase were used to imitate this environment and then exposed to stress. During transition from exponential growth to stationary phase, growth becomes unbalanced especially in laboratory systems, i.e. the synthesis of different macromolecules and cell constituents do not slow down synchronically (Nyström, 2004). Thus, stationary phase is an operational definition and does not describe a specific and fixed physiological state or response of the bacteria. It is more or less a change in physiology due to, e.g. phosphate limitation or accumulation of toxic waste products. Beside the changes in morphologies of bacteria, the gene expression pattern could be altered in stationary phase. In consequence, transcriptome analyses were run with ribonucleic acid (RNA) extracted from early stationary growth phase. In the present study, two different *P. aeruginosa* isolates were exposed to water matrices containing quite different compositions. Waste water from the influent of a municipal waste water treatment plants (WWTP) is composed of complex mixtures of xenobiotics like antibiotics, other pharmaceuticals, biocide etc., whereas tap water, in opposite, contains very low level of organic matter (including xenobiotics) as a result of the intensive drinking water conditioning processes at water-works. We analysed the transcriptional responses from two *P. aeruginosa* isolates: the antibiotic sensitive strain PA30 and the multi-resistant strain PA49. Both *P. aeruginosa* strains did neither show any differences in growth in diluted brain heart infusion (BHI) or BM2 broth nor in yields of extracted total RNA after exposure in tap water or waste water.

### Genome analyses

Large fractions of the *P. aeruginosa* genome belong to the highly conserved core genome containing only few highly variable genes (Dötsch *et al.*, 2010), while most of the genetic variation between species is restricted to the so-called *accessory genome* organized in various *regions of genomic plasticity* (RGPs) (Mathee *et al.*, 2008). Most of these RGPs represent mobile elements originating from horizontal gene transfer and include transposons, phages, plasmids and genomic islands, which are a major source of resistance genes (Battle *et al.*, 2009; Kung *et al.*, 2010; Klockgether *et al.*, 2011). The large amount of homology between the core regions of different *P. aeruginosa* strains enabled us to employ the genomic sequences of strain PAO1 chromosome and a selection of genomic islands as a blueprint for *de novo* assembly. The resulting draft genomes consist of 207 contigs with a total length of 6.77 Mb for the strain PA30 and 269 contigs with 7.01 Mb for strain PA49 respectively (Table S1).

An alignment of the contigs with *P. aeruginosa* reference strain PAO1 showed a huge overlap of 95.8% for PA30 and 96.4% for PA49 (Fig. 1; Table S2), reflecting the highly conserved character of the *P. aeruginosa* core genome. Comparing the contigs with the genome islands that were used in the alignment process revealed a distinct pattern of accessory genomic elements for the two strains covering large fractions of the various genomic islands (Fig. 1; Table S2). Strain PA30 contains full length or near-full length sequences of PAGI-5 to PAGI-11, larger fractions of PAGI-1 and PAGI-2 and several regions of PAGI-3, whereas only insignificant fractions of the remaining genomic elements occurred. In case of the multi-resistant strain PA49, all genomic islands except the smaller PAGI-9 to PAGI-11 were covered at varying percentages (Table S2). The scattered distribution of regions within the genomic islands that actually showed homology with PA49 contigs may be partially explained by incomplete sequence assembly. However, the fact that both the contigs and the genomic island reference sequences contained a large amount of non-overlapping regions (data not shown) suggests that at least in some cases, the accessory elements found in PA30 and PA49 only partially contain sequences that are homologous to the genomic islands and also include a substantial amount of new and previously uncharacterized sequences.

Since the genomes of PA30 and PA49 were nearly completely covered, the sequence types according to the multi-locus sequence typing (MLST) scheme by Curran and colleagues (2004) could be determined, enabling a phylogenetic classification of the two strains. As demonstrated by the phylogenetic tree (Fig. 2), PA30 and PA49 are members of the lineage that includes the type strain PAO1 and some recently sequenced strains. The question about their origin is open, since the sampling sites

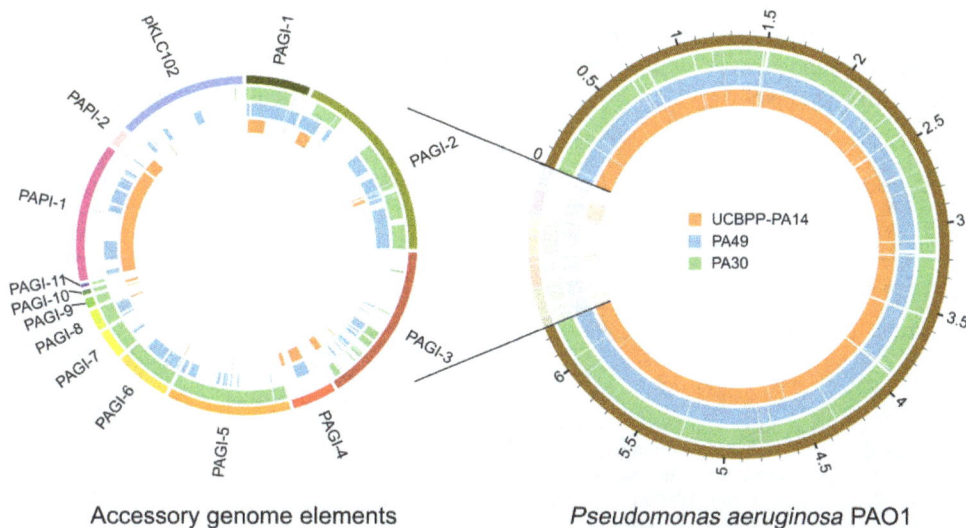

Accessory genome elements       *Pseudomonas aeruginosa* PAO1

**Fig. 1.** Coverage of genomic reference sequences by the newly assembled genomes. The reference sequences that were used for the hybrid *de novo* assembly are displayed on the outer circles (diagram to the right) with an additional display of the accessory elements alone (missing the PAO1 chromosome, diagram to the left). Regions that are covered by the contigs of strains PA30 and PA49 or overlap with the chromosome of another reference strain PA14 are highlighted by coloured areas in the concentric inner circles as specified by the color legend.

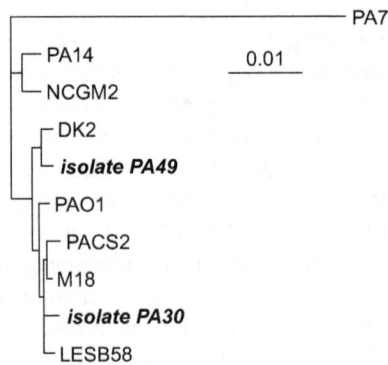

**Fig. 2.** Multi-locus sequence typing phylogenetic tree. Phylogenetic relations of the two newly sequenced strains PA30 and PA49 (bold) with eight previously published genomes of *P. aeruginosa*. The phylogenetic tree is based on seven genes that are commonly used for MLST scheme by Curran and colleagues (2004).

were influenced by hospital and housing waste waters. Selective pressures like the presence of antibiotics and other environmental criteria are a general concept that refers to many factors that create an evolutionary landscape and allow organisms with novel mutations or newly acquired characteristics to survive and proliferate (Kümmerer, 2009). There is evidence that even in subinhibitory concentration, antibiotics or other xenobiotics may still exert their impact on microbial communities (Goh *et al.*, 2002; Davies *et al.*, 2006). The direct link between antibiotics or heavy metal ions and development/selection of resistance mechanisms is obvious and manifold described (Seiler and Berendonk, 2012; Rizzo *et al.*, 2013). The impact of other harsh environmental conditions on resistance activities, recombination and horizontal gene transfer remains to be determined. Long-term effects of environmental exposure to low levels of antibiotics like these present in surface waters or in the outflow of sewage plants are also still unknown.

The prediction of protein coding sequences (CDS) yielded for both strains a comparatively large number of genes, about 99% of which were successfully annotated according to their best-hit BLAST alignments (Table S1). The vast majority of the predicted genes were found in both strains and also in the PAO1 reference genome (5262 genes), representing the conserved core genome of *P. aeruginosa*. Regarding the development of multi-drug resistance, *P. aeruginosa* is known for its high intrinsic resistance that is caused by a combination of low membrane permeability, efflux pumps and resistance genes encoded in the core genome (Nikaido, 2001; Schweizer, 2003), together with the potential to develop high-level resistance by accumulation of small mutations (Fajardo *et al.*, 2008; Dötsch *et al.*, 2009; Martinez *et al.*, 2009; Alvarez-Ortega *et al.*, 2010; Breidenstein *et al.*, 2011; Bruchmann *et al.*, 2013). However, the most obvious cause of multi-drug resistance is the acquisition of resistance genes by horizontal gene transfer (Davies and Davies, 2010). Therefore, we performed a blast search of the predicted genes of the two strains in the Comprehensive Antibiotic Resistance Database (CARD) (McArthur *et al.*, 2013) and scanned both genomes for genetic variations of intrinsic resistance determinants. In a previous work, strain PA49 was found to be resistant towards the antibiotics gentamicin (GM), amikacin (AN), azlocillin (AZ), ceftazidime (CAZ), piperacillin/tazobactam (PT), ciprofloxacin (CIP) and imipenem (IPM) (Schwartz *et al.*, 2006). Searching its genome for resistance determinants revealed the presence of one aminoglycoside acetyltransferase of the AAC(6')-type, two aminoglycoside adenylyltransferases of type ANT(2") and ANT(3") and one VIM metallo-beta-lactamase (Table 1). Two additional genes were annotated as beta-lactamases in PA49 only by the BLAST search in the National Center for Biotechnology Information (NCBI) non-redundant (nr) protein database but not found in the CARD database

**Table 1.** Comparison of antibiotic resistance determinants found in the genomes of PA30 and PA49. Identifiers state PAO1 gene IDs or RefSeq Accession where applicable. Genotypes refer to presence or absence or genes or specific alleles with 'wt' indicating the genotype found in the reference strains PAO1.

| Gene ID/accession | Gene name | Resistance type | PA30 genotype | PA49 genotype | Affected antibiotics[a] |
|---|---|---|---|---|---|
| gi\|32470063 | aac(6')-Ib | AAC(6') | – | Present | Aminoglycosides (GM, AN) |
| gi\|378773997 | aadB | ANT(2") | – | Present | Aminoglycosides (GM, AN) |
| gi\|489251134 | blaVIM-2 | VIM | – | Present | Beta-lactams (AZ, CAZ, PT) |
| gi\|88853419 | aadA10 | ANT(3") | – | Present | Aminoglycosides |
| gi\|489211498 | – | AmpC | – | Present | Beta-lactams (AZ, CAZ, PT) |
| gi\|489217979 | – | Metallo-beta-lactamase | – | Present | Beta-lactams (AZ, CAZ, PT) |
| gi\|407937916 | – | Metallo-beta-lactamase | Present | – | Beta-lactams (AZ, CAZ, PT) |
| PA3168 | gyrA | Target modification | wt | T83I | Fluoroquinolones (CIP) |
| PA4964 | parC | Target modification | wt | S87L | Fluoroquinolones (CIP) |
| PA0958 | oprD | Decreased permeability | Multiple SNPs | Multiple SNPs, frameshift insertion CC at position 279 | Carbapenems (IPM) |

**a.** Abbreviations indicate specific antibiotics. AN, amikacin; PT, Piperacillin + Tazobactam. AZ, azlocillin; CAZ, ceftazidime; CIP, ciprofloxacin; GM, gentamicin; IPM, imipenem.

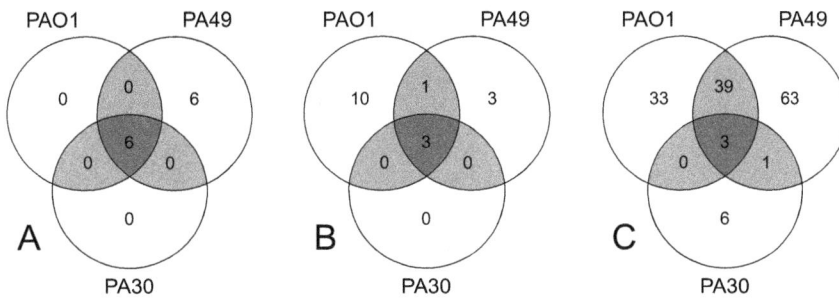

**Fig. 3.** Comparison of specific gene classes found in the genomes of PA30 and PA49 with type strain PAO1. The circles of this Venn Diagram contain the numbers of genes that were predicted from the genome sequence of the two newly sequenced strains, in comparison with the known genes of the PAO1 reference genome. PAO1 genome annotation was taken from www.pseudomonas.com (Winsor et al., 2011).
A. Genes involved in antibiotic resistance (excluding efflux pumps).
B. Genes involved in metal ion resistance.
C. Gene involved in genetic mobility – transposases, integrases, recombinases and conjugation-related proteins.

(Fig. 3A). Taken together, these genes confer resistance towards a wide range of aminoglycosides and beta-lactam antibiotics, explaining the resistance towards GM, AN, AZ, CAZ and PT. Fluoroquinolones like CIP target the DNA gyrase and Topoisomerase IV enzyme complexes, and high-level resistance towards these antibiotics is often caused by sequence variations of the two subunits GyrA (gyrase) and ParC (topoisomerase) (Ruiz, 2003) and indeed, both proteins contained a single amino acid exchange in the resistance determining region (Table 1). These two mutations represent the most common type of variations found in fluoroquinolone resistant isolates of *P. aeruginosa* and have recently been shown to be sufficient for the development of high-level resistance towards CIP (Bruchmann *et al.*, 2013). Finally, a frameshift mutation in the outer membrane porin OprD was found that is

likely to cause misfolding or decreased functionality of the protein. Defective mutations of OprD are known to cause resistance towards carbapenems including IPM in combination with intrinsic beta-lactamases and efflux pumps (Pirnay *et al.*, 2002). Of note, the strain PA30 that is sensitive towards all these antibiotics did not contain any known horizontally acquired resistance genes and harboured wild-type alleles of the target genes *gyrA*, *parC* and *oprD* (Table 1). In summary, these results provide a comprehensive explanation for the resistance phenotype covering all the antibiotics that were tested, since all resistance determining genes and alleles (besides the ones intrinsic to *P. aeruginosa*) were exclusively found in PA49 (Fig. 3A; Table 2).

Both PA30 and PA49 harbour a set of genes involved in metal ion resistance that are not found in the

**Table 2.** Expression of genes related to antibiotic resistance (excluding efflux pumps).

| Gene ID/accession | Product | PA30 | | PA49 | |
| --- | --- | --- | --- | --- | --- |
| | | T | W | T | W |
| gil489182848 | Acriflavin resistance protein | n.p.[a] | n.p. | 524 | 416 |
| gil496684660 | Bleomycin resistance protein | n.p. | n.p. | 1808 | 1737 |
| gil489211498 | Beta-lactamase | n.p. | n.p. | 1028 | 435 |
| gil489251134 | VIM-1 protein | n.p. | n.p. | 35 957 | 39 477 |
| gil32470063 | Hypothetical protein | n.p. | n.p. | 50 184 | 48 460 |
| gil88853419 | Aminoglycoside-modifying enzyme | n.p. | n.p. | 4358 | 3280 |
| gil378773997 | 2.-aminoglycoside nucleotidyltransferase | n.p. | n.p. | 7275 | 5761 |
| gil496684693 | Glyoxalase/bleomycin resistance protein/dioxygenase | n.p. | n.p. | 362 | 1547 |
| gil160901172 | Acriflavine resistance protein B | n.p. | n.p. | 468 | 408 |
| gil489217979 | Beta-lactamase | n.p. | n.p. | 109 | 64 |
| PA0706 | Chloramphenicol acetyltransferase | 399 | 304 | 48 | 73 |
| gil489208693 | Fusaric acid resistance protein | 25 | 18 | 30 | 18 |
| gil116049746 | Beta-lactamase | 107 | 64 | 272 | 131 |
| PA1129 | Probable fosfomycin resistance protein | 82 | 54 | 33 | 11 |
| PA5514 | Probable beta-lactamase | 258 | 285 | 191 | 422 |
| PA1959 | Bacitracin resistance protein | 385 | 925 | 110 | 219 |
| PA4110 | Beta-lactamase precursor AmpC | 694 | 382 | 101 | 74 |
| PA4119 | Aminoglycoside 3'-phosphotransferase type IIb | 113 | 177 | 37 | 40 |
| PA5159 | Multi-drug resistance protein | 79 | 66 | 65 | 120 |
| PA1858 | Streptomycin 3"-phosphotransferase | 145 | 131 | 31 | 52 |
| gil407937916 | beta-lactamase domain-containing protein | 233 | 201 | n.p. | n.p. |

**a.** n.p. – gene sequence is not present in this strain.
Identifiers state PAO1 gene IDs or RefSeq accession where applicable. Transcriptional activity in tap water (T) and waste water (W) was normalized for the two strains independently and is shown as normalized pseudocounts.

**Table 3.** Expression of genes related to metal tolerance.

| Gene ID/accession | Product | PA30 | | PA49 | |
| --- | --- | --- | --- | --- | --- |
| | | T | W | T | W |
| gil489181230 | Mercuric reductase | n.p.[a] | n.p. | 695 | 892 |
| gil134047226 | MerP | n.p. | n.p. | 190 | 376 |
| gil410691713 | Hg(II)-responsive transcriptional regulator MerR | n.p. | n.p. | 626 | 2339 |
| gil498493737 | Mercuric reductase | 515 | 520 | n.p. | n.p. |
| gil134047116 | MerT | 65 | 53 | 132 | 339 |
| gil152989484 | Mercuric resistance operon regulatory protein | 239 | 310 | n.p. | n.p. |
| PA2065 | Copper resistance protein A precursor | 253 | 95 | 654 | 20 |
| PA2064 | Copper resistance protein B precursor | 114 | 59 | 238 | 12 |
| PA0950 | Probable arsenate reductase | 472 | 615 | 268 | 451 |
| gil498490899 | Arsenical resistance protein ArsH | 102 | 139 | n.p. | n.p. |
| gil493249841 | Arsenate reductase | 75 | 97 | n.p. | n.p. |
| gil493264869 | Arsenic transporter | 119 | 196 | n.p. | n.p. |
| gil493249844 | ArsR family transcriptional regulator | 486 | 235 | n.p. | n.p. |
| gil495544062 | Copper oxidase | 1420 | 203 | n.p. | n.p. |
| gil493265806 | Copper resistance protein B | 612 | 100 | n.p. | n.p. |
| gil386717894 | Copper resistance protein C | 358 | 141 | n.p. | n.p. |
| gil489188126 | Copper resistance protein CopD | 717 | 392 | n.p. | n.p. |

**a.** n.p. – gene sequence is not present in this strain.
Identifiers state PAO1 gene IDs or RefSeq accession where applicable. Transcriptional activity in tap water (T) and waste water (W) was normalized for the two strains independently and is shown as normalized pseudocounts.

*P. aeruginosa* core genome. Strain PA30 contains several genes encoding resistance genes related to copper, mercury and arsenic/arsenate, while most of the genes could not be found in PA49 (Fig. 3B; Table 3).

The extent of the accessory genomes found in PA30 and PA49 point towards a high incidence of horizontal gene transfer in the evolutionary history of these strains. Therefore, we also searched the annotated genomes for genes associated with genomic mobility, mostly classified as recombinases, transposases, integrases or conjugative elements. Since mobile genetic elements *per definition* belong to the accessory genome, it is not surprising that nearly all genes associated with genetic mobility that were found in PA30 and PA49 are not present in the genome of PAO1 (Fig. 3C). Both strains contain a large number of mobility genes (75 in PA30, 103 in PA49) (Table 4).

*Transcriptome analyses*

In order to investigate the impact of waste water and tap water on the transcriptional activities, we performed RNA sequencing on both the sensitive and multi-resistant *P. aeruginosa* strain. The *de novo* assembled genomes were used as references for the mapping of reads obtained from RNA sequencing. In total, 95% of the reads mapped to the genome reference, which is comparable to results for RNA sequencing of known genomes and indicates a high quality and completeness of the two assembled genomes (Table S3). Between 1.5 and 3.6 million reads mapped uniquely to coding regions, yielding a median read count per gene of 54 to 130 and was sufficient for an in depth analysis.

Both strains, PA30 and PA49, were exposed to tap water and waste water, and differential gene expression was analysed between the different water matrices as well as between the two strains. Upon exposure to waste water, 222 genes were at least fourfold differentially expressed in strain PA30 (94 upregulated, 128 downregulated) as compared with tap water exposure (Table S4). Most of the differentially expressed genes encode for hypothetical proteins. Investigating whether any functional groups of genes were significantly overrepresented among the differentially expressed genes, we performed an enrichment analysis of gene ontology (GO) terms. Genes that were associated with 'copper ion binding' (GO:0005507) and 'potassium-transporting ATPase activity' (GO:0008556) were significantly overrepresented with six (out of 17) and three (out of three) genes being differentially expressed respectively. In strain PA49, 144 (51 upregulated, 93 downregulated, Table S5) gene showed differential expression upon exposure to the different water matrices, but no significant enrichment of GO terms was observed. A comparison of the expression of orthologous genes between PA30 and PA49 revealed a differential in the expression of 32 genes in tap water (e.g. some phenazine biosynthesis genes and a potassium-transporting ATPase, kdpABC, were upregulated in PA30), while only five gene coding for hypothetical proteins were found to be differentially expressed in waste water. This low number of differentially expressed genes between the two strains indicates a high similarity in their response to these specific environments.

The four horizontally acquired antibiotic resistance genes found in strain PA49 (Table 1) were transcriptionally active independent from the water matrix and

**Table 4.** Expression of genes related to genetic mobility and horizontal gene transfer.

| Gene ID/accession | Product | PA30 | | PA49 | |
|---|---|---|---|---|---|
| | | T | W | T | W |
| gil386063806 | Conjugal transfer protein TrbJ | n.p.[a] | n.p. | 4 | 7 |
| gil446855879 | Conjugal transfer protein TrbE | n.p. | n.p. | 58 | 33 |
| gil446855879 | Conjugal transfer protein TrbE | n.p. | n.p. | 26 | 21 |
| gil386063808 | Conjugal transfer protein TrbC | n.p. | n.p. | 10 | 16 |
| gil51492563 | TnpA | n.p. | n.p. | 481 | 609 |
| gil66045904 | TnpR resolvase | n.p. | n.p. | 2370 | 3179 |
| gil19352419 | Putative transposase | n.p. | n.p. | 3570 | 2090 |
| gil498491672 | Excisionase family DNA binding domain-containing protein | n.p. | n.p. | 263 | 432 |
| gil496684679 | Conjugal transfer coupling protein TraG | n.p. | n.p. | 22 | 8 |
| gil190573650 | Transposase | n.p. | n.p. | 218 | 206 |
| gil190572552 | Transposase IS3 | n.p. | n.p. | 189 | 156 |
| gil497209012 | Transposase | n.p. | n.p. | 0 | 0 |
| gil495918616 | Integrase | n.p. | n.p. | 187 | 126 |
| gil37958842 | Putative transposase | n.p. | n.p. | 202 | 452 |
| gil386063806 | Conjugal transfer protein TrbJ | n.p. | n.p. | 29 | 5 |
| gil446923490 | Conjugal transfer protein TrbL | n.p. | n.p. | 66 | 40 |
| gil410690825 | Transposase | n.p. | n.p. | 837 | 2156 |
| gil319765135 | Transposase TniA | n.p. | n.p. | 874 | 1657 |
| gil491446119 | TniQ | n.p. | n.p. | 448 | 546 |
| gil490385992 | Transposase | n.p. | n.p. | 26 | 30 |
| gil497303964 | Integrating conjugative element protein | n.p. | n.p. | 4 | 6 |
| gil497081867 | Conjugative transfer region protein | n.p. | n.p. | 2 | 2 |
| gil497303936 | Integrating conjugative element protein | n.p. | n.p. | 0 | 0 |
| gil497081861 | Integrating conjugative element protein | n.p. | n.p. | 13 | 11 |
| gil493535069 | Integrating conjugative element protein. PFL_4709 family | n.p. | n.p. | 19 | 19 |
| gil497081686 | Integrase family protein | n.p. | n.p. | 223 | 259 |
| gil133756449 | TniA | n.p. | n.p. | 498 | 866 |
| gil289064112 | TniQ transposition protein | n.p. | n.p. | 57 | 45 |
| gil410609201 | Transposase | n.p. | n.p. | 2789 | 2351 |
| gil446985433 | Integrase | n.p. | n.p. | 0 | 1 |
| PA4797 | Probable transposase | n.p. | n.p. | 1294 | 879 |
| gil496684684 | Conjugal transfer protein TrbE | n.p. | n.p. | 28 | 27 |
| gil496684684 | Conjugal transfer protein TrbE | n.p. | n.p. | 13 | 15 |
| gil496684684 | Conjugal transfer protein TrbE | n.p. | n.p. | 7 | 12 |
| gil496684685 | Conjugal transfer protein TrbJ | n.p. | n.p. | 1 | 0 |
| gil496684685 | Conjugal transfer protein TrbJ | n.p. | n.p. | 14 | 14 |
| gil330503729 | Transposase IS4 | n.p. | n.p. | 91 | 80 |
| gil496684687 | Conjugal transfer protein TrbL | n.p. | n.p. | 79 | 43 |
| gil256367798 | Transposase | n.p. | n.p. | 15674 | 13666 |
| gil496684681 | Conjugal transfer protein TrbB | n.p. | n.p. | 1 | 1 |
| gil116050177 | Transposase Tn4652 | n.p. | n.p. | 596 | 487 |
| gil116050169 | Recombinase | n.p. | n.p. | 553 | 1182 |
| gil152984407 | TnpT protein | n.p. | n.p. | 180 | 266 |
| gil12698413 | Transposase B | n.p. | n.p. | 50 | 71 |
| gil498341535 | Integrase [Pseudomonas fragi] | n.p. | n.p. | 453 | 462 |
| gil489250021 | Conjugal transfer protein TrbL | n.p. | n.p. | 1 | 0 |
| gil496684681 | Conjugal transfer protein TrbB | n.p. | n.p. | 1 | 2 |
| gil496684682 | Conjugal transfer protein TrbC | n.p. | n.p. | 0 | 2 |
| gil497207593 | Integrase | n.p. | n.p. | 3508 | 2908 |
| gil489182829 | Conjugative transfer protein Trbl | n.p. | n.p. | 87 | 59 |
| gil386063802 | Conjugative transfer protein TrbG | n.p. | n.p. | 25 | 15 |
| gil386063802 | Conjugative transfer protein TrbG | n.p. | n.p. | 29 | 14 |
| gil386063803 | Conjugal transfer protein TrbF | n.p. | n.p. | 22 | 3 |
| gil386063804 | Conjugal transfer protein TrbL | n.p. | n.p. | 4 | 0 |
| gil505461140 | Shufflon-specific recombinase | n.p. | n.p. | 339 | 328 |
| gil497074269 | Integrating conjugative element protein pill. pfgi-1 | n.p. | n.p. | 7 | 5 |
| gil485834156 | Transposase | n.p. | n.p. | 632 | 428 |
| gil446195994 | Transposase | n.p. | n.p. | 374 | 259 |
| gil496684690 | Conjugal transfer protein Trbl | n.p. | n.p. | 14 | 9 |
| gil496684690 | Conjugal transfer protein Trbl | n.p. | n.p. | 15 | 23 |
| gil496684689 | Conjugal transfer protein TrbG | n.p. | n.p. | 8 | 14 |
| gil495242526 | Conjugal transfer protein TrbF | n.p. | n.p. | 11 | 4 |
| gil17547297 | Conjugal transfer protein TrbL | n.p. | n.p. | 1 | 0 |
| gil3688518 | Putative transposase | n.p. | n.p. | 968 | 1026 |
| gil491446843 | Integrase | n.p. | n.p. | 48 | 67 |
| gil152987984 | Conjugal transfer protein TrbD | n.p. | n.p. | 2 | 5 |
| gil493518183 | Conjugal transfer protein TrbC | n.p. | n.p. | 3 | 1 |
| gil489201924 | Conjugal transfer protein TrbB | n.p. | n.p. | 3 | 1 |
| gil330824177 | Conjugal transfer protein TrbB | n.p. | n.p. | 8 | 2 |
| gil489194682 | Conjugal transfer protein TraG | n.p. | n.p. | 6 | 8 |
| gil92112121 | TnpA transposase | n.p. | n.p. | 2857 | 1238 |
| gil505462488 | Transposase mutator family protein | n.p. | n.p. | 483 | 234 |
| gil472324076 | Site-specific recombinase XerC | n.p. | n.p. | 187 | 201 |
| gil512557653 | Transposase | 55 | 51 | 57 | 30 |
| gil489232756 | Transposase component | 7 | 8 | 13 | 5 |
| gil495332841 | Transposase | 159 | 157 | n.p. | n.p. |

**Table 4.** *Cont.*

| Gene ID/accession | Product | PA30 | | PA49 | |
|---|---|---|---|---|---|
| | | T | W | T | W |
| gil410693866 | Transposase of ISThsp18. IS1182 family | 2730 | 2154 | n.p. | n.p. |
| gil392420288 | Integrating conjugative element ParB | 28 | 75 | 4 | 11 |
| gil330824345 | Integrating conjugative element protein | 83 | 168 | 11 | 25 |
| gil330824346 | Integrase | 31 | 67 | 36 | 24 |
| gil512557590 | Integrating conjugative element protein PilL. PFGI-1 class | 8 | 6 | n.p. | n.p. |
| gil490375364 | Integrating conjugative element protein | 15 | 35 | n.p. | n.p. |
| gil339493279 | Conjugal transfer protein TraG | 109 | 96 | n.p. | n.p. |
| gil498491306 | Integrating conjugative element protein | 6 | 2 | 9 | 9 |
| gil497303938 | Integrating conjugative element membrane protein | 3 | 8 | 7 | 2 |
| gil392420336 | Conjugal transfer protein | 8 | 9 | n.p. | n.p. |
| gil330824396 | Integrating conjugative element protein | 26 | 22 | n.p. | n.p. |
| gil330824398 | Integrating conjugative element protein | 13 | 18 | n.p. | n.p. |
| gil497303929 | Conjugative transfer ATPase | 73 | 44 | 32 | 19 |
| gil498491292 | Integrating conjugative element protein | 39 | 22 | 4 | 16 |
| gil410471275 | Phage-related integrase | 442 | 603 | n.p. | n.p. |
| PA3738 | Integrase/recombinase XerD | 177 | 231 | 122 | 275 |
| gil489224124 | Integrase | 1148 | 1068 | n.p. | n.p. |
| gil489224128 | Integrase | 1072 | 1106 | n.p. | n.p. |
| gil498248343 | Transposase | 124 | 98 | n.p. | n.p. |
| gil496762510 | Conjugal transfer protein TrbJ | 359 | 253 | n.p. | n.p. |
| gil489229539 | Conjugal transfer protein TrbL | 417 | 610 | n.p. | n.p. |
| gil489229536 | Conjugal transfer protein TrbJ | 479 | 538 | n.p. | n.p. |
| gil489229532 | Integrase | 634 | 593 | n.p. | n.p. |
| gil157420224 | TnpA | 0 | 0 | 1224 | 1485 |
| gil489225709 | Integrase | 230 | 160 | 137 | 63 |
| gil353334486 | Incl1 plasmid conjugative transfer ATPase PilQ | 241 | 176 | 84 | 59 |
| gil148807320 | Site-specific recombinase | 348 | 392 | n.p. | n.p. |
| gil152985999 | Conjugal transfer protein TraG | 565 | 416 | 207 | 167 |
| gil490477548 | Conjugal transfer protein | 98 | 44 | 19 | 16 |
| gil498491413 | Conjugative transfer ATPase | 510 | 328 | 218 | 122 |
| gil386064726 | Tyrosine recombinase XerC | 431 | 444 | 237 | 196 |
| gil386056644 | Integrase | 71 | 85 | 44 | 61 |
| gil446985433 | Integrase | 109 | 80 | 172 | 178 |
| gil163856484 | Transposase | 115 | 68 | n.p. | n.p. |
| gil496684672 | Conjugal transfer protein TraF peptidase | 9 | 2 | n.p. | n.p. |
| gil187940137 | Phage integrase family protein | 187 | 135 | n.p. | n.p. |
| gil121595234 | Conjugal transfer coupling protein TraG | 36 | 26 | 97 | 110 |
| gil386063809 | Conjugal transfer protein TrbB | 14 | 12 | 12 | 21 |
| gil492686504 | Conjugal transfer protein Trbc | 5 | 4 | n.p. | n.p. |
| gil387129929 | Conjugal transfer protein TrbD | 2 | 2 | 6 | 3 |
| gil497202778 | Conjugal transfer protein TrbE | 55 | 58 | 47 | 31 |
| gil489201921 | Conjugal transfer protein TrbJ | 47 | 28 | 11 | 7 |
| gil491446728 | Conjugal transfer protein TrbL | 869 | 474 | 16 | 8 |
| gil152984687 | Transposase | 709 | 388 | n.p. | n.p. |
| gil489221152 | Integrase | 645 | 569 | n.p. | n.p. |
| gil489184774 | Integrase | 352 | 296 | n.p. | n.p. |
| gil512563762 | P-type conjugative transfer protein TrbL | 15 | 3 | n.p. | n.p. |
| gil94310290 | conjugal transfer protein TrbF | 17 | 22 | 23 | 4 |
| gil512587667 | P-type conjugative transfer protein TrbG | 31 | 31 | 20 | 14 |
| gil489200991 | Conjugative transfer protein Trbl | 76 | 67 | 38 | 55 |
| gil490275574 | IS5 Transposase. Partial | 796 | 610 | n.p. | n.p. |
| gil218891103 | Integrase | 362 | 278 | 111 | 97 |
| gil392983596 | Integrase | 3054 | 1923 | 476 | 858 |
| gil489349534 | Transposase IS3 | 327 | 133 | n.p. | n.p. |
| gil187939496 | Transposase | 453 | 382 | 560 | 641 |
| PA1534 | Recombination protein RecR | 390 | 500 | 231 | 278 |
| PA5280 | Site-specific recombinase Sss | 320 | 320 | 112 | 113 |
| gil148807435 | Phage integrase | 115 | 87 | n.p. | n.p. |
| gil148807434 | Phage integrase | 371 | 303 | n.p. | n.p. |
| gil148807431 | Site-specific recombinase | 598 | 366 | 488 | 488 |
| gil489214965 | Integrase | 999 | 937 | 226 | 488 |
| gil148807465 | Phage integrase | 121 | 139 | n.p. | n.p. |
| gil392420677 | Transposase IS5 | 420 | 266 | n.p. | n.p. |
| gil94311234 | Integrase | 83 | 88 | 12 | 4 |
| gil493264878 | Conjugal transfer protein TraG | 81 | 100 | 101 | 58 |
| gil489180720 | Conjugal transfer protein | 6 | 1 | 2 | 2 |
| gil386717922 | Conjugal transfer protein | 76 | 77 | 54 | 53 |
| gil489215831 | Integrase | 347 | 368 | 1107 | 3025 |
| gil498490909 | Integrating conjugative element relaxase. PFGI-1 class | 212 | 126 | 181 | 129 |
| gil493264934 | Transposase IS204 | 332 | 292 | n.p. | n.p. |
| gil493265821 | Transposase | 303 | 257 | n.p. | n.p. |
| gil489180871 | Integrase | 521 | 582 | 181 | 240 |

a. n.p. – gene sequence is not present in this strain.
Identifiers state PAO1 gene IDs or RefSeq accession where applicable. Transcriptional activity in tap water (T) and waste water (W) was normalized for the two strains independently and is shown as normalized pseudocounts.

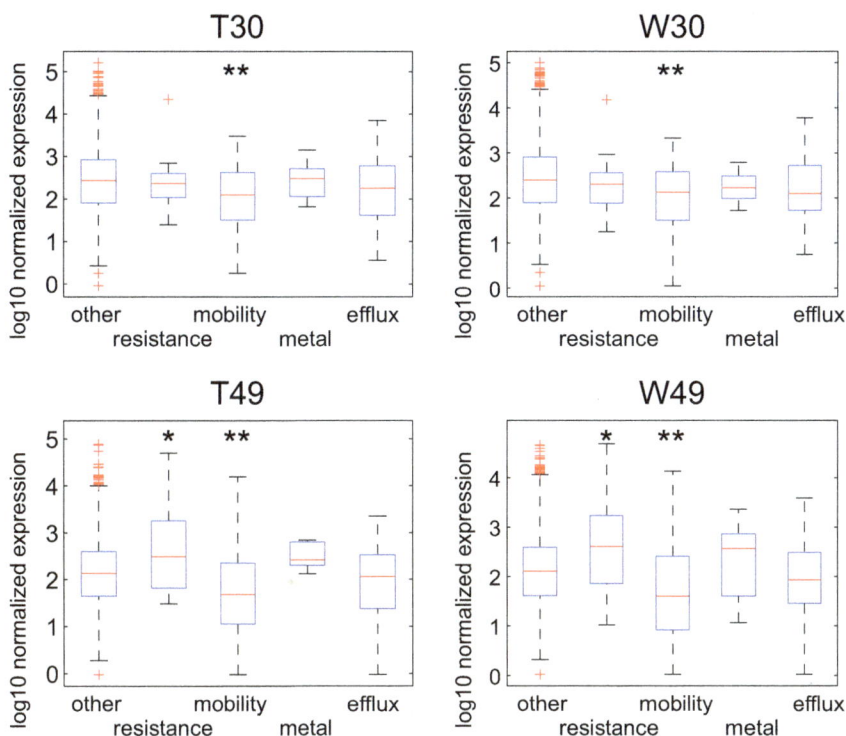

**Fig. 4.** Expression of specific genes in strain PA30 and PA49 in different water matrices. Absolute gene expression values are depicted as box plots for the different samples of strain PA30 and PA49 cultivated in tap water (T) and waste water (W). Genes were manually selected by their functional classification as *resistance* (associated with modification and deactivation of antibiotics), *mobility* (associated with horizontal gene transfer and recombination), *metal* (associated with heavy metal tolerance), *efflux* (associated with multi-drug efflux pumps) or *other* (not included in any other class). Asterisks indicate a significant difference in the medians of the particular gene class and the *other* genes determined with the Mann–Whitney–Wilcoxon test (*$P < 0.05$; **$P < 0.001$).

therefore most likely are a main cause of the observed resistance towards a wide spectrum of aminoglycoside and beta-lactam antibiotics (Table 2). Genes associated with antibiotic resistance (not including multi-drug efflux pumps) showed a higher average expression as compared with the rest of the genome in PA49 (Fig. 4), which is obviously a result of the generally high expression of horizontally acquired resistance genes (Table 2). This tendency was independent from the water matrix and not found in the transcriptome of PA30 (Fig. 4), which lacks such additional resistance genes (Fig. 2). A common cause of antibiotic resistance in *P. aeruginosa* is the overexpression of multi-drug efflux pumps (usually termed 'Mex' pumps). Indeed, the genes encoding the MexCD-OprJ efflux pump were overexpressed in PA49 in response to waste water (Table 5 and Table S5). This pump system can confer resistance towards a broad spectrum of antibiotics (Poole *et al.*, 1996) and thus may further contribute to the multi-resistance phenotype of PA49. The induced expression of this efflux pump specifically in waste water is indicating a specific stimulation presumably by one or multiple of antibiotics found in the used waste water or via so far unknown waste water components. However, since the expression of specific resistance genes and presence of resistance-related target mutations already sufficiently explains the broad resistance phenotype in PA49 (Table 1), the exact contribution of a MexCD-OprJ overexpression remains unclear. It should be again pointed out that the expression of

resistance genes (with the exception of MexCD-OprJ) in PA49 was independent from the water matrix. Similarly, the antibiotic sensitive strain PA30 does not show any difference in expression of the intrinsic resistance determinants. Thus, the exposure of both strains to polluted waste water and oligotrophic tap water resulted in similar expression profiles of resistance genes. It seems to be obvious that changes in environmental milieus result in rather unspecific transcriptional responses than selected and stimuli-specific gene regulation.

A small set of genes associated with heavy metal tolerance was also found in the genomes of PA30 and PA49 (Fig. 3B; Table 3). However, no differential expression in the two water matrices was detected. Comparing the average expression of these genes with genes not related to metal tolerance also showed no general difference, independent of strain background and water matrix (Fig. 4).

Waste waters are already known to stimulate genetic transfer due to the sublethal noxa of pharmaceutical residues (e.g. antibiotics, heavy metal ions) or other xenobiotics. But, the expression of mobile genetic element was also found to be induced after exposure to tap water. Here, physiological shifts to oligotrophic habitats and/or starvation might be responsible for the genetic activities and might contribute to horizontal gene transfer, as discussed in Davies and colleagues (2006). The genomic analysis identified a large number of mobile genomic elements in the genomes of both PA30 and PA49 (Fig. 3C;

**Table 5.** Expression of genes encoding (putative) multi-drug efflux pumps (MEX pumps).

| Gene ID/accession | Product | PA30 | | | PA49 |
|---|---|---|---|---|---|
| | | T | W | T | W |
| gil15596434 | Multi-drug resistance efflux pump | n.p.[a] | n.p. | 8 | 6 |
| gil491814602 | Macrolide efflux protein | n.p. | n.p. | 69 | 85 |
| gil386336514 | RND transporter | n.p. | n.p. | 104 | 91 |
| gil307130252 | RND efflux membrane fusion protein | n.p. | n.p. | 29 | 42 |
| PA1237 | Probable multi-drug resistance efflux pump | 9 | 6 | 10 | 0 |
| PA3719 | Anti-repressor for MexR, ArmR | 20 | 15 | 10 | 33 |
| PA4990 | SMR multi-drug efflux transporter | 103 | 79 | 21 | 15 |
| PA4374 | Probable Resistance-Nodulation-Cell Division (RND) efflux membrane fusion protein precursor | 499 | 444 | 233 | 251 |
| PA4375 | Probable Resistance-Nodulation-Cell Division (RND) efflux transporter | 1024 | 643 | 439 | 293 |
| PA3522 | Probable Resistance-Nodulation-Cell Division (RND) efflux transporter | 1090 | 102 | 381 | 82 |
| PA3523 | Probable Resistance-Nodulation-Cell Division (RND) efflux membrane fusion protein precursor | 355 | 56 | 140 | 27 |
| gil489214414 | Outer membrane component of multi-drug efflux pump. partial | 0 | 0 | 1 | 1 |
| gil15596434 | Multi-drug resistance efflux pump | 0 | 0 | 10 | 0 |
| PA0427 | Major intrinsic multiple antibiotic resistance efflux outer membrane protein OprM precursor | 2722 | 2351 | 991 | 1137 |
| PA0426 | Resistance-Nodulation-Cell Division (RND) multi-drug efflux transporter MexB | 7087 | 5994 | 2277 | 3917 |
| PA0425 | Resistance-Nodulation-Cell Division (RND) multi-drug efflux membrane fusion protein MexA precursor | 3129 | 2318 | 1098 | 1904 |
| PA0424 | Multi-drug resistance operon repressor MexR | 1305 | 520 | 331 | 321 |
| PA1238 | Probable outer membrane component of multi-drug efflux pump | 4 | 7 | 0 | 3 |
| PA1435 | Probable Resistance-Nodulation-Cell Division (RND) efflux membrane fusion protein precursor | 44 | 40 | 16 | 23 |
| PA1436 | Probable Resistance-Nodulation-Cell Division (RND) efflux transporter | 158 | 181 | 70 | 61 |
| PA0158 | Resistance-Nodulation-Cell Division (RND) triclosan efflux transporter, TriC | 2826 | 2341 | 1241 | 539 |
| PA0157 | Resistance-Nodulation-Cell Division (RND) triclosan efflux membrane fusion protein, TriB | 398 | 433 | 200 | 118 |
| PA0156 | Resistance-Nodulation-Cell Division (RND) triclosan efflux membrane fusion protein, TriA | 530 | 529 | 367 | 177 |
| PA2019 | Resistance-Nodulation-Cell Division (RND) multi-drug efflux membrane fusion protein precursor | 939 | 1323 | 79 | 70 |
| PA2018 | Resistance-Nodulation-Cell Division (RND) multi-drug efflux transporter | 3969 | 4876 | 321 | 240 |
| PA2528 | Probable Resistance-Nodulation-Cell Division (RND) efflux membrane fusion protein precursor | 508 | 452 | 349 | 690 |
| PA2527 | Probable Resistance-Nodulation-Cell Division (RND) efflux transporter | 603 | 529 | 345 | 572 |
| gil489252367 | Multi-drug transporter | 23 | 35 | 16 | 8 |
| PA2526 | Probable Resistance-Nodulation-Cell Division (RND) efflux transporter | 504 | 542 | 594 | 449 |
| gil489215922 | Resistance-Nodulation-Cell Division (RND) efflux transporter. partial | 31 | 28 | n.p. | n.p. |
| PA2522 | Outer membrane protein precursor CzcC | 80 | 17 | 116 | 15 |
| PA2521 | Resistance-Nodulation-Cell Division (RND) divalent metal cation efflux membrane fusion protein CzcB precursor | 184 | 47 | 147 | 20 |
| PA2520 | Resistance-Nodulation-Cell Division (RND) divalent metal cation efflux transporter CzcA | 369 | 77 | 430 | 51 |
| PA2495 | Multi-drug efflux outer membrane protein OprN precursor | 91 | 111 | 13 | 6 |
| PA2494 | Resistance-Nodulation-Cell Division (RND) multi-drug efflux transporter MexF | 1021 | 767 | 126 | 82 |
| PA2493 | Resistance-Nodulation-Cell Division (RND) multi-drug efflux membrane fusion protein MexE precursor | 110 | 106 | 35 | 61 |
| PA1541 | Probable drug efflux transporter | 38 | 104 | 148 | 1088 |
| PA5160 | Drug efflux transporter | 225 | 208 | 191 | 128 |
| PA4206 | Probable Resistance-Nodulation-Cell Division (RND) efflux membrane fusion protein precursor | 49 | 54 | 11 | 80 |
| PA4207 | Probable Resistance-Nodulation-Cell Division (RND) efflux transporter | 175 | 146 | 58 | 99 |
| PA4597 | Multi-drug efflux outer membrane protein OprJ precursor | 42 | 142 | 28 | 197 |
| PA4598 | Resistance-Nodulation-Cell Division (RND) multi-drug efflux transporter MexD | 153 | 297 | 83 | 964 |
| PA4599 | Resistance-Nodulation-Cell Division (RND) multi-drug efflux membrane fusion protein MexC precursor | 16 | 66 | 31 | 583 |
| PA3676 | Probable Resistance-Nodulation-Cell Division (RND) efflux transporter | 385 | 503 | 141 | 233 |
| PA3677 | Probable Resistance-Nodulation-Cell Division (RND) efflux membrane fusion protein precursor | 24 | 72 | 23 | 70 |

**a.** n.p. – gene sequence is not present in this strain.
Identifiers state PAO1 gene IDs or RefSeq accession where applicable. Transcriptional activity in tap water (T) and waste water (W) was normalized for the two strains independently and is shown as normalized pseudocounts.

Table 4). The genes that can be directly associated with horizontal gene transfer and recombination (recombinases, integrases, transposases and genes related to conjugative transfer) were mostly found to be actively expressed in both strains and independent from the water matrix. On average, these 'mobility genes' were expressed on a lower level than the 'other' genes of the genome (Fig. 4). However, their expression is insensitive to the strain background and to the water matrix.

In conclusion, the multi-drug resistance of strain PA49 can be attributed to the presence and expression of genes encoding a set of antibiotic-modifying enzymes located both in the core genome and on mobile genetic elements that were presumably acquired by horizontal gene transfer. Thus, the multi-drug resistant phenotype of PA49 seems directly linked with this set of resistance determinants. The impact of one overexpressed efflux pump being induced in waste water on the resistance characteristics of PA49 is so far an open question. Both, the antibiotic resistant and the sensitive strain, showed similar transcriptomic responses to the different water matrices but no strain-specific stress responses to both matrices (with exception to one efflux pump).

## Experimental procedures

### Isolation and cultivation of P. aeruginosa strains PA30 and PA49

Bacterial strains were enriched and isolated from a German waste water treatment plant compartment as described in a previous study (Schwartz et al., 2006). For routine culturing, bacteria were grown on agar plates containing BM2 minimal medium (Yeung et al., 2009) supplemented with 15 g $l^{-1}$ agar (Merck, Darmstadt, Germany). For overnight cultures, a colony from the agar plate was inoculated in BM2 minimal medium as well as BHI (Merck, Darmstadt, Germany) broth (1:4 diluted) and incubated at 37°C. The growth behavior of the strains was observed by diluting overnight cultures to an optical density (OD) of 0.1 in BM2 and BHI medium, incubation at 37°C with gentle agitation for a time span of 24 h and monitoring the OD over time for each strain (Infinite 200 PRO, Tecan, Männedorf, Switzerland). No difference in growth behavior between the two isolates was observed in BM2 and BHI broth respectively (data not shown).

### Antibiotic susceptibility testing PA30 and PA49

Resistance characterization for GM (10 µg disc$^{-1}$), CIP (5 µg disc$^{-1}$), IPM (10 µg disc$^{-1}$), CAZ (30 µg disc$^{-1}$), AN (30 µg disc$^{-1}$), AZ (75 µg disc$^{-1}$) and PT (100/10 µg disc$^{-1}$) was evaluated using agar diffusion test according to Clinical Laboratory Standards Institute (CLSI) guidelines, wherein the zone of growth inhibition on Miller Hinton agar (Merck) was measured after 18 h incubation at 37°C. The P. aeruginosa strain PA30 was found to be sensitive for GM, CIP, IPM, CAZ, AN, AZ and PT. In contrast, PA49 was found to be resistant against all mentioned antibiotics (Schwartz et al., 2006).

### Incubation in tap water and waste water

Distinct colonies of each strain were inoculated in 25 ml BHI medium (Merck, Darmstadt, Germany) diluted 1:4 with distilled water in a 50 ml sterile tube (Falcon, Nürtingen, Germany) and incubated on a shaker at 37°C at 100 rpm overnight. A volume of 2.5 ml of this overnight culture was used to inoculate 25 ml of 1:4 diluted BHI medium and incubated on a shaker at 37°C at 100 rpm. At an optical density ($OD_{600nm}$) of 1.0 (early stationary growth phase), bacterial suspension were pelleted at 5000 g at 20°C for 15 min. Pellets were re-suspended in 20 ml sterile tap water (T) or sterile filtered waste water (W) collected from the influent of a municipal WWTP. The OD of these suspensions with PA30 and PA49 were adjusted at 0.5. The samples were incubated on a shaker (80 rpm) at 22°C for 3 h.

The tap water conditioned from groundwater at the municipal waterworks met the requirements of the German drinking water guideline. The average total organic carbon value was measured as 0.9 mg $l^{-1}$. The chemical and physical characteristics of the final conditioned drinking water are listed in Jungfer et al. (2013; see reference waterworks).

The used waste water originated from the effluent of a municipal waste water treatment plant of a city with 445 000 inhabitants and is equipped with a conventional three treatment process (nitrification, denitrification, phosphor elimination). Chemical analyses demonstrated the presences of different classes of antibiotics (e.g. clarithromycin, roxithromycin, erythromycin, sulfamethoxazol, and trimehoprim) in a range of 0.5–1.5 µg $l^{-1}$ (unpublished data).

### DNA extraction and purification

Previous to the DNA extraction 25 ml BHI was inoculated with a single colony of PA30 and PA49, respectively, and cultivated at 37°C and 150 rpm on a rotary shaker until ODs reached 1.0 value. An aliquot of 5 ml of each culture was pelleted at 3000 g for 10 min. Subsequent DNA extraction was performed according to the protocol of QIAGEN Genomic-tip 100/G kit system (Qiagen, Germany). The concentration and purity of the obtained DNA was determined using the NanoDrop 1000 Spectrophotometer (Thermo Scientific, Germany). The quality of the genomic DNA was also controlled by agarose gel electrophoresis.

### RNA extraction and purification

Ribonucleic isolation of the samples was performed in quadruplicates that were pooled before sequencing. One millilitre of each of the four independent bacterial suspensions (T or W) was mixed with 1 ml of RNA protect (Qiagen, Hilden, Germany) and incubated for 5 min at room temperature. The bacteria were pelleted at 12.000 g for 10 min, and the supernatant was discarded. Prior to RNA extraction from bacteria, four replicate cultures from parallel experiments (tap water and waste water) from each type (PA30 and PA49) were combined. Ribonucleic acid isolation was performed using the RNeasy extraction kit (Qiagen, Hildern, Germany) according to the manufacturer's protocol, and the RNA was eluted in 50 µl RNase-free water. To eliminate residual DNA contamination, the RNA was treated with TURBO

Desoxyribonuclease (DNase, Ambion Inc., Kaufungen, Germany). Five microlitres of 10× TURBO DNase buffer and 1 µl of TURBO DNase were added to 50 µl RNA solution and incubated at 37°C for 30 min. Desoxyribonuclease inactivation reagent (5 µl) was added to the RNA solution and incubated under occasional mixing for 5 min. The sample was centrifuged at 10 000 rpm for 1.5 min, and the RNA was transferred to a new tube, and RNA concentration was measured in triplicate using the Nanodrop ND1000 spectrophotometer (PeqLab Biotechnology GmbH, Erlangen, Germany). The integrities of all RNA samples were tested using the Agilent 2100 Bioanalyzer (Agilent Technologies Sales & Services GmbH & Co.KG, Waldbronn, Germany).

### Removal of the ribosomal RNA

Removal of ribosomal RNA (rRNA) was performed with each sample. Fourteen microlitres of total RNA was mixed with 1 µl of RNase inhibitor SUPERase IN (Ambion). Ribosomal RNA was removed with the MICROBExpress KIT (Ambion) according to the manufacturer's protocol. Purified RNA was re-suspended in 25 µl TE buffer (1 mM EDTA, 10 mM Tris, pH: 8.0). The resulting purified mRNA yields were quantified with Nanodrop ND1000.

### Library preparation and Illumina sequencing

Deoxyribonucleic acid sequencing libraries were produced from 1 µg of genomic DNA and RNA libraries from 50 ng of rRNA depleted RNA, following the recommendations of the TruSeq DNA and TruSeq RNA protocols (Illumina) respectively. Briefly, the quality and quantity of ribosomal depleted RNA were assessed with the Bioanalyzer 2100 (Agilent), and the RNA-seq libraries were fragmented chemically, purified with AMPure XP beads (Beckman Coulter) and ligated to adapters with specific DNA barcode for each sample following the Illumina protocol. For DNA-seq libraries, genomic DNA was sheared to 200 bp fragments by sonication with a Covaris S2 instrument using the following settings: peak incidence power 175 W, duty factor 10%, cycle per burst 200, time 430 s. Sizes and concentrations of both RNA and DNA sequencing libraries were determined on a Bioanalyzer 2100 (DNA1000 chips, Agilent). Paired-end sequencing (2 × 50 bp) was performed on two lanes on a Hiseq1000 (Illumina) platform using TruSeq PE Cluster KIT v3 – cBot – HS and TruSeq SBS KIT v3 – HS. Cluster detection and base calling were performed using RTAV1.13 and quality of reads assessed with CASAVA v1.8.1 (Illumina). The sequencing resulted in at least 40 million pairs of 50 nt long reads for each sample, with a mean Phred quality score > 35 (Tables S1 and S3). These sequence data have been submitted to the GenBank Sequence Read Archive and are available under the accession numbers SRP041029 (PA30 genome), SRP041030 (PA49 genome), SRP041150 (PA30 transcriptomes) and SRP041151 (PA49 transcriptomes).

### Genome assembly and annotation

Raw sequence reads were trimmed and filtered using the fastq-mcf tool of the EA-UTILS software package (Aronesty, 2011) with a Phred quality cut-off of 20 and the appropriate Illumina TruSeq adapter sequences to detect and remove adapters. The genomes of strains PA30 and PA49 were assembled independently using the idba_hybrid assembler (Peng et al., 2012) with a range of k-mer sizes from 20 to 50 bp (step size 10 bp). The genome reference used to guide the hybrid assembly included the full genomic sequence of P. aeruginosa PAO1 and the sequences of 13 genomic islands and one plasmid (Table S2). To confirm the resulting contigs, the filtered and trimmed reads were mapped against the contigs using BOWTIE 2 (Langmead and Salzberg, 2012), and the read coverage of each contig was calculated by dividing the number of reads mapping to that contig by its length. Contigs with a length of less than 250 bp or with a read coverage of less than one read per base pair were discarded for being too short and/or potentially artefacts. Thereby, 207 (out of 775) and 269 (out of 887) contigs remained for the genomes of PA30 and PA49 respectively (Table S1). The overlap of the contigs with the reference sequences was determined by multi-sequence alignment using the PROMER function of the MUMMER 3.0 software package (Kurtz et al., 2004).

The filtered contigs were scanned for protein coding genes using METAGENEMARK (Trimble et al., 2012) with default parameters yielding 6572 and 6781 genes for PA30 and PA49 respectively (Table S1). The genes were annotated in two steps by using a BLASTP search (Camacho et al., 2009) first against the P. aeruginosa PAO1 genome with protein sequences taken from the Pseudomonas genome database (Winsor et al., 2011) and then for the remaining unidentified protein sequences against the database of nr protein sequences available for download from the NCBI FTP site (ftp://ftp.ncbi.nlm.nih.gov/blast/db). Thereby, 65 and 97 of the predicted genes of PA30 and PA49 remained unidentified. Additionally, the predicted protein sequences were compared by BLASTP to the CARD (McArthur et al., 2013) to identify putative resistance genes.

To enable direct comparisons between the genomes of PA30 and PA49, the orthologous genes were determined by reciprocal alignment of the protein sequences using BLAST. Genes in both genomes that reciprocally yielded the highest BLAST score for each other in the alignment with at least 80% sequence identity were considered to be orthologs.

### Analysis of gene expression

Raw reads generated from complementary DNA were trimmed and filtered using the fastq-mcf tool of the EA-UTILS software package with a Phred quality cut-off of 20 and the appropriate Illumina TruSeq adapter sequences to detect and remove adapters. Gene expression was determined by mapping the reads to the newly assembled and contigs using BOWTIE 2 and counting the reads that uniquely overlapped with the annotated genes. Differential gene expression was analysed using the R-package DESEQ (Anders and Huber, 2010). Gene were considered to be differentially expressed, when the absolute $\log_2$ fold change was greater than 2 (equivalent to fourfold upregulation or downregulation) with a P-value (adjusted for multiple hypothesis testing) below 0.05.

A GO enrichment analysis was performed using the BLAST2GO software (Conesa *et al.*, 2005; Götz *et al.*, 2008), testing for the enrichment of GO terms in the set of differentially expressed genes with a false discovery rate (FDR) of less than 0.05.

*Multilocus Sequence Typing*

The sequence types of the strains PA30 and PA49 were determined following the multilocus sequence typing (MLST) scheme described by Curran and colleagues (2004), by comparing the sequences of seven variable genes commonly found in *P. aeruginosa* (*acsA*, *aroE*, *guaA*, *mutL*, *nuoD*, *ppsA*, *trpE*) with the public online database PubMLST, http://www.pubmlst.org (Jolley and Maiden, 2010). The sequences of the seven MLST regions were concatenated for both strains and aligned with the concatenated MLST sequences of eight previously published genomes (*P. aeruginosa* strains PAO1, PA14, PA7, PASC2, LESB58, NCGM2, DK2 and M18, all taken from the *Pseudomonas* genome database http://www.pseudomonas.com (Winsor *et al.*, 2011) ) using CLUSTALW (Larkin *et al.*, 2007). A phylogenetic tree was constructed from the alignment using CLUSTALW and TREEVIEW (Page, 1996).

**Conflict of Interest**

None declared.

**References**

Alvarez-Ortega, C., Wiegand, I., Olivares, J., Hancock, R.E., and Martinez, J.L. (2010) Genetic determinants involved in the susceptibility of *Pseudomonas aeruginosa* to beta-lactam antibiotics. *Antimicrob Agents Chemother* **54**: 4159–4167.

Anders, S., and Huber, W. (2010) Differential expression analysis for sequence count data. *Genome Biol* **11**: R106.

Aronesty, E. (2011) *ea-utils: command-line tools for processing biological sequencing data.* [WWW document]. URL http://code.google.com/p/ea-utils.

Battle, S.E., Rello, J., and Hauser, A.R. (2009) Genomic islands of *Pseudomonas aeruginosa*. *FEMS Microbiol Lett* **290**: 70–78.

Bernier, S.P., Lebeaux, D., DeFrancesco, A.S., Valomon, A., Soubigou, G., Coppee, J.Y., *et al.* (2013) Starvation, together with the SOS response, mediates high biofilm-specific tolerance to the fluoroquinolone ofloxacin. *PLoS Genet* **9**: e1003144.

Breidenstein, E.B., de la Fuente-Nunez, C., and Hancock, R.E. (2011) *Pseudomonas aeruginosa*: all roads lead to resistance. *Trends Microbiol* **19**: 419–426.

Bruchmann, S., Dötsch, A., Nouri, B., Chaberny, I.F., and Häussler, S. (2013) Quantitative contributions of target alteration and decreased drug accumulation to *Pseudomonas aeruginosa* fluoroquinolone resistance. *Antimicrob Agents Chemother* **57**: 1361–1368.

Camacho, C., Coulouris, G., Avagyan, V., Ma, N., Papadopoulos, J., Bealer, K., and Madden, T.L. (2009) BLAST+: architecture and applications. *BMC Bioinformatics* **10**: 421.

Conesa, A., Götz, S., Garcia-Gomez, J.M., Terol, J., Talon, M., and Robles, M. (2005) Blast2GO: a universal tool for annotation, visualization and analysis in functional genomics research. *Bioinformatics* **21**: 3674–3676.

Curran, B., Jonas, D., Grundmann, H., Pitt, T., and Dowson, C.G. (2004) Development of a multilocus sequence typing scheme for the opportunistic pathogen *Pseudomonas aeruginosa*. *J Clin Microbiol* **42**: 5644–5649.

Davies, J., and Davies, D. (2010) Origins and evolution of antibiotic resistance. *Microbiol Mol Biol Rev* **74**: 417–433.

Davies, J., Spiegelman, G.B., and Yim, G. (2006) The world of subinhibitory antibiotic concentrations. *Curr Opin Microbiol* **9**: 445–453.

Donlan, R.M., and Costerton, J.W. (2002) Biofilms: survival mechanisms of clinically relevant microorganisms. *Clin Microbiol Rev* **15**: 167–193.

Dötsch, A., Becker, T., Pommerenke, C., Magnowska, Z., Jänsch, L., and Häussler, S. (2009) Genome-wide identification of genetic determinants of antimicrobial drug resistance in *Pseudomonas aeruginosa*. *Antimicrob Agents Chemother* **53**: 2522–2531.

Dötsch, A., Klawonn, F., Jarek, M., Scharfe, M., Blöcker, H., and Häussler, S. (2010) Evolutionary conservation of essential and highly expressed genes in *Pseudomonas aeruginosa*. *BMC Genomics* **11**: 234.

Fajardo, A., Martinez-Martin, N., Mercadillo, M., Galan, J.C., Ghysels, B., Matthijs, S., *et al.* (2008) The neglected intrinsic resistome of bacterial pathogens. *PLoS ONE* **3**: e1619.

Flemming, H.C., and Wingender, J. (2001) Relevance of microbial extracellular polymeric substances (EPSs) – part I: structural and ecological aspects. *Water Sci Technol* **43**: 1–8.

Goh, E.B., Yim, G., Tsui, W., McClure, J., Surette, M.G., and Davies, J. (2002) Transcriptional modulation of bacterial gene expression by subinhibitory concentrations of antibiotics. *Proc Natl Acad Sci USA* **99**: 17025–17030.

Götz, S., Garcia-Gomez, J.M., Terol, J., Williams, T.D., Nagaraj, S.H., Nueda, M.J., *et al.* (2008) High-throughput functional annotation and data mining with the Blast2GO suite. *Nucleic Acids Res* **36**: 3420–3435.

Jolley, K.A., and Maiden, M.C. (2010) BIGSdb: scalable analysis of bacterial genome variation at the population level. *BMC Bioinformatics* **11**: 595.

Jungfer, C., Friedrich, F., Varela-Villarreal, J., Brändle, K., Gross, H.J., Obst, U., and Schwartz, T. (2013) Drinking water biofilms on copper and stainless steel exhibit specific molecular responses towards different disinfection regimes at waterworks. *Biofouling* **29**: 891–907.

Klockgether, J., Cramer, N., Wiehlmann, L., Davenport, C.F., and Tümmler, B. (2011) *Pseudomonas aeruginosa* genomic structure and diversity. *Front Microbiol* **2**: 150–167.

Kung, V.L., Ozer, E.A., and Hauser, A.R. (2010) The accessory genome of *Pseudomonas aeruginosa*. *Microbiol Mol Biol Rev* **74**: 621–641.

Kurtz, S., Phillippy, A., Delcher, A.L., Smoot, M., Shumway, M., Antonescu, C., and Salzberg, S.L. (2004) Versatile and open software for comparing large genomes. *Genome Biol* **5:** R12.

Kümmerer, K. (2009) Antibiotics in the aquatic environment – a review – part II. *Chemosphere* **75:** 435–441.

Kwon, D.H., and Lu, C.D. (2006) Polyamines induce resistance to cationic peptide, aminoglycoside, and quinolone antibiotics in *Pseudomonas aeruginosa* PAO1. *Antimicrob Agents Chemother* **50:** 1615–1622.

Langmead, B., and Salzberg, S.L. (2012) Fast gapped-read alignment with Bowtie 2. *Nat Methods* **9:** 357–359.

Larkin, M.A., Blackshields, G., Brown, N.P., Chenna, R., McGettigan, P.A., McWilliam, H., *et al.* (2007) Clustal W and Clustal X version 2.0. *Bioinformatics* **23:** 2947–2948.

McArthur, A.G., Waglechner, N., Nizam, F., Yan, A., Azad, M.A., Baylay, A.J., *et al.* (2013) The comprehensive antibiotic resistance database. *Antimicrob Agents Chemother* **57:** 3348–3357.

Martinez, J.L., Fajardo, A., Garmendia, L., Hernandez, A., Linares, J.F., Martinez-Solano, L., and Sanchez, M.B. (2009) A global view of antibiotic resistance. *FEMS Microbiol Rev* **33:** 44–65.

Mathee, K., Narasimhan, G., Valdes, C., Qiu, X., Matewish, J.M., Koehrsen, M., *et al.* (2008) Dynamics of *Pseudomonas aeruginosa* genome evolution. *Proc Natl Acad Sci USA* **105:** 3100–3105.

Merlin, C., Bonot, S., Courtois, S., and Block, J.C. (2011) Persistence and dissemination of the multiple-antibiotic-resistance plasmid pB10 in the microbial communities of wastewater sludge microcosms. *Water Res* **45:** 2897–2905.

Morrison, A.J., and Wenzel, R.P. (1984) Epidemiology of infections due to *Pseudomonas aeruginosa*. *Rev Infect Dis* **6:** 627–642.

Nikaido, H. (2001) Preventing drug access to targets: cell surface permeability barriers and active efflux in bacteria. *Semin Cell Dev Biol* **12:** 215–223.

Nyström, T. (2004) Stationary-phase physiology. *Annu Rev Microbiol* **58:** 161–181.

Page, R.D. (1996) TreeView: an application to display phylogenetic trees on personal computers. *Comput Appl Biosci* **12:** 357–358.

Pedró, L., Baños, R.C., Aznar, S., Madrid, C., Balsalobre, C., and Juárez, A. (2011) Antibiotics shaping bacterial genome: deletion of an IS91 flanked virulence determinant upon exposure to subinhibitory antibiotic concentrations. *PLoS ONE* **6:** e27606.

Peng, Y., Leung, H.C., Yiu, S.M., and Chin, F.Y. (2012) IDBA-UD: a de novo assembler for single-cell and metagenomic sequencing data with highly uneven depth. *Bioinformatics* **28:** 1420–1428.

Pirnay, J.P., De Vos, D., Mossialos, D., Vanderkelen, A., Cornelis, P., and Zizi, M. (2002) Analysis of the *Pseudomonas aeruginosa* oprD gene from clinical and environmental isolates. *Environ Microbiol* **4:** 872–882.

Poole, K., Gotoh, N., Tsujimoto, H., Zhao, Q., Wada, A., Yamasaki, T., *et al.* (1996) Overexpression of the *mexC-mexD-oprJ* efflux operon in *nfxB*-type multidrug-resistant strains of *Pseudomonas aeruginosa*. *Mol Microbiol* **21:** 713–724.

Rizzo, L., Manaia, C., Merlin, C., Schwartz, T., Dagot, C., Ploy, M.C., *et al.* (2013) Urban wastewater treatment plants as hotspots for antibiotic resistant bacteria and genes spread into the environment: a review. *Sci Total Environ* **447:** 345–360.

Ruiz, J. (2003) Mechanisms of resistance to quinolones: target alterations, decreased accumulation and DNA gyrase protection. *J Antimicrob Chemother* **51:** 1109–1117.

Schreiber, K., Krieger, R., Benkert, B., Eschbach, M., Arai, H., Schobert, M., and Jahn, D. (2007) The anaerobic regulatory network required for *Pseudomonas aeruginosa* nitrate respiration. *J Bacteriol* **189:** 4310–4314.

Schwartz, T., Volkmann, H., Kirchen, S., Kohnen, W., Schon-Holz, K., Jansen, B., and Obst, U. (2006) Real-time PCR detection of *Pseudomonas aeruginosa* in clinical and municipal wastewater and genotyping of the ciprofloxacin-resistant isolates. *FEMS Microbiol Ecol* **57:** 158–167.

Schweizer, H.P. (2003) Efflux as a mechanism of resistance to antimicrobials in *Pseudomonas aeruginosa* and related bacteria: unanswered questions. *Genet Mol Res* **2:** 48–62.

Seiler, C., and Berendonk, T.U. (2012) Heavy metal driven co-selection of antibiotic resistance in soil and water bodies impacted by agriculture and aquaculture. *Front Microbiol* **3:** 399–408.

Stover, C.K., Pham, X.Q., Erwin, A.L., Mizoguchi, S.D., Warrener, P., Hickey, M.J., *et al.* (2000) Complete genome sequence of *Pseudomonas aeruginosa* PAO1, an opportunistic pathogen. *Nature* **406:** 959–964.

Trautmann, M., Lepper, P.M., and Haller, M. (2005) Ecology of *Pseudomonas aeruginosa* in the intensive care unit and the evolving role of water outlets as a reservoir of the organism. *Am J Infect Control* **33:** 41–49.

Trimble, W.L., Keegan, K.P., D'Souza, M., Wilke, A., Wilkening, J., Gilbert, J., and Meyer, F. (2012) Short-read reading-frame predictors are not created equal: sequence error causes loss of signal. *BMC Bioinformatics* **13:** 183.

Wang, H., Masters, S., Hong, Y., Stallings, J., Falkinham, J.O., Edwards, M.A., and Pruden, A. (2012) Effect of disinfectant, water age, and pipe material on occurrence and persistence of *Legionella*, mycobacteria, *Pseudomonas aeruginosa*, and two amoebas. *Environ Sci Technol* **46:** 11566–11574.

Winsor, G.L., Lam, D.K., Fleming, L., Lo, R., Whiteside, M.D., Yu, N.Y., *et al.* (2011) *Pseudomonas* Genome Database: improved comparative analysis and population genomics capability for *Pseudomonas* genomes. *Nucleic Acids Res* **39:** D596–D600.

Yeung, O., Law, S.P., and Yau, M. (2009) Treatment generalization and executive control processes: preliminary data from Chinese anomic individuals. *Int J Lang Commun Disord* **44:** 784–794.

## Supporting information

Additional Supporting Information may be found in the online version of this article at the publisher's web-site:

**Table S1.** Genome sequencing and assembly statistics.
**Table S2.** Coverage of references sequences. Percentages indicate the fraction of the references sequences that overlapped with the newly assembled contigs. See also Fig. 1.
**Table S3.** Ribonucleic acid sequencing statistics.
**Table S4.** MS Excel table file (.xslx) of genes that were differentially expressed comparing the transcriptomes of isolate PA30 in waste water and tap water. The file contains two table sheets listing the genes that were upregulated or downregulated upon exposure to waste water as compared with exposure to tap water.

**Table S5.** MS Excel table file (.xslx) of genes that were differentially expressed comparing the transcriptomes of isolate PA49 in waste water and tap water. The file contains two table sheets listing the genes that were upregulated or downregulated upon exposure to waste water as compared with exposure to tap water.

# 12

# Multicopy integration of mini-Tn7 transposons into selected chromosomal sites of a *Salmonella* vaccine strain

Karen Roos, Esther Werner and Holger Loessner*

*Bacterial Vaccines and Immune Sera, Department of Veterinary Medicine, Paul Ehrlich Institute, Langen 63225, Germany*

## Summary

Chromosomal integration of expression modules for transgenes is an important aspect for the development of novel *Salmonella* vectors. Mini-Tn7 transposons have been used for the insertion of one such module into the chromosomal site *attTn7*, present only once in most Gram-negative bacteria. However, integration of multiple mini-Tn7 copies might be suitable for expression of appropriate amounts of antigen or combination of different modules. Here we demonstrate that integration of a 9.6 kb mini-Tn7 harbouring the luciferase *luxCDABE* (*lux*) occurs at the natural *attTn7* site and simultaneously other locations of the *Salmonella* chromosome, which were engineered using λ-Red recombinase to contain one or two additional artificial *attTn7* sites (*a-attTn7*). Multicopy integration even at closely spaced *attTn7* sites was unexpected in light of the previously reported distance-dependent Tn7 target immunity. Integration of multiple copies of a mini-Tn7 containing a *gfp* cassette resulted in increasing green fluorescence of bacteria. Stable consecutive integration of two mini-Tn7 encoding *lacZ* and *lux* was achieved by initial transposition of *lacZ*-mini-Tn7, subsequent chromosomal insertion of *a-attTn7* and a second round of transposition with *lux*-mini-Tn7. Mini-Tn7 thus constitutes a versatile method for multicopy integration of expression cassettes into the chromosome of *Salmonella* and possibly other bacteria.

*For correspondence. E-mail holger.loessner@pei.de

**Funding Information** This work was supported by internal funding of the Paul Ehrlich Institute.

## Introduction

Engineering of whole bacterial genomes has advanced rapidly in the last decade (Carr and Church, 2009; Feher *et al.*, 2012). However, combination of multiple functions and their orchestration remains challenging. Serial modification of bacterial genomes, such as chromosomal integration of large expression modules in a rapid, targeted and stable manner is difficult to achieve, especially in bacteria other than *Escherichia coli*. These difficulties also hamper the development of new recombinant live-attenuated bacterial vaccines such as *Salmonella enterica* ssp. (*Salmonella*) vector vaccines and the exploration of their potential as vector platform. Current strategies for rational design of these vaccines are based on a detailed understanding of bacteria–host interaction and the specific requirements for the induction of protective immune responses against a targeted pathogen (Galen *et al.*, 2009; Hegazy and Hensel, 2012; Roland and Brenneman, 2013; Wang *et al.*, 2013b; Galen and Curtiss, 2014). One approach for *Salmonella* vaccine strain development is the removal of genes in order to attenuate virulence, reduce metabolic burden and genetically stabilize or redirect intracellular trafficking of bacteria. Another strategy is based on programming of the bacteria for the delivery of a heterologous cargo, such as proteinaceous antigens or DNA vaccines. In this case, expression modules for one or several heterologous antigens in addition to other factors, such as lysis determinants, secretion system components, regulatory factors or adjuvant molecules, have to be stably introduced into the respective candidate strain. By combining these strategies, a vaccine strain was engineered to contain 13 gene deletions and two chromosomal integrated expression cassettes in addition to the plasmid-based DNA vaccine, yielding a self-destructing *Salmonella* vector system for efficient DNA vaccine delivery in mice (Kong *et al.*, 2012). Although this provides an important proof of principle for *Salmonella* vector development, the recombinase-based suicide vector technology used to achieve these precise deletion/deletion–insertion mutations is cumbersome (Kang *et al.*, 2002).

Alternatively, λ-Red recombineering has become a frequently used method for targeted gene deletions and chromosomal insertions of small DNA fragments (Sawitzke et al., 2007; Sharan et al., 2009). The polymerase chain reaction (PCR)-based version of this method (Datsenko and Wanner, 2000) allows deletion of chromosomal genes by transformation of an amplified DNA fragment, containing a selectable marker flanked by ~ 50 bp extensions homologous to the chromosomal target sequence. In the presence of λ-Red recombinase, the specified chromosomal sequence is then replaced by the selectable marker gene in the recipient strain, which is subsequently removed. However, only relatively short additional heterologous DNA sequences can be chromosomally integrated this way due to a decrease of recombination efficiency for fragments exceeding ~ 1500 bp (Datsenko and Wanner, 2000; Kuhlman and Cox, 2010). Chromosomal integration of larger fragments can still be achieved by λ-Red when long flanking homology regions are employed (Yu et al., 2011; Dharmasena et al., 2013; Sabri et al., 2013; Wang et al., 2013a), but such fragments cannot easily be generated by PCR. This obstacle can be overcome by a two-step procedure, in which λ-Red recombineering is combined with a second method. For example, λ-Red was used in the first step for the integration of a so-called landing pad into a selected chromosomal locus, which is a DNA fragment harbouring a selection marker flanked by I-SceI recognition sites and short sequences homologous to the integration cassette (Kuhlman and Cox, 2010). Subsequently, this integration cassette is transformed into bacteria, excised by I-SceI and recombined into the I-SceI-cleaved landing pad region, allowing the targeting of a 7 kb cassette into selected loci of the E. coli chromosome.

Mobile elements, such as transposon or phage-derived systems, integrate either randomly or site specifically into the bacterial genome and therefore do not allow deliberate targeting (Choi and Kim, 2009; Akhverdyan et al., 2011; Murphy, 2012; Loeschcke et al., 2013). Tn7-derived transposons, so-called mini-Tn7, have been frequently used for site-specific, single-copy integration of large DNA fragments into the chromosome of Gram-negative bacteria including Salmonella (Bao et al., 1991; Yan and Meyer, 1996; Choi et al., 2005; Loessner et al., 2007; Kvitko et al., 2013). In few bacteria, e.g. Burkholderia spp. or Proteus mirabilis, two or three natural sites have been found, which can be targeted by mini-Tn7 (Choi and Schweizer, 2006; Choi et al., 2006; 2014). In its natural form, Tn7 is a 14 kb transposable element encoding genes tnsABCDE of the transposition machinery and additional antibiotic resistance genes (Lichtenstein and Brenner, 1982; Peters and Craig, 2001). Recognition of its attachment site attTn7, located downstream of gene glmS in most bacteria including Salmonella spp., is mediated by

the DNA-binding protein TnsD that recruits TnsC, an ATP-dependent DNA-binding protein, thereby linking target recognition with activation of transposase TnsAB (Bainton et al., 1993; Choi et al., 2014). Orientation-specific insertion of Tn7 is mediated by the asymmetric left and right transposon ends, Tn7L and Tn7R respectively. Both ends contain multiple TnsB binding sites. The presence of these sites in the chromosome as a consequence of transposon integration was described to confer so-called target immunity impeding further transposition events by a mechanism in which bound TnsB inactivates TnsC (Arciszewska et al., 1989; DeBoy and Craig, 1996; Stellwagen and Craig, 1997). This immunity effect was shown to be active at positions of the E. coli chromosome, which are located up to 190 kb apart from TnsB binding sites but integration into a site located 1900 kb away was not inhibited (DeBoy and Craig, 1996).

In this work, we have combined λ-Red recombineering with chromosomal integration of mini-Tn7 into the genome of the attenuated S. enterica serovar Typhimurium (S. Typhimurium) vaccine strain SL7207. We demonstrate that λ-Red-mediated integration of one or two artificial attTn7 sites (a-attTn7) into chromosomal loci of choice allows subsequent integration of multiple mini-Tn7 copies simultaneously or consecutively into the bacterial genome. Chromosomal integration of one, two or three copies of a mini-Tn7 containing a gfp expression cassette allowed us to modulate GFP expression levels. Furthermore, consecutive integration of two different mini-Tn7 harbouring lacZ and lux, respectively, gave rise to a multifunctional strain.

## Results

Mini-Tn7 transposons are frequently used for single-copy integration of DNA fragments into chromosomes of Gram-negative bacteria, with few exceptions (Choi et al., 2005; Crepin et al., 2012). However, consecutive integration of multiple mini-Tn7 copies into additional attTn7 sites present in the E. coli chromosome was until now thought to be suppressed in a distance-dependent manner by Tn7 target immunity (Arciszewska et al., 1989; DeBoy and Craig, 1996; Stellwagen and Craig, 1997; Choi et al., 2014). Here we investigated the possibility of simultaneous integration of two mini-Tn7 copies into attTn7 as well as an additional a-attTn7 site inserted at genomic positions at varying distances apart from attTn7 in the S. Typhimurium vaccine strain SL7207. The a-attTn7 was derived from the S. Typhimurium strain SL1344 genomic sequence and contains 105 bp of the 3′ end of glmS, 140 bp intergenic region and 45 bp of the 3′ end of gene SL1344_3827. The glmS sequence harbours the entire TnsD binding site and the intergenic region attTn7 downstream of glmS (Mitra et al., 2010). We successively

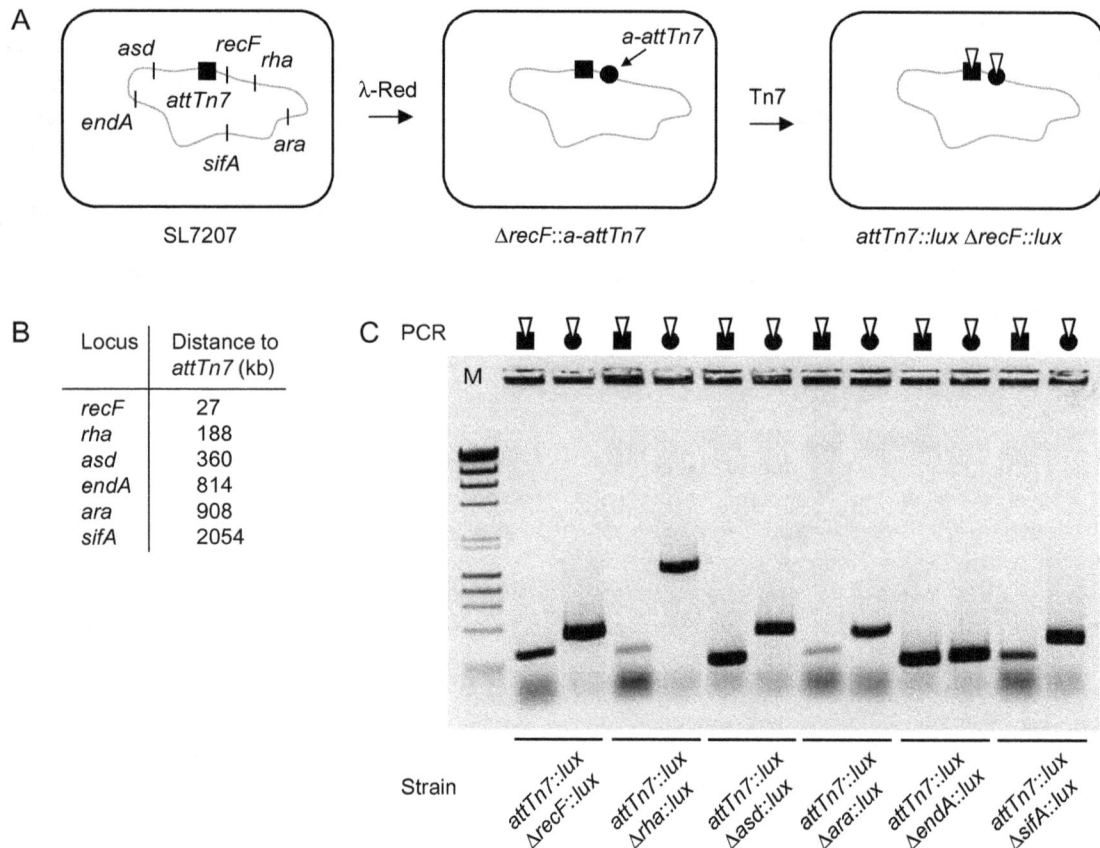

**Fig. 1.** Integration of two *lux*-mini-Tn7 into *attTn7* and in addition one *a-attTn7* site inserted into selected chromosomal loci of
*S.* Typhimurium.
A. Strain SL7207 was modified in two steps for each of the indicated genes. λ-Red-mediated deletion and concurrent integration of *a-attTn7*
(black circle) is depicted for *recF*. Strains harbouring the natural *attTn7* site (black square) and in addition *a-attTn7* were used for transposition
with *lux*-mini-Tn7 9.6 kb in size (white inversed triangle).
B. Approximate distances between *attTn7* and the newly introduced *a-attTn7* sites in respective strains.
C. Integration of *lux*-mini-Tn7 into *attTn7* or *a-attTn7* (indicated by combined symbols) was confirmed by the bioluminescent phenotype (data
not shown) and by colony PCR with primers homologous to sequences of the right end of *lux*-mini-Tn7 (Tn7R) and the neighbouring genomic
location (Supporting Information Table S1). The expected band sizes for *lux*-mini-Tn7 integration into *attTn7* is 332 bp, for integrations into
*a-attTn7* sites in mutant strains are *recF* 586 bp, *rha* 1523 bp, *asd* 629 bp, *endA* 526 bp, *ara* 378 bp and *sifA* 543 bp respectively. Δ indicates
a gene deletion and double colon indicates a chromosomal insertion. PCR data for representative clones are shown.

replaced chromosomal genes *ara*, *asd*, *endA*, *recF*, *rha*
and *sifA* of strain SL7207 with *a-attTn7* one at a time by
λ-Red-mediated homologous recombination (Fig. 1A).
The targeted loci were located at varying distances below
or above 190 kb apart from *attTn7* (Fig. 1B). We then
subjected the six different mutant strains carrying
*attTn7* as well as one additional *a-attTn7* site to transposition with *lux*-mini-Tn7, which contains a constitutive
*lux*-cassette and a kanamycin marker. Bioluminescent
*Salmonella* colonies with chromosomal insertions of *lux*-
mini-Tn7 were selected on LB plates containing streptomycin and kanamycin. Colony PCR of all mutant strains
revealed that integration of *lux*-mini-Tn7 occurred into
both integration sites (Fig. 1C). This was even the case for
strains carrying *a-attTn7* in the *recF* and the *rha* loci,
which are less than 190 kb away from *attTn7*. Therefore,

we did not observe immunity-related inhibition of *lux*-mini-
Tn7 integration into either one of the two transposon
integration sites present in the genome of all tested
mutant strains.

To investigate simultaneous integration of *lux*-mini-Tn7
into additional chromosomal locations, *a-attTn7* sites
were integrated into *ara* and *asd* loci by two subsequent
rounds of λ-Red-mediated gene replacement (Fig. 2A).
The intermediate strain harbouring three Tn7 integration
sites was then again subjected to transposition with *lux*-
mini-Tn7 (Fig. 2A). In this strain the three Tn7 integration
sites were spaced more than 190 kb apart from each
other. PCR analysis of colonies obtained from selective
plates revealed that mini-Tn7 integration occurred consistently into all three transposon integration sites in
all tested colonies (Fig. 2B). Such clones could also be

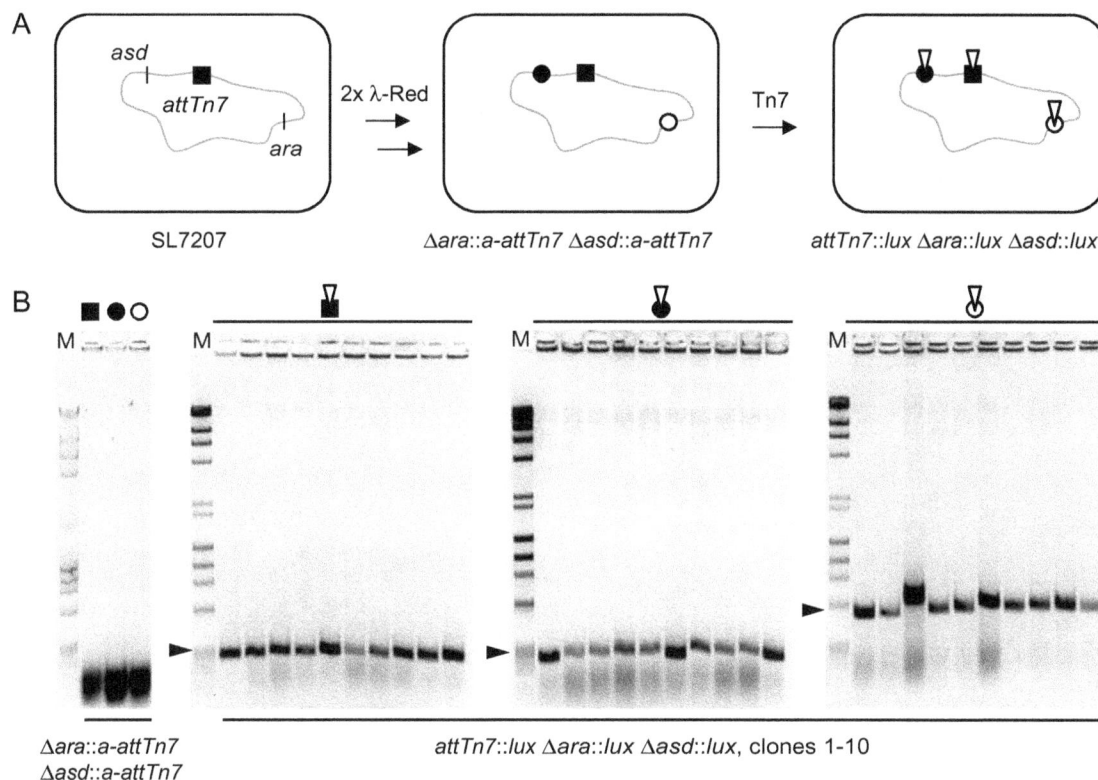

**Fig. 2.** Consistency of simultaneous transposition of *lux*-mini-Tn7 into three loci of the *S.* Typhimurium chromosome.
A. A SL7207 derivative with the natural *attTn7* site (black square) and two additional *a-attTn7* sites (black and white circles) was generated by two rounds of λ-Red-mediated recombination linked with chromosomal integration of the *a-attTn7* sequence each time.
B. Strain SL7207Δ*ara::a-attTn7*Δ*asd::a-attTn7* harbouring three sites for Tn7 transposition was used for transposition with *lux*-mini-Tn7 (white inversed triangle). *lux*-mini-Tn7 integration into the three sites was confirmed by colony PCR for 10 bioluminescent clones selected on LB plates containing streptomycin and kanamycin. The strain harbouring three empty Tn7 transposition sites was used as control template. The band size for *lux*-mini-Tn7 integration into *attTn7* is 332 bp and for integrations into *a-attTn7* are *ara* 378 bp and *asd* 629 bp, band positions are indicated by black arrow heads.

identified due to the bioluminescent phenotype alone by plating a dilution series of the triple mating culture on LB plates containing only streptomycin. The mean frequency of identified positive colonies from three experimental repetitions was $(4.5 \pm 1.7) \times 10^{-5}$, indicating that high efficiency of multicopy integration can even be achieved with a mini-Tn7 devoid of an antibiotic resistance marker if phenotypical identification of colonies is possible.

To determine if chromosomal integration of multiple copies of a *gfp*-mini-Tn7 allows a stepwise, copy number-dependent increase in the expression level of a transgene, we integrated one, two or three copies of a GFP expression cassette into SL7207. Towards this, either one or two additional *a-attTn7* sites were initially recombined into the *ara* and *asd* loci, and then a *gfp*-mini-Tn7 was used for transposition of respective strains. Colonies containing insertions within the present Tn7 integration sites were identified by colony PCR (Fig. 3A). These clones were then grown in liquid medium and green fluorescence of cells was analysed by flow cytometry upon induction of GFP synthesis by addition of

L-arabinose. The level of green fluorescence intensity measured correlated with the copy number of chromosomally integrated *gfp*-mini-Tn7 at selected loci, while cultures grown in the absence of L-arabinose displayed only low GFP fluorescence (Fig. 3B and Supporting Information Fig. S1).

In addition to the integration of multiple copies of the same mini-Tn7, we also investigated the consecutive integration of two different mini-Tn7 carrying *lacZ* and *lux* into strain SL7207. In a first round of transposition, either *lacZ*-mini-Tn7 or *lux*-mini-Tn7 was integrated into the native Tn7 recognition site *attTn7*. Thereafter, the strain harbouring *lacZ*-mini-Tn7 was subjected to λ-Red-mediated integration of *a-attTn7* into the *rha* locus. Transposition of *lux*-mini-Tn7 into this strain gave rise to bacteria harbouring both mini-Tn7 encoding *lacZ* and *lux* (Fig. 3C). Phenotypical analysis of these strains confirmed the expression of *lacZ*, *lux* or both transgenes (Fig. 3D). To assess the stability of *lacZ* and *lux* expression cassettes we evaluated colonization and foreign gene expression of the recombinant *S.* Typhimurium

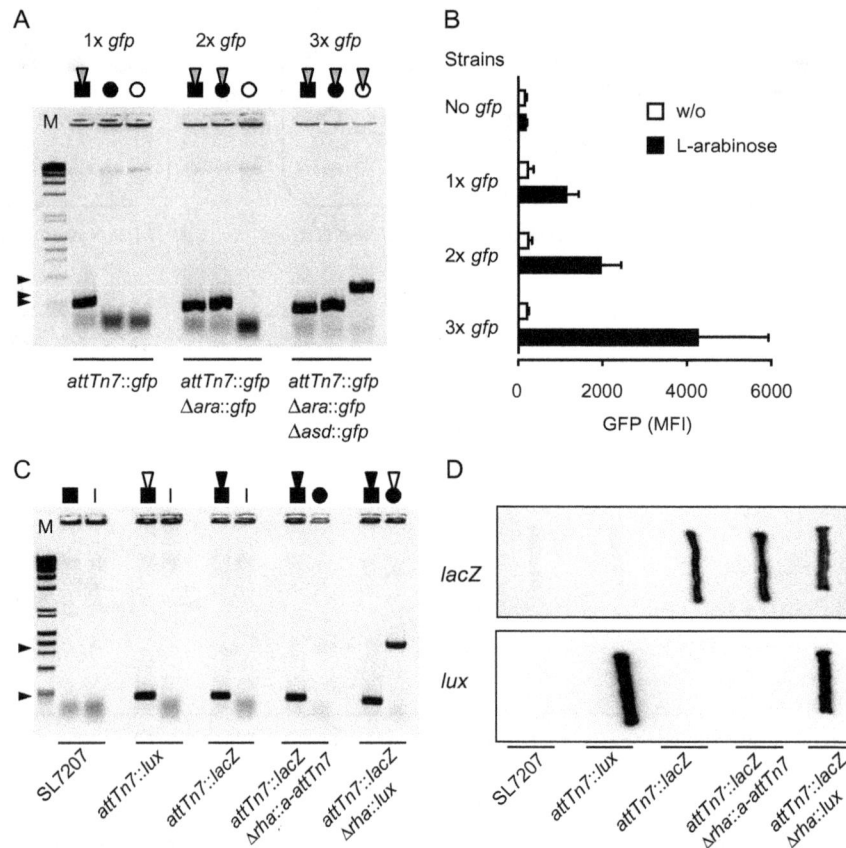

**Fig. 3.** Modulation of GFP synthesis by chromosomal integration of either one, two or three copies of *gfp*-mini-Tn7 and consecutive chromosomal integration of two different mini-Tn7 encoding *lacZ* and *lux* into *S.* Typhimurium strain SL7207.

A. Strain SL7207 with the natural *attTn7* site (black square) or derivatives with one or two additional a *attTn7* sites integrated into *ara* and *asd* loci (black and white circles) were used for transposition with *gfp*-mini-Tn7 harbouring the L-arabinose inducible *gfp* expression cassette (grey inversed triangle). *gfp*-mini-Tn7 integration into each site was verified by colony PCR and expected bands were obtained at 332 bp for integration into *attTn7*, 378 bp for integration into *a-attTn7* at the *ara* locus and 629 bp for integration into *a-attTn7* at the *asd* locus. Bands are indicated by black arrow heads.

B. Strains SL7207 and derivatives with one, two or three chromosomal copies of the *gfp* cassette were grown in the absence or presence of L-arabinose prior to flow cytometric analysis. Depicted mean fluorescence intensities (MFI) represent three experimental repetitions and error bars indicate standard deviation.

C. Strain SL7207 was used for transposition with *lacZ*-mini-Tn7 or *lux*-mini-Tn7 (black and white inversed triangles). Recipient strains carry either *lacZ* or *lux* within *attTn7* (black square) and are designated *attTn7::lacZ* and *attTn7::lux* respectively. Subsequently, the *rha* operon of strain *attTn7::lacZ* was replaced by *a-attTn7* (white square) giving rise to strain *attTn7::lacZΔrha::a-attTn7*. This strain was employed for another transposition round with *lux*-mini-Tn7 yielding strain *attTn7::lacZΔrha::lux*. Colony PCR verified transposon integrations into the chromosome of each strain. Band sizes for integration into *attTn7* are 332 bp and into *a-attTn7* at the *rha* locus 1523 bp.

D. *lacZ* expression of bacteria was observed on X-Gal plates by blue dye formation and *lux* expression by detection of bioluminescence from the same plate.

strains in BALB/c mice. Upon oral administration, tissue colonization of bioluminescent SL7207 bacteria peaks around day 7 after oral administration (Burns-Guydish *et al.*, 2005). Groups of five mice were orally inoculated with the original strain SL7207 and derivatives *attTn7::lux* and *attTn7::lacZΔrha::lux*. The *in vivo* colonization course of the two bioluminescent strains was followed by non-invasive *in vivo* imaging for 9 days (Supporting Information Fig. S2A). Mice were then sacrificed, and Peyer's patches, mesenteric lymph nodes and spleens were harvested for determination of colony-forming units (Supporting Information Fig. S2B). We

found that all tested strains readily colonized the analysed organs, even though the strains expressing additional transgenes yielded lower numbers, indicating an attenuating effect. However, all engineered bacteria retained *lacZ* and *lux* expression as confirmed by phenotypical examination of bacterial colonies recovered from organs (Supporting Information Fig. S2B), illustrating that consecutive transposition of different mini-Tn7 is a feasible approach for stable chromosomal integration of various heterologous expression cassettes into *Salmonella* and should also be considered for other Gram-negative bacteria.

## Discussion

Recombinant *Salmonella* vector vaccines stimulate humoral and cell-mediated immune responses in mucosal and systemic compartments to self-antigens as well as to heterologous antigens derived from infectious agents or malignant tissues (Galen *et al.*, 2009; Curtiss *et al.*, 2010; Paterson *et al.*, 2010; Hegazy and Hensel, 2012; Roland and Brenneman, 2013; Toussaint *et al.*, 2013; Wang *et al.*, 2013b). The potency of such a vaccine depends on its ability to induce high-level production of the respective antigens once bacteria have reached the immune inductive sites. Therefore, the introduction of appropriately tailored antigen expression cassettes is an important step during construction of a recombinant *Salmonella* vector vaccine. The use of a self-replicating plasmid is one option for the introduction of such cassettes. Such plasmids are designed to assure stable retention by the recipient bacteria and to maintain colonization of immune inductive tissues (Bauer *et al.*, 2005; Galen *et al.*, 2010; Xin *et al.*, 2012). However, this situation is difficult to achieve with multicopy plasmids harbouring expression cassettes for large gene clusters, but these limitations can be overcome by integrating such cassettes into the bacterial chromosome (Husseiny and Hensel, 2008; Yu *et al.*, 2011; Dharmasena *et al.*, 2013).

A variety of transposons have been used as gene delivery tools for bacteria and eukaryotic cells (Choi and Kim, 2009; Ivics and Izsvak, 2011). Tn7 is the hallmark transposon for site-specific single-copy integration into the chromosome of Gram-negative bacteria (Peters and Craig, 2001). Here we investigated the possibility of integrating multiple copies of a mini-Tn7 into the chromosome of the S. Typhimurium strain SL7207 containing several *attTn7* sites. Towards this, we used λ-Red recombineering for the insertion of initially one *a-attTn7* site into selected positions of the *Salmonella* chromosome located at varying distances from *attTn7*. Using a 9.6 kb mini-Tn7 with a constitutive luciferase expression cassette, we obtained for each position transposon insertions at both sites *attTn7* and *a-attTn7*. Surprisingly, this was independent from the distance between the sites, which was as small as 27 kb in the case of the *recF* locus and as large as 2054 kb in the case of the *sifA* locus. Integration of three *lux*-mini-Tn7 copies into *attTn7* and two additional *a-attTn7* sites located in the *asd* and *ara* loci occurred with the same consistency. To our knowledge, this is the first report of integration of multiple mini-Tn7 copies into deliberately selected loci of the bacterial chromosome down to a distance below 190 kb between integration sites, which was previously considered inefficient due to Tn7 target immunity (DeBoy and Craig, 1996). Chromosomal integration of multiple mini-Tn7 copies has been reported to occur naturally in *Burkholderia* ssp. and *P. mirabilis*, which

carry two or three natural integration sites in their bacterial chromosomes (Choi and Schweizer, 2006; Choi *et al.*, 2006; 2008). For example in *Burkholderia pseudomallei* strain 1026b three natural *attTn7* sites are present, which are located downstream of three *glmS* genes (Choi *et al.*, 2008). Mini-Tn7 integration occurs in > 65% of events as single-copy integrations downstream of gene *glmS2*. The other two *attTn7* sites downstream of genes *glmS1* and *glmS3* were targeted by single-copy integration of mini-Tn7 at much lower rates. Double insertions of a mini-Tn7 in one cell occur at a rate of 10–20%, but triple insertions at once were only rarely observed (Choi *et al.*, 2008). The reason of the low double and triple mini-Tn7 insertion frequency in this strain is unknown. Tn7 target immunity would not be expected to be the obvious reason as *attTn7* sites on chromosome 1 are 1,1 Mb apart from each other, and one *attTn7* site is located on chromosome 2. In *P. mirabilis* strain HI4320, Choi and Schweizer (2006) always observed simultaneous mini-Tn7 insertion into two sites, the *attTn7* site downstream of *glmS* and one site located in *carA*, when bacteria were grown in rich medium. These sites are spaced 724 kb and would also not expected to be affected by Tn7 target immunity. However, in our work with *Salmonella* we observed a frequency of 100% of simultaneous mini-Tn7 insertions at two or three *attTn7* sites, even in the case of closely spaced *attTn7* sites 27 or 188 kb apart.

Designing *Salmonella* vector vaccines for the immunization against heterologous pathogens has so far required a number of precise chromosomal gene deletions and module integrations in parallel. The combination of λ-Red-mediated gene deletion with mini-Tn7 integration reported here has the potential to considerably facilitate this process. As target loci for mini-Tn7 integration, we specifically selected genes, which have been previously removed from *Salmonella* vaccine strains as such sites should be suited for the stable integration of heterologous expression cassettes. Δ*asd* strains were used as hosts for plasmids stabilized by the complementing essential *asd* gene, referred to as balanced lethal systems (Nakayama *et al.*, 1988). In addition, the Δ*asd* mutation is present in strains undergoing a so-called delayed type of bacterial lysis (Kong *et al.*, 2008). Δ*ara* and Δ*rha* mutations have been introduced into vaccine strains in order to prevent degradation of L-arabinose or L-rhamnose when such sugars are used as inductors of gene expression (Kong *et al.*, 2008). Δ*recE* strains were shown to display a reduced capacity for recombination and therefore are suited for stable maintenance of dual-plasmid systems in vaccine strains (Zhang *et al.*, 2011; Xin *et al.*, 2012). Δ*sifA* strains were employed for the cytosolic delivery of heterologous antigens or DNA vaccines due to their ability to escape the phagosomal compartment inside host cells (Brumell *et al.*, 2002). When we used a *gfp*-mini-Tn7, integration of one, two or three copies into chromosomal loci *attTn7*, Δ*ara::a-attTn7* and Δ*asd::a-attTn7*, respectively, correlated with a stepwise increase of GFP fluorescence demonstrating a stepwise modulation of the expression level by targeted multicopy integration of a mini-Tn7.

Live bacterial vaccine or vector construction often necessitates chromosomal integration of expression cassettes for different transgenes (Galen *et al.*, 2009; Curtiss *et al.*, 2010; Hegazy and Hensel, 2012; Wang *et al.*, 2013b). Therefore, we investigated the consecutive introduction of two different mini-Tn7 harbouring reporter genes *lacZ* or *lux* into a strain carrying *attTn7* and *a-attTn7* respectively. The recombinant bacteria containing both modules were readily obtained by initial transposition of *lacZ*-mini-Tn7, followed by λ-Red-mediated replacement of the *rha* locus with *a-attTn7* and subsequent transposition of the *lux*-mini-Tn7. Importantly, *lacZ* and *lux* expression was stably retained during colonization of mouse tissues upon oral administration. However, instability of such modules might potentially be caused by endogenous recombinases acting on homologous sequences such as attTn7/a-attTn7, but even more so between multiple copies of the same mini-Tn7. This problem might be circumvented by the removal of host recombinases, reduction of sequence homologies and appropriate chromosomal positioning of insertion sequences (Zhang *et al.*, 2011). Bioluminescence *in vivo* imaging and subsequent recovery and quantification of bacteria from mice indicated that the strains containing mini-Tn7 insertions colonized tissues to a somewhat lower extent possibly due to an attenuating effect of constitutive expression of *lacZ* and *lux*. However, combination of our approach with strategies for the inducible expression of heterologous factors may be able to resolve such effects (Hohmann *et al.*, 1995).

In conclusion, we consider λ-Red recombinase-mediated positioning of additional Tn7 attachment sites in the bacterial genome followed by mini-Tn7 transposition a versatile method for the construction of multifunctional *Salmonella* vaccines and vectors. Further work should be devoted to answer questions such as (i) how many mini-Tn7 copies can be reliably integrated and (ii) will consecutive insertions in close proximity be compromised by Tn7 target immunity. The ability of researchers to construct multifunctional vector strains may help to overcome known limitations of present recombinant *Salmonella* vaccines and eventually allow the development of a new generation of efficacious and safe *Salmonella* vaccines for clinics (Galen *et al.*, 2009; Wang *et al.*, 2013b). Moreover, such novel *Salmonella* vector systems also constitute promising delivery systems for other medical interventions, such as tumour therapy (Forbes, 2010; Leschner and Weiss, 2010).

## Experimental procedures

### Bacterial strains and growth conditions

*Salmonella enterica* serovar Typhimurium strain SL7207 (Δ*aroA*, Δ*hisG*) was originally obtained from Bruce Stocker (Hoiseth and Stocker, 1981) and kindly provided by Siegfried Weiss (Helmholtz Center for Infection Research, Braunschweig). Chromosomal modification of this strain was mediated by λ-Red recombinase or transposition of a mini-Tn7 (see below). Strain derivatives are described in the Results section. *Escherichia coli* strains Top10 and Pir2 (Life Technologies) were used as hosts for cloning, and *E. coli* strain SM10λ*pir* (Simon *et al.*, 1983) was used for mobilization of mini-Tn7 and the Tn7 helper plasmid. Bacteria were grown on LB agar plates or in LB medium supplemented with 100 μg ml⁻¹ ampicillin, 30 μg ml⁻¹ kanamycin, 30 μg ml⁻¹ streptomycin, 2 mg ml⁻¹ L-arabinose, 50 μg ml⁻¹ diaminopimelic acid or 30 μg ml⁻¹ 5-Bromo-4-chloro-3-indolyl-β-D-galactopyranoside (X-Gal), where appropriate. LB medium base and supplements were purchased from Carl Roth.

### Plasmid constructions

A 300 bp sequence harbouring the *attTn7* integration site (*a-attTn7*) was derived from the genomic sequence of strain *S.* Typhimurium SL1344 obtained from NCBI GenBank (accession number FQ312003, position 4090151–4090450 bp). This sequence including flanking EcoRI restriction sites was synthesized and sequenced (Geneart, Life Technologies). To apply λ-Red recombination for chromosomal integration of *a-attTn7*, the template plasmid pKD4 (Datsenko and Wanner, 2000) was modified. The synthetic DNA was cleaved with EcoRI and inserted into the NdeI site of plasmid pKD4 after ends were made compatible with Klenow enzyme. The plasmid containing the insert in the same orientation as the kanamycin resistance gene was named pKR31a. This plasmid was subsequently used as template for the generation of PCR products with homology ends for λ-Red recombination. Primer pairs for the replacement of *ara*, *asd*, *endA*, *recF*, *rha* and *sifA* are listed in Supporting Information Table S1. Plasmids pKD46 (Datsenko and Wanner, 2000) and pCP20 (Cherepanov and Wackernagel, 1995) were used for the expression of λ-Red (*exo*, *bet* and *gam* genes) and FLP recombinase respectively. Expression cassettes for the bioluminescence operon *luxCDABE* (*lux*) derived from *Photorhabdus luminescence*, the green fluorescence protein mutant 2 variant (*gfp*) (Cormack *et al.*, 1996) and β-galactosidase of *E. coli* (*lacZ*) were cloned into a derivative of the mini-Tn7 pUX-BF5 (Bao *et al.*, 1991). The constitutive *lux* expression cassette was originally subcloned from plasmid pLite201 (Voisey and Marincs, 1998) into plasmid pHL300 (Loessner *et al.*, 2009). This plasmid was cut with BspLUII, blunt ended and again digested with BsrGI. The 1307 bp partial *lux* fragment was subsequently inserted into the mini-Tn7 transposon plasmid pHL289 (Loessner *et al.*, 2007). For this, plasmid pHL289 was opened with NheI, blunt ended and thereafter cut with BsrGI. The product was designated pHL305 (*lux*-mini-Tn7). *gfp* linked to the L-arabinose inducible promoter P_BAD (Loessner *et al.*, 2008) was cloned into a derivative of pUX-BF5 giving rise to plasmid pKR49 (*gfp*-mini-Tn7). *lacZ* was amplified from plasmid pCMVβ with *lacZ* forward and reverse primers (Supporting Information Table S1). The PCR fragment was digested with XbaI and HindIII and inserted into the same restriction sites of plasmid pHL222 (Loessner *et al.*, 2006) giving rise to plasmid pHL325. *lacZ* in conjunction with the constitutive β-lactamase promoter of *E. coli* was obtained from this plasmid by cleavage with SmaI and HindIII and inserted into the mini-Tn7 plasmid pKR50 opened with SalI and MluI after ends were blunted. The construct was designated pKR61 (*lacZ*-mini-Tn7). Plasmid pUX-BF13 containing Tn7 transposon genes *tnsABCDE* was used as helper plasmid for transposition of mini-Tn7 (Bao *et al.*, 1991). Primers used for confirmation of chromosomal integrations of mini-Tn7 are listed in Supporting Information Table S1.

### λ-Red recombinase-mediated gene replacement

λ-Red recombinase-mediated gene replacement was carried out as previously described (Datsenko and Wanner, 2000). Briefly, PCR products harbouring ~ 40 bp end sequences homologous to the respective *S.* Typhimurium target genes, a kanamycin resistance marker flanked by FLP recombinase recognition sites, and *a-attTn7*, so-called Red knock out (ko) fragments, were amplified with pKR31a as template and so-called Red ko primer (Supporting Information Table S1). Strain SL7207 harbouring plasmid pKD46 was grown in liquid LB medium supplemented with ampicillin at 30°C and 200 r.p.m. agitation up to an optical density at 600 nm (OD600) of ~ 0.4. At this time point λ-Red recombinase expression was induced by addition of L-arabinose, and 1 h later electro-competent cells were prepared by washing cells three times in ice-cold distilled water. An amount of approximately 200 ng of the Red ko fragment was electroporated into cells with the Gen Pulser Xcell System (Bio-Rad) at 2.5 kV, 400 Ohm and 25 μF using a 0.2 cm cuvette. SL7207 mutant clones were selected on media plates containing kanamycin and streptomycin at 37°C. The kanamycin resistance gene was then removed by action of FLP recombinase as previously described (Datsenko and Wanner, 2000).

### Transposition of mini-Tn7 into S. Typhimurium

Transposon-mediated site-specific integration into the chromosome of strain SL7207 was carried out according to the method of Bao *et al.* (1991). Briefly, triple mating cultures consisting of the helper strain *E. coli* SM10λpir harbouring pUX-BF13, *E. coli* SM10λpir harbouring a mini-Tn7 plasmid and the *S.* Typhimurium strain SL7207 were incubated on non-selective medium plates at 30°C for 24 h. Thereafter, bacteria were recovered and plated on medium plates containing streptomycin, kanamycin or chloramphenicol respectively. Site-specific chromosomal integration was verified by colony PCR (see Supporting Information). Of note, plasmid pUX-BF13 mediates synthesis of TnsABCDE proteins. TnsABC + D proteins are sufficient for target-specific chromosomal insertion of mini-Tn7 into *attTn7* (Waddell and Craig, 1988). In contrast TnsABC + E mediated mini-Tn7 insertion occurs at non-*attTn7* sites, but only on conjugal plasmids (Wolkow *et al.*, 1996). Therefore, we do not expect

interference of TnsE expression with the chromosomal mini-Tn7 insertions we observe in this work.

### Flow cytometry

Bacteria were grown at 37°C and 200 r.p.m. agitation up to OD600 of ~ 0.6. Then *gfp* expression was induced by addition of L-arabinose and 1 h later bacteria were pelleted, suspended in phosphate-buffered saline and subsequently analysed on a BD Accuri C6 flow cytometer (Becton Dickinson) as previously described (Loessner *et al.*, 2006). An appropriate scatter gate was used to distinguish bacteria from other particles. Data were acquired and analysed with BD CSampler software (Becton Dickinson).

### Animal work and ethics statement

Experimental procedures for the work with mice are provided as Supporting Information. All animal experiments were in compliance with the German Animal Welfare Act and approved by the competent authority (Regierungspraesidium Darmstadt).

## Acknowledgements

We would like to thank Franziska Domaschka for technical assistance, Siegfried Weiss (Helmholtz Center for Infection Research, Braunschweig) for providing of materials, and Veronika von Messling (Paul Ehrlich Institute, Langen) and Zoltan Ivics (Paul Ehrlich Institute, Langen) for critical reading of the manuscript. K.R. received a Paul Ehrlich Scholarship for her doctoral studies.

## Conflict of interest

None declared.

## References

Akhverdyan, V.Z., Gak, E.R., Tokmakova, I.L., Stoynova, N.V., Yomantas, Y.A., and Mashko, S.V. (2011) Application of the bacteriophage Mu-driven system for the integration/amplification of target genes in the chromosomes of engineered Gram-negative bacteria – mini review. *Appl Microbiol Biotechnol* **91**: 857–871.

Arciszewska, L.K., Drake, D., and Craig, N.L. (1989) Transposon Tn7. cis-Acting sequences in transposition and transposition immunity. *J Mol Biol* **207**: 35–52.

Bainton, R.J., Kubo, K.M., Feng, J.N., and Craig, N.L. (1993) Tn7 transposition: target DNA recognition is mediated by multiple Tn7-encoded proteins in a purified in vitro system. *Cell* **72**: 931–943.

Bao, Y., Lies, D.P., Fu, H., and Roberts, G.P. (1991) An improved Tn7-based system for the single-copy insertion of cloned genes into chromosomes of gram-negative bacteria. *Gene* **109**: 167–168.

Bauer, H., Darji, A., Chakraborty, T., and Weiss, S. (2005) *Salmonella*-mediated oral DNA vaccination using stabilized eukaryotic expression plasmids. *Gene Ther* **12**: 364–372.

Brumell, J.H., Tang, P., Zaharik, M.L., and Finlay, B.B. (2002) Disruption of the *Salmonella*-containing vacuole leads to increased replication of *Salmonella enterica* serovar Typhimurium in the cytosol of epithelial cells. *Infect Immun* **70:** 3264–3270.

Burns-Guydish, S.M., Olomu, I.N., Zhao, H., Wong, R.J., Stevenson, D.K., and Contag, C.H. (2005) Monitoring age-related susceptibility of young mice to oral *Salmonella enterica* serovar Typhimurium infection using an *in vivo* murine model. *Pediatr Res* **58:** 153–158.

Carr, P.A., and Church, G.M. (2009) Genome engineering. *Nat Biotechnol* **27:** 1151–1162.

Cherepanov, P.P., and Wackernagel, W. (1995) Gene disruption in *Escherichia coli*: TcR and KmR cassettes with the option of Flp-catalyzed excision of the antibiotic-resistance determinant. *Gene* **158:** 9–14.

Choi, K.H., and Kim, K.J. (2009) Applications of transposon-based gene delivery system in bacteria. *J Microbiol Biotechnol* **19:** 217–228.

Choi, K.H., and Schweizer, H.P. (2006) Mini-Tn7 insertion in bacteria with secondary, non-*glmS*-linked *attTn7* sites: example *Proteus mirabilis* HI4320. *Nat Protoc* **1:** 170–178.

Choi, K.H., Gaynor, J.B., White, K.G., Lopez, C., Bosio, C.M., Karkhoff-Schweizer, R.R., *et al.* (2005) A Tn7-based broad-range bacterial cloning and expression system. *Nat Methods* **2:** 443–448.

Choi, K.H., DeShazer, D., and Schweizer, H.P. (2006) Mini-Tn7 insertion in bacteria with multiple *glmS*-linked *attTn7* sites: example *Burkholderia mallei* ATCC 23344. *Nat Protoc* **1:** 162–169.

Choi, K.H., Mima, T., Casart, Y., Rholl, D., Kumar, A., Beacham, I.R., *et al.* (2008) Genetic tools for select-agent-compliant manipulation of *Burkholderia pseudomallei*. *Appl Environ Microbiol* **74:** 1064–1075.

Choi, K.Y., Spencer, J.M., and Craig, N.L. (2014) The Tn7 transposition regulator TnsC interacts with the transposase subunit TnsB and target selector TnsD. *Proc Natl Acad Sci USA* **111:** E2858–E2865.

Cormack, B.P., Valdivia, R.H., and Falkow, S. (1996) FACS-optimized mutants of the green fluorescent protein (GFP). *Gene* **173:** 33–38.

Crepin, S., Harel, J., and Dozois, C.M. (2012) Chromosomal complementation using Tn7 transposon vectors in *Enterobacteriaceae*. *Appl Environ Microbiol* **78:** 6001–6008.

Curtiss, R., III, Xin, W., Li, Y., Kong, W., Wanda, S.Y., Gunn, B., *et al.* (2010) New technologies in using recombinant attenuated *Salmonella* vaccine vectors. *Crit Rev Immunol* **30:** 255–270.

Datsenko, K.A., and Wanner, B.L. (2000) One-step inactivation of chromosomal genes in *Escherichia coli* K-12 using PCR products. *Proc Natl Acad Sci USA* **97:** 6640–6645.

DeBoy, R.T., and Craig, N.L. (1996) Tn7 transposition as a probe of cis interactions between widely separated (190 kilobases apart) DNA sites in the *Escherichia coli* chromosome. *J Bacteriol* **178:** 6184–6191.

Dharmasena, M.N., Hanisch, B.W., Wai, T.T., and Kopecko, D.J. (2013) Stable expression of *Shigella sonnei* form I O-polysaccharide genes recombineered into the chromo-

some of live *Salmonella* oral vaccine vector Ty21a. *Int J Med Microbiol* **303:** 105–113.

Feher, T., Burland, V., and Posfai, G. (2012) In the fast lane: large-scale bacterial genome engineering. *J Biotechnol* **160:** 72–79.

Forbes, N.S. (2010) Engineering the perfect (bacterial) cancer therapy. *Nat Rev Cancer* **10:** 785–794.

Galen, J.E., and Curtiss, R., III (2014) The delicate balance in genetically engineering live vaccines. *Vaccine* **32:** 4376–4385.

Galen, J.E., Pasetti, M.F., Tennant, S., Ruiz-Olvera, P., Sztein, M.B., and Levine, M.M. (2009) *Salmonella enterica* serovar Typhi live vector vaccines finally come of age. *Immunol Cell Biol* **87:** 400–412.

Galen, J.E., Wang, J.Y., Chinchilla, M., Vindurampulle, C., Vogel, J.E., Levy, H., *et al.* (2010) A new generation of stable, nonantibiotic, low-copy-number plasmids improves immune responses to foreign antigens in *Salmonella enterica* serovar Typhi live vectors. *Infect Immun* **78:** 337–347.

Hegazy, W.A., and Hensel, M. (2012) *Salmonella enterica* as a vaccine carrier. *Future Microbiol* **7:** 111–127.

Hohmann, E.L., Oletta, C.A., Loomis, W.P., and Miller, S.I. (1995) Macrophage-inducible expression of a model antigen in *Salmonella typhimurium* enhances immunogenicity. *Proc Natl Acad Sci USA* **92:** 2904–2908.

Hoiseth, S.K., and Stocker, B.A. (1981) Aromatic-dependent *Salmonella typhimurium* are non-virulent and effective as live vaccines. *Nature* **291:** 238–239.

Husseiny, M.I., and Hensel, M. (2008) Construction of highly attenuated *Salmonella enterica* serovar Typhimurium live vectors for delivering heterologous antigens by chromosomal integration. *Microbiol Res* **163:** 605–615.

Ivics, Z., and Izsvak, Z. (2011) Nonviral gene delivery with the sleeping beauty transposon system. *Hum Gene Ther* **22:** 1043–1051.

Kang, H.Y., Dozois, C.M., Tinge, S.A., Lee, T.H., and Curtiss, R., III (2002) Transduction-mediated transfer of unmarked deletion and point mutations through use of counterselectable suicide vectors. *J Bacteriol* **184:** 307–312.

Kong, W., Wanda, S.Y., Zhang, X., Bollen, W., Tinge, S.A., Roland, K.L., *et al.* (2008) Regulated programmed lysis of recombinant *Salmonella* in host tissues to release protective antigens and confer biological containment. *Proc Natl Acad Sci USA* **105:** 9361–9366.

Kong, W., Brovold, M., Koeneman, B.A., Clark-Curtiss, J., and Curtiss, R., III (2012) Turning self-destructing *Salmonella* into a universal DNA vaccine delivery platform. *Proc Natl Acad Sci USA* **109:** 19414–19419.

Kuhlman, T.E., and Cox, E.C. (2010) Site-specific chromosomal integration of large synthetic constructs. *Nucleic Acids Res* **38:** e92.

Kvitko, B.H., McMillan, I.A., and Schweizer, H.P. (2013) An improved method for *oriT*-directed cloning and functionalization of large bacterial genomic regions. *Appl Environ Microbiol* **79:** 4869–4878.

Leschner, S., and Weiss, S. (2010) *Salmonella*-allies in the fight against cancer. *J Mol Med* **88:** 763–773.

Lichtenstein, C., and Brenner, S. (1982) Unique insertion site of Tn7 in the *E. coli* chromosome. *Nature* **297:** 601–603.

Loeschcke, A., Markert, A., Wilhelm, S., Wirtz, A., Rosenau, F., Jaeger, K.E., *et al.* (2013) TREX: a universal tool for the transfer and expression of biosynthetic pathways in bacteria. *ACS Synth Biol* **2:** 22–33.

Loessner, H., Endmann, A., Rohde, M., Curtiss, R., III, and Weiss, S. (2006) Differential effect of auxotrophies on the release of macromolecules by *Salmonella enterica* vaccine strains. *FEMS Microbiol Lett* **265:** 81–88.

Loessner, H., Endmann, A., Leschner, S., Westphal, K., Rohde, M., Miloud, T., *et al.* (2007) Remote control of tumour-targeted *Salmonella enterica* serovar Typhimurium by the use of L-arabinose as inducer of bacterial gene expression in vivo. *Cell Microbiol* **9:** 1529–1537.

Loessner, H., Endmann, A., Leschner, S., Bauer, H., Zelmer, A., zur Lage, S., *et al.* (2008) Improving live attenuated bacterial carriers for vaccination and therapy. *Int J Med Microbiol* **298:** 21–26.

Loessner, H., Leschner, S., Endmann, A., Westphal, K., Wolf, K., Kochruebe, K., *et al.* (2009) Drug-inducible remote control of gene expression by probiotic *Escherichia coli* Nissle 1917 in intestine, tumor and gall bladder of mice. *Microbes Infect* **11:** 1097–1105.

Mitra, R., McKenzie, G.J., Yi, L., Lee, C.A., and Craig, N.L. (2010) Characterization of the TnsD-*attTn7* complex that promotes site-specific insertion of Tn7. *Mob DNA* **1:** 18.

Murphy, K.C. (2012) Phage recombinases and their applications. *Adv Virus Res* **83:** 367–414.

Nakayama, K., Kelly, S.M., and Curtiss, R., III (1988) Construction of an Asd + expression-cloning vector: stable maintenance and high level expression of cloned genes in a *Salmonella* vaccine strain. *Biotechnology (N Y)* **6:** 693–697.

Paterson, Y., Guirnalda, P.D., and Wood, L.M. (2010) *Listeria* and *Salmonella* bacterial vectors of tumor-associated antigens for cancer immunotherapy. *Semin Immunol* **22:** 183–189.

Peters, J.E., and Craig, N.L. (2001) Tn7: smarter than we thought. *Nat Rev Mol Cell Biol* **2:** 806–814.

Roland, K.L., and Brenneman, K.E. (2013) *Salmonella* as a vaccine delivery vehicle. *Expert Rev Vaccines* **12:** 1033–1045.

Sabri, S., Steen, J.A., Bongers, M., Nielsen, L.K., and Vickers, C.E. (2013) Knock-in/Knock-out (KIKO) vectors for rapid integration of large DNA sequences, including whole metabolic pathways, onto the *Escherichia coli* chromosome at well-characterised loci. *Microb Cell Fact* **12:** 60.

Sawitzke, J.A., Thomason, L.C., Costantino, N., Bubunenko, M., Datta, S., and Court, D.L. (2007) Recombineering: *in vivo* genetic engineering in *E. coli*, *S. enterica*, and beyond. *Methods Enzymol* **421:** 171–199.

Sharan, S.K., Thomason, L.C., Kuznetsov, S.G., and Court, D.L. (2009) Recombineering: a homologous recombination-based method of genetic engineering. *Nat Protoc* **4:** 206–223.

Simon, R., Priefer, U.B., and Puhler, A. (1983) A broad host range mobilization system for in vivo genetic engineering: transposon mutagenesis in gram negative bacteria. *Bio/Technology* **1:** 784–791.

Stellwagen, A.E., and Craig, N.L. (1997) Avoiding self: two Tn7-encoded proteins mediate target immunity in Tn7 transposition. *EMBO J* **16:** 6823–6834.

Toussaint, B., Chauchet, X., Wang, Y., Polack, B., and Le, G.A. (2013) Live-attenuated bacteria as a cancer vaccine vector. *Expert Rev Vaccines* **12:** 1139–1154.

Voisey, C.R., and Marincs, F. (1998) Elimination of internal restriction enzyme sites from a bacterial luminescence (*luxCDABE*) operon. *Biotechniques* **24:** 58.

Waddell, C.S., and Craig, N.L. (1988) Tn7 transposition: two transposition pathways directed by five Tn7-encoded genes. *Genes Dev* **2:** 137–149.

Wang, J.Y., Harley, R.H., and Galen, J.E. (2013a) Novel methods for expression of foreign antigens in live vector vaccines. *Hum Vaccin Immunother* **9:** 1558–1564.

Wang, S., Kong, Q., and Curtiss, R., III (2013b) New technologies in developing recombinant attenuated *Salmonella* vaccine vectors. *Microb Pathog* **58:** 17–28.

Wolkow, C.A., DeBoy, R.T., and Craig, N.L. (1996) Conjugating plasmids are preferred targets for Tn7. *Genes Dev* **10:** 2145–2157.

Xin, W., Wanda, S.Y., Zhang, X., Santander, J., Scarpellini, G., Ellis, K., *et al.* (2012) The Asd(+)-DadB(+) dual-plasmid system offers a novel means to deliver multiple protective antigens by a recombinant attenuated *Salmonella* vaccine. *Infect Immun* **80:** 3621–3633.

Yan, Z.X., and Meyer, T.F. (1996) Mixed population approach for vaccination with live recombinant *Salmonella* strains. *J Biotechnol* **44:** 197–201.

Yu, B., Yang, M., Wong, H.Y., Watt, R.M., Song, E., Zheng, B.J., *et al.* (2011) A method to generate recombinant *Salmonella typhi* Ty21a strains expressing multiple heterologous genes using an improved recombineering strategy. *Appl Microbiol Biotechnol* **91:** 177–188.

Zhang, X., Wanda, S.Y., Brenneman, K., Kong, W., Zhang, X., Roland, K., *et al.* (2011) Improving *Salmonella* vector with *rec* mutation to stabilize the DNA cargoes. *BMC Microbiol* **11:** 31.

## Supporting information

Additional Supporting Information may be found in the online version of this article at the publisher's web-site:

**Fig. S1.** Modulation of GFP synthesis in *S.* Typhimurium strain SL7207 by chromosomal integration of either one, two or three copies of *gfp*-mini-Tn7 (Fig. 3A,B). Strains SL7207 and derivatives with one, two or three chromosomal copies of the *gfp* cassette were induced with L-arabinose and subsequently analysed by flow cytometry. The histogram shows data of one representative experiment.

**Fig. S2.** Consecutive chromosomal integration of two different mini-Tn7 encoding *lacZ* and *lux* into *S.* Typhimurium and stable maintenance of both modules by bacteria during colonization of mice.

A. $10^9$ bacteria of strains SL7207, *attTn7::lux* and *attTn7:lacZΔrha::lux* (Fig. 3C,D) were orally administered to BALB/c mice ($n = 5$ each group). The colonization course of bioluminescent strains harbouring *lux* was followed by non-invasive *in vivo* imaging. Bioluminescence intensities of abdominal regions of mice are expressed as means of radiance and error bars indicate standards deviations.

B. At day 9 post-infection (p.i.) Peyer's patches (PP), mesenteric lymph nodes (MLN) and spleens were harvested ($n = 5$ each group), and bacterial content of organs was determined by plating dilution series of tissue homogenates on X-Gal plates. Data are presented as means of cfu with error bars indicating standard deviations. Stability of *lacZ* and *lux* cassettes was judged based on the blue and bioluminescent colony phenotype.

**Table S1.** Primers used in this study.

**Appendix S1.** Experimental procedures.

# 13

# Genetic characterization of caffeine degradation by bacteria and its potential applications

Ryan M. Summers,[1†] Sujit K. Mohanty,[2†] Sridhar Gopishetty[3] and Mani Subramanian[2,3]*

[1]Department of Chemical and Biological Engineering, The University of Alabama, Tuscaloosa, AL 35487, USA.

[2]Department of Chemical and Biochemical Engineering and [3]Center for Biocatalysis and Bioprocessing, The University of Iowa, Coralville, IA 52241, USA.

## Summary

The ability of bacteria to grow on caffeine as sole carbon and nitrogen source has been known for over 40 years. Extensive research into this subject has revealed two distinct pathways, N-demethylation and C-8 oxidation, for bacterial caffeine degradation. However, the enzymological and genetic basis for bacterial caffeine degradation has only recently been discovered. This review article discusses the recent discoveries of the genes responsible for both N-demethylation and C-8 oxidation. All of the genes for the N-demethylation pathway, encoding enzymes in the Rieske oxygenase family, reside on 13.2-kb genomic DNA fragment found in *Pseudomonas putida* CBB5. A nearly identical DNA fragment, with homologous genes in similar orientation, is found in *Pseudomonas* sp. CES. Similarly, genes for C-8 oxidation of caffeine have been located on a 25.2-kb genomic DNA fragment of *Pseudomonas* sp. CBB1. The C-8 oxidation genes encode enzymes similar to those found in the uric acid metabolic pathway of *Klebsiella pneumoniae*. Various biotechnological applications of these genes responsible for bacterial caffeine degradation, including bio-decaffeination, remediation of caffeine-contaminated environments, production of chemical and fuels and development of diagnostic tests have also been demonstrated.

*For correspondence. E-mail mani-subramanian@uiowa.edu

**Funding Information** This research was supported by University of Iowa Research Funds.

## Introduction

Caffeine (1,3,7-trimethylxanthine) and related methylxanthines are natural purine alkaloids found in many plants around the world (Ashihara and Crozier, 1999). These compounds are hypothesized to serve as natural insecticides, and have been shown to protect the plants from insects and other predators (Nathanson, 1984; Hollingsworth *et al.*, 2002). Other possible reasons for biosynthesis of caffeine include inhibition of plant matter (Waller, 1989) and improved pollination (Wright *et al.*, 2013).

Methylxanthines are often consumed by humans in foods and beverages, including chocolate, coffee and tea. Coffee is a major worldwide agricultural commodity, with millions of metric tons produced and distributed globally each year (Summers *et al.*, 2014). In addition to the food industry, caffeine and related methylxanthines are also used in pharmaceuticals as stimulants, diuretics, bronchodilators, vasodilators and in the treatment and/or prevention of axial myopia, glaucoma and macular degeneration (Stavric, 1988a,b,c; Trier *et al.*, 1999; Dash and Gummadi, 2006b; Daly, 2007).

Although bacterial caffeine degradation has been studied since the 1970s, very little was known concerning the enzymes and genes responsible for caffeine degradation until recently. Several excellent reviews have summarized the bacterial caffeine catabolic pathways (Mazzafera, 2004; Dash and Gummadi, 2006a,b; Gummadi *et al.*, 2012), including recent developments (Gopishetty *et al.*, 2012). In this review, we focus on the discoveries made since the publication of these reviews, with specific emphasis on the genes responsible for caffeine degradation in bacteria.

## Caffeine metabolism

To date, over 35 bacterial strains that are capable of degrading caffeine have been isolated and reported (Table 1). While there is some diversity among the types

**Table 1.** Characterized bacterial strains capable of degrading caffeine.

| Organism | Location isolated | Catabolic pathway | References |
|---|---|---|---|
| *Pseudomonas putida* strain 40 | California, USA | *N*-demethylation | Woolfolk, 1975 |
| *Pseudomonas putida* C1 | Germany | *N*-demethylation | Blecher and Lingens, 1977 |
| *Pseudomonas putida* C3024 | Netherlands | N.R. | Middelhoven and Bakker, 1982 |
| *Pseudomonas putida* WS | Germany | *N*-demethylation | Glück and Lingens, 1987 |
| *Pseudomonas* sp. No. 6 | Japan | *N*-demethylation | Asano *et al.*, 1993 |
| *Pseudomonas putida* No. 352 | Japan | *N*-demethylation | Asano *et al.*, 1994 |
| *Serratia marcescens* | Brazil | *N*-demethylation | Mazzafera *et al.*, 1996 |
| *Pseudomonas putida* ATCC 700097 | California, USA | *N*-demethylation | Ogunseitan, 1996 |
| *Klebsiella* and *Rhodococcus* | India | C-8 oxidation | Madyastha and Sridhar, 1998 |
| *Pseudomonas putida* (8 strains) | Brazil | N.R. | Yamaoka-Yano and Mazzafera, 1998 |
| *Pseudomonas fluorescens* | Brazil | N.R. | Yamaoka-Yano and Mazzafera, 1998 |
| Coryneform (4 strains) | Brazil | N.R. | Yamaoka-Yano and Mazzafera, 1998 |
| *Acinetobacter* sp. (3 strains) | Brazil | N.R. | Yamaoka-Yano and Mazzafera, 1998 |
| *Flavobacterium* sp. (2 strains) | Brazil | N.R. | Yamaoka-Yano and Mazzafera, 1998 |
| *Moraxella* sp. | Brazil | N.R. | Yamaoka-Yano and Mazzafera, 1998 |
| *Pseudomonas putida* IF-3 | Japan | *N*-demethylation | Koide *et al.*, 1996 |
| *Pseudomonas putida* L | Brazil | *N*-demethylation | Yamaoka-Yano and Mazzafera, 1999 |
| *Pseudomonas putida* KD6 | N.R. | *N*-demethylation | Sideso *et al.*, 2001 |
| *Alcaligenes* sp. | Canada | C-8 oxidation | Mohapatra *et al.*, 2006 |
| *Pseudomonas putida* NCIM 5235 | India | *N*-demethylation | Dash and Gummadi, 2006a |
| *Pseudomonas* sp. CBB1 | Iowa, USA | C8-oxidation | Yu *et al.*, 2008 |
| *Pseudomonas alcaligenes* CFR 1708 | India | N.R. | Sarath Babu *et al.*, 2005 |
| *Alcaligenes fecalis* T1 | India | N.R. | Sarath Babu *et al.*, 2005 |
| *Acetobacter* sp. T3 | India | N.R. | Sarath Babu *et al.*, 2005 |
| *Pseudomonas putida* CBB5 | Iowa, USA | *N*-demethylation | Yu *et al.*, 2009 |
| *Pseudomonas* sp. CES | Iowa, USA | *N*-demethylation | Yu *et al.*, 2014 |

N.R., not reported.

of bacteria isolated, the majority are *Pseudomonas*, primarily *Pseudomonas putida*. Caffeine-degrading bacteria are geographically dispersed, and have been found in coffee fields (Yamaoka-Yano and Mazzafera, 1998), wastewater streams (Ogunseitan, 1996) and garden soil (Blecher and Lingens, 1977; Yu *et al.*, 2008; 2009; 2014).

Metabolic studies with these caffeine-degrading bacterial isolates have revealed only two catabolic pathways: *N*-demethylation and C-8 oxidation. The *N*-demethylation pathway appears to be the most common, as it has been observed in over 80% of reported isolates where metabolism has been characterized. In both pathways, bacteria break caffeine down to carbon dioxide and ammonia to harvest energy and cellular building blocks.

During *N*-demethylation, the caffeine molecule is sequentially *N*-demethylated to form xanthine (Fig. 1A). Each of the three methyl groups is removed with incorporation of molecular oxygen to produce one formaldehyde and one water molecule per reaction. Theobromine (3,7-dimethylxanthine) is the major metabolite formed from the first step in the pathway, with small amounts of paraxanthine (1,7-dimethylxanthine) also reported in some strains (Yamaoka-Yano and Mazzafera, 1999; Yu *et al.*, 2009; 2014). The second step of the pathway is the $N_3$-demethylation of theobromine or the $N_1$-demethylation of paraxanthine to form 7-methylxanthine. 7-Methylxathine is further $N_7$-demethylated to form

xanthine. Finally, xanthine is converted to uric acid, which enters normal purine catabolic pathway.

Theophylline (1,3-dimethylxanthine) has not been reported as a metabolite of caffeine in bacteria. However, it is the first major metabolite of caffeine in fungi (Hakil *et al.*, 1998), and is further degraded via *N*-demethylation to 3-methylxanthine and xanthine. Although the bacterium *P. putida* CBB5 does not produce theophylline from caffeine, it has been reported to degrade theophylline by *N*-demethylation (Yu *et al.*, 2009). Both 3-methylxanthine (major product) and 1-methylxanthine (minor product) are formed from theophylline in CBB5 and are further *N*-demethylated to form xanthine, as in the caffeine catabolic pathway.

Some of the metabolites formed during the bacterial *N*-demethylation of caffeine also undergo C-8 oxidation to form their corresponding uric acids (Blecher and Lingens, 1977; Yamaoka-Yano and Mazzafera, 1999). In most strains, this involves the formation of 3,7-dimethyluric acid, 1,7-dimethyluric acid and 7-methyluric acid from theobromine, paraxanthine and 7-methylxanthine respectively. These methyluric acids are not formed during caffeine *N*-demethylation in *P. putida* CBB5. However, CBB5 converts approximately 25% of the entire theophylline metabolite pool to 1,3-dimethyluric acid, 1-methyluric acid and 3-methyluric acid. There is no evidence that these methyluric acids are further metabolized (Yu *et al.*, 2009).

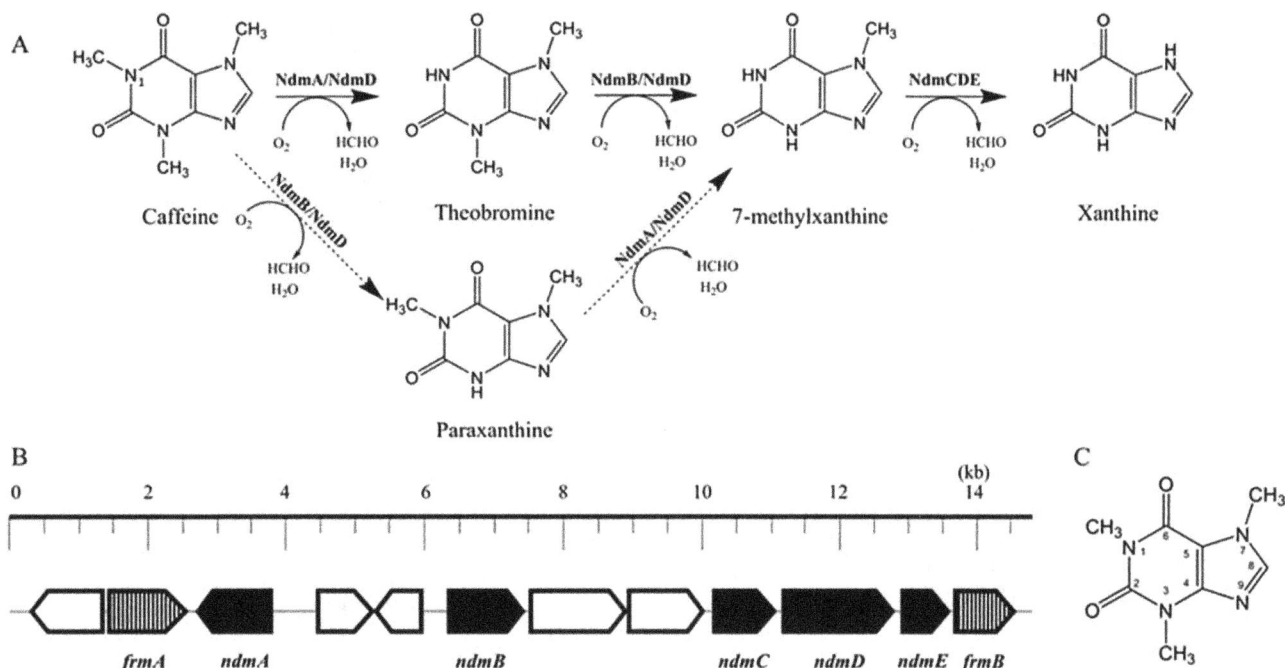

**Fig. 1.** Proposed caffeine $N$-demethylation pathway (A) and map of associated genes (B) in *Pseudomonas putida* CBB5. The dashed arrows in part (A) represent a minor pathway, accounting for 1–2% of metabolized caffeine. NdmA/*ndmA* = $N_1$-demethylase specific for $N_1$-methyl group of caffeine; NdmD/*ndmD* = reductase; NdmB/*ndmB* = $N_3$-demethylase specific for $N_3$-methyl group of theobromine; NdmCDE = protein complex containing $N_7$-demethylase specific for $N_7$-demethylation of 7-methylxanthine; *ndmC* = NdmC gene; *ndmE* = NdmE gene; *frmA* = glutathione-dependent formaldehyde dehydrogenase; *frmB* = $S$-formylglutathione hydrolase. (C) A numbered structure of the caffeine molecule.

C-8 oxidation involves the oxidation of caffeine to form 1,3,7-trimethyluric acid (TMU), which is further degraded by a pathway homologous to the uric acid metabolic pathway (Fig. 2). This pathway has been observed in both mixed cultures (Madyastha and Sridhar, 1998) and pure bacterial isolates (Mohapatra *et al.*, 2006; Yu *et al.*, 2008; Mohanty *et al.*, 2012). 1,3,7-trimethyluric acid is further metabolized to sequentially form 1,3,7-trimethyl-5-hydroxyisouric acid (TM-HIU), 3,6,8-trimethyl-2-oxo-4-hydroxy-4-carboxy-5-ureidoimidazoline (TM-OHCU) and 3,6,8-trimethylallantoin (TMA) (Mohanty *et al.*, 2012). Further degradation of TMA has not yet been fully characterized. However, it is believed that only S-(+)-TMA is formed enzymatically and its degradation proceeds through trimethylallantoic acid (TMAA) before being mineralized to glyoxylic acid, dimethylurea and monomethylurea (Madyastha and Sridhar, 1998; Mohanty, 2013). These latter compounds are then assumed to enter the central metabolic cycles of the bacterial cell.

### Genetic basis for caffeine degradation

Although bacterial degradation of caffeine has been studied for over 40 years, very little was discovered concerning the enzymes involved in the metabolism. Similarly, the genetics of bacterial caffeine metabolism were unknown. Recent work has revealed the nature of enzymes involved in both $N$-demethylation and C-8 oxidation pathways (Asano *et al.*, 1994; Madyastha *et al.*, 1999; Yamaoka-Yano and Mazzafera, 1999; Yu *et al.*, 2008; 2009; Summers *et al.*, 2011; Mohanty, *et al.*, 2012). The gene sequences for these enzymes have also been elucidated, reported and deposited in the GenBank database (Mohanty *et al.*, 2012; Summers *et al.*, 2012; 2013). Currently, the only known bacterial genes responsible for caffeine metabolism are those that are directly responsible for catabolism. Other genes encoding proteins for caffeine uptake, chemotaxis and regulation of caffeine-degrading enzymes have not yet been reported.

### N-demethylation

Many of the earliest works on caffeine $N$-demethylase enzymes indicated that they are labile in nature and quickly lost activity during purification (Hohnloser *et al.*, 1980; Glück and Lingens, 1988; Sideso *et al.*, 2001; Beltran *et al.*, 2006). Glück and Lingens (1988) reported partial purification of a 7-methylxanthine demethylase from *P. putida* WS that was not active towards any other

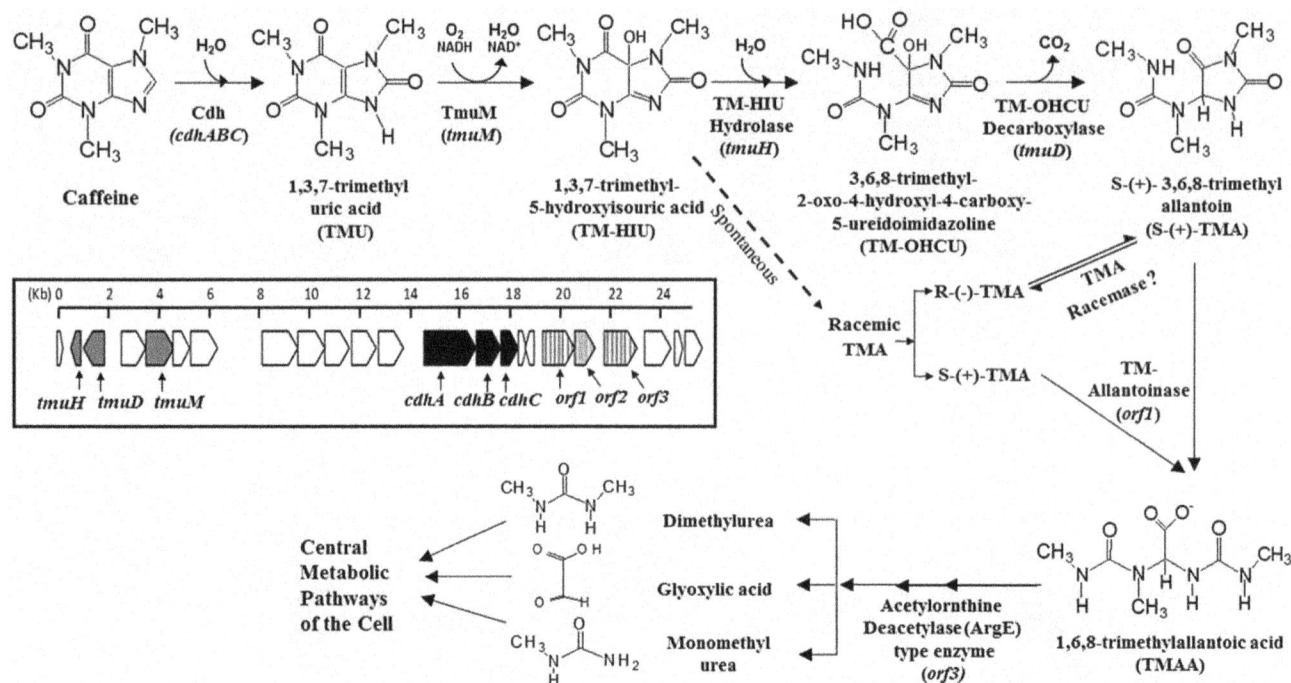

**Fig. 2.** Proposed caffeine C-8 oxidation pathway and associated genes in *Pseudomonas* sp. CBB1. The solid arrows represent enzymatic steps, while the dashed arrow represents the spontaneous degradation of 1,3,7-trimethyl-5-hydroxyisouric acid (TM-HIU) to racemic TMA. Inset: map of the 25.2 kb caffeine gene cluster on the CBB1 genome containing the entire C-8 oxidation pathway genes. Cdh/ *cdhABC* = trimeric caffeine dehydrogenase; TmuM/*tmuM* = trimethyluric acid monooxygenase; *tumH* = putative TM-HIU hydrolase; *tmuD* = putative TM-OHCU decarboxylase; *orf1* = putative trimethylallantoinase; *orf2, orf3* = putative genes responsible for catabolism of 1,6,8-trimethylallantoic acid to simpler metabolites such as dimethylurea, monomethylurea and glyoxylic acid.

methylxanthine. This enzyme lost activity within 12 h and could not be completely purified. Asano and colleagues (1994) observed two distinct *N*-demethylase fractions eluting from an ion-exchange chromatography column loaded with cell extracts of *P. putida* No. 352. One protein fraction was active towards caffeine, resulting in formation of theobromine. The second fraction converted theobromine to 7-methylxanthine. These results indicate that caffeine is *N*-demethylated by at least three methylxanthine-specific enzymes.

Recently, the five enzymes (NdmABCDE) that catalyse the entire caffeine *N*-demethylation pathway in *P. putida* CBB5 were purified and characterized (Summers *et al.*, 2011; 2013). These same enzymes are also responsible for metabolism of theophylline in CBB5. In addition, the gene sequences of all five enzymes (*ndmABCDE*) were determined from a 13.2-kb CBB5 genomic DNA fragment (Fig. 1B) and deposited in the GenBank database (Summers *et al.*, 2012; 2013). NdmA and NdmB are Rieske [2Fe-2S] monooxygenases with a non-heme iron at the active site (Summers *et al.*, 2012). NdmA specifically removes the $N_1$-methyl group from caffeine, paraxanthine, theophylline and 1-methylxanthine to form theobromine, 7-methylxanthine, 3-methylxanthine and xanthine respectively. Similarly, NdmB is an $N_3$-

specific demethylase, converting caffeine, theobromine, theophylline and 3-methylxanthine to paraxanthine, 7-methylxanthine, 1-methylxanthine and xanthine respectively.

Both NdmA and NdmB are entirely dependent upon NdmD (Fig. 1A), which is a partner reductase that transfers electrons from nicotinamide adenine dinucleotide (NADH) to power the reaction (Summers *et al.*, 2011). NdmD is a redox-dense protein, containing two [2Fe-2S] clusters, one FMN binding domain and one NADH binding domain. When NdmD is expressed with NdmC and NdmE, the three proteins form a large protein complex that catalyses the $N_7$-demethylation of 7-methylxanthine to xanthine (Summers *et al.*, 2013). NdmC is also a non-heme iron monooxygenase but does not contain a Rieske [2Fe-2S] cluster, as do NdmA and NdmB. NdmE is a glutathione-*S*-transferase-like protein that is postulated to have a structural role in aligning the extra Rieske cluster found on NdmD with the NdmC subunit to catalyse the last *N*-demethylation step.

Among the Rieske oxygenase family, NdmCDE protein complex is very unique (Summers *et al.*, 2013). The reductase, NdmD, is the only reported Rieske reductase that contains an extra [2Fe-2S] cluster. The NdmCDE protein complex is also the first reported instance in which the Rieske domain of the oxygenase is split off and fused to the reductase. Also, this was the first reported instance in which a glutathione-S-transferase-like protein is absolutely required for solubility and activity of a Rieske oxygenase. Interestingly, a BLAST search using the *ndmCDE* genes as queries identified an additional 18 organisms with homologous *ndmCDE* clusters coding for enzymes of unknown function (Summers *et al.*, 2013). The *ndmCDE* genes clustered most closely with homologues from *Janthinobacter* sp. Marseille (accession no. NC_009659.1), *Klebsiella pneumoniae* subsp. *pneumoniae* WGLW2 (accession no. NZ_JH930420.1) and *Pseudomonas* sp. TJI-51 (accession no. NZ_AEWE01000207.1). Other bacteria containing *ndmCDE* homologues predominantly belong to the *Sinorhizobium* and *Mesorhizobium* genera. Although the functions of these homologues are unknown, the widespread dissemination of *ndmCDE* homologues indicates that similar caffeine-degrading enzymes may be found in bacteria other than those in the *Pseudomonas* genera.

Enzymes and genes homologous to NdmABCDE were also found in caffeine-grown *Pseudomonas* sp. CES cells (Yu *et al.*, 2014). The sequences of these homologous enzymes revealed high similarity to the enzymes (80–90% identity) and genes (72–77% identity) in *P. putida* CBB5. The similarity between the caffeine-degrading enzymes in *P. putida* CBB5 and *Pseudomonas* sp. CES indicate that these N-demethylase enzymes may be conserved among many bacteria that metabolize caffeine *via* N-demethylation. The *ndmABCDE* genes in *P. putida* CBB5 and *Pseudomonas* sp. CES are encoded on the genomic DNA. In contrast, Dash and Gummadi (2006a) reported that N-demethylation in *Pseudomonas* sp. NCIM 5235 is encoded on a 12 kb plasmid, although the gene sequences have not yet been reported. Other caffeine N-demethylase genes may yet be discovered in the future, which will only increase our understanding of how bacteria respond to many other N-methylated compounds in the environment.

Flanking the *ndmABCDE* genes on the CBB5 chromosome are two genes homologous to those known to catalyse the conversion of formaldehyde to formic acid, *frmA* and *frmB* (Fig. 1B). Formaldehyde production during N-demethylation has been detected many times (Blecher and Lingens, 1977; Glück and Lingens, 1988; Summers *et al.*, 2011; 2012; 2013). Thus, the genetic basis of caffeine N-demethylation, including the cellular utilization of formaldehyde produced as a by-product has been substantiated.

## C-8 oxidation

To date, three enzymes catalyzing the C-8 oxidation of caffeine have been purified and characterized. An 85-kDa caffeine oxidase was purified from a mixed culture of *Klebsiella* sp. and *Rhodococcus* sp. (Madyastha *et al.*, 1999). Mohapatra and colleagues (2006) discovered a 65-kDa caffeine oxidase in *Alcaligenes* sp. Both caffeine oxidase enzymes displayed low activity towards theobromine, theophylline and a few other theobromine analogues. A heterotrimeric caffeine dehydrogenase (Cdh) enzyme was discovered in *Pseudomonas* sp. CBB1 (Yu *et al.*, 2008), which catalysed C-8 oxidation of caffeine to form TMU (Fig. 2). This 158-kDa protein was a novel quinone-dependent oxidoreductase (EC 1.17.5.2) that exhibited no activity with $NAD(P)^+$. The second step of the C-8 oxidation pathway, conversion of TMU to TM-HIU, was catalysed by a 43-kDa NADH-dependent trimethyluric acid monooxygenase (TmuM) also isolated from CBB1 (Mohanty *et al.*, 2012). TM-HIU generated by TmuM was found to be unstable and spontaneously degraded to racemic TMA. However, the two step enzymatic transformation of TM-HIU is expected to yield S-(+)-TMA via TM-OHCU in biological systems, analogous to the uric acid metabolic pathway (Ramazzina *et al.*, 2006).

In *Pseudomonas* sp. CBB1, several genes, including genes encoding Cdh and TmuM, have been identified in a 25.2-kb caffeine gene cluster of the CBB1 genome (Fig. 2 inset). A BLASTX-based sequence analysis of Cdh genes (*cdhABC*) revealed significant homology with other heterotrimeric enzymes such as xanthine dehydrogenase, hydratase/alcohol dehydrogenase, aldehyde oxidase and carbon monoxide dehydrogenase. The high similarity between *cdhABC* and their homologues further facilitated in associating these genes with their respective cofactor-binding subunits (Mohanty *et al.*, 2012). Similarly, the *tmuM* gene showed significant homology with FAD (Flavin adenine dinucleotide)-containing aromatic-ring hydroxylases. In particular, *tmuM* showed a high degree of similarity with *hpxO* gene (encoding HpxO, a FAD-dependant uric acid oxidase) from *K. pneumoniae*. Further, a protein homology model of TmuM based on HpxO led to a better understanding of the changes in the active site pocket of this enzyme and its specificity towards methyluric acids (Mohanty *et al.*, 2012).

Two other genes in this cluster, *tmuH* and *tmuD* (Fig. 2 inset), are proposed to encode putative enzymes (TM-HIU hydroxylase and TM-OHCU decarboxylase, respectively) of the C-8 oxidation pathway (Mohanty *et al.*, 2012) based on their homology to enzymes of the uric acid metabolic pathway in *K. pneumoniae* (French and Ealick, 2010; 2011; Mohanty, 2013). The presence of these putative enzymes in CBB1 suggests enzymatic formation of S-(+)-TMA from TM-HIU via TM-OHCU in the C-8 oxidation pathway (Mohanty *et al.*, 2012). *Klebsiella pneumoniae* also contains an allantoin racemase to covert (R)-allantoin, formed by spontaneous degradation of hydroxyisouric acid, to (S)-allantoin (French *et al.*, 2011). Mohanty and colleagues (2012) proposed a similar conversion of (R)-(-)-TMA, formed by spontaneous degradation of TM-HIU, to (S)-(+)-TMA by a TMA racemase. However, none of the genes in the 25.2-kb CBB1 genetic cluster showed any homology to allantoin racemase (Mohanty, 2013). Thus, there yet remain undiscovered genes encoding enzymes active in the caffeine C-8 oxidation pathway.

Further degradation of S-(+)-TMA to glyoxylic acid, dimethylurea and monomethylurea, as suggested by Madyastha and Sridhar (1998), is poorly understood. Mohanty and colleagues (2012) have identified an open reading frame (*orf1*) in the 25.2-kb caffeine gene-cluster in the CBB1 genome, which encodes a putative trimethylallantoinase homologous to allantoinases that hydrolyse allantoin to allantoic acid (Fig. 2). This suggests that S-(+)-TMA is further hydrolysed into TMAA in the C-8 oxidation pathway. Subsequent formation of glyoxylic acid, dimethylurea and monomethylurea, which then enter the central metabolic pathway for total mineralization, may occur by hydrolysis of non-peptide C-N bonds of TMAA (Mohanty, 2013).

### C-8 oxidation of N-*demethylated metabolites*

There are several reports concerning the C-8 oxidation of N-demethylated metabolites. Woolfolk (1985) discovered a xanthine dehydrogenase capable of oxidizing xanthine, 1-methylanthine and 3-methylxanthine in *P. putida* 40. Yamaoka-Yano and Mazzafera (1999) reported that *P. putida* L contains a broad-specificity xanthine oxidase responsible for C-8 oxidation of theobromine, paraxanthine and 7-methylxanthine. The purified oxidase also oxidized xanthine, 3-methylxanthine and theophylline. *Pseudomonas putida* CBB5 also contains a broad-substrate xanthine dehydrogenase that was partially purified from cell extracts and found to be active towards theophylline, 1-methylxanthine, 3-methylxanthine and xanthine (Yu *et al.*, 2009). In all cases, the methyluric acids formed from N-demethylated metabolites accumulate in the growth media and are not further degraded. In addition, none of the methyluric acids were utilized as sole carbon and nitrogen sources. Thus, the reason for their production is currently unknown, although it is likely that they are simply the result of a broad-specificity xanthine dehydrogenase that was partially purified from cell extracts of CBB5 (Yu *et al.*, 2009). To our knowledge, these are the only three reports that describe the enzymes responsible for C-8 oxidation of N-demethylated metabolites in bacteria. Currently, it is unknown whether these enzymes are specific for N-demethylated metabolites or are simply the general xanthine dehydrogenase with broad specificity. No bacterial gene has been asso-

ciated with the C-8 oxidation activity of N-demethylated caffeine metabolites to date.

## Applications of bacterial caffeine degradation

An understanding of the genes involved in bacterial caffeine degradation may open up several new biotechnological applications. Some of these include biological decaffeination of coffee, tea and caffeinated plant matter, environmental remediation of soils and waters with high caffeine concentrations, synthesis of alkylxanthines and alkyl uric acids for use as chemicals or pharmaceuticals and development of a rapid diagnostic test to detect caffeine and related methylxanthines.

### Bio-decaffeination

Bio-decaffeination of coffee and tea using whole microbial cells or enzymes has been discussed for a number of years (Kurtzman and Schwimmer, 1971; Sideso *et al.*, 2001; Beltran *et al.*, 2006; Gopishetty *et al.*, 2012). *Pseudomonas putida* CBB5 can completely decaffeinate coffee and tea extracts, while *Pseudomonas* sp. CBB1 has also been used to decaffeinate tea extracts (Gopishetty *et al.*, 2012). In terms of relative efficacy, strain CBB5 used the N-demethylation pathway to degrade a higher amount of caffeine in a shorter amount of time than did strain CBB1 through the C-8 oxidation pathway. An immobilized mixed culture of *Klebsiella* sp. and *Rhodococcus* sp. was also used to decaffeinate tea extract *via* C-8 oxidation under both batch and continuous processes (Summers *et al.*, 2014). Overall, the N-demethylation pathway appears to be more efficient than C-8 oxidation for use in the microbial decaffeination of coffee. However, the use of bacterial cells for bio-decaffeination of beverages may not be feasible due to the potential for release of endotoxins.

Alternately, use of purified caffeine-degrading enzymes (either soluble or immobilized) may provide a viable alternative (Beltran *et al.*, 2006; Gopishetty *et al.*, 2012) to eliminate endotoxin problems. The *ndmABCDE* genes could be cloned into *Escherichia coli* for large-scale recombinant enzyme production in order to carry out bio-decaffeination of beverages. Another approach is to clone caffeine-degrading genes into *Saccharomyces cerevisiae* or another generally regarded as safe (GRAS) organism for whole-cell bio-decaffeination (Gopishetty *et al.*, 2012), thus circumventing the endotoxin problem. Through this method, optimized genetic cassettes could be transformed into the GRAS organism, creating an enhanced caffeine-degrading strain.

A greater opportunity for microbial bio-decaffeination may be in the decaffeination of coffee and tea wastes. During processing of coffee, millions of metric tons of waste are generated each year (Brand *et al.*, 2000). This waste has a high nutritional content, including a carbohydrate content of 45–60% (dry weight) (Summers *et al.*, 2014). Unfortunately, a caffeine concentration in the waste greater than 1% makes it unsuitable as animal feed or as a biofuel feedstock. Treatment of coffee waste with caffeine-degrading microorganisms (either wild type or recombinant) may transform the waste into a valuable by-product, rather than a waste stream.

*Environmental remediation*

There are many routes by which caffeine enters the environment, where it can exhibit toxic effects on the surrounding plants, insects and microbes (Nathanson, 1984; Waller, 1989). In coffee and tea fields, fallen leaves, stems and seeds decompose, releasing caffeine into the soil. Solid and liquid wastes from coffee and tea processing plants also contain high levels of caffeine, which enter soil and groundwater. The widespread use of caffeine in foods, beverages and pharmaceuticals lead to high levels of caffeine in human wastewater streams, as well. In fact, caffeine can be used as an anthropogenic marker for wastewater contamination in the environment (Buerge *et al.*, 2003). In all of these cases, either wild-type or recombinant caffeine-degrading microorganisms may be of use in removing caffeine from contaminated environments.

*Chemical production*

While caffeine is a relatively inexpensive molecule, many of the metabolites formed by both *N*-demethylation and C-8 oxidation of caffeine and their analogues are high-value chemicals. Many of these chemicals have great potential in the pharmaceutical and cosmetic industries. Uric acid and methyluric acids are antioxidants (Nishida, 1991; Schlotte *et al.*, 1998), and 8-oxomethylxanthines may be used in treatments for obesity, skin cosmetics and anti-dandruff products (Simic and Jovanovic, 1989). Methylxanthines have been used as diuretics, bronchodilators, antioxidants and asthma control (Lee, 2000; Daly, 2007).

Most methylxanthines and methyluric acids are difficult to synthesize chemically because selective alkylation of each nitrogen atom is difficult to achieve (Taylor *et al.*, 1961; Shamim *et al.*, 1989; Gopishetty *et al.*, 2012). The recent discovery of genes encoding bacterial caffeine-degrading enzymes may help to facilitate synthesis of these high-value chemicals. The *ndmABCDE* genes catalyse specific *N*-demethylation of alkylxanthines, which leave a specific methyl group open for chemical derivatization. Caffeine dehydrogenase from *Pseudomonas* sp. CBB1 can oxidize caffeine to TMU, and displays some activity towards other methylxanthines, as well.

Currently, there is only one report of methylxanthine production from caffeine using engineered cells (Summers *et al.*, 2014). The genes *ndmA* and *ndmD* were cloned into *E. coli*, resulting in a bacterial strain that was able to effectively convert caffeine to theobromine. A second *E. coli* strain was constructed to convert theobromine to 7-methylxanthine using *ndmB* and *ndmD* genes. This preliminary report demonstrated the feasibility of specific methylxanthine production from caffeine.

*Diagnostics*

Indiscriminate introduction of caffeine in food and beverages has led to growing concern among caffeine-sensitive consumers and the US Federal Drug Administration, giving rise to an ever-increasing demand for a suitable 'in-home' test to detect caffeine. Recently, a Cdh enzyme-based colorimetric test was developed. This test was rapid and sensitive enough to detect caffeine in beverages, including coffee, soft drinks and nursing mother's milk, within minutes (Mohanty *et al.*, 2014). Based on the type of dye (electron acceptor) used, the test developed a bright colour upon exposure to caffeine even at 1–5 ppm level (Fig. 3). The test could successfully detect caffeine in samples with a wide range of pH and variations, with milk and sugar, or with other active pharmaceutical ingredients. Thus, this test is now deemed to be highly suitable for further development into an 'in-home' type strip-based test (Mohanty *et al.*, 2014).

The *N*-demethylase genes were also used to addict *E. coli* to caffeine (Quandt *et al.*, 2013), resulting in a whole-cell caffeine biosensor. Genes *ndmABCD* and an *ndmE* homologue were cloned into an *E. coli* guanine auxotroph. In order to grow, the engineered cells were required to convert caffeine to xanthine, which was further converted to guanine. This engineered *E. coli* strain was then used to accurately determine the caffeine concentration of various beverages. While impractical for an at-home caffeine diagnostic test, this *E. coli* strain could find use detecting caffeine in a laboratory setting or in environment samples.

**Conclusion**

Research for over 40 years has uncovered two distinct caffeine metabolic pathways in bacteria: *N*-demethylation and C-8 oxidation. While there are a couple of reports on the enzymes involved in these processes, work was stalled due to lack of knowledge concerning the genetics of bacterial caffeine degradation. The recent discovery of bacterial genes responsible for metabolism of caffeine, by

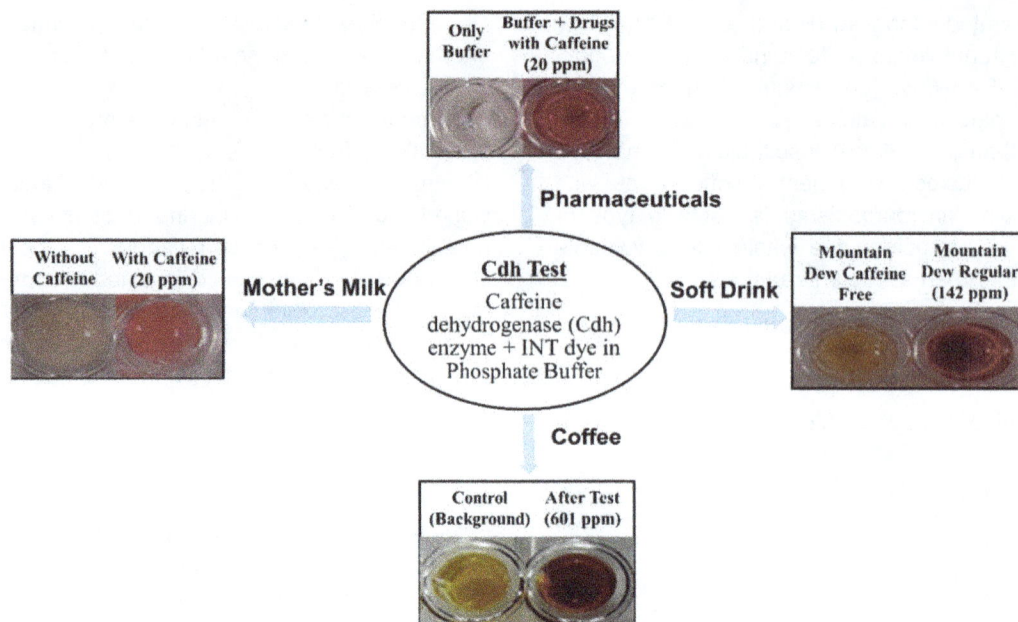

**Fig. 3.** Caffeine dehydrogenase (Cdh)-based caffeine detection. Iodonitrotetrazolium chloride (INT) dye in the presence of Cdh results in shades of red to detect caffeine in test samples such as nursing mothers' milk (left, with 20 ppm caffeine), pharmaceuticals (top, tablets dissolved and diluted to 20 ppm caffeine), soft drinks (right) and brewed coffee (bottom).

both *N*-demethylation and C-8 oxidation routes, opens numerous potential biotechnological applications. These novel genes and enzymes may be of great use in home diagnostic tests, remediation of caffeine-contaminated environments and production of chemicals, pharmaceuticals, animal feed and biofuels.

## Conflict of interest

None declared.

## References

Asano, Y., Komeda, T., and Yamada, H. (1993) Microbial production of theobromine from caffeine. *Biosci Biotechnol Biochem* **57:** 1286–1289.

Asano, Y., Komeda, T., and Yamada, H. (1994) Enzymes involved in theobromine production from caffeine by *Pseudomonas putida* No. 352. *Biosci Biotechnol Biochem* **58:** 2303–2304.

Ashihara, H., and Crozier, A. (1999) Biosynthesis and metabolism of caffeine and related purine alkaloids. *Adv Bot Res* **30:** 117–205.

Beltran, J.G., Leask, R.L., and Brown, W.A. (2006) Activity and stability of caffeine demethylases found in *Pseudomonas putida* IF-3. *Biochem Eng J* **31:** 8–13.

Blecher, R., and Lingens, F. (1977) The metabolism of caffeine by a *Pseudomonas putida* strain. *Hoppe-Seyler's Z Physiol Chem* **358:** 807–817.

Brand, D., Pandey, A., Roussos, S., and Soccol, C. (2000) Biological detoxification of coffee husk by filamentous fungi using a solid state fermentation system. *Enzyme Microb Technol* **27:** 127–133.

Buerge, I.J., Poiger, T., Muller, M.D., and Buser, H.R. (2003) Caffeine, an anthropogenic marker for wastewater contamination of surface waters. *Environ Sci Technol* **37:** 691–700.

Daly, J.W. (2007) Caffeine analogs: biomedical impact. *Cell Mol Life Sci* **64:** 2153–2169.

Dash, S.S., and Gummadi, S.N. (2006a) Biodegradation of caffeine by *Pseudomonas* sp. NCIM 5235. *Res J Microbiol* **1:** 115–123.

Dash, S.S., and Gummadi, S.N. (2006b) Catabolic pathways and biotechnological applications of microbial caffeine degradation. *Biotechnol Lett* **28:** 1993–2002.

French, J.B., and Ealick, S.E. (2010) Structural and mechanistic studies on *Klebsiella pneumoniae* 2-oxo-4-hydroxy-4-carboxy-5-ureidoimidazoline decarboxylase. *J Biol Chem* **285:** 35446–35454.

French, J.B., and Ealick, S.E. (2011) Structural and kinetic insights into the mechanism of 5-hydroxyisourate hydrolase from *Klebsiella pneumoniae*. *Acta Crystallogr* **D67:** 671–677.

French, J.B., Neau, D.B., and Ealick, S.E. (2011) Characterization of the structure and function of *Klebsiella pneumoniae* allantoin racemase. *J Mol Biol* **410:** 447–460.

Glück, M., and Lingens, F. (1987) Studies on the microbial production of theobromine and heteroxanthine from caffeine. *Appl Microbiol Biotechnol* **25:** 334–340.

Glück, M., and Lingens, F. (1988) Heteroxanthine-demethylase, a new enzyme in the degradation of caffeine by *Pseudomonas putida*. *Appl Microbiol Biotechnol* **28:** 59–62.

Gopishetty, S.R., Louie, T.M., Yu, C.L., and Subramanian, M. (2012) Microbial degradation of caffeine, methylxanthines, and its biotechnological applications. In *Microbial Biotechnology: Methods and Applications*. Thatoi, H.N., and Mishra, B.B. (eds). New Delhi, India: Narosa Publishing House Pvt, pp. 44–67.

Gummadi, S.N., Bhavya, B., and Ashock, N. (2012) Physiology, biochemistry and possible applications of microbial caffeine degradation. *Appl Microbiol Biotechnol* **93**: 545–554.

Hakil, M., Denis, S., Viniegra-Gonzalez, G., and Augur, C. (1998) Degradation and product analysis of caffeine and related dimethylxanthines by filamentous fungi. *Enzyme Microb Technol* **22**: 355–359.

Hohnloser, W., Osswal, B., and Lingens, F. (1980) Enzymological aspects of caffeine demethylation and formaldehyde oxidation by *Pseudomonas putida* C1. *Hoppe-Seyler's Z Physiol Chem* **361**: 1763–1766.

Hollingsworth, R.G., Armstrong, J.W., and Campbell, E. (2002) Pest control: caffeine as a repellent for slugs and snails. *Nature* **417**: 915–916.

Koide, Y., Nakane, S., and Imai, Y. (1996) Caffeine demethylate gene-containing DNA fragment and microbial process for producing 3-methyl-7-alkylxanthine. United States Patent US5550041.

Kurtzman, R.H., and Schwimmer, S. (1971) Caffeine removal from growth media by microorganisms. *Experiencia* **127**: 481–482.

Lee, C. (2000) Antioxidant ability of caffeine and its metabolites based on the study of oxygen radical absorbing capacity and inhibition of LDL peroxidation. *Clin Chim Acta* **295**: 141–154.

Madyastha, K.M., and Sridhar, G.R. (1998) A novel pathway for the metabolism of caffeine by a mixed culture consortium. *Biochem Biophys Res Commun* **249**: 178–181.

Madyastha, K.M., Sridhar, G.R., Vadiraga, B.B., and Madhavi, Y.S. (1999) Purification and partial characterization of caffeine oxidase – A novel enzyme from a mixed culture consortium. *Biochem Biophys Res Commun* **263**: 460–464.

Mazzafera, P. (2004) Catabolism of caffeine in plants and microorganisms. *Front Biosci* **9**: 1348–1359.

Mazzafera, P., Olsson, O., and Sandberg, G. (1996) Degradation of caffeine and related methylxanthines by *Serratia marcescens* isolated from soil under coffee cultivation. *Microb Ecol* **31**: 199–207.

Middelhoven, W.J., and Bakker, C.M. (1982) Degradation of caffeine by immobilized cells of *Pseudomonas putida* strain C 3024. *Eur J Appl Microbiol Biotechnol* **15**: 214–217.

Mohanty, S.K. (2013) A. Genetic characterization of the caffeine C-8 oxidation pathway in *Pseudomonas* Sp. CBB1. B. Validation of caffeine dehydrogenase as a suitable enzyme for a rapid caffeine diagnostic test. PhD diss., University of Iowa, 2013.

Mohanty, S.K., Yu, C.L., Das, S., Louie, T.M., Gakhar, L., and Subramanian, M. (2012) Delineation of the caffeine C-8 oxidation pathway in *Pseudomonas* sp. strain CBB1 via characterization of a new trimethyluric acid monooxygenase and genes involved in trimethyluric acid metabolism. *J Bacteriol* **194**: 3872–3882.

Mohanty, S.K., Yu, C.L., Gopishetty, S., and Subramanian, M. (2014) Validation of caffeine dehydrogenase from *Pseudomonas* sp. strain CBB1 as a suitable enzyme for a rapid caffeine detection and potential diagnostic test. *J Agric Food Chem* **62**: 7939–7946.

Mohapatra, B.R., Harris, N., Nordin, R., and Mazumder, A. (2006) Purification and characterization of a novel caffeine oxidase from *Alcaligenes* species. *J Biotechnol* **125**: 319–327.

Nathanson, J.A. (1984) Caffeine and related methylxanthines: possible naturally occurring pesticides. *Science* **226**: 184–187.

Nishida, Y. (1991) Inhibition of lipid peroxidation by methylated analogues of uric acid. *J Pharm Pharmacol* **43**: 885–887.

Ogunseitan, O.A. (1996) Removal of caffeine in sewage by *Pseudomonas putida*: implications for water pollution index. *World J Microbiol Biotechnol* **12**: 251–256.

Quandt, E.M., Hammerling, M.J., Summers, R.M., Otoupal, P.B., Slater, R.N., Alnahhas, A., *et al.* (2013) Decaffeination and measurement of caffeine content by addicted *Escherichia coli* with a refactored N-demethylation operon from *Pseudomonas putida* CBB5. *ACS Synth Biol* **2**: 301–307.

Ramazzina, I., Folli, C., Secchi, A., Berni, R., and Percudani, R. (2006) Completing the uric acid degradation pathway through phylogenetic comparison of whole genomes. *Nat Chem Biol* **2**: 144–148.

Sarath Babu, V.R., Patra, S., Thakur, M.S., Karanth, N.G., and Varadaraj, M.C. (2005) Degradation of caffeine by *Pseudomonas Alcaligenes* CFR 1708. *Enzyme Microb Technol* **37**: 617–624.

Schlotte, V., Sevanian, A., Hochstein, P., and Weithmann, K.U. (1998) Effect of uric acid and chemical analogues on oxidation of human low density lipoprotein in vitro. *Free Radic Biol Med* **25**: 839–847.

Shamim, M.T., Ukena, D., Padgett, W.L., and Daly, J.W. (1989) Effects of 8-phenyl and 8-cycloalkyl substituents on the activity of mono-, di, and trisubstituted alkylxanthines with substitution at the 1-, 3-, and 7-positions. *J Med Chem* **32**: 1231–1237.

Sideso, O.F.P., Marvier, A.C., Katerelos, N.A., and Goodenough, P.W. (2001) The characteristics and stabilization of a caffeine demethylase enzyme complex. *Int J Food Sci Technol* **36**: 693–698.

Simic, M.G., and Jovanovic, S.V. (1989) Antioxidation mechanisms of uric acid. *J Am Chem Soc* **111**: 5778–5782.

Stavric, B. (1988a) Methylxanthines: toxicity to humans. 1. Theophylline. *Food Chem Toxicol* **26**: 541–565.

Stavric, B. (1988b) Methylxanthines: toxicity to humans. 2. Caffeine. *Food Chem Toxicol* **26**: 645–662.

Stavric, B. (1988c) Methylxanthines: toxicity to humans. 3. Theobromine, paraxanthine, and the combined effects of methylxanthines. *Food Chem Toxicol* **26**: 725–733.

Summers, R.M., Louie, T.M., Yu, C.L., and Subramanian, M. (2011) Characterization of a broad specificity non-haem iron N-demethylase from *Pseudomonas putida* CBB5 capable of utilizing several purine alkaloids as sole carbon and nitrogen source. *Microbiology* **157**: 583–592.

Summers, R.M., Louie, T.M., Yu, C.L., Gakhar, L., Louie, K.C., and Subramanian, M. (2012) Novel, highly specific *N*-demethylases enable bacteria to live on caffeine and related purine alkaloids. *J Bacteriol* **194:** 2041–2049.

Summers, R.M., Seffernick, J.L., Quandt, E.M., Yu, C.L., Barrick, J.E., and Subramanian, M.V. (2013) Caffeine junkie: an unprecedented glutathione *S*-transferase-dependent oxygenase required for caffeine degradation by *Pseudomonas putida* CBB5. *J Bacteriol* **195:** 3933–3939.

Summers, R.M., Gopishetty, S., Mohanty, S.K., and Subramanian, M. (2014) New genetic insights to consider coffee waste as feedstock for fuel, feed, and chemicals. *Cent Eur J Chem* **12:** 1271–1279.

Taylor, E.C., Barton, J.W., and Paudler, W.W. (1961) Studies in purine chemistry. X. Some derivatives of 9-aminopuries. *J Org Chem* **26:** 4961–4967.

Trier, K., Olsen, E.B., Kobayashi, T., and Ribel-Madsen, S.M. (1999) Biochemical and ultrastructural changes in rabbit sclera after treatment with 7-methylxanthine, theobromine, acetazolamide, or L-ornithine. *Br J Ophthalmol* **83:** 1370–1375.

Waller, G.R. (1989) Biochemical frontiers of allelopathy. *Biol Plant* **31:** 418–447.

Woolfolk, C.A. (1975) Metabolism of *N*-methylpurines by a *Pseudomonas putida* strain isolated by enrichment on caffeine as the sole source of carbon and nitrogen. *J Bacteriol* **123:** 1088–1106.

Woolfolk, C.A. (1985) Purification and properties of a novel ferricyanide-linked xanthine dehydrogenase from Pseudomonas putida 40. *J Bacteriol* **163:** 600–609.

Wright, G.A., Baker, D.D., Palmer, M.J., Stabler, D., Mustard, J.A., Power, E.F., *et al.* (2013) Caffeine in floral nectar enhances a pollinator's memory of reward. *Science* **339:** 1202–1204.

Yamaoka-Yano, D.M., and Mazzafera, P. (1998) Degradation of caffeine by *Pseudomonas putida* isolated from soil. *Allelopathy J* **5:** 23–34.

Yamaoka-Yano, D.M., and Mazzafera, P. (1999) Catabolism of caffeine and purification of xanthine oxidase for methyluric acid production in *Pseudomonas putida* L. *Rev Microbiol* **30:** 62–70.

Yu, C.L., Kale, Y., Gopishetty, S., Louie, T.M., and Subramanian, M. (2008) A novel caffeine dehydrogenase in *Pseudomonas* sp. strain CBB1 oxidizes caffeine to trimethyluric acid. *J Bacteriol* **190:** 772–776.

Yu, C.L., Louie, T.M., Summers, R., Kale, Y., Gopishettey, S., and Subramanian, M. (2009) Two distinct pathways for metabolism of theophylline and caffeine are coexpressed in *Pseudomonas putida* CBB5. *J Bacteriol* **191:** 4624–4632.

Yu, C.L., Summers, R., Li, Y., Mohanty, S., Subramanian, M., and Pope, M. (2014) Rapid identification and quantitative validation of a caffeine-degrading pathway in *Pseudomonas* sp. CES. *J Proteome Res* **14:** 95–106.

# 14

# Meta-barcoded evaluation of the ISO standard 11063 DNA extraction procedure to characterize soil bacterial and fungal community diversity and composition

Sebastien Terrat,[1†] Pierre Plassart,[1†] Emilie Bourgeois,[2†] Stéphanie Ferreira,[3] Samuel Dequiedt,[1] Nathalie Adele-Dit-De-Renseville,[3] Philippe Lemanceau,[2] Antonio Bispo,[4] Abad Chabbi,[5] Pierre-Alain Maron[1,2] and Lionel Ranjard[1,2*]

[1]INRA, UMR1347 Agroécologie, Plateforme GenoSol, Dijon, France.
[2]INRA, UMR1347 Agroécologie, Dijon, France.
[3]Equipe R&D Santé-Environnement, Campus Pasteur, Genoscreen, Lille, France.
[4]Service Agriculture et Forêt, ADEME, Angers Cedex 01, France.
[5]INRA-URP3F, Lusignan, France.

## Summary

This study was designed to assess the influence of three soil DNA extraction procedures, namely the International Organization for Standardization (ISO-11063, GnS-GII and modified ISO procedure (ISOm), on the taxonomic diversity and composition of soil bacterial and fungal communities. The efficacy of each soil DNA extraction method was assessed on five soils, differing in their physico-chemical characteristics and land use. A meta-barcoded pyrosequencing approach targeting 16S and 18S rRNA genes was applied to characterize soil microbial communities. We first observed that the GnS-GII introduced some heterogeneity in bacterial composition between replicates. Then, although no major difference was observed between extraction procedures for soil bacterial diversity, we saw that the number of fungal genera could be underestimated by the ISO-11063. In particular, this procedure underestimated the detection in several soils of the genera *Cryptococcus*, *Pseudallescheria*, *Hypocrea* and *Plectosphaerella*, which are of ecological interest. Based on these results, we recommend using the ISOm method for studies focusing on both the bacterial and fungal communities. Indeed, the ISOm procedure provides a better evaluation of bacterial and fungal communities and is limited to the modification of the mechanical lysis step of the existing ISO-11063 standard.

*For correspondence. E-mail ranjard@dijon.inra.fr

**Funding Information** This work involving technical facilities at the GenoSol platform of the infrastructure ANAEE-France received a grant from the French state through the National Agency for Research under the program 'Investments for the Future' (reference ANR-11-INBS-0001), as well as grants from the Regional Council of Burgundy. This work was also supported by the European Commission within EcoFINDERS project (FP7-264465).

## Introduction

During the last three decades, the challenge to better characterize soil microbial communities has led to the development of culture-independent techniques that are well suited to deciphering the huge diversity of soil microbes as they provide access to previously hidden genetic resources (Martin-Laurent *et al.*, 2001). These methods are based essentially on the direct extraction and characterization of soil DNA. In this context, most efforts have been devoted to optimizing the soil DNA extraction procedure in order to obtain suitable representative extracts for quantitative and qualitative characterization of the microbial communities (Roesch *et al.*, 2007; Rajendhran and Gunasekaran, 2008; Terrat *et al.*, 2012). These efforts led to the development of various homemade DNA extraction protocols and even commercial kits (Zhou *et al.*, 1996; Martin-Laurent *et al.*, 2001; Delmont *et al.*, 2011a; Terrat *et al.*, 2012). However, each method had its own advantages and potential biases, leading to variations in DNA representativeness and consequently to effects on soil microbial assessments, making comparisons between studies impossible (Zhou *et al.*, 1996; Martin-Laurent *et al.*, 2001; Terrat *et al.*, 2012). To deal with this issue, Delmont and colleagues (2011b) suggested that several soil sampling and DNA extraction strategies should be combined to access the whole soil microbial metagenome in terms of species richness. However, this approach is clearly not applicable or relevant to wide-scale studies, where time and cost constraints make the need to use a standardized single DNA extraction procedure obvious (Dequiedt *et al.*, 2011).

In this context, a standardized 'ISO-11063: Soil quality – Method to directly extract DNA from soil' was developed and validated by independent laboratories to efficiently recover bacterial DNA from various soil samples (Philippot et al., 2010; Petric et al., 2011). However, archaeal and fungal groups also constitute a significant proportion of the soil microbial biodiversity and are key organisms for soil processes. In a previous study, we tested the sensitivity of the ISO-11063 method for the detection of these groups (Plassart et al., 2012). Briefly, three different procedures were compared on five soils with contrasting land-use and physico-chemical properties: (i) the ISO-11063 standard; (ii) a modified ISO procedure (ISOm) that includes a particular mechanical lysis step (a FastPrep®-24 lysis step instead of the recommended bead beating using a mini bead-beater cell disruptor); and (iii) a custom procedure called GnS-GII, which also includes the FastPrep®-24 mechanical lysis step. This evaluation revealed that the ISO-11063 procedure yielded significantly less overall microbial DNA, (corroborated by measurement of the bacterial, archaeal and fungal densities by real-time PCR), whatever the soil is (Plassart et al., 2012). Furthermore, the analysis of fungal communities' structure with terminal restriction fragment length polymorphism (T-RFLP) patterns showed that the two non-ISO methods clearly outperformed the ISO-11063 method, leading to more significant variations because of soil type and management. Finally, one major conclusion of this study was that the non-ISO methods provided a better representativeness of soil DNA mainly due to use of the FastPrep®-24 bead-beating system, achieving lysis of the majority of cells with tough walls and particularly fungal cells, more efficiently than the usual bead beating (Ranjard et al., 2010; Rousk et al., 2010; Yarwood et al., 2010; Dequiedt et al., 2011; Plassart et al., 2012). Nevertheless, this comparative study was carried out using classical molecular approaches, i.e., quantitative PCR and community DNA fingerprinting through T-RFLP. Nowadays, high throughput sequencing technologies (e.g. 454 or Illumina) are readily available to assess microbial diversity with greater precision by obtaining hundreds of thousands of ribosomal rRNA gene sequences from a single metagenomic DNA (Roesch et al., 2007; Will et al., 2010; Maron et al., 2011). Nonetheless, the DNA extraction techniques previously described has never been evaluated with these new technologies, in terms of efficiency and representativeness, despite their widespread use in soil microbial diversity studies.

In the present study, the same three DNA extraction procedures, coupled with high throughput sequencing technology, were evaluated to identify a technique suitable to characterize the diversity and composition of bacterial and fungal communities simultaneously. The guideline standard ISO-11063, the custom GnS-GII and a custom DNA extraction procedure derived from the ISO-11063 standard (ISOm), were used to extract template DNA from five different soils with contrasting land-use and physico-chemical properties (Plassart et al., 2012). A meta-barcoded pyrosequencing technique, targeting the 16S and 18S rRNA genes, was used to characterize bacterial and fungal communities' richness [based on the number of operational taxonomic units (OTUs) and genera detected], diversity (using Shannon and Evenness indices) and composition (taxonomic affiliation of OTUs). We also measured the phylogenetic distance between sets of OTUs in a phylogenetic tree using the UNIFRAC method to determine whether bacterial and fungal community compositions were influenced by the DNA extraction procedures.

## Results and discussion

Since the development of molecular tools to study soil microbial communities, it has been largely demonstrated that the characterization of these communities might be influenced by the method used to recover soil metagenomic DNA (Delmont et al., 2011b; Terrat et al., 2012). It is consequently essential to test the representativeness of soil DNA extraction methods in terms of bacterial and fungal organisms, which constitute a major part of the soil microbial community. Here, the efficacy of three soil DNA extraction methods (ISO-11063, ISOm and GnS-GII) was assessed on five soils with different physico-chemical characteristics and land use (Table 1) using a meta-barcoded pyrosequencing technique targeting bacterial and fungal communities. This approach was chosen because it is a recently developed powerful technique widely used for detailed phylogenetic and taxonomic surveys of microbial communities (Roesch et al., 2007; Rousk et al., 2010; Will et al., 2010; Lienhard et al., 2013a).

*Influence of soil DNA extraction procedure on bacterial richness and diversity*

Bacterial rRNA gene sequences were successfully amplified by PCR and sequenced from all soils using each of the three DNA extraction procedures (Table 2). After bioinformatic filters, 2322 high-quality reads per sample were kept, analyzed and taxonomically identified using a curated database derived from SILVA (Quast et al., 2013) (Table 3). Rarefaction curves of bacterial richness demonstrated that our sequencing depth allowed accurate description of the bacterial community diversity in each soil sample studied (Supporting Information Fig. S1).

No significant differences were found between the three DNA extraction methods for the number of bacterial genera detected, the number of bacterial OTUs or for the

**Table 1.** Origin, physical and chemical parameters of the five French soils used.

| Soil | Collection site | Origin | Clay | Fine loam | Coarse loam | Fine sand | Coarse sand | Organic carbon | Total N | C/N | CaCO$_3$ | pH |
|------|-----------------|--------|------|-----------|-------------|-----------|-------------|----------------|---------|------|----------|------|
| C | Agricultural Site (Champdotre, Burgundy) | Crop soil | 504 | 180 | 145 | 73 | 98 | 24.9 | 2.8 | 9 | 102 | 7.75 |
| E | INRA Experimental Site (Epoisses, Burgundy) | Crop soil | 392 | 320 | 228 | 34 | 26 | 16.5 | 1.65 | 10 | 2 | 7 |
| F | Forest Observatory Plot (La Mailleraye-sur-Seine, Normandy) | Forest soil | 101 | 167 | 205 | 217 | 310 | 103.3 | 3.1 | 34 | < 1 | 3.8 |
| L | INRA Experimental Site SOERE-ACBB (Lusignan, Poitou) | Grassland | 175 | 369 | 304 | 73 | 79 | 13.2 | 1.33 | 9.92 | < 1 | 6.6 |
| R | INRA Experimental Site (Pierrelaye, Ile-de-France) | Crop soil | 79 | 66 | 44 | 315 | 496 | 50.2 | 2.16 | 23.3 | 22 | 7.5 |

Clay, fine loam, coarse loam, fine sand and coarse sand, organic carbon, total N and calcium carbonate are given in mg g$^{-1}$. Originally published and extracted from (Plassart et al., 2012).

Shannon and Evenness indices in any of the soils (Table 2). This means that neither the mechanical lysis step (using a mini bead-beater cell disruptor or the FastPrep®-24) nor the complete DNA extraction procedures had a significant effect on the evaluation of bacterial diversity parameters by meta-barcoding for the wide range of soil types and land uses tested (Table 1).

On the other hand, soil type did have an impact on bacterial richness and diversity indices, as significant differences were highlighted between soils, whatever the DNA extraction procedure (Table 2). Indeed, F and L soils (respectively the sandy acidic forest soil and the loamy grassland soil) were significantly different ($P < 0.001$) based on the number of OTUs, Shannon and Evenness indices. More precisely, the F soil had the lowest richness (number of OTUs and genera) and diversity, with for example a Shannon index of about 4.1 against 5.6 for the L soil (Table 2). This observation can be linked to particular physico-chemical characteristics, because the F soil had a pH of 3.8 and a C/N ratio of 34 (Dequiedt et al., 2011; Lienhard et al., 2013b). Several studies have high-lighted that bacterial richness had a positive correlation with soil pH (Fierer and Jackson, 2006; Lauber et al., 2008; Terrat et al., 2012) and a negative correlation with C/N ratio (Kuramae et al., 2012). Indeed, a high C/N ratio is generally typical of a large recalcitrant organic matter content that is unfavourable for bacterial growth (Boer et al., 2005). However, the sandy crop soil R, also harbouring a C/N ratio of the same magnitude (23.3), holds a greater richness of OTUs and genera than the forest soil (Table 2). This might partly be due to either the high sand content (Table 1), which increases soil microscale heterogeneity and stimulates the bacterial richness (Chau et al., 2011) or an alkaline pH (7.5) favouring bacterial richness (Fierer and Jackson, 2006; Lauber et al., 2008; Terrat et al., 2012).

Altogether, our results confirmed that bacterial diversity and richness can be strongly linked to soil characteristics and especially soil pH, organic matter and texture (Fierer and Jackson, 2006; Lauber et al., 2008; Kuramae et al., 2012; Terrat et al., 2012; Lienhard et al., 2013b). All DNA extraction procedures tested gave enough and similar

**Table 2.** Bacterial richness and diversity indices of the five soils used.

| | | Number of genera | OTUs (95% of similarity) | Shannon | Evenness |
|---|------|------------------|--------------------------|---------|----------|
| C | GnS | 207.67 (± 5.73) 2 | 485.67 (± 49.13) 1,2 | 5.17 (± 0.13) 1,2 | 0.84 (± 0.01) 1,2 |
| | ISO | 207.33 (± 4.19) 1,2 | 524.33 (± 15.11) 1,2 | 5.15 (± 0.01) 1,2 | 0.82 (± 0.00) 1,2 |
| | ISOm | 205.50 (± 0.5) 1,2 | 521.5 (± 2.5) 1,2 | 5.24 (± 0.03) 1,2 | 0.84 (± 0.01) 1,2 |
| E | GnS | 205.67 (± 7.59) 2 | 522.67 (± 50.37) 1,2 | 5.16 (± 0.12) 1,2 | 0.82 (± 0.01) 1,2 |
| | ISO | 194.00 (± 5.89) 1,2 | 545.33 (± 31.48) 1,2 | 5.22 (± 0.06) 1,2 | 0.83 (± 0.00) 1,2 |
| | ISOm | 200.00 (± 7.48) 1,2 | 498.67 (± 21.64) 1,2 | 5.16 (± 0.02) 1,2 | 0.80 (± 0.00) 1,2 |
| F | GnS | 102.67 (± 11.14) 1 | 329.33 (± 50.31) 1 | 4.12 (± 0.26) 1 | 0.71 (± 0.03) 1 |
| | ISO | 111.00 (± 1.63) 1 | 358.33 (± 31.56) 1 | 4.33 (± 0.21) 1 | 0.74 (± 0.03) 1 |
| | ISOm | 97.33 (± 2.87) 1 | 281.33 (± 12.5) 1 | 3.99 (± 0.13) 1 | 0.71 (± 0.02) 1 |
| L | GnS | 234.33 (± 14.27) 1,2 | 658.3 (± 22.48) 2 | 5.58 (± 0.04) 2 | 0.86 (± 0.00) 2 |
| | ISO | 232.67 (± 2.87) 2 | 668.67 (± 38.69) 2 | 5.68 (± 0.06) 2 | 0.87 (± 0.00) 2 |
| | ISOm | 231.67 (± 11.09) 1,2 | 692 (± 50.34) 2 | 5.62 (± 0.09) 2 | 0.86 (± 0.01) 2 |
| R | GnS | 219.00 (± 9.80) 2 | 561.67 (± 72.67) 1,2 | 5.31 (± 0.15) 1,2 | 0.84 (± 0.01) 1,2 |
| | ISO | 223.33 (± 6.13) 2 | 653.33 (± 39.35) 1,2 | 5.63 (± 0.07) 1,2 | 0.87 (± 0.00) 1,2 |
| | ISOm | 231.00 (± 6.98) 2 | 653.67 (± 33.89) 1,2 | 5.5 (± 0.04) 1,2 | 0.85 (± 0.00) 1,2 |

The means were calculated with three replicates per soil (C, E, F, L and R) and procedure (ISO, ISOm and GnS-GII), and the standard errors of the means are indicated in parentheses. Significant differences between soils for the same procedure are indicated with numbers (1 – 1,2 – 2).

**Table 3.** Bioinformatic parameters and databases used in the analysis of bar-coded pyrosequencing results.

| Step | Parameter | Targeted rDNA Gene | |
|---|---|---|---|
| | | 16S | 18S |
| Preprocessing | Length threshold | 370 | 300 |
| | Number of ambiguities tolerated | 0 | 0 |
| | Detection of proximal primer sequence | Complete and perfect | Complete and perfect |
| | Detection of distal primer sequence | No | Perfect, but potentially incomplete |
| Clustering | Chosen level of similarity (%) | 95 | 95 |
| | Ignoring differences in homopolymer lengths | Yes | Yes |
| Filtering | Chosen clustering similarity threshold | 95 | 95 |
| | Used taxonomic database | SILVA (r114) | SILVA (r111) |
| | Chosen taxonomic level | Phylum | Phylum |
| | Similarity or confidence threshold (%) | 90 | 85 |
| Homogenization | High-quality reads kept for each sample | 2322 | 4378 |
| Taxonomy | Used taxonomic database | SILVA (r114) | SILVA (r111) |
| | Method or tool of comparison | USEARCH | MEGABLAST |
| | Similarity or confidence threshold (%) | 80 | 80 |
| Analysis | Chosen level of similarity (%) | 95 | 95 |
| | Ignoring differences in homopolymer lengths | Yes | Yes |
| | Computation of a UNIFRAC distance matrix | Yes | Yes |

sensitivity to detect changes between indigenous bacterial communities of soils differing by their characteristics and management. These data also support the idea that a study limited to these diversity indices could not be sufficient to determine whether a DNA extraction procedure is more powerful than another to describe soil bacterial communities and that it might be completed by a more detailed bacterial community composition analysis.

### Influence of soil DNA extraction procedure on fungal richness and diversity

Using the same DNA extracts as for the bacterial analysis (three DNA extraction procedures applied to five soils), 18S rRNA gene sequences were successfully amplified and sequenced from all samples (Table 4). Homogenized high-quality reads (4378 per sample) were then analyzed using taxonomically dependent and independent analyses to determine fungal richness and diversity (Table 3). As for bacteria, the rarefaction curves of fungal richness confirmed that the number of high-quality reads allowed accurate description of the fungal community diversity in each soil sample studied (Supporting Information Fig. S1).

With regard to the number of detected genera, the numbers of OTUs and the computed indices (Shannon and Evenness), significant differences among the three DNA extraction procedures were recorded only for the L soil (Table 4), in which a lower number of fungal genera was significantly detected using the ISO procedure ($P = 0.003$). Moreover, in all the other soils but F, the

**Table 4.** Fungal richness and diversity indices of the five French soils used.

| | | Number of *genera* | OTUs (95% of similarity) | Shannon | Evenness |
|---|---|---|---|---|---|
| C | GnS | 116.00 (± 11.43) *1* | 350.33 (± 42.32) *1,2* | 3.72 (± 0.05) *1,2* | 0.64 (± 0.01) *1,2* |
| | ISO | 92.33 (± 15.69) *1* | 273.67 (± 63.67) *1* | 3.31 (± 0.33) *1* | 0.59 (± 0.04) *1,2* |
| | ISOm | 118.67 (± 9.67) *1* | 340.67 (± 68.23) *1,2* | 3.73 (± 0.15) *1,2* | 0.64 (± 0.01) *1,2* |
| E | GnS | 128.00 (± 13.74) *1* | 287.33 (± 30.58) *1* | 3.39 (± 0.08) *1,2* | 0.6 (± 0.01) *1,2* |
| | ISO | 108.33 (± 9.81) *1* | 239.33 (± 31.54) *1* | 3.34 (± 0.22) *1* | 0.61 (± 0.03) *1,2* |
| | ISOm | 125.67 (± 6.55) *1* | 289 (± 33.66) *1* | 3.54 (± 0.18) *1,2* | 0.63 (± 0.02) *1,2* |
| F | GnS | 129.67 (± 8.34) *1* | 249.67 (± 11.15) *1* | 2.98 (± 0.05) *1* | 0.54 (± 0.01) *1* |
| | ISO | 136.00 (± 7.79) *2* | 312 (± 35.36) *1* | 3.19 (± 0.12) *1* | 0.56 (± 0.01) *1* |
| | ISOm | 140.33 (± 10.14) *1* | 267 (± 24.91) *1* | 3.27 (± 0.18) *1* | 0.59 (± 0.03) *1* |
| L | GnS | 127.33 (± 9.29) *a.1* | 416.33 (± 89.46) *2* | 4.05 (± 0.21) *2* | 0.67 (± 0.01) *2* |
| | ISO | 89.67 (± 5.79) *b.1* | 353.33 (± 51.45) *1* | 3.89 (± 0.09) *1* | 0.66 (± 0.03) *1,2* |
| | ISOm | 129.00 (± 6.48) *a.1* | 382.33 (± 71.82) *2* | 3.74 (± 0.49) *2* | 0.63 (± 0.06) *2* |
| R | GnS | 141.33 (± 13.82) *1* | 407.33 (± 84.94) *2* | 3.9 (± 0.2) *1,2* | 0.65 (± 0.01) *1,2* |
| | ISO | 111.00 (± 11.00) *1,2* | 399.00 (± 30.00) *2* | 4.14 (± 0.01) *1* | 0.69 (± 0.01) *2* |
| | ISOm | 135.00 (± 12.68) *1* | 407 (± 67.38) *2* | 3.94 (± 0.12) *1,2* | 0.66 (± 0.01) *1,2* |

The means were calculated with three replicates per soil (C, E, F, L and R) and procedure (ISO, ISOm and GnS-GII), and the standard errors of the means are indicated in parentheses. Significant differences between procedures for the same soil are indicated by letters (*a, b*), and significant differences between soils for the same procedure are indicated with numbers (1 – 1,2 – 2).

number of genera recovered followed the same trend, with lower values detected for the ISO, than what was observed for the two other procedures. These genera missed by the ISO demonstrate that fungal diversity can be skewed using this procedure. As the main difference among the ISO and the two other procedures is the soil-grinding step; we can hypothesize that the traditional bead-beating system is not sufficient to lyze some fungal cells. Indeed, many fungi have cell walls that impede lysis and the recovery of nucleic acids (Fredricks et al., 2005). The mechanical lysis step of the ISOm and GnS-GII procedures was strongly optimized in terms of type and size of the glass beads as well as in terms of the strength and duration of grinding using the FastPrep®-24 (Terrat et al., 2012).

When fungal diversity was compared between soils, significant differences in the Shannon and Evenness indices ($P < 0.05$) and in the number of OTUs ($P < 0.1$) were observed whatever the DNA extraction procedure (Table 4). More precisely, the acidic forest soil F harboured the lowest richness and diversity, and the alkaline sandy crop soil R the highest (Table 4). These differences could be explained by several soil physico-chemical parameters, namely their contrasting pH (3.8 against 7.75), but also their C/N ratio (34 against 23.3) (Table 1). Although extreme environments like acidic soils may provide suitable biotopes for fungi (Baker and Banfield, 2003; Butinar et al., 2005), the lowest richness and diversity was detected in the acidic forest soil F, indicating that other physico-chemical parameters can limit fungal communities. Thus, a high C/N ratio is typical of soil systems with a low rate of organic matter degradation because of the presence of a high proportion of recalcitrant organic matter (Kuramae et al., 2012). Strickland and Rousk (2010) demonstrated in a previous study that the optimal C/N for fungi is expected to range from 5 to 15; i.e. closer to the C/N of the sandy crop soil R than to the ratio of the forest soil F, which has a higher carbon content. Focusing on the number of fungal genera recovered by the three DNA extraction procedures, only the ISO allowed the detection of significant differences between soils. This finding has to be seriously questioned because we demonstrated in the previous paragraph that the ISO underestimates the number of fungal genera.

*Influence of soil DNA extraction procedure on bacterial community composition*

The bacterial community composition in the five soils was compared by computing the UNIFRAC distances on a phylogenetic tree (Lozupone and Knight, 2005). In addition to analyzing the phylogenetic distances, we also compared the bacterial communities' compositions based on the relative abundance of the bacterial genera detected in the samples (Fig. 1). Due to the size and variability of the genus table, only the most highly represented bacterial genera in the samples (i.e. only those for which the sum of the relative abundances of the genus in all samples was higher than 5%) were identified and mapped.

The clustering of soil bacterial communities indicated that replicates from the same soil were more similar to each other than replicates from other soils, whatever the DNA extraction procedure (Fig. 1). This observation demonstrated the good reproducibility between replicates for each type of soil DNA extraction procedure even if, surprisingly, two GnS-GII replicates from soils L and R seemed to be erroneously clustered. More precisely, two main clusters were identified, sorting the samples from the acidic forest soil F (which hosted a very different bacterial composition) apart from the four other soils (Fig. 1). Four sub-clusters could also be defined, each one grouping samples from each of the four soils and confirming that the studied soils hosted distinct bacterial communities, as already demonstrated by DNA fingerprinting approach (Plassart et al., 2012). This observation demonstrated the good reproducibility between replicates for ISO and ISOm procedures. However, even if clustering revealed that soil type had a more important effect on bacterial composition than the DNA extraction procedure, it is interesting to note that this latter could induce significant variations (Fig. 1). For all soils (except the forest soil F), the bacterial diversity profiles resulting from the ISOm and GnS-GII DNA extraction procedures grouped together (i.e. were not discriminated by the UNIFRAC analysis), but were different from those obtained with the ISO-11063 procedure (Fig. 1). These observations confirm the influence of soil DNA extraction procedure on soil bacterial composition and especially the clear distinction between ISO-11063 and the two other procedures, potentially explained by differences in the soil-grinding methods (as discussed above for fungal richness and diversity). These differences were also confirmed by a more detailed analysis of bacterial composition (Fig. 1, subcells A–C). For example, the genus *Brevundimonas* was more detected ($P < 0.05$) with the ISO-11063 procedure than with the two others in the clayey crop soil C (Fig. 1, subcell A), as were the genera *Massilia, Pseudospirillum, Herbaspirillum, Enterobacter, Thermomonas* and *Lysobacter*. Similarly, the genus *Polaromonas* was more detected in the sandy crop soil R (Fig. 1, subcell C), but not in the other soils. On the contrary, the genera *Clostridium, Nitrosospira, Microvirga* and *Pseudonocardia* were respectively less detected ($P < 0.05$) with ISO-11063 than with the ISOm and GnS-GII procedures in soils C, E, L and R (Fig. 1, subcell B). Because the genera *Clostridium* or *Pseudonocardia* are known to be potentially recalcitrant to mechanical lysis, because of their spore-forming ability (Kaewkla and Franco, 2011; Yang

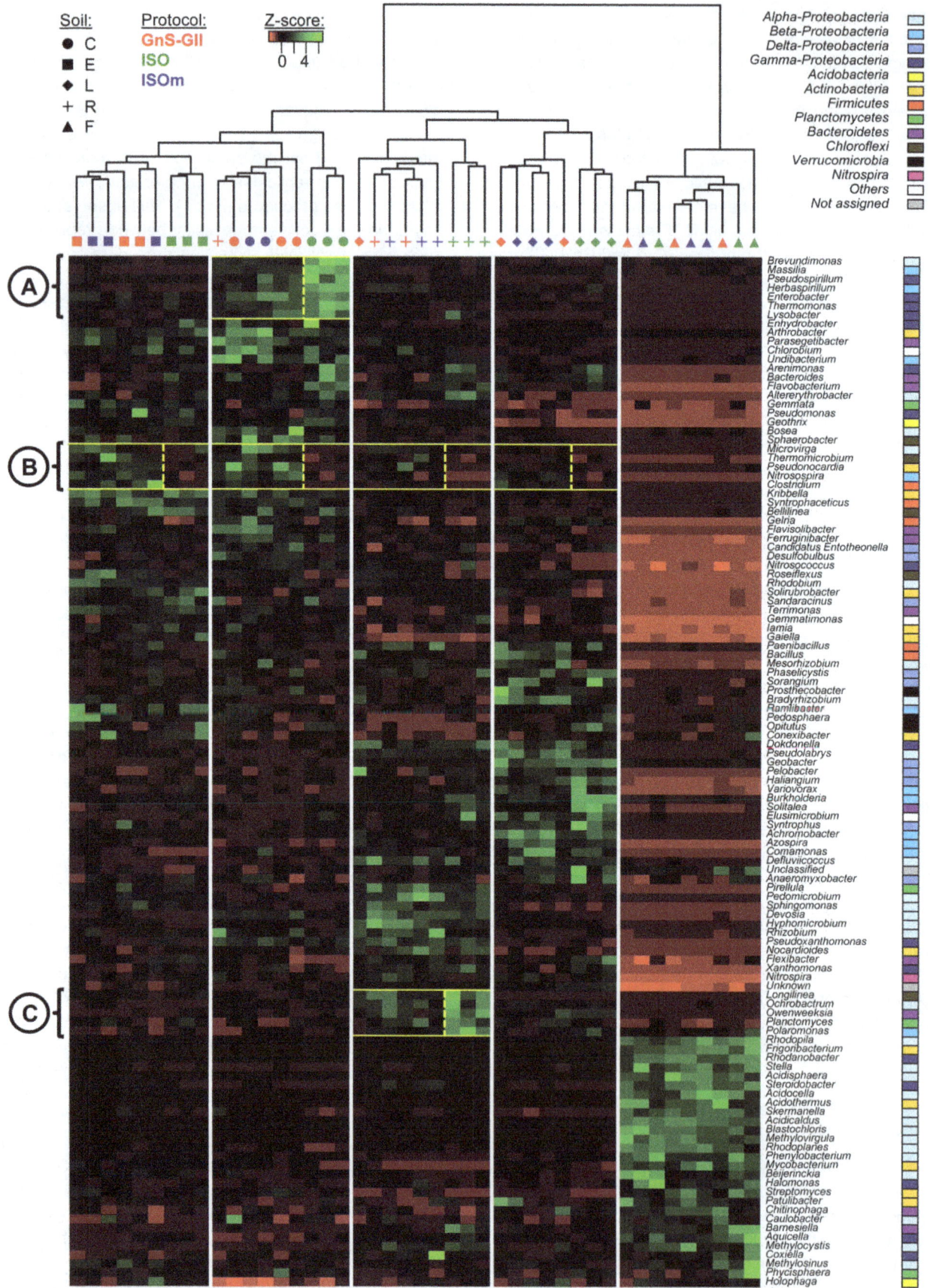

**Fig. 1.** Heat map comparison of the dominant bacterial genera detected in soils according to extraction procedures. The five different soils (C, E, F, L, R) were organized based on the UPGMA dendrogram of UNIFRAC distances (weighted and normalized) between soil samples according to the three DNA extraction procedures (ISO-11063, GnS-GII and ISOm). The legend shows the Z-scores (relative abundances are expressed as median centred Z-scores between all samples, and the colours scaled to standard deviations). Subcells A, B and C in the heat map have been highlighted by yellow squares and numbered to identify significant differences in the relative abundance of particular bacterial genera according to DNA extraction procedure.

and Ponce, 2011), their lower detection with the ISO-11063 procedure may be explained by the less efficient mechanical lysis (bead beating) of this procedure, compared with the two others, which are based on FastPrep®-24 grinding (Plassart et al., 2012; Terrat et al., 2012).

### Influence of soil DNA extraction procedure on fungal community composition

As with the bacterial communities, the fungal communities' composition in all soils was compared by using the UNIFRAC distances and determining the most highly represented fungal genera in the samples (Fig. 2). The UNIFRAC dendrogram revealed a better discrimination of fungal composition between soils than between DNA extraction methods, demonstrating a good reproducibility between replicates for all procedures (Fig. 2). Moreover, as for the bacterial communities, the same clustering organization was obtained for fungal communities, revealing a significant distinction between the forest soil and the other soils. This observation corroborates other studies in which soil characteristics (e.g. pH, texture, C/N) were shown to impact fungal community diversity and composition (Rousk et al., 2010; Strickland and Rousk, 2010; McGuire et al., 2013). The fungal populations in this acidic soil clearly differed from those of the other soils, with a dominance of the *Basidiomycota* phylum (e.g. genera *Sebacina, Boletus, Pleurotus* or *Hericium*), which is common in forest soils (Buée et al., 2009).

For soils F, L and R, the patterns of the fungal communities resulting from the ISO-11063 procedure were discriminated from those obtained with the two non-ISO protocols (Fig. 2). This observation evidenced that in these soils, the fungal community compositions detected with the ISO-11063 differed from those detected with the non-ISO procedures. More precisely, in the loamy grassland soil L, several *genera* (e.g. *Knufia* and *Diversispora*) were more detected ($P < 0.1$) with the ISO-11063 protocol (Fig. 2, subcell A). However, this positive impact of the ISO-11063 procedure was only visible for this particular soil. On the contrary, the genera *Myrothecium, Cryptococcus, Glomerella* and *Plectosphaerella* were respectively less detected ($P < 0.05$) with the ISO-11063 protocol than with the other methods in soils C, E, L and R (Fig. 2, subcell B), as *Pseudallescheria* in soils C, E and R, and *Hypocrea* in soils E and L ($P < 0.1$). This would be of great importance in ecological studies as some of these

genera (e.g. *Cryptococcus, Pseudallescheria, Hypocrea* and *Plectosphaerella*) are saprotrophic fungi known to play key roles in organic matter turnover (Martínez et al., 2003; Jaklitsch et al., 2005; Buée et al., 2009; McGuire et al., 2013). Moreover, the genus *Pseudallescheria*, which has been found in compost-amended or heavily hydrocarbon-polluted soils, can be used as an indicator of soil disturbance (April et al., 1998). Therefore, the ISO-induced underrepresentation of these genera could lead to a misinterpretation of the functioning of an ecosystem.

This difference in community composition is, together with the lower number of fungal genera recovered with the ISO described earlier, a clue indicating that the ISO procedure may not be the most appropriate to investigate soil fungal communities. Besides, these differences are thought to be due to the less efficient mechanical lysis of soil with the ISO-11063 procedure; the classical system seems not to break open as many cells as the FastPrep®-24 bead-beating system, particularly in the case of fungal cells with tough walls. This is why the ISOm and GnS-GII methods are thought to be more efficient at extracting fungal DNA from different types of soils. This conclusion strengthens the idea that the physical lysis step is of crucial importance in a soil DNA extraction procedure (Feinstein et al., 2009; İnceoğlu et al., 2010; Delmont et al., 2011b). This finding is in agreement with previous comparisons of these DNA extraction procedures based on quantitative PCR and community DNA fingerprinting (Plassart et al., 2012).

### Conclusion

In the context of modern microbial ecology, where investigations to describe the whole soil microbiota in numerous samples are carried out on a very large scale, the importance of using a single, standardized soil DNA extraction procedure is paramount. Among the three DNA extraction procedures evaluated in this study, the GnS-GII introduced some heterogeneity in bacterial composition between replicates, and the ISO-11063 DNA caused an underrepresentation of several fungal groups of ecological interest. Therefore, the ISOm procedure provides a better snapshot of bacterial and fungal communities.

### Experimental procedures

#### Soil samples

Five soils were chosen for their contrasting land-use and physico-chemical characteristics (Table 1) (Plassart et al.,

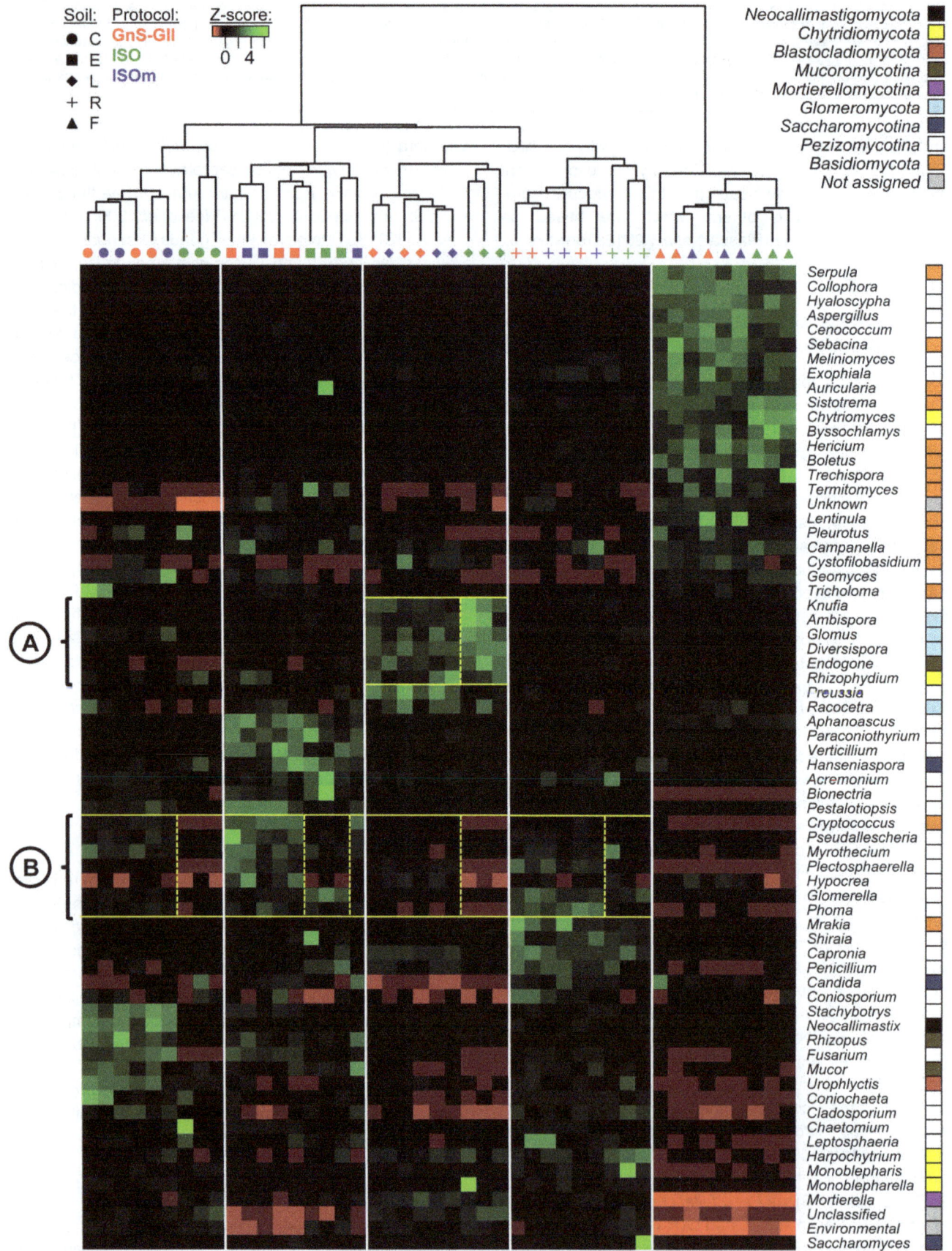

**Fig. 2.** Heat map comparison of the dominant fungal genera detected in soils according to extraction procedures. The five different soils (C, E, F, L, R) were organized based on the UPGMA dendrogram of UNIFRAC distances (weighted and normalized) between soil samples according to the three DNA extraction procedures (ISO-11063, GnS-GII and ISOm). The legend shows the Z-scores (relative abundances are expressed as median centred Z-scores between all samples, and the colours scaled to standard deviations). Subcells A and B in the heat map have been highlighted by yellow squares and numbered to identify significant differences in the relative abundance of particular fungal genera according to DNA extraction procedure.

◀ ─────────────────────────────────────────────────

2012). All necessary permits were obtained from the respective land owners (INRA, ADEME and private owners). For each soil, three independent replicates were collected at a depth of 20 cm [fully described in (Plassart *et al.*, 2012)]. Physico-chemical characteristics (pH, texture, organic carbon, total N and $CaCO_3$) were analyzed, using international standard procedures, by the Soil Analysis Laboratory at INRA (Arras, France, http://www.lille.inra.fr/las).

### Soil DNA extraction, purification and quantification

Three different procedures were tested: the GnS-GII protocol, the ISO-11063 standard and the ISOm. All three procedures are adapted to extract DNA from 1 g of soil (dry weight) and have already been described by Plassart and colleagues (2012).

*ISO-11063 procedure.* This protocol is a version of the ISO-11063 standard (Martin-Laurent *et al.*, 2001; Petric *et al.*, 2011). Soil was added to a bead-beating tube containing 2 g of glass beads of 106 μm diameter and eight glass beads of 2 mm diameter. Each soil sample was mixed with a solution of 100 mM Tris-HCl (pH 8), 100 mM EDTA (pH 8), 100 mM NaCl, 2% (w/v) polyvinylpyrrolidone (40 g mol$^{-1}$) and 2% (w/v) sodium dodecyl sulfate. The tubes were then shaken for 30 s at 1600 r.p.m. in a mini bead-beater cell disruptor (Mikro-Dismembrator, Braun Biotech International), then incubated for 10 min at 70°C and centrifuged at 14,000$g$ for 1 min. After removing the supernatant, proteins were precipitated, with 1/10 volume of 3 M sodium acetate prior to centrifugation (14,000$g$ for 5 min at 4°C). Finally, nucleic acids were precipitated by adding 1 volume of ice-cold isopropanol. The DNA pellets obtained after centrifugation (14,000$g$ for 5 min at 4°C) were washed with 70% ethanol (full details are described in (Martin-Laurent *et al.*, 2001; Philippot *et al.*, 2010; Petric *et al.*, 2011).

*ISOm procedure.* This protocol is a modified version of ISO-11063 standard as it includes a different mechanical lysis step (FastPrep® bead-beating instead of the recommended bead beating). Soil was added to 15 ml of Falcon tube containing 2.5 g of 1.4 mm diameter ceramic beads, 2 g of 106 μm diameter silica beads and four glass beads of 4 mm diameter. Each soil sample was mixed with a solution of 100 mM Tris-HCl (pH 8), 100 mM EDTA (pH 8), 100 mM NaCl, 2% (w/v) polyvinylpyrrolidone (40 g mol$^{-1}$) and 2% (w/v) sodium dodecyl sulfate. The tubes were then shaken for 3 × 30 s at 4 m sec$^{-1}$ in a FastPrep®-24 (MP-Biomedicals, NY, USA), before incubation for 10 min at 70°C and centrifugation at 14,000$g$ for 1 min. After removing the supernatant, proteins were precipitated with 1/10 volume of 3 M sodium acetate prior to centrifugation (14,000$g$ for 5 min at 4°C). Finally, nucleic acids were precipitated by adding 1 volume of ice-

cold isopropanol. The DNA pellets obtained after centrifugation (14,000$g$ for 5 min at 4°C) were washed with 70% ethanol.

*GnS-GII procedure.* This DNA extraction procedure was initially developed and optimized by the GenoSol platform (Terrat *et al.*, 2012). Soil was added to 15 ml of Falcon tube containing 2.5 g of 1.4 mm diameter ceramic beads, 2 g of 106 μm diameter silica beads and four glass beads of 4 mm diameter. Each soil sample was mixed with a solution of 100 mM Tris-HCl (pH 8), 100 mM EDTA (pH 8), 100 mM NaCl, 2% (w/v) and 2% (w/v) sodium dodecyl sulfate. The tubes were then shaken for 3 × 30 s at 4 m sec$^{-1}$ in a FastPrep®-24 (MP-Biomedicals, NY, USA), before incubation for 30 min at 70°C and centrifugation at 7,000$g$ for 5 min at 20°C. After removing the supernatant, proteins were precipitated with 1/10 volume of 3 M sodium acetate prior to centrifugation (14,000$g$ for 5 min at 4°C). Finally, nucleic acids were precipitated by adding 1 volume of ice-cold isopropanol. The DNA pellets obtained after centrifugation (14,000$g$ for 5 min at 4°C) were washed with 70% ethanol.

*Purification and quantification procedure.* As the DNA purification step is not part of the evaluated protocols to avoid additional biases among the three procedures and only compare the extraction step, all crude soil DNA extracts were purified and quantified using the same procedure (Ranjard *et al.*, 2003; Plassart *et al.*, 2012). Briefly, 100 μl aliquots of crude DNA extracts were loaded onto PVPP (polyvinylpolypyrrolidone) Microbiospin minicolumns (Bio-Rad) and centrifuged for 4 min at 1000$g$ and 10°C. Eluates were then collected and purified for residual impurities using the Geneclean Turbo kit (MP-Biomedicals, NY, USA). Purified DNA extracts were quantified using the PicoGreen staining Kit (Molecular Probes, Paris, France).

### Pyrosequencing of 16S and 18S rRNA gene sequences

Microbial diversity was determined for each biological replicate and for each soil (C, F, E, L and R) by 454 pyrosequencing of ribosomal genes. A 16S rRNA gene fragment with sequence variability and appropriate size (about 450 bases) for 454 pyrosequencing was amplified using the primers F479 (5′-CAGCMGCYGCNGTAANAC-3′) and R888 (5′-CCGYCAATTCMTTTRAGT-3′) (Supporting Information Table S1 for *in silico* match analysis, Terrat *et al.* 2014). For each sample, 5 ng of DNA were used for a 25 μl of PCR conducted under the following conditions: 94°C for 2 min, 35 cycles of 30 s at 94°C, 52°C for 30 s and 72°C for 1 min, followed by 7 min at 72°C. The PCR products were purified using a MinElute gel extraction kit (Qiagen, Courtaboeuf, France) and quantified using the PicoGreen staining Kit (Molecular Probes, Paris, France). Similarly, an 18S rRNA gene fragment of about 350 bases was amplified using the primers FR1 (5′-ANCCATTCAATCGGTANT-3′) and FF390 (5′-CGATAACGAACGAGACCT-3′) (Prevost-Boure *et al.*, 2011) under the following PCR conditions: 94°C for 3 min, 35 cycles of 1 min at 94°C, 52°C for 1 min and 72°C for 1 min, followed by 5 min at 72°C. A second PCR of nine cycles was then conducted twice for each sample under similar PCR conditions with purified PCR products and 10 base pair multiplex identifiers added to the primers at 5′ position to specifically identify each sample and avoid PCR bias. Finally, the duplicate PCR products were pooled, purified and quantified as previously described. Pyrosequencing was then carried out on a GS FLX Titanium (Roche 454 Sequencing System) by Genoscreen (Lille, France).

### Bioinformatic analysis of 16S and 18S rRNA gene sequences

Bioinformatic analyses were done using the GnS-PIPE initially developed by the Genosol platform (INRA, Dijon, France) (Terrat *et al.*, 2012) and recently optimized. The parameters chosen for each bioinformatic step can be found in Table 3. First, all the 16S and 18S raw reads were sorted according to the multiplex identifier sequences. The raw reads were then filtered and deleted based on (i) their length, (ii) their number of ambiguities (Ns) and (iii) their primer(s) sequence(s). A PERL program was then applied for rigorous dereplication (i.e. clustering of strictly identical sequences). The dereplicated reads were then aligned using INFERNAL alignment (Cole *et al.*, 2009), and clustered into OTU using a PERL program that groups rare reads to abundant ones, and does not count differences in homopolymer lengths. A filtering step was then carried out to check all single singletons (reads detected only once and not clustered, which might be artefacts, such as PCR chimeras) based on the quality of their taxonomic assignments. Finally, in order to compare the data sets efficiently and avoid biased community comparisons, the reads retained were homogenized by random selection closed to the lowest dataset.

The retained high-quality reads were used for (i) taxonomy-independent analyses, determining several diversity and richness indices using the defined OTU composition at the genus level and (ii) taxonomy-based analysis using similarity approaches against dedicated reference databases from SILVA (Quast *et al.*, 2013) (see Table 3). The raw data sets are available on the European Bioinformatics Institute database system under project accession number PRJEB4825.

### Statistical analyses

The effects of the DNA extraction procedure on bacterial and fungal diversities were tested by analysis of variance (multiple paired comparisons). The effects of the DNA extraction procedure on bacterial and fungal community compositions were assessed by Kruskal–Wallis tests. All statistical analyses were performed under XLSTAT software (Addinsoft®). The bacterial and fungal communities from all samples were also compared by using UNIFRAC (Lozupone and Knight, 2005), based on the 16S and 18S phylogenetic trees computed with FASTTREE (Price *et al.*, 2010).

### Acknowledgements

Thanks to the National Research Infrastructure 'Agroécosystèmes, Cycles Biogéochimique et Biodiversité (SOERE-ACBB http://www.soere-acbb.com/fr/) for providing support during the sampling campaign and making available the physico-chemical soil properties data. The funders [National Agency for Research (through 'Investments for the Future' program), Regional Council of Burgundy and European Commission) had no role in the study design, data collection and analysis, decision to publish, or preparation of the manuscript.

### Conflict of Interest

None declared.

### References

April, T.M., Abbott, S.P., Foght, J.M., and Currah, R.S. (1998) Degradation of hydrocarbons in crude oil by the ascomycete *Pseudallescheria boydii* (*Microascaceae*). *Can J Microbiol* **44**: 270–278.

Baker, B.J., and Banfield, J.F. (2003) Microbial communities in acid mine drainage. *FEMS Microbiol Ecol* **44**: 139–152.

Boer, W.d., Folman, L.B., Summerbell, R.C., and Boddy, L. (2005) Living in a fungal world: impact of fungi on soil bacterial niche development. *FEMS Microbiol Rev* **29**: 795–811.

Buée, M., Reich, M., Murat, C., Morin, E., Nilsson, R.H., Uroz, S., and Martin, F. (2009) 454 Pyrosequencing analyses of forest soils reveal an unexpectedly high fungal diversity. *New Phytol* **184**: 449–456.

Butinar, L., Santos, S., Spencer-Martins, I., Oren, A., and Gunde-Cimerman, N. (2005) Yeast diversity in hypersaline habitats. *FEMS Microbiol Lett* **244**: 229–234.

Chau, J.F., Bagtzoglou, A.C., and Willig, M.R. (2011) The effect of soil texture on richness and diversity of bacterial communities. *Environ Forensics* **12**: 333–341.

Cole, J.R., Wang, Q., Cardenas, E., Fish, J., Chai, B., Farris, R.J., *et al.* (2009) The Ribosomal Database Project: improved alignments and new tools for rRNA analysis. *Nucleic Acids Res* **37**: D141–D145.

Delmont, T.O., Robe, P., Cecillon, S., Clark, I.M., Constancias, F., Simonet, P., *et al.* (2011a) Accessing the soil metagenome for studies of microbial diversity. *Appl Environ Microbiol* **77**: 1315–1324.

Delmont, T.O., Robe, P., Clark, I., Simonet, P., and Vogel, T.M. (2011b) Metagenomic comparison of direct and indirect soil DNA extraction approaches. *J Microbiol Methods* **86**: 397–400.

Dequiedt, S., Saby, N.P.A., Lelievre, M., Jolivet, C., Thioulouse, J., Toutain, B., *et al.* (2011) Biogeographical patterns of soil molecular microbial biomass as influenced by soil characteristics and management. *Glob Ecol Biogeogr* **20**: 641–652.

Feinstein, L.M., Sul, W.J., and Blackwood, C.B. (2009) Assessment of bias associated with incomplete extraction of microbial DNA from soil. *Appl Environ Microbiol* **75**: 5428–5433.

Fierer, N., and Jackson, R.B. (2006) The diversity and bio-geography of soil bacterial communities. *Proc Natl Acad Sci USA* **103:** 626–631.

Fredricks, D.N., Smith, C., and Meier, A. (2005) Comparison of six DNA extraction methods for recovery of fungal DNA as assessed by quantitative PCR comparison of six DNA extraction methods for recovery of Fungal DNA as assessed by quantitative PCR. *J Clin Microbiol* **43:** 5122–5128.

İnceoğlu, Ö., Hoogwout, E.F., Hill, P., and van Elsas, J.D. (2010) Effect of DNA extraction method on the apparent microbial diversity of soil. *Appl Environ Microbiol* **76:** 3378–3382.

Jaklitsch, W.M., Komon, M., Kubicek, C.P., and Druzhinina, I.S. (2005) *Hypocrea voglmayrii* sp. nov. from the Austrian Alps represents a new phylogenetic clade in *Hypocrea/Trichoderma*. *Mycol* **97:** 1365–1378.

Kaewkla, O., and Franco, C.M.M. (2011) *Pseudonocardia eucalypti* sp. nov., an endophytic actinobacterium with a unique knobby spore surface, isolated from roots of a native Australian eucalyptus tree. *Int J Syst Evol Microbiol* **61:** 742–746.

Kuramae, E.E., Yergeau, E., Wong, L.C., Pijl, A.S., van Veen, J.A., and Kowalchuk, G.A. (2012) Soil characteristics more strongly influence soil bacterial communities than land-use type. *FEMS Microbiol Ecol* **79:** 12–24.

Lauber, C.L., Strickland, M.S., Bradford, M.A., and Fierer, N. (2008) The influence of soil properties on the structure of bacterial and fungal communities across land-use types. *Soil Biol Biochem* **40:** 2407–2415.

Lienhard, P., Terrat, S., Mathieu, O., Levêque, J., Prévost-Bouré, N.C., Nowak, V., *et al.* (2013a) Soil microbial diversity and C turnover modified by tillage and cropping in Laos tropical grassland. *Environ Chem Lett* **11:** 1–8.

Lienhard, P., Tivet, F., Chabanne, A., Dequiedt, S., Lelievre, M., Sayphoummie, S., *et al.* (2013b) No-till and cover crops shift soil microbial abundance and diversity in Laos tropical grasslands. *Agron Sustain Dev* **33:** 375–384.

Lozupone, C., and Knight, R. (2005) UniFrac: a new phylogenetic method for comparing microbial communities. *Appl Environ Microbiol* **71:** 8228–8235.

McGuire, K.L., Payne, S.G., Palmer, M.I., Gillikin, C.M., Keefe, D., Kim, S.J., *et al.* (2013) Digging the New York City skyline: soil fungal communities in green roofs and city parks. *PLoS ONE* **8:** e58020.

Maron, P.-A., Mougel, C., and Ranjard, L. (2011) Soil microbial diversity: methodological strategy, spatial overview and functional interest. *C R Biol* **334:** 403–411.

Martínez, M., López-Solanilla, E., Rodríguez-Palenzuela, P., Carbonero, P., and Díaz, I. (2003) Inhibition of plant-pathogenic fungi by the barley cystatin Hv-CPI (Gene Icy) is not associated with its cysteine-proteinase inhibitory properties. *Mol Plant Microbe Interact* **16:** 876–883.

Martin-Laurent, F., Philippot, L., Hallet, S., Chaussod, R., Germon, J.C., Soulas, G., and Catroux, G. (2001) DNA extraction from soils: old bias for new microbial diversity analysis methods. *Appl Environ Microbiol* **67:** 2354–2359.

Petric, I., Philippot, L., Abbate, C., Bispo, A., Chesnot, T., Hallin, S., *et al.* (2011) Inter-laboratory evaluation of the ISO standard 11063 'Soil quality – Method to directly extract DNA from soil samples'. *J Microbiol Methods* **84:** 454–460.

Philippot, L., Abbate, C., Bispo, A., Chesnot, T., Hallin, S., Lemanceau, P., *et al.* (2010) Soil microbial diversity: an ISO standard for soil DNA extraction. *J Soils Sediments* **10:** 1344–1345.

Plassart, P., Terrat, S., Thomson, B., Griffiths, R., Dequiedt, S., Lelievre, M., *et al.* (2012) Evaluation of the ISO standard 11063 DNA extraction procedure for assessing soil microbial abundance and community structure. *PLoS ONE* **7:** e44279.

Prevost-Boure, N.C., Christen, R., Dequiedt, S., Mougel, C., Lelievre, M., Jolivet, C., *et al.* (2011) Validation and application of a PCR primer set to quantify fungal communities in the soil environment by real-time quantitative PCR. *PLoS ONE* **6:** e24166.

Price, M.N., Dehal, P.S., and Arkin, A.P. (2010) FastTree 2 – approximately maximum-likelihood trees for large alignments. *PLoS ONE* **5:** e9490.

Quast, C., Pruesse, E., Yilmaz, P., Gerken, J., Schweer, T., Yarza, P., *et al.* (2013) The SILVA ribosomal RNA gene database project: improved data processing and web-based tools. *Nucleic Acids Res* **41:** D590–D596.

Rajendhran, J., and Gunasekaran, P. (2008) Strategies for accessing soil metagenome for desired applications. *Biotechnol Adv* **26:** 576–590.

Ranjard, L., Lejon, D.P.H., Mougel, C., Schehrer, L., Merdinoglu, D., and Chaussod, R. (2003) Sampling strategy in molecular microbial ecology: influence of soil sample size on DNA fingerprinting analysis of fungal and bacterial communities. *Environ Microbiol* **5:** 1111–1120.

Ranjard, L., Dequiedt, S., Jolivet, C., Saby, N.P.A., Thioulouse, J., Harmand, J., *et al.* (2010) Biogeography of soil microbial communities: a review and a description of the ongoing French national initiative. *Agron Sustain Dev* **30:** 359–365.

Roesch, L.F.W., Fulthorpe, R.R., Riva, A., Casella, G., Hadwin, A.K.M., Kent, A.D., *et al.* (2007) Pyrosequencing enumerates and contrasts soil microbial diversity. *ISME J* **1:** 283–290.

Rousk, J., Baath, E., Brookes, P.C., Lauber, C.L., Lozupone, C., Caporaso, J.G., *et al.* (2010) Soil bacterial and fungal communities across a pH gradient in an arable soil. *ISME J* **4:** 1340–1351.

Strickland, M.S., and Rousk, J. (2010) Considering fungal: bacterial dominance in soils – methods, controls, and eco-system implications. *Soil Biol Biochem* **42:** 1385–1395.

Terrat, S., Christen, R., Dequiedt, S., Lelievre, M., Nowak, V., Regnier, T., *et al.* (2012) Molecular biomass and MetaTaxogenomic assessment of soil microbial communities as influenced by soil DNA extraction procedure. *Microb Biotechnol* **5:** 135–141.

Terrat, S., Dequiedt, S., Horrigue, W., Lelievre, M., Cruaud, C., Saby, N., *et al.* (2014) Improving soil bacterial taxa-area relationships assessment using DNA meta-barcoding. *Heredity*. In press.

Will, C., Thürmer, A., Wollherr, A., Nacke, H., Herold, N., Schrumpf, M., *et al.* (2010) Horizon-specific bacterial community composition of German grassland soils, as revealed by pyrosequencing-based analysis of 16S rRNA genes. *Appl Environ Microbiol* **76:** 6751–6759.

Yang, W.-W., and Ponce, A. (2011) Validation of a *Clostridium* endospore viability assay and analysis of Greenland ices and Atacama desert soils. *Appl Environ Microbiol* **77:** 2352–2358.

Yarwood, S., Bottomley, P., and Myrold, D. (2010) Soil microbial communities associated with Douglas-fir and red alder stands at high- and low-productivity forest sites in Oregon, USA. *Microb Ecol* **60:** 606–617.

Zhou, J., Bruns, M.A., and Tiedje, J.M. (1996) DNA recovery from soils of diverse composition. *Appl Environ Microbiol* **62:** 316–322.

## Supporting information

Additional Supporting Information may be found in the online version of this article at the publisher's web-site:

**Fig. S1.** Rarefaction curves of bacterial and fungal OTUs detected in soils according to extraction procedures.
**Table S1.** Detailed hit frequencies (%) of the in silico analysis of the F479/R888 primer set for Bacteria, Archaea and Eukaryota.

# 15

# A novel strain of *Cellulosimicrobium funkei* can biologically detoxify aflatoxin B$_1$ in ducklings

Lv-Hui Sun,[1†] Ni-Ya Zhang,[1†] Ran-Ran Sun,[1] Xin Gao,[1] Changqin Gu,[1] Christopher Steven Krumm[2] and De-Sheng Qi[1]*

[1]*Department of Animal Nutrition and Feed Science, College of Animal Science and Technology, Huazhong Agricultural University, Wuhan, Hubei 430070, China.*
[2]*Department of Animal Science, Cornell University, Ithaca, NY 14853, USA.*

## Summary

Two experiments were conducted to screen microorganisms with aflatoxin B$_1$ (AFB$_1$) removal potential from soils and to evaluate their ability in reducing the toxic effects of AFB$_1$ in ducklings. In experiment 1, we screened 11 isolates that showed the AFB$_1$ biodegradation ability, and the one exhibited the highest AFB$_1$ removal ability (97%) was characterized and identified as *Cellulosimicrobium funkei* (*C. funkei*). In experiment 2, 80 day-old Cherry Valley ducklings were divided into four groups with four replicates of five birds each and were used in a 2 by 2 factorial trial design, in which the main factors included administration of AFB$_1$ versus solvent and *C. funkei* versus solvent for 2 weeks. The AFB$_1$ treatment significantly decreased the body weight gain, feed intake and impaired feed conversion ratio. AFB$_1$ also decreased serum albumin and total protein concentration, while it increased activities of alanine aminotransferase and aspartate aminotransferase and liver damage in the ducklings. Supplementation of *C. funkei* alleviated the adverse effects of AFB$_1$ on growth performance, and provided protective effects on the serum biochemical indicators, and decreased hepatic injury in the ducklings. Conclusively, our results suggest that the novel isolated *C. funkei* strain could be used to mitigate the negative effects of aflatoxicosis in ducklings.

*For correspondence. E-mail qds@mail.hzau.edu.cn

**Funding Information** This project was supported by the Chinese Natural Science Foundation projects (31072058 and 31272479), Fundamental Research Funds for the Central Universities (2011QC050 and 2013BQ059), and Hubei Provincial Natural Science Foundation (2013CFA010).

## Introduction

Aflatoxins (AF) are secondary fungal metabolites that are largely produced by the fungi *Aspergillus flavus* and *Aspergillus parasiticus* (Diaz *et al.*, 2002). Among the various dangerous AF and their metabolites, aflatoxin B$_1$ (AFB$_1$) is the most toxic mycotoxin, having harmful hepatotoxic, mutagenic, carcinogenic and teratogenic effects on many species of livestock. It is also classified as a group one carcinogen [International Agency for Research on Cancer (IARC), 1987]. Unfortunately, AFB$_1$ can easily contaminate various types of crops and is a very prevalent contaminant of maize-based food and feed all over the world (Wu and Guclu, 2012; Hamid *et al.*, 2013). The feed contaminated by AFB$_1$ can pose serious problems to the health and productivity of livestock and can therefore cause significantly economic losses (Rawal *et al.*, 2010; Wu and Guclu, 2012).

Several physical and chemical detoxification methods used to control AFB$_1$ have been to some extent successful, while most of them have major disadvantages including nutrients loss and high costs, which limited their practical applications (Varga *et al.*, 2010; Jard *et al.*, 2011). Thus, scientists have come to favor the biological method, which is utilization of microorganisms and/or their enzymatic products to remove AF through microbial binding and/or degradation of mycotoxins into less toxic compounds, giving a characterization of specific, efficient and environmentally sound detoxification (Wu *et al.*, 2009; Guan *et al.*, 2011).

Many studies have shown that AFB$_1$ can be biologically detoxified by various species of microorganisms, including fungi, such as *Pleurotus ostreatus* (Motomura *et al.*, 2003), *Trametes versicolor* (Zjalic *et al.*, 2006), yeast such as *Trichosporon mycotoxinivorans* (Molnar *et al.*, 2004) and *Saccharomyces cerevisiae* (Pizzolitto *et al.*, 2013), and bacteria, such as lactic acid bacteria (Bagherzadeh Kasmani *et al.*, 2012; Nikbakht Nasrabadi *et al.*, 2013), *Stenotrophomonas maltophilia* (Guan *et al.*, 2008), *Myxococcus fulvus* (Guan *et al.*, 2010) and *Rhodococcus species* (Cserháti *et al.*, 2013). Unfortunately, few of these microorganisms, their metabolites and/or degradation products have been utilized in animal feed due to a lack of information on the mechanisms of detoxification, the

**Table 1.** Ability of AFB$_1$ biodegradation by screened isolates[a].

| Isolate[b] | AFB$_1$ biodegradation (%) |
|---|---|
| T$_1$-1 | 84.9 ± 4.0 |
| T$_1$-2 | 41.7 ± 4.1 |
| T$_1$-3 | 86.3 ± 7.3 |
| T$_1$-4 | 24.9 ± 3.8 |
| T$_2$-1 | 75.9 ± 1.8 |
| T$_3$-1 | 51.5 ± 6.9 |
| T$_3$-2 | 81.2 ± 4.9 |
| T$_3$-3 | 87.1 ± 4.0 |
| T$_3$-4 | 20.4 ± 4.3 |
| T$_3$-5 | 94.2 ± 2.4 |
| T$_3$-6 | 70.1 ± 4.5 |

**a.** Values are expressed as means ± SD ($n = 5$).
**b.** Isolates are screened from soil samples using coumarin as the only carbon source.

efficiency and stability of detoxification under different oxygen, pH or bile conditions, as well as their potential side-effects needs to be further investigated.

The objective of this study was to screen novel AFB$_1$ biodegradation microorganisms that could be applied in feed industry. Because AFB$_1$ is a furanocoumarin derivative which has a similar chemical structure to polycyclic aromatic hydrocarbons (PAHs) (Yu *et al.*, 2004; Guan *et al.*, 2008), we therefore hypothesized that the microorganism having biodegradable activity on PAHs maybe also have the same effect towards AFB$_1$. We therefore chose soil samples around petroleum factories, which were contaminated with PAHs for the screening of the microorganisms that could biodegrade AFB$_1$ (Pampanin and Sydnes, 2013). Since coumarin is the basic chemical structure of AFB$_1$, along with the relatively safe and inexpensive characterization, we used medium-containing coumarin as the only carbon source to screen AFB$_1$-biodegrading microorganisms by replicating the well-established method previously conducted (Guan *et al.*, 2008). Since duckling is extremely sensitive to the toxic effects of AFB$_1$ (Shi *et al.*, 2013), and thus it was used to evaluate the detoxification effects of the isolate in this study. In this study, we successfully obtained the isolate

*Cellulosimicrobium funkei* T$_3$-5 (*C. funkei*), which exhibited excellent AFB$_1$ biodegradation ability both *in vitro* and *in vivo*. Our findings suggested a feasible approach for a safe and efficient method to control AFB$_1$ levels in the animal feed industry.

## Results

### Experiment 1

*Screening for AFB$_1$ biodegradation microbes.* A total of 11 strains were isolated from three soil samples by coumarin medium, and all of them showed various degrees of ability to reduce concentrations of AFB$_1$ in the liquid medium after 72 h incubation at 37°C (Table 1), which were calculated from the high-performance liquid chromatography (HPLC) results. The chromatograms of HPLC analysed results were shown in Fig. S2. Among the 11 screened strains, seven showed the potential of reducing AFB$_1$ in the medium over 70%, and the isolate T$_3$-5 was the most effective strain with an observed 94.2% AFB$_1$ reduction in the medium (Table 1).

*Identification of isolate T$_3$-5.* Microscopic morphological results showed that isolate T$_3$-5 is a gram-positive bacterium, which appeared as circular, yellow, a smooth surface and an entire edge after 18 h of incubation at 37°C on the Luria Bertani (LB) agar (Fig. S1). Physiological and biochemical studies showed that the T$_3$-5 strain was able to utilize most oligosaccharides including glucose, maltose, D-xylose, galactose, D-sorbitol, D-raffinose and as well as sucrose as a sole carbon source. The T$_3$-5 strain could also hydrolyse cellulose, gelatin and Tween 80, but not amylum (Table 2). The 16S rDNA sequencing result of the isolate T$_3$-5 [National Center for Biotechnology Information (NCBI) GenBank: KM032184] showed 99% deoxyribonucleic acid (DNA) sequence homology to that of *Cellulosimicrobium funkei* listed by NCBI (GenBank: NR_042937.1, Fig. 1). Taken together, the morphological, physiological, biochemical

**Table 2.** Biochemical and physiological characteristics of *C. funkei* T$_3$-5.

| Characteristic | Result | Characteristic | Result | Characteristic | Result |
|---|---|---|---|---|---|
| Acid fermentation of: | | Phosphatidylcholine | − | 10% NaCl | − |
| Glucose | + | Tween 80 | − | Acid medium | − |
| Maltose | + | Amylum | − | Other test: | |
| D-Xylose | + | Cellulose | + | Motility | − |
| Galactose | + | Xylan | − | Catalase test | + |
| D-Sorbitol | − | Nitrate reduction | + | Methyl red test | + |
| D-Raffinose | + | Organic acid | + | Urease test | + |
| Sucrose | + | Yeast cell | − | V-P test | − |
| Glycerol | + | Growth on: | | Oxidase test | − |
| Hydrolysis of: | | 2%, 5% NaCl | + | Congo red tolerance | + |
| Gelatin liquefaction | + | 7% NaCl | (+) | | |

+, Positive; −, negative; (+), weakly positive.

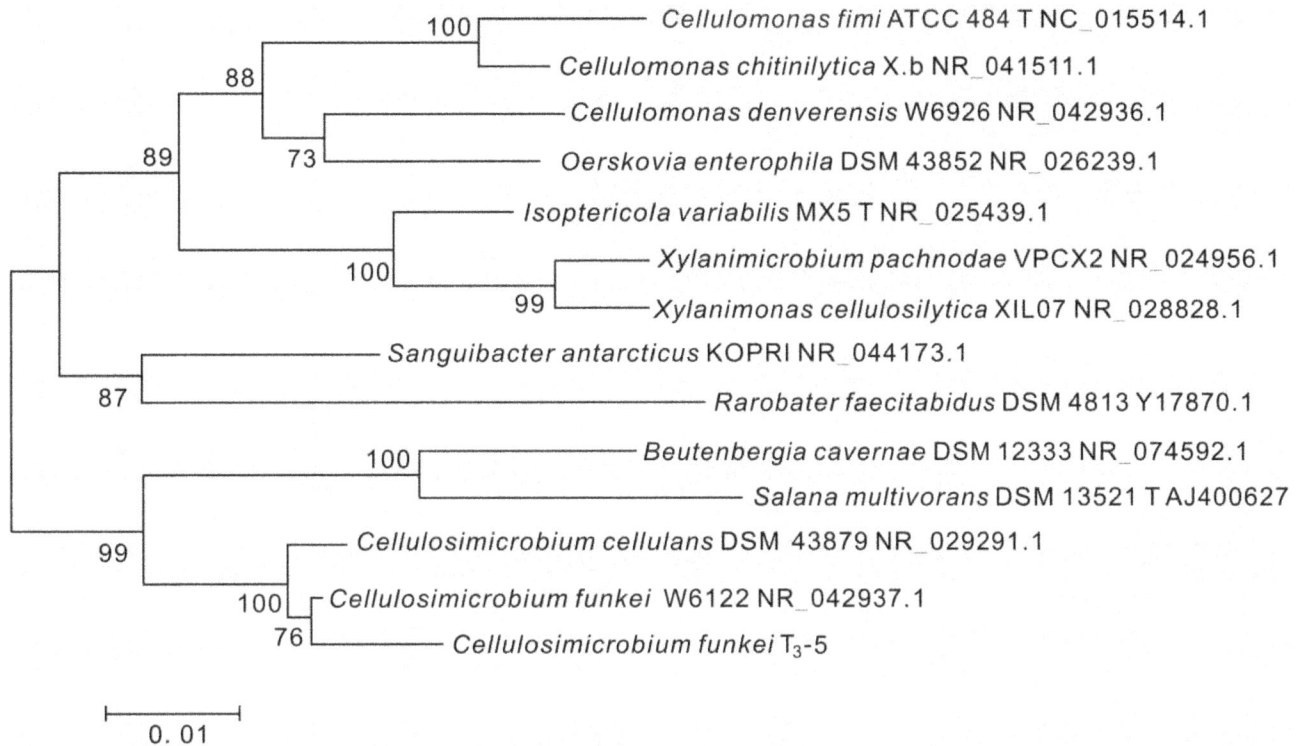

**Fig. 1.** Neighbour-joining phylogenetic tree based on 16S rDNA gene sequences showing the relationships among the species of the genus *Cellulomonas* and related specifies. Bootstrap values calculated for 1000 replications are indicated. Bar, 2 substitutions per 100 nucleotides. Accession numbers from Genbank are given in brackets.

data as well as the NCBI blast results suggest that the isolate $T_3$-5 belonged to *C. funkei*, which is an aerobic and facultatively anaerobic gram-positive bacterium. The above strain is deposited at China Center for Type Culture Collection (CCTCC) in Wuhan University, and has a preservation number of CCTCC NO: M 2013564.

*Characterization of AFB$_1$ biodegradation by* C. funkei $T_3$-5. The culture supernatant of *C. funkei* $T_3$-5 showed the strongest ($P < 0.05$) AFB$_1$ biodegradation ability compared with viable cell and cell extract, which removed 97%, 20% and 16% AFB$_1$ after 72 h incubation respectively (Fig. 2A). These results indicated that the activity of AFB$_1$ biodegradation occurred primarily within the culture supernatant of *C. funkei*. The AFB$_1$ biodegradation ability of the culture supernatant of *C. funkei* was decreased ($P < 0.05$) 54% and 56% after treated with proteinase K with or without SDS, respectively, while it was only slightly decreased by heat treatment (Fig. 2B).

*Experiment 2*

*Performance.* Non-significant differences in initial body weight were observed among the four groups (Table 3). After 2 weeks of experimental treatments, the growth performance was significantly affected by oral administra-

tion of AFB$_1$ and *C. funkei* or their interactions (Table 3). Compared with the control, the final body weight (BW), overall daily BW gain and overall daily feed intake of ducklings were decreased ($P < 0.05$) 42%, 49% and 38%, along with increased ($P < 0.05$) 23% overall feed/gain ratio by AFB$_1$ administration respectively. Although the AFB$_1$ + *C. funkei* group showed the similar trend to AFB$_1$

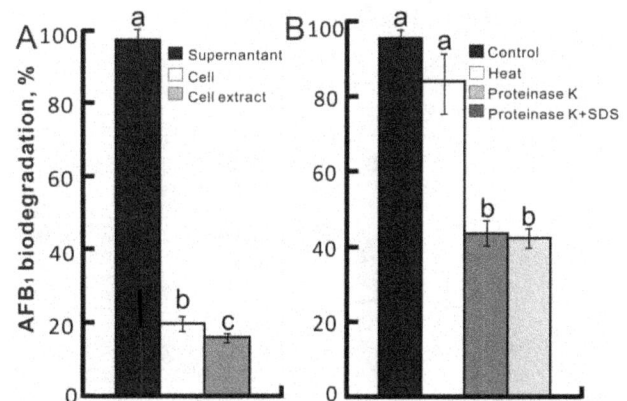

**Fig. 2.** Ability of AFB$_1$ biodegradation by culture supernatant, cell and cell extract of *C. funkei* after 72 h fermentation (A); and culture supernatant of *C. funkei* was determined by pretreating the supernatant by proteinase K with or without SDS or heat respectively (B). Values are means ± SD, $n = 5$. Bars without a common letter differ, $P < 0.05$.

**Table 3.** Effects of administration of $AFB_1$ and *C. funkei* on growth performance in ducklings[a].

|  | Control | $AFB_1$[a] | *C. funkei*[c] | $AFB_1$ + *C. funkei* |
|---|---|---|---|---|
| week 0 BW, g | 119.3 ± 1.3 | 118.8 ± 1.0 | 118.5 ± 0.6 | 118.8 ± 0.5 |
| week 1 BW, g | 372.5 ± 30.0[a] | 309.4 ± 7.0[c] | 364.7 ± 16.2[a] | 336.6 ± 16.4[b] |
| week 2 BW, g | 786.9 ± 61.2[a] | 458.5 ± 61.7[b] | 858.8 ± 42.6[a] | 527.8 ± 63.3[b] |
| week 2 BW gain, g/day | 47.7 ± 4.3[a] | 24.3 ± 4.4[b] | 52.9 ± 3.0[a] | 29.2 ± 4.5[b] |
| week 2 feed intake, g/day | 89.7 ± 9.0[a] | 55.5 ± 5.7[b] | 90.3 ± 3.1[a] | 55.1 ± 7.8[b] |
| week 2 feed/gain, g/g | 1.88 ± 0.04[b] | 2.32 ± 0.25[a] | 1.71 ± 0.03[c] | 1.89 ± 0.16[b] |

**a.** Values are expressed as means ± SD ($n = 5$), and means with different superscript letters differ ($P < 0.05$).
**b.** Each duckling oral administrated 100 μg $AFB_1$/kg BW per day.
**c.** Each duckling oral administrated *C. funkei* at $10^8$ cfu/per day.

group that decreased ($P < 0.05$) the final BW (33%), overall daily BW gain (39%) and overall daily feed intake (39%) of ducklings respectively, administration of *C. funkei* prevented ($P < 0.05$) the loss in final feed conversion and BW (9%) at the first week. In addition, administration of *C. funkei* alone reduced ($P < 0.05$) the overall feed/gain ratio (9%), and did not affect the other growth performance parameters, when compared with the control. No mortality due to $AFB_1$ administration was found in this study.

*Serum biochemistry and liver histology.* The results showed that the serum biochemical and histological parameters were significantly affected by administration of $AFB_1$ and *C. funkei* or their interactions (Table 4). The $AFB_1$ administration led to increased ($P < 0.05$) activity of aspartate aminotransferase (AST; 404% and 867%) and alanine aminotransferase (ALT; 82% and 282%), along with decreased ($P < 0.05$) concentration of total protein (TP; 34% and –) and albumin (ALB; 44% and –) in the serum of ducklings at the first and second week respectively. Strikingly, $AFB_1$ + *C. funkei* group decreased ($P < 0.05$) the activity of AST (43 and 44%) and ALT (28 and 20%), along with increased ($P < 0.05$) concentration

of TP (22% and –) and ALB (42% and –) in serum of ducklings at the first and second week, respectively, compared with the $AFB_1$ group (Table 4). Furthermore, the histological analysis results showed $AFB_1$ administration-induced hepatic injury, such as vacuolar degeneration, necrosis and bile duct hyperplasia at the first week and increased liver damage on the second week. Notably, $AFB_1$ + *C. funkei* group alleviated the liver damage was observed in the $AFB_1$ group (Fig. 3).

## Discussion

The two most novel findings from the present study were: (i) we successfully screened a novel $AFB_1$ biodegradation microorganism *C. funkei* $T_3$-5, and (ii) oral administration of *C. funkei* effectively alleviated the adverse effects induced by $AFB_1$ in the ducklings. The *C. funkei* is a gram-positive, aerobic and facultatively anaerobic, non-spore-forming rod or coccus-shaped bacterium of the genus *Cellulosimicrobium*, consistent with previously reported (Brown *et al.*, 2006). *In vivo*, *C. funkei* demonstrated effective $AFB_1$ biodegradation ability that 97% of $AFB_1$ can be removed after 72 h incubation. Interestingly, these reported values were much higher than those of previously reported various microorganisms, such as *Rhodococcus erythropolis* (67%, Alberts *et al.*, 2006), *Flavobacterium aurantiacum* (74.5%, Smiley and Draughon, 2000), *Mycobacterium strain* (80%, Hormisch *et al.*, 2004), *M. fulvus* (81%, Guan *et al.*, 2010) and *S. maltophilia* (83%, Guan *et al.*, 2008). Strikingly, Cserháti and colleagues (2013) found that several *Rhodococcus* species displayed more than 97% $AFB_1$-degrading ability, along with effective degrading ability to other common mycotoxins also, which also offered a promising strategy to control mycotoxins. Moreover, our results implied that the compounds biodegrading $AFB_1$ were mainly within the fermentation supernatant of *C. funkei* rather than in its viable cell and cell extract. Notably, the $AFB_1$ biodegradation activity of fermentation supernatant of *C. funkei* was decreased more than 50% after treated with proteinase K or plus SDS, which is similar to the $AFB_1$ biodegradation by the culture supernatant of *F. aurantiacum* and

**Table 4.** Effects of administration of $AFB_1$ and *C. funkei* on serum biochemical parameters in ducklings[a].

|  | Control | $AFB_1$[b] | *C. funkei*[c] | $AFB_1$ + *C. funkei* |
|---|---|---|---|---|
| week 1 |  |  |  |  |
| ALT, U/l | 46.5 ± 5.8[c] | 234.8 ± 48.7[a] | 48.3 ± 7.8[c] | 132.8 ± 26.7[b] |
| AST, U/l | 68.3 ± 8.4[c] | 124.5 ± 5.4[a] | 68.8 ± 11.3[c] | 89.8 ± 8.1[b] |
| TP, g/l | 29.4 ± 1.2[a] | 19.4 ± 1.8[c] | 28.9 ± 0.8[a] | 23.6 ± 1.0[b] |
| ALB, g/l | 13.5 ± 0.4[a] | 7.6 ± 0.8[c] | 13.4 ± 0.7[a] | 10.8 ± 0.1[b] |
| week 2 |  |  |  |  |
| ALT, U/l | 35.3 ± 10.2[c] | 341.0 ± 52.0[a] | 39.00 ± 11.0[c] | 190.0 ± 43.3[b] |
| AST, U/l | 53.5 ± 8.7[c] | 204.5 ± 44.7[a] | 54.3 ± 8.3[c] | 162.8 ± 30.9[b] |
| TP, g/l | 29.9 ± 1.7 | – | 30.0 ± 0.3 | – |
| ALB, g/l | 14.0 ± 0.9 | – | 13.9 ± 0.2 | – |

**a.** Values are expressed as means ± SD ($n = 5$), and means with different superscript letters differ ($P < 0.05$).
**b.** Each duckling oral administrated 100 μg $AFB_1$/kg BW per day.
**c.** Each duckling oral administrated *C. funkei* at $10^8$ cfu/per day.
–, Undetectable.

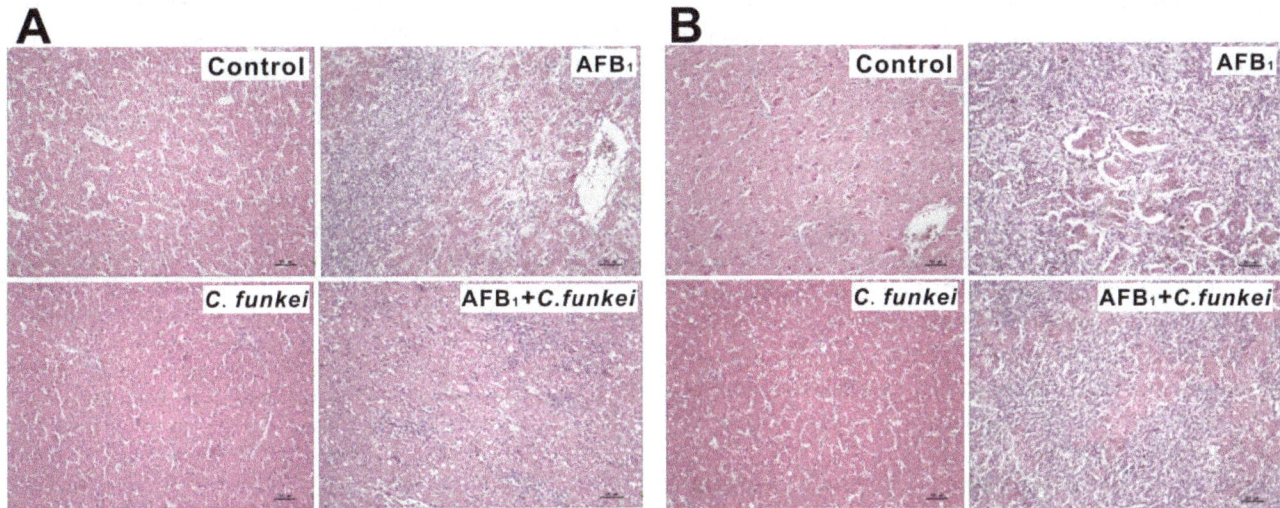

**Fig. 3.** Photomicrographs of hepatic sections stained with haematoxylin and eosin (40× magnification) of ducklings from different treatment groups on (A) week 1 and (B) week 2 respectively.

*S. maltophilia* reported earlier, and indicated the active ingredient could therefore be protein or perhaps an enzyme (Alberts *et al.*, 2006; Guan *et al.*, 2008). Furthermore, the AFB$_1$ biodegradation activity was positively correlated with the protein content from the fermentation supernatant of *C. funkei* by ammonium sulfate precipitation (Table S1) which provides further evidence that the active ingredient could be protein (Callejón *et al.*, 2014; Duong-Ly and Gabelli, 2014). However, heating may cause denaturation of proteins, strikingly, the AFB$_1$ biodegradation activity of fermentation supernatant of *C. funkei* was only slightly affected by heating. It may be interpreted by (i) the protein involved in the AFB$_1$ biodegradation is heat resistant (Wang *et al.*, 2007), and (ii) the fermentation supernatant may contain cell wall, and heating may increase permeability of their external layer and lead to the increasing availability of the otherwise hidden binding sites for AFB$_1$ (Shetty *et al.*, 2007). Taken together, our results revealed that the active ingredient of AFB$_1$ biodegradation in the fermentation supernatant of *C. funkei* could be enzyme and other active ingredients such as, cell wall. However, systematic identification of the active ingredients in the fermentation supernatant of *C. funkei* and its detoxification mechanism is still needed to be explored in the future.

Administration of AFB$_1$ reduced the growth rate and efficiency of feed utilization of ducklings, which were in accordance with those in previous studies (Pasha *et al.*, 2007). The negative effects of AFB$_1$ on feed intake, BW gain and the feed conversion have been associated with anorexia, reluctance and inhibition of protein synthesis and lipogenesis (Bagherzadeh Kasmani *et al.*, 2012). The present study showed that supplementation of

*C. funkei* at dose of $10^8$ cfu/day prevented the loss in feed conversion throughout. Meanwhile, although supplementation of *C. funkei* had no significant effect on the final BW, while improved body weight was observed during the first week. The beneficial effects of supplementation of *C. funkei* could be due to (i) toxin biodegraded by *C. funkei in vivo* was evidenced by a 97% removal of AFB$_1$, and (ii) the *C. funkei* could hydrolyse cellulose and improved the cellulose utilization. In addition, no adverse effects in productivity parameters were found between ducklings in control group and the experimental group administered *C. funkei* alone, indicating that *C. funkei* was non-toxic and safe. Similar results were obtained from recent studies that *Nocardia corynebacteroides* and *S. cerevisiae* can partly detoxify chicken feed contaminated with AFB$_1$ (Tejada-Castañeda *et al.*, 2008; Pizzolitto *et al.*, 2013).

Activities of serum enzymes such as ALT and AST, and concentrations of serum TP and ALB have been described as valuable parameters of hepatic injury and function (Bagherzadeh Kasmani *et al.*, 2012; Lv *et al.*, 2014). Administration of AFB$_1$ alone increased ALT and AST activity, along with decreased TP and ALB concentrations compared with the control diet. These outcomes were consistent with previous studies, which provided evidence that liver injury was induced by AFB$_1$ (Bagherzadeh Kasmani *et al.*, 2012; Shi *et al.*, 2013). Results obtained from the present study showed that serum biochemical changes could be ameliorated by *C. funkei* administration. Moreover, histopathological changes in the livers of ducklings exposed to AFB$_1$ were similar to those reported on avian aflatoxicosis (Denli *et al.*, 2009). Administration of *C. funkei* showed stronger protective effect on the histopathological changes on the first week, but consistent with the growth performance results, was unable to prevent liver injury on the second week. This may be due to (i) the duckling-ingested $AFB_1$ was added along with the increase in BW, while the *C. funkei* ($10^8$ cfu/day) dosage was not changed, and (ii) the toxic potency of $AFB_1$ was increased due to the prolonged exposure times (Centoducati *et al.*, 2009) and finally beyond the detoxification capacity of *C. funkei*. Since ducklings exposure to $AFB_1$ from the naturally contaminated feed is usually much lower (at least 10 times) than the administration of $AFB_1$ at 100 µg/kg BW per day in our study (Yang *et al.*, 2012), supplementation of *C. funkei* may therefore exert better protective effects on aflatoxicosis in practice.

Although *C. funkei* has showed potent $AFB_1$ biodegradation capability and safety *in vivo* study, directly using this microbe as a feed additive seems challenged by the fact that *C. funkei* is an opportunistic pathogen (Petkar *et al.*, 2011). Therefore, our ongoing research was focus in two directions: (i) exploring the mechanism of $AFB_1$ biodegradation by *C. funkei*, which try to separate the enzyme and/or other active ingredients such as cell wall that could biodegradation the $AFB_1$ and (ii) using *C. funkei* alone or with other microbial to do the solid-state fermentation on rapeseed meal and cottonseed meal to improve crude fiber digestibility, reduce $AFB_1$ contents and produce $AFB_1$ biodegradation active ingredients in these feedstuffs.

In summary, the *C. funkei* isolated in the present study, exhibited significant improvements in the capabilities of biodegradation of $AFB_1$ *in vitro*. Moreover, an *in vivo* study verified its $AFB_1$ biodegradation activity in ducklings with regard to partial improvement growth performance, serum biochemistry, hepatotoxicity and histopathology of livers. Additionally, the *in vivo* study showed that administration of *C. funkei* at $10^8$ cfu/day was non-toxic and safe to administer to ducklings. Overall, these findings suggest that the use of *C. funkei* in $AFB_1$-contaminated feed offers a new strategy to reduce the adverse effects of aflatoxicosis in ducklings.

## Experimental procedures

### Experiment 1

*Soil samples and $AFB_1$ biodegradation microorganism isolation.* Three soil samples designated as $T_1$, $T_2$ and $T_3$, were collected around the factories of SINOPEC Wuhan Company, Shandong Jinqiao Coal Mine Company and Shandong Yankuang International Coking Company. All these samples were air-dried at room temperature. Microorganisms that could use coumarin as the only carbon source were then isolated from soil samples by using a standard procedure with minor modifications (Guan *et al.*, 2008). Single colonies that were able to grow on the coumarin

(Sigma Chemical Co., Bellefonte, USA) plate were selected for $AFB_1$ biodegradation activity analysis according to the protocol described by Guan and colleagues (2008) with minor modifications. Initially, candidate isolates were cultured at 37°C in LB medium for 72 h, and then 950 µl fermented supernatant was taken and mixed with 50 µl 10 µg/ml $AFB_1$ solution (Sigma Chemical Co., Bellefonte, USA) in a sterilized centrifuge tube, and then biodegradation tests were conducted at 37°C for 72 h. Finally, the reaction solution was centrifuged at 10 000 *g* at 4°C for 10 min to remove cells and the supernatant, and then it was collected for $AFB_1$ quantification (Guan *et al.*, 2008). The $AFB_1$ concentration was determined by HPLC (Teniola *et al.*, 2005) with a minor modification. $AFB_1$ was extracted three times with chloroform from liquid cultures and cell-free extracts. The chloroform was evaporated under nitrogen gas, and the samples were dissolved in methanol, filtered by 0.22 µm filters for HPLC analysis. HPLC analysis was performed on a Shimadazu LC-20A binary gradient liquid chromatography equipped with a 5 µm × 4.6 mm × 250 mm C-18 reverse-phase column (ZORBAX Eclipse XDB-C18, Agilent). The mobile phase was acetonitrile/methanol/water (1:1:2, v/v/v) at a flow rate of 1 ml/min, and the sample temperature was set at 30°C. $AFB_1$ was measured by UV (365 nm.) detector (Shimadazu SPD-20A). The sterilized LB medium alone substituted fermentation supernatant incubated with $AFB_1$ solution was used as the negative control.

*Identification of the $AFB_1$ biodegradation isolate.* Total DNA was extracted from the $AFB_1$ biodegradation isolate using TIANamp Bacterial DNA Kit (Tiangen Biotech, Beijing, China) according to the manufacturer's instructions. The forward primer (27f: 5′-GAGAGTTTGATCCTGGCTCAG-3′) and the reverse primer (1492r: 5′-CTACGGCTACCTTGT TACGA-3′) were used to amplify the 16S ribosomal (r)DNA (Minerdi *et al.*, 2012). After the amplified 16S rDNA fragment was purified using the Gel Extraction Kit, it was ligated into the pMD18-T vector, and transformed into the *Escherichia coli* JM109 strain by calcium chloride activation (Dagert and Ehrlich, 1979). The positive colonies were selected for DNA sequencing (Tsingke, Wuhan, China). The obtained DNA sequence and NCBI GenBank-derived sequences were aligned using the CLUSTALX program (Thompson *et al.*, 1997). Neighbour-joining phylogenetic tree and bootstrap values were analysed by the MEGA program (Tamura *et al.*, 2013). Physiological and biochemical tests were carried out following the method described by Holt and colleagues (1994).

*Characterization of $AFB_1$ biodegradation activity of C. funkei $T_3$-5.* After *C. funkei* $T_3$-5 grew at 37°C in LB medium for 72 h, then the cell, cell extract and fermentation supernatant were prepared as previously described (Guan *et al.*, 2008), and their $AFB_1$ biodegradation ability was tested as described before. Specifically, supernatant was obtained by centrifuging fermentation supernatant at 12 000 *g* at 4°C for 20 min; cell was collected after being centrifuged at 12 000 *g* at 4°C for 20 min and washed twice with phosphate buffer (50 mM; pH 7.0); cell extract was produced by using ultrasonic cell disintegrator on ice, and the suspension was centrifuged at 12 000 *g* for 20 min at 4°C, and then it was filtered by 0.22 µm pore size sterile cellulose pyrogen free filters. Since the main active ingredients for $AFB_1$ biodegradation were found within the fermentation supernatant, further assessment was conducted through *in vivo* experiments. The $AFB_1$ biodegradation stability of fermentation supernatant of *C. funkei* was determined by the residual activity after the supernatant was treated by proteinase K (0.5 mg/ml) with or without SDS (5.0%) at 37°C for 6 h, or boiled at 100 °C for 15 min respectively. The untreated fermentation supernatant of *C. funkei* was used as the positive control.

### Experiment 2

*Ducklings, treatments and samples collection.* Our animal protocol was approved by the Institutional Animal Care and Use Committee of Huazhong Agricultural University, China. A total of 80 day-old Cherry Valley commercial ducklings were randomly divided into four treatment groups with four replicates of five birds each. The trial was arranged in a 2 by 2 factorial design that included oral administration of $AFB_1$ or solvent and *C. funkei* or solvent respectively. All birds were allowed free access to a similar corn-soybean meal diet (Shi *et al.*, 2013) and distilled water ad libitum. The $LD_{50}$ of $AFB_1$ in duckling is 2.8 mg/kg BW (Yunus *et al.*, 2011), and we chose 100 µg/kg BW since the dose of subchronic toxicity test was chosen between 1/10-1/50 $LD_{50}$ (Jin *et al.*, 2008). After 3 days of acclimation, each group was administered an oral dose of $AFB_1$ [dissolved in 1.0% dimethylsulphoxide (DMSO)] at 100 µg/kg BW or an equivalent amount of sterile DMSO, along with an administration of 1 ml $10^8$ cfu/ml *C. funkei* or an equivalent amount of sterile LB medium per day respectively. The administration continued for 2 week. Birds were monitored mortality daily, along with body weight and feed intake measured weekly. Meanwhile, four birds from each treatment group were slaughtered weekly to collect blood and liver for the preparation of serum, and liver histological tissue samples were prepared as previously described (Shi *et al.*, 2013; Sun *et al.*, 2013).

*Serum biochemical and histological analysis.* The serum activities of ALT and AST, along with concentrations of TP and ALB were determined in serum samples. Analyses of the serum samples were measured by an automatic biochemistry analyser (Beckman Synchron CX4 PRO, CA, USA). The liver tissues were fixed in 10% neutral buffered formalin and processed for paraffin embedding, sectioned at 5 µm and stained with haematoxylin and eosin, by standard procedure (Pizzolitto *et al.*, 2013). Liver sections from all birds were microscopically examined.

*Statistical analysis.* Data generated from experiment 1 were analysed by one-way ANOVA to test the main effects of $AFB_1$ biodegradation activity of *C. funkei*. Data generated from experiment 2 were analysed by two-way ANOVA to test the main effects of administration $AFB_1$ and *C. funkei*. The Bonferroni *t*-test was followed for multiple mean comparisons if there was a main effect. All analyses were conducted using SAS 8.2 (SAS Institute). Data were presented as means ± SD, and significance level was set at $P < 0.05$.

## Acknowledgements

We thank Weiche Wu for his technical assistance.

## Conflict of interest

Lv-Hui Sun, Ni-Ya Zhang, Ran-Ran Sun, Xin Gao, Changqin Gu, Christopher Steven Krumm and De-Sheng Qi have no conflicts of interest.

## Reference

Alberts, J.F., Engelbrecht, Y., Steyn, P.S., Holzapfel, W.H., and van Zyl, W.H. (2006) Biological degradation of aflatoxin B1 by Rhodococcus erythropolis cultures. *Int J Food Microbiol* **109**: 121–126.

Bagherzadeh Kasmani, F., Karimi Torshizi, M.A., Allameh, A., and Shariatmadari, F. (2012) A novel aflatoxin-binding Bacillus probiotic: performance, serum biochemistry, and immunological parameters in Japanese quail. *Poult Sci* **91**: 1846–1853.

Brown, J.M., Steigerwalt, A.G., Morey, R.E., Daneshvar, M.I., Romero, L.J., and McNeil, M.M. (2006) Characterization of clinical isolates previously identified as Oerskovia turbata: proposal of Cellulosimicrobium funkei sp. nov. and emended description of the genus Cellulosimicrobium. *Int J Syst Evol Microbiol* **56**: 801–804.

Callejón, S., Sendra, R., Ferrer, S., and Pardo, I. (2014) Identification of a novel enzymatic activity from lactic acid bacteria able to degrade biogenic amines in wine. *Appl Microbiol Biotechnol* **98**: 185–198.

Centoducati, G., Santacroce, M.P., Lestingi, A., Casalino, E., and Crescenzo, G. (2009) Characterization of the cellular damage induced by Aflatoxin B1 in sea bream (Sparus aurata Linnaeus, 1758) hepatocytes. *Ital J Anim Sci* **8**: 848–850.

Cserháti, M., Kriszt, B., Krifaton, C., Szoboszlay, S., Háhn, J., Tóth, S., *et al.* (2013) Mycotoxin-degradation profile of Rhodococcus strains. *Int J Food Microbiol* **166**: 176–185.

Dagert, M., and Ehrlich, S.D. (1979) Prolonged incubation in calcium chloride improves the competence of Escherichia coli cells. *Gene* **6**: 23–28.

Denli, M., Blandon, J.C., Guynot, M.E., Salado, S., and Perez, J.F. (2009) Effects of dietary AflaDetox on performance, serum biochemistry, histopathological changes, and aflatoxin residues in broilers exposed to aflatoxin B(1). *Poult Sci* **88**: 1444–1451.

Diaz, D.E., Hagler, W.M., Jr, Hopkins, B.A., and Whitlow, L.W. (2002) Aflatoxin binders I: in vitro binding assay for aflatoxin B1 by several potential sequestering agents. *Mycopathologia* **156**: 223–226.

Duong-Ly, K.C., and Gabelli, S.B. (2014) Salting out of proteins using ammonium sulfate precipitation. *Methods Enzymol* **541**: 85–94.

Guan, S., Ji, C., Zhou, T., Li, J., Ma, Q., and Niu, T. (2008) Aflatoxin B(1) degradation by Stenotrophomonas maltophilia and other microbes selected using coumarin medium. *Int J Mol Sci* **9**: 1489–1503.

Guan, S., Zhao, L., Ma, Q., Zhou, T., Wang, N., Hu, X., and Ji, C. (2010) In vitro efficacy of Myxococcus fulvus ANSM068 to biotransform aflatoxin $B_1$. *Int J Mol Sci* **11**: 4063–4079.

Guan, S., Zhou, T., Yin, Y.L., Xie, M.Y., Ruan, Z., and Young, J.C. (2011) Microbial strategies to control aflatoxins in food and feed. *World Mycotoxin J* **4**: 413–424.

Hamid, A.S., Tesfamariam, I.G., Zhang, Y., and Zhang, Z.G. (2013) Aflatoxin B1-induced hepatocellular carcinoma in developing countries: geographical distribution, mechanism of action and prevention. *Oncol Lett* **5**: 1087–1092.

Holt, J.G., Krieg, N.R., Sneath, P.H., Staley, J.T., and Williams, S.T. (1994) *Bergey's Manual of Determinative Bacteriology*, 9th edn. Baltimore, MD, USA: Williams & Witkins Baltimore.

Hormisch, D., Brost, I., Kohring, G.W., Giffhorn, F., Kroppenstedt, R.M., Stackebrandt, E., *et al.* (2004) Mycobacterium fluoranthenivorans sp. nov., a fluoranthene and aflatoxin B1 degrading bacterium from contaminated soil of a former coal gas plant. *Syst Appl Microbiol* **27**: 653–660.

International Agency for Research on Cancer (IARC) (1987) *IARC Monographs on the Evaluation of Carcinogenic Risk of Chemicals to Humans; Overall Evaluation of Carcinogenicity: An Updating of IARC Monographs*, Vol. **1–42**, Supplement 7. Lyon, France: IARC, p. 59.

Jard, G., Liboz, T., Mathieu, F., Guyonvarc'h, A., and Lebrihi, A. (2011) Review of mycotoxin reduction in food and feed: from prevention in the field to detoxification by adsorption or transformation. *Food Addit Contam Part A Chem Anal Control Expo Risk Assess* **28**: 1590–1609.

Jin, Y.E., Yuan, H., Yuan, L.Y., Liu, J.H., Tu, D., and Jiang, K. (2008) Study on acute and subchronic toxicities of avermectin microcapsule in Rats. *Prog in Vet Med* **29**: 27–30.

Lv, L.X., Hu, X.J., Qian, G.R., Zhang, H., Lu, H.F., Zheng, B.W., *et al.* (2014) Administration of *Lactobacillus salivarius* LI01 or *Pediococcus pentosaceus* LI05 improves acute liver injury induced by D-galactosamine in rats. *Appl Microbiol Biotechnol* **98**: 5619–5632.

Minerdi, D., Zgrablic, I., Sadeghi, S.J., and Gilardi, G. (2012) Identification of a novel Baeyer–Villiger monooxygenase from *Acinetobacter radioresistens*: close relationship to the Mycobacterium tuberculosis prodrug activator EtaA. *Microb Biotechnol* **5**: 700–716.

Molnar, O., Schatzmayr, G., Fuchs, E., and Prillinger, H. (2004) Trichosporon mycotoxinivorans sp. nov., a new yeast species useful in biological detoxification of various mycotoxins. *Syst Appl Microbiol* **27**: 661–671.

Motomura, M., Toyomasu, T., Mizuno, K., and Shinozawa, T. (2003) Purification and characterization of an aflatoxin degradation enzyme from Pleurotus ostreatus. *Microbiol Res* **158**: 237–242.

Nikbakht Nasrabadi, E., Jamaluddin, R., Abdul Mutalib, M.S., Khaza'ai, H., Khalesi, S., and Mohd Redzwan, S. (2013) Reduction of aflatoxin level in aflatoxin-induced rats by the activity of probiotic *Lactobacillus casei* strain Shirota. *J Appl Microbiol* **114**: 1507–1515.

Pampanin, D.M., and Sydnes, M.O. (2013) Polycyclic aromatic hydrocarbons a constituent of petroleum: presence and influence in the aquatic environment. In *Hydrocarbon*. Vladimir, K., and Kolesnikov, A. (eds). Rijeka, Croatia: InTech, pp. 83–118.

Pasha, T.N., Farooq, M.U., Khattak, F.M., Jabbar, M.A., and Khan, A.D. (2007) Effectiveness of sodium bentonite and two commercial products as aflatoxin absorbents in diets for broiler chickens. *Anim Feed Sci Technol* **132**: 103–110.

Petkar, H., Li, A., Bunce, N., Duffy, K., Malnick, H., and Shah, J.J. (2011) Cellulosimicrobium funkei: first report of infection in a nonimmunocompromised patient and useful phenotypic tests for differentiation from *Cellulosimicrobium cellulans* and *Cellulosimicrobium terreum*. *J Clin Microbiol* **49**: 1175–1178.

Pizzolitto, R.P., Armando, M.R., Salvano, M.A., Dalcero, A.M., and Rosa, C.A. (2013) Evaluation of *Saccharomyces cerevisiae* as an antiaflatoxicogenic agent in broiler feedstuffs. *Poult Sci* **92**: 1655–1663.

Rawal, S., Kim, J.E., and Coulombe, R., Jr (2010) Aflatoxin B₁ in poultry: toxicology, metabolism and prevention. *Res Vet Sci* **89**: 325–331.

Shetty, P.H., Hald, B., and Jespersen, L. (2007) Surface binding of aflatoxin B1 by *Saccharomyces cerevisiae* strains with potential decontaminating abilities in indigenous fermented foods. *Int J Food Microbiol* **113**: 41–46.

Shi, F., Seng, X.L., Tang, H.Q., Zhao, S.M., Deng, Y., Jin, R.W., and Li, Y.L. (2013) Effect of low levels of aflatoxin B1 on performance, serum biochemistry, hepatocyte apoptosis and liver histopathological changes of cherry valley ducks. *J Anim Vet Adv* **12**: 1126–1130.

Smiley, R.D., and Draughon, F.A. (2000) Preliminary evidence that degradation of aflatoxin B1 by *Flavobacterium aurantiacum* is enzymatic. *J Food Prot* **63**: 415–418.

Sun, L.H., Li, J.G., Zhao, H., Shi, J., Huang, J.Q., Wang, K.N., *et al.* (2013) Porcine serum can be biofortified with selenium to inhibit proliferation of three types of human cancer cells. *J Nutr* **143**: 1115–1122.

Tamura, K., Stecher, G., Peterson, D., Filipski, A., and Kumar, S. (2013) MEGA6: molecular evolutionary genetics analysis version 6.0. *Mol Biol Evol* **30**: 2725–2729.

Tejada-Castañeda, Z.I., Avila-Gonzalez, E., Casaubon-Huguenin, M.T., Cervantes-Olivares, R.A., Vásquez-Peláez, C., Hernández-Baumgarten, E.M., and Moreno-Martínez, E. (2008) Biodetoxification of aflatoxin-contaminated chick feed. *Poult Sci* **87**: 1569–1576.

Teniola, O.D., Addo, P.A., Brost, I.M., Färber, P., Jany, K.D., Alberts, J.F., *et al.* (2005) Degradation of aflatoxin B(1) by cell-free extracts of *Rhodococcus erythropolis* and *Mycobacterium fluoranthenivorans* sp. nov. DSM44556(T). *Int J Food Microbiol* **105**: 111–117.

Thompson, J.D., Gibson, T.J., Plewniak, F., Jeanmougin, F., and Higgins, D.G. (1997) The CLUSTAL_X windows interface: flexible strategies for multiple sequence alignment aided by quality analysis tools. *Nucleic Acids Res* **25**: 4876–4882.

Varga, J., Kocsubé, S., Péteri, Z., Vágvölgyi, C., and Tóth, B. (2010) Physical and biological approaches to prevent ochratoxin induced toxicoses in humans and animals. *Toxins (Basel)* **2**: 1718–1750.

Wang, J.L., Ruan, H., Zhang, H.F., Zhang, Q., Zhang, H.B., He, G.Q., and Shen, S.R. (2007) Characterization of a thermostable and acidic-tolerable beta-glucanase from aerobic fungi *Trichoderma koningii* ZJU-T. *J Food Sci* **72**: C452–C456.

Wu, F., and Guclu, H. (2012) Aflatoxin regulations in a network of global maize trade. *PLoS ONE* **7**: e45151.

Wu, Q., Jezkova, A., Yuan, Z., Pavlikova, L., Dohnal, V., and Kuca, K. (2009) Biological degradation of aflatoxins. *Drug Metab Rev* **41:** 1–7.

Yang, J., Bai, F., Zhang, K., Bai, S., Peng, X., Ding, X., *et al.* (2012) Effects of feeding corn naturally contaminated with aflatoxin B1 and B2 on hepatic functions of broilers. *Poult Sci* **91:** 2792–2801.

Yu, J., Chang, P.K., Ehrlich, K.C., Cary, J.W., Bhatnagar, D., Cleveland, T.E., *et al.* (2004) Clustered pathway genes in aflatoxin biosynthesis. *Appl Environ Microbiol* **70:** 1253–1262.

Yunus, A.W., Razzazi-Fazeli, E., and Bohm, J. (2011) Aflatoxin B(1) in affecting broiler's performance, immunity, and gastrointestinal tract: a review of history and contemporary issues. *Toxins (Basel)* **3:** 566–590.

Zjalic, S., Reverberi, M., Ricelli, A., Mario Granito, V., Fanelli, C., and Adele Fabbri, A. (2006) Trametes versicolor: a possible tool for aflatoxin control. *Int J Food Microbiol* **107:** 243–249.

## Supporting information

Additional Supporting Information may be found in the online version of this article at the publisher's web-site:

**Fig. S1.** Morphology of (A) colony and (B) gram staining of *C. funkei* $T_3$-5.

**Fig. S2.** The selected chromatogram of HPLC, (A) 25 ug/kg $AFB_1$ standard; (B) negative control; (C) positive control; (D) after $AFB_1$ biodegradation by *C. funkei*. $AFB_1$ biodegradation (%) = $(C_{peak\ area} - D_{peak\ area})/C_{peak\ area} \times 100\%$.

**Table S1.** Ability of $AFB_1$ biodegradation by the protein from the culture supernatant of *C. funkei* by ammonium sulfate precipitation[1].

# 16

# Hfq regulates antibacterial antibiotic biosynthesis and extracellular lytic-enzyme production in *Lysobacter enzymogenes* OH11

Gaoge Xu,[1] Yuxin Zhao,[1] Liangcheng Du,[2]
Guoliang Qian[1]** and Fengquan Liu[1,3]*
[1]*College of Plant Protection, Nanjing Agricultural University, China/Key Laboratory of Integrated Management of Crop Diseases and Pests (Nanjing Agricultural University), Ministry of Education, Nanjing 210095, China.*
[2]*Department of Chemistry, University of Nebraska-Lincoln, Lincoln, NE 68588, USA.*
[3]*Institute of Plant Protection, Jiangsu Academy of Agricultural Science, Nanjing 210014, China.*

## Summary

***Lysobacter enzymogenes* is an important biocontrol agent with the ability to produce a variety of lytic enzymes and novel antibiotics. Little is known about their regulatory mechanisms. Understanding these will be helpful for improving biocontrol of crop diseases and potential medical application. In the present study, we generated an *hfq* (encoding a putative ribonucleic acid chaperone) deletion mutant, and then utilized a new genomic marker-free method to construct an *hfq*-complemented strain. We showed for the first time that Hfq played a pleiotropic role in regulating the antibacterial antibiotic biosynthesis and extracellular lytic enzyme activity in *L. enzymogenes*. Mutation of *hfq* significantly increased the yield of WAP-8294A2 (an antibacterial antibiotic) as well as the transcription of its key**

biosynthetic gene, *waps1*. However, inactivation of *hfq* almost abolished the extracellular chitinase activity and remarkably decreased the activity of both extracellular protease and cellulase in *L. enzymogenes*. We further showed that the regulation of *hfq* in extracellular chitinase production was in part through the impairment of the secretion of chitinase A. Collectively, our results reveal the regulatory roles of *hfq* in antibiotic metabolite and extracellular lytic enzymes in the underexplored genus of *Lysobacter*.

For correspondence. *E-mail fqliu20011@sina.com
**E-mail glqian@njau.edu.cn

**Funding Information** This study was supported by National Natural Science Foundation of China (31371981), Program for New Century Excellent Talents in University of Ministry of Education of China (NCET-13-0863), Fundamental Research Funds for the Central Universities (No. KYZ201205 and KYTZ201403), Special Fund for Agro-Scientific Research in the Public Interest (No. 201203034; 201003004), the National High Technology Research and Development Program ('863' Program) of China (2011AA10A205), Modern Agricultural Industry Technology System (No.CARS-29-09), Zhongshan Scholars of Nanjing Agricultural University National "948" program (2014-Z24). Research in the Du lab is supported in part by NIH R01AI097260 and Nebraska Research Initiatives.

## Introduction

*Lysobacter* is a genus in the family of Xanthomonadaceae and is one of the most ubiquitous environmental microorganisms (Christensen and Cook, 1978). *Lysobacter enzymogenes* of the genus is the best characterized species. This species is known for its ability to produce a variety of extracellular lytic enzymes, including chitinase, cellulase and protease (Kobayashi *et al.*, 2005). These enzymes are able to destroy the cell wall of pathogenic fungi and oomycetes and are highly linked to biocontrol activity of *L. enzymogenes* against crop pathogens (Zhang and Yuen, 2000; Palumbo *et al.*, 2003). Previously, efforts had been devoted to cloning the genes encoding the lytic enzymes (Epstein and Wensink, 1988; Zhang and Yuen, 2000; Palumbo *et al.*, 2003), but little is known about the regulation of the lytic-enzyme production or secretion in *L. enzymogenes*, except the global regulator Clp (Kobayashi *et al.*, 2005; Wang *et al.*, 2014).

In addition to producing abundant lytic enzymes, *L. enzymogenes* produces diverse bioactive natural products and is recently emerging as a new source of antibiotics, such as the antibacterial WAP-8294A2 (Kato *et al.*, 1998; Zhang *et al.*, 2011; Xie *et al.*, 2012). WAP-8294A2 is a cyclic lipodepsipeptide with a strong activity against Gram-positive bacteria, including methicillin-resistant *Staphylococcus aureus* (Zhang *et al.*, 2011). However, *L. enzymogenes* produces a low yield of this antibiotic under the common growth condition (Zhang *et al.*, 2011). Meanwhile, the complex chemical structure of WAP-8294A2 makes it extremely challenging for chemical synthesis. We have recently identified the gene cluster responsible for the biosynthesis of WAP-8294A2 in

*L. enzymogenes* (Zhang *et al.*, 2011). Among the gene, *waps1*, encoding a typical non-ribosomal peptide synthetase, is a key gene for WAP-8294A2 biosynthesis (Zhang *et al.*, 2011). The molecular biosynthetic mechanism for WAP-8294A2 production has been proposed (Zhang *et al.*, 2011). Little is known, however, about the molecular mechanisms that regulate the biosynthesis of this antibiotic. Understanding the regulatory mechanism for the production of the antibiotic WAP-8294A2 and the lytic enzymes (e.g. chitinase) is important because it could lead to new genetic approaches to improvement of biocontrol of crop disease and potential medical application.

Hfq is a protein serving as a conserved ribonucleic acid (RNA) chaperon and was first characterized as a host factor for phage Qβ RNA replication (Franze de Fernandez *et al.*, 1968). It is widely distributed in bacterial annotated genomic databases (Caswell *et al.*, 2012; Wang *et al.*, 2012). Hfq can bind AU-rich sequence of target messenger RNA (mRNA) and facilitate the pairing interaction between mRNA and small RNA (Wang *et al.*, 2012), which suggests that this protein is a global post-transcriptional regulator in most cases. However, recent studies also showed that Hfq was able to directly bind deoxyribonucleic acid (DNA) (Updegrove *et al.*, 2010; Geinguenaud *et al.*, 2011), even tRNA (Lee and Feig, 2008) and proteins (Butland *et al.*, 2005) to modulate the transcriptional expression of target genes, which indicates Hfq also can function as a transcriptional regulator in some bacterial species. Hfq was shown to play critical roles in diverse animal bacterial pathogens and plant-associated bacteria, such as *Escherichia coli, Salmonella, Sinorhizobium meliloti, Staphylococcus* and *Pseudomonas*; the *hfq* mutant exhibited pleiotropic phenotypes in these bacterial species, including decreased growth rate (Fantappiè *et al.*, 2009; Chambers and Bender, 2011), increased sensitivity to various environmental stressors (Kadzhaev *et al.*, 2009; Schiano *et al.*, 2010) and attenuated virulence (Geng *et al.*, 2009; Ramos *et al.*, 2011). The role of *hfq* in a number of bacterial biological control agents had been investigated. For example, Hfq was found to regulate antibiotic production in *Pseudomonas aeruginosa* M18, a rhizobacterium bacterium that can efficiently inhibit soil-borne phytopathogenic fungi (Wang *et al.*, 2012). In *Pseudomonas fluorescens* 2P24, *hfq* is involved in the colonization, biofilm formation, antibiotic synthesis and quorum sensing signal production (Wu *et al.*, 2010). However, essentially nothing is known about the role of *hfq* in any of *Lysobacter* species.

In the present study, we identified an *hfq* homologue from the genome of strain OH11, a Chinese isolate of *L. enzymogenes*. The results show that this *hfq* plays a pleiotropic role in regulating the antibacterial antibiotic

biosynthesis and extracellular lytic enzyme activity in *L. enzymogenes*. To our knowledge, this study represents the first attempt to address the regulatory function of *hfq* in the genus of *Lysobacter*. Our findings also add an understanding of the conserved Hfq protein in different bacterial species.

## Results and discussion

### Sequence analysis of hfq in L. enzymogenes

Hfq protein was first discovered from *E. coli*, and this Hfq was considered as the model in bacteria (Franze de Fernandez *et al.*, 1968; Kajitani and Ishihama, 1991). To examine whether *L. enzymogenes* possesses an Hfq homologue, we used the *E. coli* Hfq protein (AAC43397.1) as the query sequence to perform a local BLASTP search in the draft genome of strain OH11 (Lou *et al.*, 2011). This led to the identification of an Hfq homologue (KM186922) in *L. enzymogenes*. As shown in Fig. 1, Hfq protein of *L. enzymogenes* shares 75% similarity to that of *E. coli* at the amino acid level. Moreover, the locus of *miaA-hfq-hflX-hflK-hflC* of *E. coli* was also conserved in *L. enzymogenes*. Next, we selected the reported Hfq proteins from four *Lysobacter*-related bacterial species, all belonging to the Xanthomonadaceae family, to do a sequence alignment. The result showed that Hfq of *L. enzymogenes* shared a high similarity (85–96%) to that of these taxonomically related species, including Hfq$_{xoo}$ of *Xanthomonas oryzae* pv. *oryzae* (WP_014503655.1), Hfq$_{xcc}$ of *X. campestris* pv. *campestris* (WP_011036893.1), Hfq$_{sm}$ of *Stenotrophomonas maltophilia* (WP_019183319.1) and Hfq$_{xf}$ of *Xylella fastidiosa* (WP_004085558.1). The results suggest that *L. enzymogenes* possesses a putative conserved Hfq protein.

### Generation of a set of marker free of hfq-derived strains of L. enzymogenes

In order to explore the function of *hfq* in *L. enzymogenes*, a 155-bp internal fragment of *hfq* was in-frame deleted, which led to the generation of an *hfq* mutant, named as Δ*hfq* (Fig. S1). Subsequently, a genomic marker-free strategy was utilized to construct the *hfq* complemented strain (Fig. 2A). In this *hfq* complemented strain, no exogenous antibiotic selection markers were introduced into the *hfq* mutant, which can eliminate their potential effect on the tested phenotypes of the present study in *L. enzymogenes*. For this purpose, we selected α*lp* (Wang and Qian, 2012), an α-lytic-protease encoding gene as a target for *hfq* integration. To our knowledge, this gene is not associated with the tested phenotypes of the present study, including the antibiotic WAP-8294A2

**A**

**B**

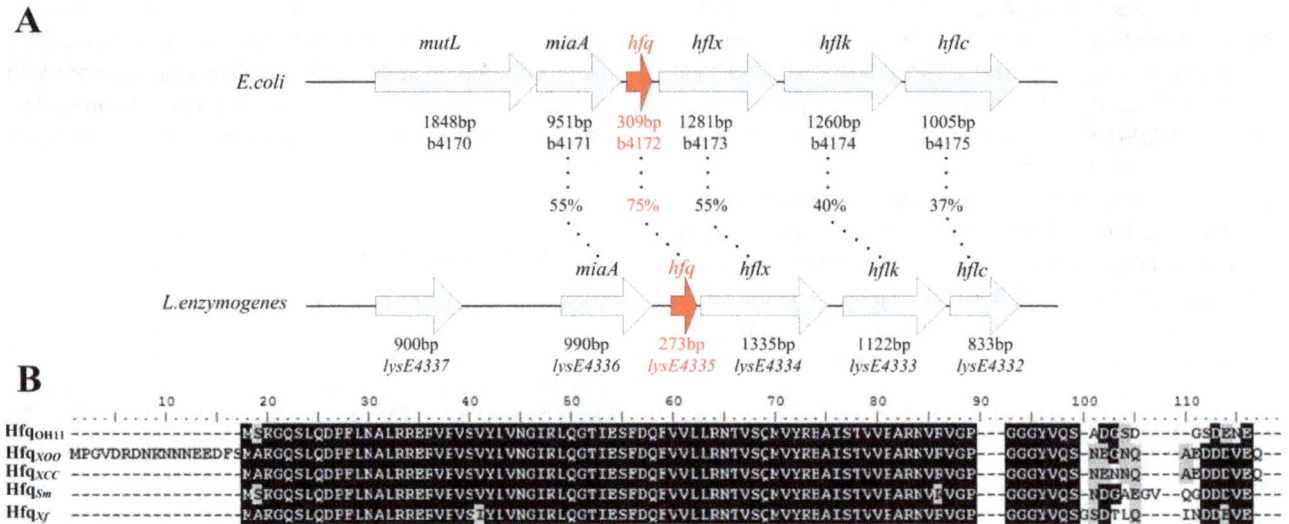

**Fig. 1.** Identification of Hfq (LysE4335) in *Lysobacter enzymogenes*.
A. Comparison of the *hfq* locus between *L. enzymogenes* strain OH11 and the well-studied *Escherichia coli*. The percentage numbers (expressed by %) represent the identity/similarity between the gene/protein homologue in *E. coli* and *L. enzymogenes* at amino-acid level. The size of each gene was presented below each arrow.
B. Sequence alignment of Hfq$_{OH11}$ of *L. enzymogenes* with other Hfq proteins from taxonomically related bacterial species, all belonging to the Xanthomonadaceae family.

production and the activity of extracellular chitinase, protease or cellulase. By using homologous recombination method as shown in Fig. 2A, we introduced the intact *hfq* gene together with its own promoter into the *αlp* genomic locus in the background of the *hfq* mutant, and generated a chromosomal marker-free *hfq*-complemented strain Δ*hfq*(*hfq*)$_{Δαlp}$ (Fig. 2B). Meanwhile, we also created single

mutation of *αlp* (Δ*αlp*) and double mutations of *hfq* and *αlp* (Δ*hfq*Δ*αlp*) as controls (Table S1). To verify the mutants, we utilized an reverse transcription polymerase chain reaction (RT-PCR) assay to examine whether the target gene (*hfq* or *αlp*) was transcribed or not in these *hfq*-derived strains. As shown in Fig. 2C, we detected the transcription of *hfq* in wild-type OH11, the *hfq*-

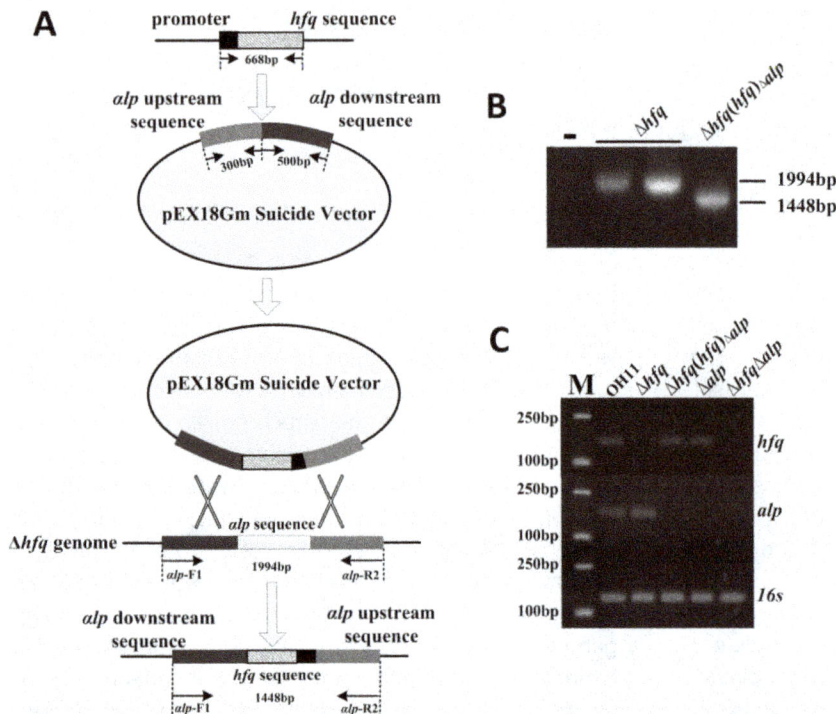

**A**

**B**

**C**

**Fig. 2.** The marker-free integration of *hfq* to generate a complementation strain of *Lysobacter enzymogenes*.
A. Physical map of the marker-free complemented strain of the *hfq* mutant.
B. PCR confirmation of the replacement of the *αlp* gene by *hfq*.
C. RT-PCR to conform the expression of the target gene (*hfq* or *αlp*) in the *hfq* mutant and its derivative strains. OH11, the wild-type strain of *L. enzymogenes*; Δ*hfq*, the *hfq* deletion mutant; Δ*hfq*(*hfq*)$_{Δαlp}$, the genomic marker-free complemented strain of the *hfq* mutant (the *αlp* gene was replaced by the *hfq* in the background of the *hfq* mutation); Δ*αlp*, the *αlp* deletion mutant; Δ*hfq*Δ*αlp*, the double mutant of *hfq* and *αlp*. *αlp*, an α-lytic-protease encoding gene of *L. enzymogenes*. '-' in B represents the blank control.

complemented strain ($\Delta hfq(hfq)_{\Delta alp}$) and the $alp$ mutant ($\Delta alp$) but not in the $hfq$ mutant ($\Delta hfq$) and the double mutant ($\Delta hfq\Delta alp$). Similarly, it was also found that $alp$ was transcribed in wild-type OH11 and the $hfq$ mutant ($\Delta hfq$) but not in the $hfq$ complemented strain, the $alp$ deletion mutant and the double mutant. These results verified the $hfq$, $alp$, double mutations and the marker-free $hfq$ complemented strains.

### Mutation of hfq caused a significant increase of the yield of WAP-8294A2 in L. enzymogenes

To test the role of $hfq$ in the regulation of the antibiotic WAP-8294A2 biosynthesis, we examined WAP-8294A2 production in the $hfq$ mutant in *L. enzymogenes*. To eliminate the potential influence of growth alteration on WAP-8294A2 production between the wild-type strain and the $hfq$ mutant, we subsequently determined the growth ability (expressed by $OD_{600\,nm}$, the optical density at 600 nm) of wild-type OH11, the $hfq$ mutant and its derivative strains in 20% TSB broth that was used to cultivate these *Lysobacter* strains for WAP-8294A2 extraction. Meanwhile, considering that inactivation of $hfq$ in several bacteria leads to an increased cell size (Tsui *et al.*, 1994; Boudry *et al.*, 2014), we then tested the role of $hfq$ in this phenotype (cell size) in *L. enzymogenes*, because if disruption of $hfq$ causes bigger cells, it will affect the cell density that is expressed by $OD_{600\,nm}$ in the present study. As shown in Fig. S2, we found that mutation of $hfq$ almost did not alter cell size compared with the wild-type strain at the two selected time points (after growth of 24 h and 48 h). These results suggest that application of $OD_{600\,nm}$ as an indicator to reflect the cell growth status between the wild-type strain and the $hfq$ mutant should be reasonable in the present study. Next, the growth rate of the wild-type strain and the $hfq$ mutant in 20% Tryptic Soy Broth (TSB) broth was determined and compared. As shown in Fig. 3, we observed that mutation of $hfq$ resulted in a changed growth pattern that was taken place in the logarithmic phase and originated the delay in reaching the stationary phase compared with that of the wild-type strain. Under the same condition, the complemented strain exhibited the wild-type growth rate. As expected, mutation of $alp$ did not alter the wild-type growth rate, indicating that $alp$ was not involved in the growth capacity in the tested medium. This finding was further verified by the growth rate of the double mutant ($\Delta hfq\Delta alp$), as this double mutant displayed a closely similar growth rate to that of the $hfq$ mutant. The results suggested that mutation of $hfq$ had a slight effect on the growth of the wild-type strain in the tested medium (20% TSB).

Next, we cultivated the tested *Lysobacter* strains in 20% TSB broth to extract WAP-8294A2, and determined its yield in each strain by high-performance liquid chroma-

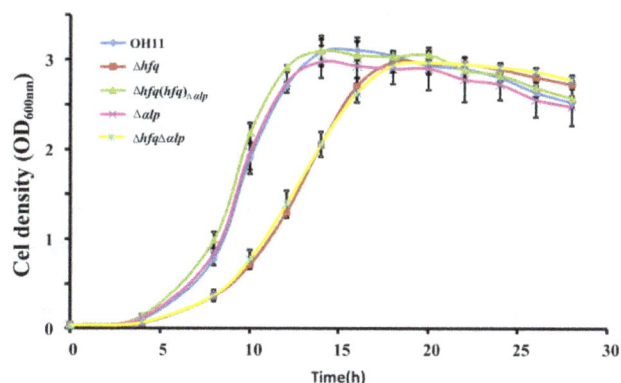

**Fig. 3.** The growth curves of various *Lysobacter* strains in 20% TSB medium. The growth level of each strain was measured by $OD_{600\,nm}$ at regular intervals (2 h or 4 h). Three replicates for each treatment/strain were used, and the experiment was performed three times. Vertical bars represent standard errors. The strain information in Fig. 3 is shown in the legend of Fig. 2.

tography (HPLC). To completely eliminate the influence of growth alteration on WAP-8294A2 production, we used the indicator of 'Antibiotic production (peak area/$OD_{600\,nm}$)' to quantitatively evaluate the ability of WAP-8294A2 production in the wild-type strain and its derivatives. Here, peak area indicates the area of WAP-8294A2 determined by HPLC, whereas $OD_{600\,nm}$ represents the cell density of the tested strains at the time point used for the extraction of WAP-8294A2 in *L. enzymogenes*. In this way, we found that mutation of $hfq$ significantly enhanced WAP-8294A2 production ($\sim$ 2.2 fold) compared with the wild-type strain, whereas the marker-free complemented strain of the $hfq$ mutant displayed the wild-type level in this ability (Fig. 4A). Given that the marker-free $hfq$ complemented strain was constructed by integration of $hfq$ into the $alp$ gene in the background of $hfq$ mutation, the parent $hfq$ and $alp$ genes were both missing in this complemented strain. Therefore, it is also possible that the restoration of WAP-8294A2 production in the marker-free $hfq$ complemented strain could be due to the effect of both missing $hfq$ and $alp$. To test this possibility, we determined the WAP-8294A2 production in the double mutant ($\Delta hfq\Delta alp$). We showed that mutation of $alp$ in the background of the $hfq$ mutant did not affect the yield of WAP-8294A2 in this mutant (the $hfq$ mutant) (Fig. 4A). This result provided supportive evidence to verify that restoration of WAP-8294A2 production in the marker-free $hfq$ complemented strain was due to the replacement of $alp$ by $hfq$ in the background of the $hfq$ mutation but not associated with the effect of the simultaneous deletion of both $hfq$ and $alp$. Furthermore, mutation of $alp$ in the background of wild-type strain did not alter WAP-8294A2 production, supporting that $alp$ was not associated with WAP-8294A2 biosynthesis. Collectively, these results indicated that $hfq$ played an important role in the negative regulation of WAP-8294A2 biosynthesis in *L. enzymogenes*.

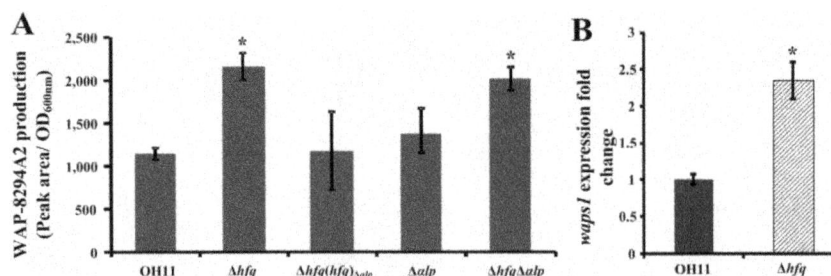

**Fig. 4.** Mutation of *hfq* significantly increased the production of WAP-8294A2 in *Lysobacter enzymogenes*.
A. Quantitative measurement of the yield of the antibacterial antibiotic WAP-8294A2 in the *hfq* mutant and its derivative strains.
B. Quantitative determination of the transcription of the critical biosynthetic gene (*waps1*) for WAP-8294A2. The strain information in Fig. 4 is shown in the legend of Fig. 2. Each column indicates the mean of three biologically independent experiments. Vertical bars represent standard errors. '*' ($P < 0.05$; *t*-test) above the bars indicate a significant difference between the wild-type strain OH11 and the *hfq* mutant.

To further verify the important role of *hfq* in WAP-8294A2 biosynthesis, the transcriptional level of *waps1* (Zhang *et al.*, 2011), the key gene responsible for WAP-8294A2 biosynthesis, was tested using quantitative (q)RT-PCR between the wild-type strain and the *hfq* mutant. Based on the result of Fig. 3, we finally collected the cells at the logarithmic phase from the wild-type strain and the *hfq* mutant at different time point corresponds to the same cell density ($OD_{600 nm} = 1.0$), respectively, because at this cell density, the gene *waps1* was previously shown to be expressed at transcriptional level in the wild-type OH11 of *L. enzymogenes* (Wang *et al.*, 2014). The results of the qRT-PCR assay showed that mutation of *hfq* caused a significant increase (~ 2.3 fold) of the transcription of *waps1* compared with the wild-type strain (Fig. 4B). This finding was consistent with the HPLC result for the WAP-8294A2 production in the *hfq* mutant (Fig. 4A), and further verified that *hfq* was involved in the negative regulation of WAP-8294A2 biosynthesis in *L. enzymogenes*.

*Mutation of hfq almost abolished extracellular chitinase activity and significantly reduced the activity of both extracellular protease and cellulase in L. enzymogenes*

In addition to WAP-8294A2 biosynthesis, we are also interested in addressing the role of *hfq* in lytic-enzyme production in *L. enzymogenes*. We therefore examined the activity of three known extracellular lytic enzymes, including chitinase, protease and cellulase on the corresponding detecting media. In the present study, the ratio of the area of hydrolytic zones divided by cell density ($OD_{600 nm}$) was used as a quantitative indicator for each enzyme activity. By this way, we found that mutation of *hfq* almost abolished extracellular chitinase activity, and significantly reduced the activity of both extracellular protease ($P < 0.01$; *t*-test) or cellulase ($P < 0.05$; *t*-test) in *L. enzymogenes* (Fig. 5). Under the same conditions, the marker-free *hfq* complemented strain restored the wild-

type level in each tested enzyme activity (Fig. 5). Furthermore, mutation of *alp* in the background of the *hfq* mutant did not influence the activity of each tested enzymes (Fig. 5). This result suggested that the activity restoration of each tested enzymes in the marker-free *hfq*

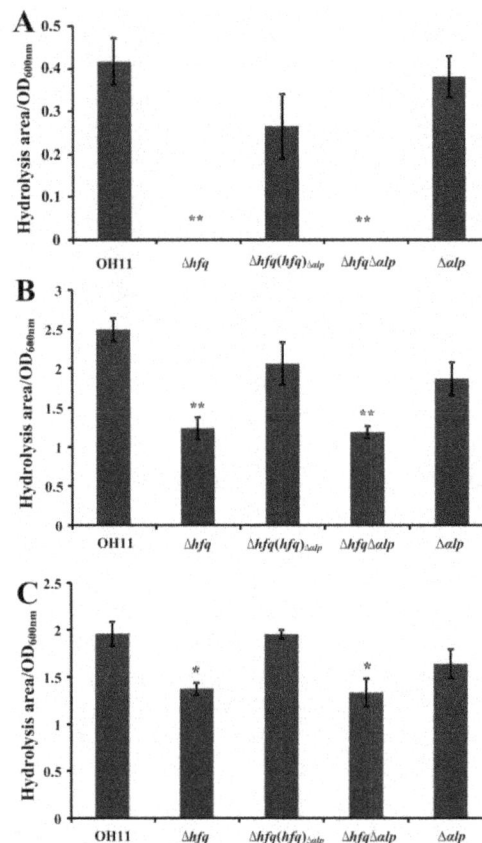

**Fig. 5.** Quantitative determination of the activity of three extracellular lytic enzymes, chitinase (A), protease (B) and cellulase (C) from various *Lysobacter enzymogenes* strains. Each column indicates the mean of three biologically independent experiments. Vertical bars represent standard errors. '*' ($P < 0.05$; *t*-test) or '**' ($P < 0.01$; *t*-test) above the bars indicate a significant difference between the wild-type strain and its derivatives. The strain information in Fig. 5 is shown in the legend of Fig. 2.

complemented strain was due to the replacement of *alp* by *hfq* in the background of the *hfq* mutation but not associated with the effect of the deletion of both *hfq* and *alp*. Moreover, mutation of *alp* in the background of the wild-type strain did not alter the activity of each tested enzyme (Fig. 5), supporting that *alp* was not involved in the regulation of the activity of these three tested enzymes in *L. enzymogenes*. Collectively, our results indicated that *hfq* played a key role in the regulation of the activity of extracellular chitinase, protease and cellulase in *L. enzymogenes*.

*Mutation of hfq inhibited the secretion of chitinase A*

The deficiency of the *hfq* mutant in extracellular chitinase production promotes us to focus the mechanism(s) by which *hfq* modulates this activity in *L. enzymogenes*. For this purpose, we first investigated whether the deficiency of the *hfq* mutant in extracellular chitinase activity might be associated with its growth reduction, as we observed that the final cell density of the *hfq* mutant was significantly reduced (~ 2.3 fold) compared with that of the wild-type strain when the initial inoculated cell concentration was the same (Fig. S3). Subsequently, the experimental evidence presented here eliminated this possibility, because we clearly found that even though the *hfq* mutant possessed a significantly increased cell density (*P* < 0.05; *t*-test) compared with the wild-type strain (when the initial inoculated cell concentration of the *hfq* mutant was 10-fold higher than that of the wild-type strain) (Fig. S3), we still did not find any hydrolytic zone around the colonies of the *hfq* mutant (Fig. S3). These results provided supportive evidence to show that the deficiency of the *hfq* mutant in chitinase activity was not due to its growth reduction, which implies that other mechanism(s) may be utilized by *hfq* to modulate extracellular chitinase production in *L. enzymogenes*.

Next, we attempted to test which type(s) of chitinases was reduced in the *hfq* mutant, resulted in its deficiency in extracellular chitinase activity. For this purpose, we made a survey to detect how many chitinase-encoding genes are present in the genome of *L. enzymogenes*. This led to identification of a total of three chitinase-encoding genes, including *chiA* (Qian *et al.*, 2012), *chiB* and *chiC* (Table S3). We subsequently generated the *chiB* and *chiC* deletion mutants (Table S1). The extracellular chitinase activity of these two mutants as well as the *chiA* mutant (Qian *et al.*, 2012) was then tested. As shown in Fig. S4, we found that mutation of *chiA* completely abolished the chitinase activity, which was closely consistent with the role of the *hfq* mutant in this phenotype (Fig. 5). However, mutation of *chiB* or *chiC* did not alter the extracellular chitinase activity (Fig. S4). This finding raised the possibility that the regulation of *hfq* on extracellular chitinase

activity was probably through the alteration of *chiA* in *L. enzymogenes* under the testing conditions.

Considering the fact that Hfq usually functions as a post-transcriptional regulator (Lee and Feig, 2008; Sittka *et al.*, 2009), we then examined whether *hfq* plays a post-transcriptional regulation on *chiA*. We first constructed a plasmid (pBBR1-MCS5)-based *hfq*-complemented strain by introduction of pBBR1-MCS5 containing the *hfq* gene into the *hfq* mutant. Then, the extracellular chitinase activity was tested in this complemented strain. We showed that the plasmid (pBBR1-MCS5)-based *hfq*-complemented strain partially restored the wild-type level in extracellular chitinase activity, whereas the *hfq* mutant containing the empty vector was deficient in this function (Fig. S5). This result indicates that the plasmid (pBBR1-MCS5) containing *hfq* was functional in restoring the deficiency of the *hfq* mutant in extracellular chitinase activity. Next, a flag-tagged *chiA* sequence was cloned into the broad-host-vector pBBR1-MCS5, and transformed into the *chiA* mutant. The introduction of pBBR1-MCS5 containing the flag-tagged *chiA* restored the chitinase activity of the *chiA* mutant (Fig. S6A), supporting the correction of the construction. We then individually introduced the pBBR1-MCS5 containing the flag-tagged *chiA* into the *hfq* mutant and wild-type OH11, respectively. Next, the anti-flag antibody dependent western blot assay was performed to compare the protein content of the flag-tagged ChiA (Chitinase A) in wild-type OH11 and the *hfq* mutant. As shown in Fig. 6, we observed that the band of flag-tagged ChiA was detected both in the total cells and the supernatant of wild type, suggesting ChiA was synthesized and secreted outside the cells under the testing condition. Meanwhile, we only detected the flag-tagged ChiA band in the total cells but not in the supernatant of the *hfq* mutant. This result indicated the ChiA protein was probably synthesized in the cells of the *hfq* mutant but was unable to be secreted out of the cells or degraded once secreted in the *hfq* mutant. To verify this, we introduced the construct (pBBR1-MCS5 containing the flag-tagged *chiA*) into the *hfq* mutant and tested its chitinase

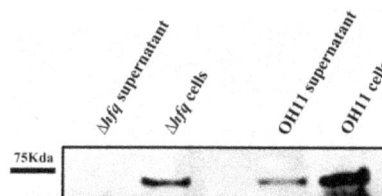

**Fig. 6.** Western blot analysis of the yield of flag-tagged ChiA (Chitinase A) in *Lysobacter enzymogenes*. The yield of flag-tagged ChiA (Chitinase A) both in the supernatant and total cell of the wild-type strain OH11 and the *hfq* mutant (Δ*hfq*) was comparatively analysed by western blot using the anti-flag antibody. The data are the representative results of three independent experiments. The expected size of ChiA protein is 71.6 Kda.

**Table 1.** Strains and plasmids used in this study.

| Strains and plasmids | Characteristics[a] | Source |
|---|---|---|
| *Lysobacter enzymogenes* | | |
| OH11 | Wild-type, Km$^R$ | Qian *et al.*, 2009 |
| Δ*hfq* | *hfq* in-frame deletion mutant, Km$^R$ | This study |
| Δ*hfq(hfq)* | Δ*hfq* harbouring plasmid *hfq*-pBBR, Gm$^R$, Km$^R$ | This study |
| Δ*hfq*(pBBR) | Δ*hfq* harbouring plasmid pBBR1-MCS5, Gm$^R$, Km$^R$ | This study |
| Δ*hfq*Δα*lp* | *hfq* and α*lp* in-frame deletion mutant, Km$^R$ | This study |
| Δ*hfq(hfq)*$_{\Delta\alpha lp}$ | *hfq* in-frame deletion mutant and the α*lp* gene was replaced by *hfq*, Km$^R$ | This study |
| Δα*lp* | α*lp* in-frame deletion mutant, Km$^R$ | Wang and Qian, 2012 |
| Δ*chiB* | *chiB* in-frame deletion mutant, Km$^R$ | This study |
| Δ*chiC* | *chiC* in-frame deletion mutant, Km$^R$ | This study |
| Δ*chiA* | *chiA* in-frame deletion mutant, Km$^R$ | Qian *et al.*, 2012 |
| Δ*chiA(chiA*-flag) | Δ*chiA* harbouring flag-tagged *chiA*-pBBR | This study |
| OH11(*chiA*-flag) | OH11 harbouring flag-tagged *chiA*-pBBR | This study |
| Δ*hfq(chiA*-flag) | Δ*hfq* harbouring flag-tagged *chiA*-pBBR | This study |
| *Escherichia coli* | | |
| TOP10 | *supE44lacU169*(Δ*lacZ*ΔM15) *hsdR17 recA lendA1gyrA96 thi-1 relA11* | Lab collection |
| Plasmids | | |
| pEX18GM | Suicide vector with a *sacB* gene, Gm$^R$ | Hoang *et al.*, 1998 |
| pBBR1-MCS5 | Broad-host- vector with a P$_{lac}$ promoter, Gm$^R$ | Kovach *et al.*, 1995 |
| *hfq*-pEX18 | pEX18GM with two flanking fragments of *hfq*, Gm$^R$, | This study |
| α*lp*-pEX18 | pEX18GM with two flanking fragments of α*lp*, Gm$^R$, | Wang and Qian, 2012 |
| *hfq*-pBBR | pBBR1-MCS5 cloned with a 668-bp fragment containing intact *hfq* and its predicted promoter | This study |
| *chiB*-pEX18 | pEX18GM with two flanking fragments of *chiB*, Gm$^R$, | This study |
| *chiC*-pEX18 | pEX18GM with two flanking fragments of *chiC*, Gm$^R$, | This study |
| *chiA*-flag-pBBR | pBBR1-MCS5 cloned with a 2403 bp fragment containing flag-tagged *chiA* | This study |

**a.** Km$^R$, Gm$^R$ = Kanamycin, Gentamicin-resistance respectively.

activity. We found that the *hfq* mutant containing this construct did not restore the chitinase activity, as no hydrolytic zones around its colonies was detected on the chitinase selective medium, whereas the *chiA* mutant and wild-type OH11 containing this construct displayed visible chitinase activity (Fig. S6B). This result provided supportive evidence to show that the effect of *hfq* on chitinase activity was probably in part through the impairment of the secretion of ChiA in *L. enzymogenes*. However, we do not know how *hfq* mediates the secretion of ChiA in *L. enzymogenes* at this time.

A recent study shows that in *Flavobacterium johnsoniae* chitinase secretion is dependent on type IX secretion system (T9SS), consisting of products of the key genes of *gldK*, *gldL*, *gldM*, *gldNO*, *sprA*, *sprE* and *sprT* (Kharade and McBride, 2014). However, we did not find the homologue of any of these T9SS-associated key genes/proteins in the genome of *L. enzymogenes* OH11, indicating that the effects of *hfq* on the secretion of Chitinase A in *L. enzymogenes* is probably not associated with T9SS and is therefore different from the finding in *F. johnsoniae*. Similar to our result, the *hfq* mutant of *Listeria monocytogenes* showed a less chitinolytic activity compared with that of wild-type strain. This report pointed out that uncharacterized *hfq*-dependent small RNAs may mediate the stimulating effect of *hfq* on the activity and/or secretion of the chitinase in *L. monocytogenes* (Nielsen *et al.*, 2011). This information provided a clue to further explore the regulatory mechanism of *hfq* on the secretion

of chitinase A by focusing on the *hfq*-dependent small RNAs in *L. enzymogenes*.

*Conclusion remarks*

The production of lytic enzymes and antibiotics is one of the distinctive characteristics of the important but underexplored biological control agent *L. enzymogenes*. The genetic determinants that regulate the biosynthesis of these factors are largely unidentified. The present study reported Hfq, a putative RNA chaperone played a pleiotropic role in the modulation of the antibacterial antibiotic WAP-8294A2, and the activity of extracellular chitinase, protease and cellulase in *L. enzymogenes* OH11. The regulation of *hfq* on extracellular chitinase production was further shown to be in part through the impairment of the secretion of chitinase A. These findings provide new insights into the role of *hfq* in bacteria. In future works, we aim to further address the molecular mechanisms of *hfq* on the regulation of WAP-8294A2 as well as the secretion of chitinase A in *L. enzymogenes*. These future works will help us further understand the signalling pathway of *hfq* in the regulation of antibiotics and lytic enzymes in *L. enzymogenes*.

**Experimental procedures**

*Bacterial strains and growth conditions*

Strains and plasmids used in this study are shown in Table 1. *Escherichia coli*, strain Top 10 was grown in LB medium at

37°C. Unless otherwise stated, *Lysobacter enzymogenes* strain OH11 (CGMCC No. 1978), and its derivative strains were grown in LB medium at 28°C. When required, appropriate antibiotics were added into the medium to a final concentration of kanamycin (Km) 50 µg ml$^{-1}$ and gentamicin (Gm) 150 µg ml$^{-1}$.

## Generation of deletion mutants of target genes in *L. enzymogenes*

The wild-type strain OH11 (CGMCC No. 1978) of *L. enzymogenes* was used as an original strain to generate the in-frame deletion mutants. Construction of gene-deletion mutants in *L. enzymogenes* was performed as described previously (Qian *et al.*, 2013). As a representative example, the scheme of the *hfq* mutant construction and molecular confirmation in *L. enzymogenes* was provided in Fig. S1 in the supplementary file. All the primers and plasmids used in the present study were provided in Table S2 and Table 1 respectively.

## Construction and confirmation of a marker-free hfq complemented strain

The *hfq* fragment with its predicted promoter (668-bp) was inserted into a pEX18Gm-*αlp* (Wang and Qian, 2012) recombinant suicide vector, which has an upstream fragment (300-bp) and a downstream fragment (500-bp) of *αlp* gene of the wild-type OH11. The final construct was transformed into the *hfq* mutant. After twice homologous recombination, the *αlp* gene was replaced by *hfq* in the genome of the wild-type OH11, which is defined as the marker-free-complemented strain of the *hfq* mutant. This marker-free *hfq*-complemented strain was verified both by PCR and RT-PCR assays. For PCR assay, the expected size of the DNA fragment from the *hfq* mutant was 1994 bp by using the primers of *αlp*-F1/R2 (Table S2). When the *αlp* gene was replaced by the *hfq* gene in the background the *hfq* mutant, the expected size of DNA fragment was 1448 bp by using the same primer pairs (*αlp*-F1/R2). For RT-PCR assay, the expression of the target gene (*hfq* or *αlp*) in the *hfq* mutant and its derivative strains was tested. The expected size of the DNA fragment amplified from the complementary DNA (cDNA) of the wild-type strain corresponds to the *hfq*, *αlp* and 16s rRNA was 158 bp, 174 bp and 147 bp respectively. The corresponding primers were shown in Table S4. In the present study, 16s rRNA was used as an internal control as described previously (Qian *et al.*, 2013; 2014).

## Observation of the cell size of L. enzymogenes by electronic microscope

The wild-type OH11 and the *hfq* deletion mutant were grown on 20% TSB with agar (TSA) plates, and their cells were collected after growth of 24 h and 48 h, respectively. These cells were further used for the analysis of electronic microscope as described previously in our lab (Chen *et al.*, 2009)

## Detection of growth curve

Various *Lysobacter* strains were cultured in LB medium at 28°C overnight. Then, 500 µl of the overnight culture for each

strain was transferred into the fresh 20% TSB broth (50 ml) to grow until the cell density expressed by OD$_{600 nm}$ reached to 1.0. Next, 1 ml of each culture was transferred again into the fresh 20% TSB broth (50 ml) to start the detection of growth curve. All inoculation broths were grown at 28°C with shaking at 200 r.p.m., and the OD$_{600 nm}$ value was determined every 2 h or 4 h until bacterial growth reached the stationary stage. Each sample involves three technical replicates and the experiment was performed three times.

## RNA extraction, qRT-PCR and RT-PCR

The wild-type OH11 and the *hfq* deletion mutant were grown on 20% TSB. The cells of the wild-type strain and the *hfq* mutant were collected at the time point, 8 h and 11 h, corresponds to the same cell density (OD$_{600 nm}$ = 1.0). Then, the total RNA was extracted from the cells of each strain using a kit with a code of R6950-01 from OMIGA Company (China). Next, the qRT-PCR and RT-PCR assays, including cDNA synthesis and PCR amplification were performed as described previously (Qian *et al.*, 2013; 2014). Primer sequences used in this assay are listed in Table S4.

## Extraction and detection of WAP-8294A2

Various *Lysobacter* strains were cultured in 20% TSB broth at 28°C until the cell density expressed by OD$_{600 nm}$ reached to 1.0. Next, 1 ml of each culture was transferred into the fresh 20% TSB broth (50 ml) for 3 days shaking culturing. Then, the extraction and HPLC analysis of WAP-8294A2 from *L. enzymogenes* were performed as described previously (Zhang *et al.*, 2011; 2014). Three replicates were used for each strain, and the experiment was performed three times.

## Quantitative determination of the activity of three extracellular lytic enzymes

A sterile filter membrane (10 mm diameter) was put on the surface of the selective plates for chitinase, protease and cellulase. The composition of each selective plates was described in previous studies (Kobayashi *et al.*, 2005; Qian *et al.*, 2013). In brief, 3 µl of bacterial culture of various *Lysobacter* strains with the same cell density (OD$_{600 nm}$ = 2.0) was spotted on the filter membrane. After 3 days of growth, the filter member was taken off from the plates. The diameters of the hydrolytic zones in the plates were measured and the corresponding hydrolytic areas were calculated. Meanwhile, the cells of each strain on the filter member were washed by 800-µL sterilized ddH$_2$O, and the cell density of each strain was measured and expressed by OD$_{600 nm}$. Finally, the ratio of the area of hydrolytic zones divided by cell density (OD$_{600 nm}$) was used as a quantitative indicator for each enzyme activity of the tested *Lysobacter* strains.

## Western blot

The flag-tagged *chiA* was amplified by PCR with the primers of *chiA*-F and *chiA*-R (flag) (Table S2), and cloned into the broad-host-vector pBBR1-MSC5 (Table 1). Then, the pBBR1-

MCS5 containing the flag-tagged *chiA* was transferred into wild-type OH11 and the *hfq* deletion mutant to generate two strains, OH11 (*chiA*-flag) and Δ*hfq* (*chiA*-flag). These strains were cultured in 50 ml of LB broth to grow until the value of OD$_{600\,nm}$ reached to 2.0. Then, the total cells and the culture supernatants of tested strains were collected respectively. Next, for cells, 1-ml RIPA buffer (CWBIO Company, China) with a code of 1713L was used to lyse cells, followed by a centrifugation (10 000 × *g* at 4°C for 10 min). These cell supernatants were collected and used for further study. For culture supernatants, they were concentrated to 1 ml using a vacuum freeze drier, and used for further studies. The following western blot assay was performed as described previously (Ansong *et al.*, 2009). The experiments were performed three times, and each involves three replicates for each treatment.

*Data submission*

The sequence data of the present study have been submitted to the National Center for Biotechnology Information Genbank under accession number KM186921(*miaA*), KM186922 (*hfq*), KM186923 (*hflx*), KM186924 (*hflk*) KM186925 (*hflc*) and KM186926 (*αlp*) respectively.

*Data analysis*

All analyses were conducted using SPSS 14.0 (SPSS Inc, Chicago, IL, USA). The hypothesis test of percentages (*t*-test, $P < 0.05$ or 0.01) was used to determine significant differences in the production of antibiotic metabolites, lytic-enzyme activity and gene expressions between the wild-type OH11 and its derivatives.

**Acknowledgement**

We thank Prof Hu Dongwei from Zhejiang University (China) for his support in electronic microscope.

**Conflict of interest**

None declared.

**References**

Ansong, C., Yoon, H., Porwollik, S., Mottaz-Brewer, H., Petritis, B.O., Jaitly, N., *et al.* (2009) Global systems-level analysis of Hfq and SmpB deletion mutants in *Salmonella*: implications for virulence and global protein translation. *PLoS ONE* **4:** e4809.

Boudry, P., Gracia, C., Monot, M., Caillet, J., Saujet, L., Hajnsdorf, E., *et al.* (2014) Pleiotropic role of the RNA chaperone protein Hfq in the human pathogen *Clostridium difficile. J Bacteriol* **196:** 3234–3248.

Butland, G., Peregrin-Alvarez, J.M., Li, J., Yang, W., Yang, X., Canadien, V., *et al.* (2005) Interaction network containing conserved and essential protein complexes in *Escherichia coli. Nature* **433:** 531–537.

Caswell, C.C., Gaines, J.M., and Roop, R.M. (2012) The RNA chaperone Hfq independently coordinates expression of the VirB type IV secretion system and the LuxR-type regulator BabR in *Brucella abortus* 2308. *J Bacteriol* **194:** 3–14.

Chambers, J.R., and Bender, K.S. (2011) The RNA chaperone Hfq is important for growth and stress tolerance in *Francisella novicida. PLoS ONE* **6:** e19797.

Chen, L., Hu, B.S., Qian, G.L., Wang, C., Yang, W.F., Han, Z.C., and Liu, F.Q. (2009) Identification and molecular characterization of twin-arginine translocation system (Tat) in *Xanthomonas oryzae pv. oryzae* strain PXO99. *Arch Microbiol* **191:** 163–170.

Christensen, P., and Cook, F.D. (1978) *Lysobacter*, a new genus of nonfruiting, gilding bacteria with a high base ratio. *Int J Syst Bacteriol* **28:** 367–393.

Epstein, D.M., and Wensink, P.C. (1988) The α-lytic protease gene of *Lysobacter enzymogenes*. The nucleotide sequence predicts a large prepro-peptide with homology to pro-peptides of other chymotrypsin-like enzymes. *J Biol Chem* **263:** 16586–16590.

Fantappiè, L., Metruccio, M.M.E., Seib, K.L., Oriente, F., Cartocci, E., Ferlicca, F., *et al.* (2009) The RNA chaperone Hfq is involved in stress response and virulence in *Neisseria meningitidis* and is a pleiotropic regulator of protein expression. *Infect Immun* **77:** 1842–1853.

Franze de Fernandez, M.T., Eoyang, L., and August, J.T. (1968) Factor fraction required for the synthesis of bacteriophage Qbeta-RNA. *Nature* **219:** 588–590.

Geinguenaud, F., Calandrini, V., Teixeira, J., Mayer, C., Liquier, J., Lavelle, C., and Arluison, V. (2011) Conformational transition of DNA bound to Hfq probed by infrared spectroscopy. *Phys Chem Chem Phys* **13:** 1222–1229.

Geng, J., Song, Y., Yang, L., Feng, Y., Qiu, Y., Li, G., *et al.* (2009) Involvement of the post-transcriptional regulator Hfq in *Yersinia pestis* virulence. *PLoS ONE* **4:** e6213.

Hoang, T.T., Karkhoff-Schweizer, R.R., Kutchma, A.J., and Schweizer, H.P. (1998) A broad-host-range Flp-FRT recombination system for site-specific excision of chromosomally-located DNA sequences: application for isolation of unmarked *Pseudomonas aeruginosa* mutants. *Gene* **212:** 77–86.

Kadzhaev, K., Zingmark, C., Golovliov, I., Bolanowski, M., Shen, H., Conlan, W., and Sjöstedt, A. (2009) Identification of genes contributing to the virulence of *Francisella tularensis* SCHU S4 in a mouse intradermal infection model. *PLoS ONE* **4:** e5463.

Kajitani, M., and Ishihama, A. (1991) Identification and sequence determination of the host factor gene for bacteriophage Q beta. *Nucleic Acids Res* **19:** 1063–1066.

Kato, A., Nakaya, S., Kokubo, N., Aiba, Y., Ohashi, Y., Hirata, H., *et al.* (1998) A new anti-MRSA antibiotic complex, WAP-8294A. I. Taxonomy, isolation and biological activities. *J Antibiot (Tokyo)* **51:** 929–935.

Kharade, S.S., and McBride, M.J. (2014) *Flavobacterium johnsoniae* chitinase ChiA is required for chitin utilization and is secreted by the type IX secretion system. *J Bacteriol* **196:** 961–970.

Kobayashi, D.Y., Reedy, R.M., Palumbo, J.D., Zhou, J.M., and Yuen, G.Y. (2005) A *clp* gene homologue belonging to the *crp* gene family globally regulates lytic enzyme production, antimicrobial activity, and biological control activity expressed by *Lysobacter enzymogenes* strain C3. *Appl Environ Microbiol* **71**: 261–269.

Kovach, M.E., Elzer, P.H., Hill, D.S., Robertson, G.T., Farris, M.A., Roop, R.M., 2nd, and Peterson, K.M. (1995) Four new derivatives of the broad-host-range cloning vector pBBR1MCS, carrying different antibiotic-resistance cassettes. *Gene* **166**: 175–176.

Lee, T., and Feig, A.L. (2008) The RNA binding protein Hfq interacts specifically with tRNAs. *RNA* **14**: 514–523.

Lou, L.L., Qian, G.L., Xie, Y.X., Hang, J.L., Chen, H.T., Zaleta-Rivera, K., *et al.* (2011) Biosynthesis of HSAF, a tetramic acid-containing macrolactam from *Lysobacter enzymogenes*. *J Am Chem Soc* **133**: 643–645.

Nielsen, J.S., Larsen, M.H., Lillebaek, E.M.S., Bergholz, T.M., Christiansen, M.H.G., Boor, K.J., *et al.* (2011) A small RNA controls expression of the chitinase ChiA in *Listeria monocytogenes*. *PLoS ONE* **6**: e19019.

Palumbo, J.D., Sullivan, R.F., and Kobayashi, D.Y. (2003) Molecular characterization and expression in *Escherichia coli* of three β-1, 3-glucanae genes from *Lysobacter enzymogenes* strain N4-7. *J Bacteriol* **185**: 4362–4370.

Qian, G.L., Hu, B.S., Jiang, Y.H., and Liu, F.Q. (2009) Identification and characterization of *Lysobacter enzymogenes* as a biological control agent against some fungal pathogens. *Agric Sci China* **8**: 68–75.

Qian, G.L., Wang, Y.S., Qian, D.Y., Fan, J.Q., Hu, B.S., and Liu, F.Q. (2012) Selection of available suicide vectors for gene mutagenesis using *chiA* (a chitinase encoding gene) as a new reporter and primary functional analysis of *chiA* in *Lysobacter enzymogenes* strain OH11. *World J Microbiol Biotechnol* **28**: 549–557.

Qian, G.L., Wang, Y.L., Liu, Y.R., Xu, F.F., He, Y.W., Du, L.C., *et al.* (2013) *Lysobacter enzymogenes* uses two distinct cell-cell signaling systems for differential regulation of secondary-metabolite biosynthesis and colony morphology. *Appl Environ Microbiol* **79**: 6604–6616.

Qian, G.L., Xu, F.F., Venturi, V., Du, L.C., and Liu, F.Q. (2014) Roles of a solo luxR in the biological control agent *Lysobacter enzymogenes* strain OH11. *Phytopathology* **104**: 224–231.

Ramos, C.G., Sousa, S.A., Grilo, A.M., Feliciano, J.R., and Leitão, J.H. (2011) The second RNA chaperone, Hfq2, is also required for survival under stress and full virulence of *Burkholderia cenocepacia* J2315. *J Bacteriol* **193**: 1515–1526.

Schiano, C.A., Bellows, L.E., and Lathem, W.W. (2010) The small RNA chaperone Hfq is required for the virulence of *Yersinia pseudotuberculosis*. *Infect Immun* **78**: 2034–2044.

Sittka, A., Sharma, C.M., Rolle, K., and Vogel, J. (2009) Deep sequencing of *Salmonella* RNA associated with heterologous Hfq proteins in vivo reveals small RNAs as a major target class and identifies RNA processing phenotypes. *RNA Biol* **6**: 266–275.

Tsui, H.C., Leung, H.C., and Winkler, M.E. (1994) Characterization of broadly pleiotropic phenotypes caused by an *hfq* insertion mutation in *Escherichia coli* K-12. *Mol Microbiol* **13**: 35–49.

Updegrove, T.B., Correia, J.J., Galletto, R., Bujalowski, W., and Wartell, R.M. (2010) *E. coli* DNA associated with iso-lated Hfq interacts with Hfq's distal surface and C-terminal domain. *Biochim Biophys Acta* **1799**: 588–596.

Wang, G.H., Huang, X.Q., Li, S.N., Huang, J.F., Wei, X., Li, Y., and Xu, Y. (2012) The RNA chaperone Hfq regulates antibiotic biosynthesis in the rhizobacterium *Pseudomonas aeruginosa* M18. *J Bacteriol* **194**: 2443–2457.

Wang, Y.S., and Qian, G.L. (2012) Deletion and primary functional analysis of α-lytic protease gene in *Lysobacter enzymogenes* OH11. *Sciencepaper Online*. URL http://www.docin.com/p-390563768.html.

Wang, Y.S., Zhao, Y.X., Zhang, J., Zhao, Y.Y., Shen, Y., Su, Z.H., *et al.* (2014) Transcriptomic analysis reveals new regulatory roles of Clp signaling in secondary metabolite biosynthesis and surface motility in *Lysobacter enzymogenes* OH11. *Appl Microbiol Biotechnol* **98**: 9009–9020.

Wu, X.G., Duan, H.M., Tian, T., Yao, N., Zhou, H.Y., and Zhang, L.Q. (2010) Effect of the *hfq* gene on 2, 4 – diacetylphloroglucinol production and the Pcol/PcoR quorum-sensing system in *Pseudomonas fluorescens* 2P24. *FEMS Microbiol Lett* **309**: 16–24.

Xie, Y., Wright, S., Shen, Y.M., and Du, L.C. (2012) Bioactive natural products from *Lysobacter*. *Nat Prod Rep* **29**: 1277–1287.

Zhang, J., Du, L.C., Liu, F.Q., Xu, F.F., Hu, B.S., Venturi, V., and Qian, G.L. (2014) Involvement of both PKS and NRPS in antibacterial activity in *Lysobacter enzymogenes* OH11. *FEMS Microbiol Lett* **355**: 170–176.

Zhang, W., Li, Y.Y., Qian, G.L., Wang, Y., Chen, H.T., Li, Y.Z., *et al.* (2011) Identification and characterization of the anti-methicillin-resistant *Staphylococcus aureus* WAP-8294A2 biosynthetic gene cluster from *Lysobacter enzymogenes* OH11. *Antimicrob Agents Chemother* **55**: 5581–5589.

Zhang, Z., and Yuen, G.Y. (2000) The role of chitinase production by *Stenotrophomonas maltophilia* Strain C3 in biological control of *Bipolaris sorokiniana*. *Phytopathology* **90**: 384–389.

## Supporting information

Additional Supporting Information may be found in the online version of this article at the publisher's web-site:

**Fig. S1.** The scheme of the *hfq* mutant construction and molecular confirmation in *Lysobacter enzymogenes*.
A. Physical map of deletion mutant construction used in this study. The 364-bp (amplified by *hfq*-F1/R1) (Table S2) and 368-bp (amplified by *hfq*-F2/R2) (Table S2) DNA fragment of *hfq* was used as 5′ and 3′ fragment for homologue recombination respectively. The internal 155-bp DNA fragment would be deleted in the *hfq* mutant. The primers *hfq*-F1/R2 (Table S2) was used for molecular confirmation of *hfq* mutant. B. PCR verification of the *hfq* mutant. An 887-bp DNA fragment was amplified from the wild-type OH11 with the primers *hfq*-F1/R2, while only a 732-bp fragment was obtained from the *hfq* mutant with the same primers due to the deletion of internal 155-bp DNA fragment. '-' in B represents the blank control. This strategy was applied into the construction of other mutants of target genes in the present study (Table S1 and Table S2).

**Fig. S2.** Comparison of the cell size between the wild-type strain and the *hfq* mutant of *Lysobacter enzymogenes*.

A. The representative result of cell size from three independent experiments between the wild-type strain (OH11) and the *hfq* mutant (Δ*hfq*) under electronic microscope at two selected time points (24 h and 48 h after growth on solid 20% TSA medium). The scale bars represent 1 μm.

B. Statistic analysis of the cell size of the wild-type strain and the *hfq* mutant. In each technological repeat, at least 15 cells of each strain were selected for analysis. The experiment was performed three times. Each column indicates the mean of three biologically independent experiments. Vertical bars represent standard errors. No significant difference ($P < 0.05$; *t*-test) in cell size between the wild-type strain and the *hfq* mutant was found.

**Fig. S3.** Further determination of extracellular chitinase activity between the wild-type strain and the *hfq* mutant of *Lysobacter enzymogenes* on solid medium.

A. The representative phenotype of extracellular chitinase activity between the wild-type OH11 and the *hfq* mutant from three independent experiments.

B. Quantitative analysis of cell density ($OD_{600 nm}$) of each strain on the surface of the filter membrane. Each column indicates the mean of three biologically independent experiments. Vertical bars represent standard errors. '**' ($P < 0.01$; *t*-test) above the bars indicate a significant difference between the wild-type strain and the *hfq* mutant. The initial inoculated cell concentration expressed by $OD_{600 nm}$ for the wild-type OH11, Δ*hfq*-a, Δ*hfq*-b was 2, 2 and 20 respectively.

**Fig. S4.** Detection of the chitinase activity of three mutants of predicted chitinase synthesis genes (Δ*chiA*, Δ*chiB*, Δ*chiC*) on chitin plate. Only the *chiA* mutant cannot hydrolyse chitin under the tested conditions. The gene information of *chiB* and *chiC* was provided in Table S3. The mutant construction and confirmation was provided in Table S1.

**Fig. S5.** Introduction of the broad-host-vector pBBR1-MCS5 containing *hfq* partially restored extracellular chitinase production of the *hfq* mutant in *Lysobacter enzymogenes*. OH11, the wild-type strain of *L. enzymogenes*; Δ*hfq*, the *hfq* deletion mutant; Δ*hfq*(pBBR), the *hfq* mutant containing the empty vector (pBBR1-MCS5); Δ*hfq*(*hfq*-pBBR), the pBBR1-MCS5-based *hfq* complemented strain. Figure S5 is the representative result of three independent experiments.

**Fig. S6.** Determination of extracellular chitinase activities of *Lysobacter* strains.

A. The construct of flag-tagged *chiA* restored the chitinase activity of the *chiA* mutant.

B. The *hfq* mutant containing the construct of flag-tagged *chiA* did not restore the chitinase activity. OH11, the wild-type strain of *L. enzymogenes*; Δ*chiA*, the *chiA* deletion mutant (Qian *et al.*, 2012); Δ*chiB*, the *chiB* deletion mutant; Δ*chiC*, the *chiC* deletion mutant; Δ*chiA*(*chiA*-flag), the *chiA* mutant containing flag-tagged *chiA*; Δ*hfq*(pBBR), the *hfq* mutant containing an original pBBR1-MSC5 vector; Δ*hfq*(*hfq*-pBBR), the *hfq* mutant containing *hfq*-pBBR complemented vector; Δ*hfq*(*chiA*-flag), the *hfq* mutant containing flag-tagged *chiA*; OH11(*chiA*-flag), the wild-type containing flag-tagged *chiA*.

**Table S1.** Mutant confirmation by PCR in this study.

**Table S2.** Primers used for in-frame deletion and complementation in this study.

**Table S3.** Three chitinase encoding genes in *Lysobacter enzymogenes*.

**Table S4.** Primers used for qRT-PCR or RT-PCR in this study.

# 17

# Production of 2-ketoisocaproate with *Corynebacterium glutamicum* strains devoid of plasmids and heterologous genes

**Michael Vogt, Sabine Haas, Tino Polen,
Jan van Ooyen\* and Michael Bott\*\***
*Institute of Bio- and Geosciences, IBG-1: Biotechnology,
Forschungszentrum Jülich, D-52425 Jülich, Germany.*

## Summary

2-Ketoisocaproate (KIC), the last intermediate in
L-leucine biosynthesis, has various medical and
industrial applications. After deletion of the *ilvE* gene
for transaminase B in L-leucine production strains of
*Corynebacterium glutamicum*, KIC became the major
product, however, the strains were auxotrophic for
L-isoleucine. To avoid auxotrophy, reduction of IlvE
activity by exchanging the ATG start codon of *ilvE* by
GTG was tested instead of an *ilvE* deletion. The resulting strains were indeed able to grow in glucose
minimal medium without amino acid supplementation, but at the cost of lowered growth rates and KIC
production parameters. The best production performance was obtained with strain MV-KICF1, which
carried besides the *ilvE* start codon exchange three
copies of a gene for a feedback-resistant 2-
isopropylmalate synthase, one copy of a gene for a
feedback-resistant acetohydroxyacid synthase and
deletions of *ltbR* and *iolR* encoding transcriptional
regulators. In the presence of 1 mM L-isoleucine,
MV-KICF1 accumulated 47 mM KIC (6.1 g l$^{-1}$) with a
yield of 0.20 mol/mol glucose and a volumetric productivity of 1.41 mmol KIC l$^{-1}$ h$^{-1}$. Since MV-KICF1 is
plasmid free and lacks heterologous genes, it is an
interesting strain for industrial application and as
platform for the production of KIC-derived compounds, such as 3-methyl-1-butanol.

For correspondence. \*E-mail j.van.ooyen@fz-juelich.de
\*\*E-mail m.bott@fz-juelich.de

**Funding Information** No funding information provided.

## Introduction

*Corynebacterium glutamicum* is the major host for
biotechnological production of amino acids, the most
important ones being the flavor enhancer L-glutamate
and the feed additive L-lysine. In the past decades,
*C. glutamicum* strains have been developed for the production of various other commercially interesting compounds (Becker and Wittmann, 2012), including organic
acids (Okino *et al.*, 2008; Litsanov *et al.*, 2012a,b;
Wieschalka *et al.*, 2013), diamines (Mimitsuka *et al.*,
2007; Kind and Wittmann, 2011; Schneider and
Wendisch, 2011) or alcohols (Inui *et al.*, 2004; Smith
*et al.*, 2010; Blombach *et al.*, 2011). Besides small molecules, also heterologous proteins can be efficiently produced with this Gram-positive bacterium (Scheele *et al.*,
2013, and references therein). Thus, *C. glutamicum* has
become a production platform in white biotechnology.
Three monographs (Eggeling and Bott, 2005; Burkovski,
2008; Yukawa and Inui, 2013) document the rapidly
increasing knowledge on this species, which is based on
the genome sequence (Ikeda and Nakagawa, 2003;
Kalinowski *et al.*, 2003) and efficient techniques for its
genetic engineering (Kirchner and Tauch, 2003).

The spectrum of amino acids produced with
*C. glutamicum* includes the essential branched-chain
amino acids (BCAAs) L-valine, L-isoleucine and L-leucine,
which are produced in quantities of up to 5000 tons per
year in a steadily growing market (Becker and Wittmann,
2012). They have different applications in the food, feed
and pharmaceutical industry (Park and Lee, 2010). The
biosynthesis pathways of the BCAAs in *C. glutamicum* are
overlapping and partly share the same precursors and
enzymes (Fig. 1). The direct precursors of L-valine,
L-isoleucine and L-leucine are 2-ketoisovalerate (KIV),
2-keto-3-methylvalerate (KMV), and 2-ketoisocaproate
(KIC) respectively. These keto acids are predominantly
transaminated to the respective amino acids by the
transaminase IlvE (Radmacher *et al.*, 2002; Marienhagen
*et al.*, 2005). Similar to their corresponding amino acids,
KIV, KMV and KIC have a variety of applications in the
medical, biological and food area, since they play an

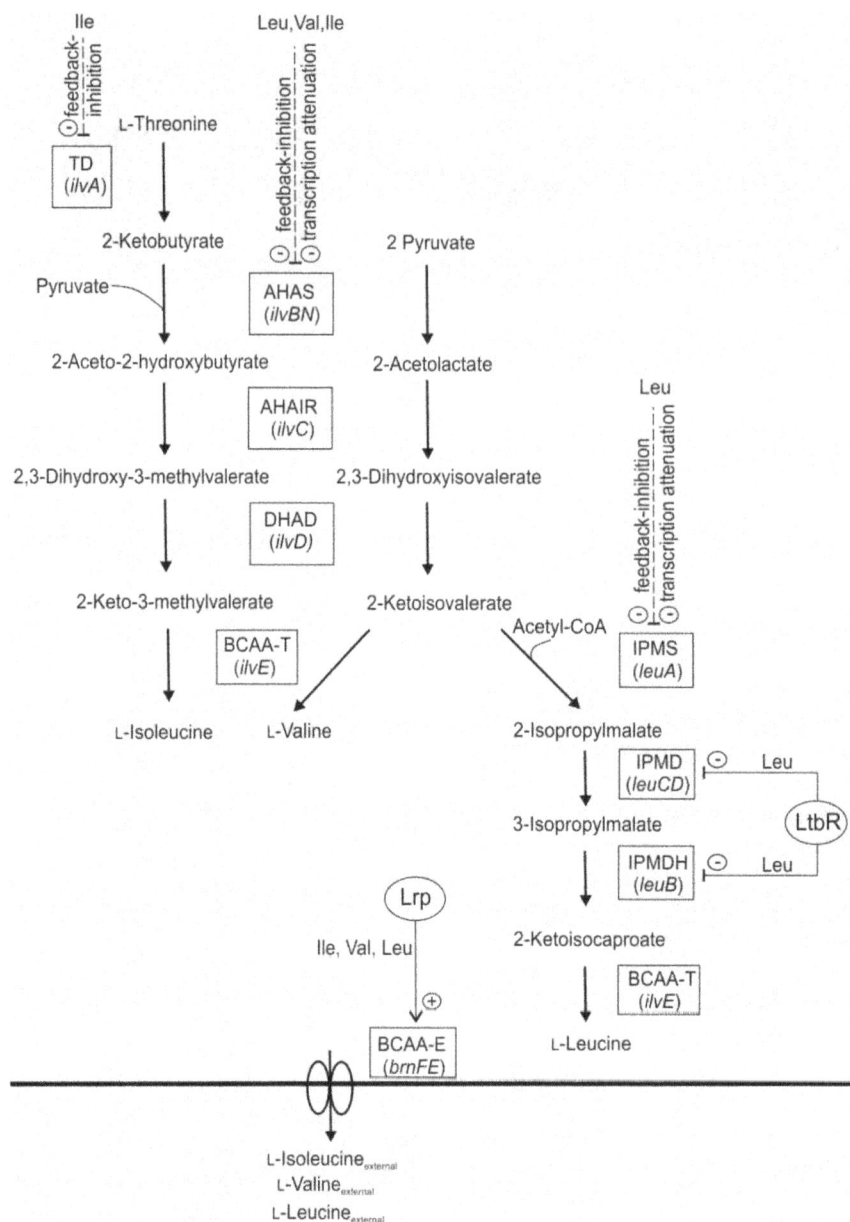

**Fig. 1.** Biosynthesis pathways and their control by various regulatory mechanisms of the three branched-chain amino acids and the respective keto acids in *C. glutamicum*. Enzymes and their corresponding genes are shown in boxes. Lines with '+' indicate activation of gene expression; '-' indicates repression of gene expression (solid lines) or transcription attenuation or feedback inhibition (dashed lines). 'Leu', 'Val' and 'Ile' indicate the presence of L-leucine, L-valine and L-isoleucine respectively. Not shown is the *avtA* gene encoding the branched-chain amino acid transaminase AvtA, which predominantly transaminates 2-ketoisovalerate to L-valine. Abbreviations: AHAIR, acetohydroxyacid isomeroreductase; AHAS, acetohydroxyacid synthase; BCAA-E, branched-chain amino acid exporter (BrnFE); BCAA-T, branched-chain amino acid transaminase IlvE; DHAD, dihydroxyacid dehydratase; IPMD, 3-isopropylmalate dehydratase; IPMDH, 3-isopropylmalate dehydrogenase; IPMS, 2-isopropylmalate synthase; Lrp, leucine-responsive regulatory protein; LtbR, leucine and tryptophane biosynthesis regulator; TD, threonine dehydratase (threonine ammonia-lyase).

important role in living organisms as regulatory factors in metabolism and key intermediates in biosynthesis (Krause *et al.*, 2010; Zhu *et al.*, 2011; Bückle-Vallant *et al.*, 2014). They are used, for example, in the therapy of chronic kidney disease patients (Aparicio *et al.*, 2012). Similar to L-leucine, KIC has anti-catabolic properties through inhibition of muscle proteolysis and provokes

enhancement of protein synthesis, especially in the skeletal muscle (Escobar *et al.*, 2010; Zanchi *et al.*, 2011). Additionally, an insulin-releasing action of KIC (Heissig *et al.*, 2005) and an inhibitory effect on glucagon release (Leclercq-Meyer *et al.*, 1979) were discussed. It has been shown that KIC can also serve as a basis for the production of the biofuel isopentanol (Cann and Liao, 2010).

KIV, KMV and KIC are mainly produced by chemical synthesis using harsh reaction conditions and multiple purification steps resulting in plenty of waste (Cooper et al., 1983). The biotechnological production of these keto acids is thus an interesting alternative. Besides a biotransformation process with *Rhodococcus opacus* using L-leucine as substrate for KIC formation (Zhu et al., 2011), fermentative processes with glucose as substrate have recently been described for the production of KIV (Krause et al., 2010) and KIC (Bückle-Vallant et al., 2014), showing that deletion of *ilvE* in certain engineered *C. glutamicum* strains results in KIC formation. Whereas these strains contained plasmids and in part heterologous genes, the *C. glutamicum* KIC production strains developed in our work are plasmid free and lack heterologous genes.

## Results and discussion

### Initial studies on KIC production using plasmid-containing strains of C. glutamicum

Based on recently developed efficient production strains of *C. glutamicum* ATCC 13032 (Abe et al., 1967) for L-leucine (Vogt et al., 2014), we intended to modify these strains for the production of KIC. The conversion of KIC to L-leucine is catalysed by the transaminase IlvE, which also converts KIV to L-valine and KMV to L-isoleucine using L-glutamate as amino donor (Radmacher et al., 2002; Marienhagen et al., 2005). An *ilvE* deletion has been reported to cause auxotrophy for L-leucine and L-isoleucine, but not for L-valine, since the transaminase AvtA also effectively converts KIV to L-valine using L-alanine as amino donor (Marienhagen et al., 2005). According to this knowledge, deletion of *ilvE* in L-leucine production strains should lead to the accumulation of KIC and potentially also KMV. In a first series of experiments, we deleted *ilvE* in the wild-type *C. glutamicum* ATCC 13032 and transformed the Δ*ilvE* mutant with plasmid pAN6-*leuA*_B018, carrying an IPTG-inducible *leuA* allele encoding a feedback-resistant 2-isopropylmalate synthase (IPMS) (Vogt et al., 2014). 2-Isopropylmalate synthase of *C. glutamicum* is strongly inhibited by L-leucine with a $K_i$ of 0.4 mM (Pátek et al., 1994) and the presence of a feedback-resistant variant is the key for L-leucine overproduction (Vogt et al., 2014). The Δ*ilvE* mutant and the Δ*ilvE* strain with plasmid pAN6-*leuA*_B018 were cultivated in 500 ml baffled Erlenmeyer flasks with 50 ml CGXII minimal medium (Keilhauer et al., 1993) with 4% (w/v) glucose, 1 mM L-leucine and 1 mM L-isoleucine at 30°C and 120 rpm on a rotary shaker. Keto acids and amino acids were quantified by high-performance liquid chromatography as described (Vogt et al., 2014). Chromosomal in-frame deletions and integrations of DNA fragments were performed by two-step homologous recombination using the vector pK19*mobsacB* (Schäfer et al., 1994) and a method described previously (Niebisch and Bott, 2001).

*Corynebacterium glutamicum* Δ*ilvE* exhibited a growth rate of $0.38 \pm 0.01$ h$^{-1}$ and excreted up to 5 mM KIV but no detectable concentrations of KIC (detection limit < 0.1 mM), whereas *C. glutamicum* Δ*ilvE* pAN6-*leuA*_B018 showed a growth rate of $0.30 \pm 0.01$ h$^{-1}$ and accumulated $37 \pm 0.7$ mM KIC in the supernatant when induced with 0.1 mM IPTG, confirming that over-expression of the *leuA* allele encoding the feedback-resistant IPMS increased metabolic flux into the leucine pathway (Fig. 1). Surprisingly, *C. glutamicum* Δ*ilvE* carrying pAN6-*leuA*_B018 also accumulated L-leucine ($12.3 \pm 0.4$ mM) and in fact was only auxotrophic for L-isoleucine, but not for L-leucine. A possible limitation of L-valine due to a high metabolic flux from KIV towards KIC was excluded for this strain since additional supplementation of L-valine did not improve growth (data not shown). Accumulation of L-leucine was also reported for other KIC-producing Δ*ilvE* strains and explained by the activity of unspecific transaminases (e.g. AlaT or AvtA) using KIC as substrate when it is present in high concentrations (Bückle-Vallant et al., 2014). Consequently, supplementation of the medium with L-leucine was omitted in the following cultivations. The formation of L-leucine as by-product additionally necessitates the presence of feedback-resistant IPMS for KIC overproduction. The results described above demonstrated that our previously described L-leucine producers (Vogt et al., 2014) can serve as basis for the construction of KIC production strains.

### Deletion of ilvE in plasmid-free L-leucine production strains

Analogous to our strategy used for L-leucine strain development (Vogt et al., 2014), we intended to construct KIC production strains devoid of plasmids, heterologous genes and auxotrophies. Depending on the composition of the medium used in the fermentation process, auxotrophies can necessitate the addition of supplements, increasing the costs of the fermentation process. Plasmids usually necessitate the addition of antibiotics to the medium, which is undesirable for production strains applied in the food and feed industry and can be prohibited by regulatory authorities (Tauch et al., 2002). Moreover, the absence of plasmids, antibiotic resistance markers and heterologous genes often results in more stable producer strains (Pátek, 2007). The use of heterologous genes is also an undesired trait for strains used in the food and feed industry. In a first attempt to construct a plasmid-free KIC producer, we deleted the *ilvE* gene in the previously constructed L-leucine producer MV-Leu20 (Table 1; Vogt et al., 2014), which contains a deletion of the *ltbR* gene, encoding a

**Table 1.** Strains and plasmids used in this study[a,b].

| Strain or plasmid | Relevant characteristics[c] | Source or reference |
|---|---|---|
| *C. glutamicum* strains | | |
| Wild type | ATCC 13032, biotin-auxotrophic | Abe and colleagues (1967) |
| Δ*ilvE* | ATCC 13032 derivative with in-frame deletion of *ilvE* | Marienhagen and colleagues (2005) |
| MV-Leu20 | Rationally designed *C. glutamicum* L-leucine producer (Δ*ltbR* Δ*leuA*::P*tuf-leuA*_B018) | Vogt and colleagues (2014) |
| MV-Leu20 Δ*ilvE* | MV-Leu20 derivative with in-frame deletion of *ilvE* | This study |
| SH-KIC20 | MV-Leu20 derivative with chromosomal replacement of ATG start codon of *ilvE* by GTG start codon | This study |
| MV-LeuF1 | Rationally designed *C. glutamicum* L-leucine producer (Δ*ltbR*::P*tuf-leuA*_B018 Δ*leuA*::P*tuf-leuA*_B018 IR(cg1121/1122)::P*tuf-leuA*_B018 Δ*iolR ilvN*_fbr) | Vogt and colleagues (2014) |
| MV-KICF1 | MV-LeuF1 derivative with chromosomal replacement of ATG start codon of *ilvE* by GTG start codon | This study |
| Δ*ilvE* Δcg0018 | Δ*ilvE* derivative with cg0018 in-frame deletion | This study |
| Δ*ilvE* Δcg1121 | Δ*ilvE* derivative with cg1121 in-frame deletion | This study |
| Δ*ilvE* Δcg1219 | Δ*ilvE* derivative with cg1219 in-frame deletion | This study |
| Δ*ilvE* Δcg1419 | Δ*ilvE* derivative with cg1419 in-frame deletion | This study |
| Δ*ilvE* Δcg1658 | Δ*ilvE* derivative with cg1658 in-frame deletion | This study |
| Δ*ilvE* Δcg2557 | Δ*ilvE* derivative with cg2557 in-frame deletion | This study |
| Δ*ilvE* Δcg2676 | Δ*ilvE* derivative with cg2676 in-frame deletion | This study |
| Δ*ilvE* Δcg3334 | Δ*ilvE* derivative with cg3334 in-frame deletion | This study |
| Δ*ilvE* Δcg1121::cg1121 | Δ*ilvE* Δcg1121 derivative with re-integrated gene cg1121 into its wild-type locus | This study |
| *E. coli* strains | | |
| DH5α | F⁻ Φ80*lacZ*ΔM15 Δ(*lacZYA-arg*F)U169 *rec*A1 *end*A1 *hsd*R17 (r$_K$⁻, m$_K$⁺) *phoA* *sup*E44 λ⁻ *thi*-1 *gyr*A96 *rel*A1 | Invitrogen (Karlsruhe, Germany) |
| Plasmids | | |
| pAN6 | Kan$^r$; *E. coli*/*C. glutamicum* shuttle vector for inducible gene expression (P$_{tac}$, *lacI*$^q$, pBL1 pUC18 *oriV*$_{E.coli}$, pBL1 *oriV*$_{C.glutamicum}$) | Frunzke and colleagues (2008) |
| pAN6-*leuA*_B018 | Kan$^r$; pAN6 derivative containing *leuA* allele coding for feedback-resistant 2-isopropylmalate synthase under control of the *tac* promoter | Vogt and colleagues (2014) |
| pAN6-*leuA*_B018-cg1121 | Kanr; pAN6-*leuA*_B018 derivative carrying additional gene coding for cg1121 along with its upstream (94 bp) and downstream (305 bp) regions | This study |
| pK19*mobsacB* | Kan$^r$; vector for allelic exchange in *C. glutamicum* (pK18 *oriV*$_{E.coli}$ *sacB lacZα*) | Schäfer and colleagues (1994) |
| pK19*mobsacB*-Δ*ilvE* | Kan$^r$, pK19*mobsacB* derivative for in-frame deletion of gene *ilvE* | Marienhagen and colleagues (2005) |
| pK19*mobsacB*-GTG-*ilvE* | Kan$^r$, pK19*mobsacB* derivative for replacement of ATG start codon of *ilvE* by GTG | This study |
| pK19*mobsacB*-Δcg0018 | Kan$^r$, pK19*mobsacB* derivative for in-frame deletion of gene coding for Cg0018 | This study |
| pK19*mobsacB*-Δcg1121 | Kan$^r$, pK19*mobsacB* derivative for in-frame deletion of gene coding for Cg1121 | This study |
| pK19*mobsacB*-Δcg1219 | Kan$^r$, pK19*mobsacB* derivative for in-frame deletion of gene coding for Cg1219 | This study |
| pK19*mobsacB*-Δcg1419 | Kan$^r$, pK19*mobsacB* derivative for in-frame deletion of gene coding for Cg1419 | This study |
| pK19*mobsacB*-Δcg1658 | Kan$^r$, pK19*mobsacB* derivative for in-frame deletion of gene coding for Cg1658 | This study |
| pK19*mobsacB*-Δcg2557 | Kan$^r$, pK19*mobsacB* derivative for in-frame deletion of gene coding for Cg2557 | This study |
| pK19*mobsacB*-Δcg2676 | Kan$^r$, pK19*mobsacB* derivative for in-frame deletion of gene coding for Cg2676 | This study |
| pK19*mobsacB*-Δcg3334 | Kan$^r$, pK19*mobsacB* derivative for in-frame deletion of gene coding for Cg3334 | This study |
| pK19*mobsacB*-cg1121 | Kan$^r$, pK19*mobsacB* derivative for re-integration of gene coding for Cg1121 into its wild-type locus | This study |

**a.** All constructed plasmids as well as chromosomal deletions and integrations in engineered strains were verified by DNA sequencing.
**b.** Plasmid constructions were performed in *E. coli* DH5α. Description of plasmid constructions and used DNA oligonucleotides (Table S1) can be found in the Supporting Information.
**c.** Kan$^r$, kanamycin resistance

repressor of the L-leucine biosynthesis genes, and a replacement of the wild-type *leuA* gene by the feedback-resistant variant *leuA*_B018 under control of the strong *tuf* promoter (Vogt *et al.*, 2014). In shake flask cultivations with CGXII medium containing 4% (w/v) glucose, MV-Leu20 accumulated about 20 mM L-leucine. When cultivated in the same medium supplemented with 1 mM L-isoleucine, the strain MV-Leu20 Δ*ilvE* accumulated 18.0 ± 1.6 mM KIC in the supernatant and formed as by-products 5.6 ± 0.3 mM L-leucine, 2.1 ± 0.5 mM KIV and 7.3 ± 1.7 mM KMV. Since KIV and KMV are substrates of the tran-

saminase IlvE (Marienhagen *et al.*, 2005), the *ilvE* deletion leads to an accumulation of these keto acids. The presumably low concentrations of L-isoleucine and L-valine in strain MV-Leu20 Δ*ilvE* may also contribute to overproduction of KIV and KMV by reducing the feedback-inhibition of threonine dehydratase (encoded by *ilvA*) by L-isoleucine (Möckel *et al.*, 1992) and of acetohydroxyacid synthase (encoded by *ilvBN*) by L-valine and L-isoleucine (Eggeling *et al.*, 1987).

As mentioned above, L-leucine formation in the absence of the transaminase IlvE is presumably due to high cytoplasmic KIC concentrations, allowing its conversion by other transaminases such as AvtA that have weak affinities for KIC (Marienhagen et al., 2005). To test this assumption, we measured the cytoplasmic KIC concentrations in the wild type and strain MV-Leu20 $\Delta ilvE$. Cells were grown in CGXII medium with 4% (w/v) glucose, harvested in the early exponential phase (optical density at 600 nm ($OD_{600} = 5$), and cytoplasmic concentrations were determined as described (Paczia et al., 2012). The internal KIC concentration of the wild type was below the detection limit of 5 μM, whereas MV-Leu20 $\Delta ilvE$ accumulated approximately 2.8 mM KIC inside the cell, corresponding to a more than 500-fold increase.

### Exchange of the ilvE start codon in plasmid-free L-leucine production strains

To avoid the L-isoleucine auxotrophy of strain MV-Leu20 $\Delta ilvE$, we intended to reduce IlvE activity to a value that was high enough to provide L-isoleucine for growth, but low enough to allow KIC overproduction. For this purpose, the ATG start codon of ilvE was exchanged against GTG, which should decrease the translation rate (Becker et al., 2010) of the ilvE transcript and thereby reduce the specific IlvE activity. The start codon exchange was performed in the L-leucine producers MV-Leu20 and MV-LeuF1 (Table 1). The latter strain contains (i) three copies of the leuA_B018 gene in the chromosome under control of the tuf promoter, two of them replacing ltbR and the native leuA gene, (ii) a deletion of iolR (Klaffl et al., 2013) for enhanced glucose uptake and (iii) a feedback-resistant acetohydroxyacid synthase encoded by ilvN_fbr (Vogt et al., 2014). The strains SH-KIC20 (from MV-Leu20) and MV-KICF1 (from MV-LeuF1) resulting from the ilvE start codon exchange (Table 1) were cultivated in CGXII medium with 4% (w/v) glucose to test for KIC accumulation in the supernatant (Fig. 2).

Without supplementation of L-isoleucine, SH-KIC20 showed a strongly reduced growth rate of $0.08 \pm 0.01$ $h^{-1}$ compared with the ancestor strain MV-Leu20 ($0.31 \pm 0.01$ $h^{-1}$). This phenotype suggests that the ilvE start codon exchange reduced the availability of BCAAs and consequently the growth rate. Strain SH-KIC20 accumulated about 19 mM KIC in 49 h, which correlates with the KIC concentration produced by MV-Leu20 $\Delta ilvE$. The slower growth of SH-KIC20 led to a lowered volumetric productivity of 0.38 mmol $l^{-1}$ $h^{-1}$ in comparison to MV-Leu20 $\Delta ilvE$ (approximately 0.8 mmol $l^{-1}$ $h^{-1}$). Strain MV-KICF1 showed a growth rate of only $0.03 \pm 0.01$ $h^{-1}$, which is 85% lower than the one of the parent strain ($\mu = 0.20 \pm 0.01$ $h^{-1}$), and yielded a maximal KIC concentration of 33 mM after 95 h (Table 2). Supplementation of the medium with 1 mM L-isoleucine enabled MV-KICF1 to reach the same growth rate ($\mu = 0.21 \pm$

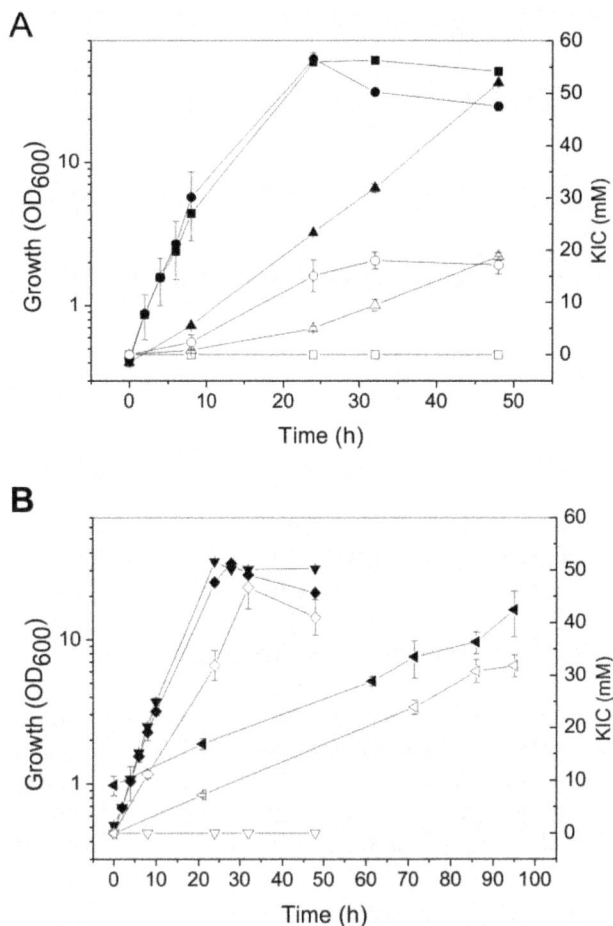

**Fig. 2.** Growth and KIC formation of different C. glutamicum strains in shake flasks with CGXII minimal medium containing 4% (w/v) glucose.
A. MV-Leu20 without supplements (growth, ■; KIC, □), MV-Leu20 $\Delta ilvE$ supplemented with 1 mM L-isoleucine (growth, ●; KIC, ○) and SH-KIC20 without supplements (growth, ▲; KIC, △).
B. MV-LeuF1 without supplements (growth, ▼; KIC, ▽), MV-KICF1 without supplements (growth, ◄; KIC, ◁) and MV-KICF1 with 1 mM L-isoleucine (growth, ◆; KIC, ◇). The data represent mean values and standard deviations obtained from three independent cultivations.

$0.01$ $h^{-1}$) as its parent MV-LeuF1 and to form $47 \pm q$ 4 mM KIC (6.1 g $l^{-1}$) after 32 h with a yield of $0.20 \pm 0.02$ mol KIC per mol of glucose and a productivity of $1.41 \pm 0.13$ mmol KIC $l^{-1}$ $h^{-1}$ (Table 2). Both SH-KIC20 and MV-KICF1 formed KIV, KMV and L-leucine as by-products (Table 2).

The results described above demonstrate that the start codon exchange for reduction of IlvE activity was successful and allowed growth and KIC accumulation without supplementation of BCAAs; however, the production parameters were lower compared with supplementation with 1 mM L-isoleucine (Table 2). A successful industrial application depends on high product yields combined with sufficient cell growth, resulting in

**Table 2.** Growth and production parameters of strains MV-KICF1 and SH-KIC20 in shake flask cultivations[a,b].

| Parameter | MV-KICF1 + 1 mM L-isoleucine | MV-KICF1 without L-isoleucine | SH-KIC20[d] without L-isoleucine |
|---|---|---|---|
| Growth rate (h⁻¹) | $0.21 \pm 0.01$ | $0.03 \pm 0.01$ | $0.08 \pm 0.01$ |
| KIC (mM) | $46.7 \pm 4.1$ | $31.8 \pm 2.1$ | $18.8 \pm 0.67$ |
| By-products[c]: | | | |
|   KIV (mM) | $13.3 \pm 2.2$ | $19.0 \pm 4.1$ | $2.6 \pm 0.3$ |
|   KMV (mM) | $8.8 \pm 1.3$ | $4.9 \pm 0.6$ | $8.7 \pm 0.1$ |
|   L-leucine (mM) | $3.0 \pm 0.2$ | $10.3 \pm 3.1$ | $4.8 \pm 0.2$ |
| Molar product yield (mol KIC per mol glucose) | $0.204 \pm 0.018$ | $0.143 \pm 0.010$ | $0.084 \pm 0.001$ |
| Volumetric productivity (mmol KIC l⁻¹ h⁻¹) | $1.41 \pm 0.13$ | $0.34 \pm 0.02$ | $0.38 \pm 0.02$ |

**a.** Cultivations were performed in 500 ml baffled shake flasks containing 50 ml CGXII minimal medium with 4% (w/v) glucose. Supplementation of L-isoleucine is indicated.
**b.** Mean values and standard deviations from three independent cultivations are shown.
**c.** Concentrations of L-valine and L-isoleucine were below 2 mM.
**d.** Cultivation of SH-KIC20 supplemented with 1 mM L-isoleucine was not tested.

competitive productivity. Therefore, the addition of L-isoleucine is still important to improve growth of the constructed strains to reach better productivity values. As an alternative approach to adjust the IlvE activity to an optimal value for prototrophic growth with simultaneous KIC production, *ilvE* gene expression could be fine-tuned by testing promoters with varying strength (Vašicová *et al.*, 1999; Hammer *et al.*, 2006).

Recently, Bückle-Vallant and colleagues (2014) described a plasmid-based *C. glutamicum* strain for the production of KIC. This strain is characterized by deletions of the genes *ilvE*, *ltbR*, *prpC1* and *prpC2*, an exchange of the two *gltA* promoters (van Ooyen *et al.*, 2011) by the mutated *dapA* promoter L1 (Vašicová *et al.*, 1999) to reduce citrate synthase activity (van Ooyen *et al.*, 2012), and plasmid-based overexpression of *ilvBNCD* and a *leuA* allele of *Escherichia coli* encoding a feedback-resistant IPMS. This strain accumulated up to 71 mM KIC when cultivated with glucose plus acetate as carbon sources and supplemented with 2 mM each of L-isoleucine and L-valine. Under cultivation conditions comparable to ours, i.e. without acetate, this strain reached KIC titres

($54 \pm 4$ mM) and yields (0.22 mol per mol of glucose) in a similar range as strain MV-KICF1 when supplemented with L-isoleucine. In comparison, the biotransformation with *Rhodococcus opacus* transcribed by Zhu and colleagues (2011) reached about 10 mM KIC using 39 mM L-leucine as substrate.

*2-Ketoisocaproate transport*

When MV-KICF1 was batch cultivated in a bioreactor as described (Vogt *et al.*, 2014) using a medium supplemented with 1 mM L-isoleucine, comparable KIC titres as in shake flasks of about 50 mM were achieved (data not shown). Interestingly, additional feeding of glucose in fed-batch experiments did not further increase the KIC concentrations and led to an arrest of cell growth and glucose consumption (data not shown). A possible explanation is a block of the L-leucine biosynthesis pathway by elevated KIC concentrations, as Bückle-Vallant and colleagues (2014) found competitive inhibition of IPMS by KIC and non-competitive inhibition of 3-isopropylmalate dehydratase by KIC. Additionally, a competitive inhibition

**Table 3.** Putative transporter genes showing increased expression in a KIC producer.

| Gene | Annotation | mRNA ratio[a] (MV-Leu20 $\Delta$*ilvE*/ Wild type) | TMH[b] |
|---|---|---|---|
| cg0018 | putative membrane protein, conserved | 3.5 | 9 |
| cg1121 | putative permease of the major facilitator superfamily | 2.2 | 7 |
| cg1219 | putative membrane protein | 3.5 | 10 |
| cg1419 | putative Na⁺-dependent transporter, bile acid:Na⁺ symporter BASS family | 7.2 | 8 |
| cg1658 | putative permease of the major facilitator superfamily | 35.6 | 12 |
| cg2557 | putative secondary Na⁺/bile acid symporter, bile acid:Na⁺ symporter BASS family | 2.5 | 8 |
| cg2676 | putative ABC-type dipeptide/oligopeptide/nickel transport system, permease component | 2.1 | 6 |
| cg3334 | putative arabinose efflux permease, MFS type | 2.0 | 12 |

**a.** Transcriptome analyses of KIC producer MV-Leu20 $\Delta$*ilvE* in comparison to the wild type were performed using DNA microarrays as described (Vogt *et al.*, 2014). Candidate transporter genes were chosen based on an mRNA ratio (MV-Leu20 $\Delta$*ilvE*/wild type) of > 2, an annotation as (putative) membrane or transporter proteins and the prediction of multiple transmembrane helices in the encoded proteins. Data represent mean values of at least two (maximum four) evaluable microarray experiments ($P$-value < 0.05).
**b.** TMH, number of transmembrane helices predicted with the SOSUI engine version 1.11.

of acetohydroxyacid synthase by KIV has been reported by Krause and colleagues (2010). To avoid a reduced flux into the leucine synthesis pathway by competitive inhibition of IPMS by KIC, the concentration of KIV within the cell needs to be increased and/or the concentration of KIC within the cell should be decreased. These concentrations are determined on one hand by the rates of synthesis and further metabolic conversion and on the other hand by the rates of export to and import from the supernatant.

Knowledge on KIC transport is very limited. Obviously, as shown by our studies and that of Bückle-Vallant and colleagues (2014), KIC can leave the cell, and previous studies by Groeger and Sahm (1987) demonstrated that KIC can enter the cell. As previously described for L-isoleucine (Zittrich and Krämer, 1994), passive diffusion, carrier-mediated uptake and carrier-mediated excretion must be considered as possibilities for an amphiphilic solute like KIC to cross the cytoplasmic membrane. Due to its similarity to L-isoleucine, it seems likely that also KIC is able to diffuse across the membrane. However, a necessity to possess carriers for KIC import or KIC export is not obvious, in contrast to the advantage of having importers for amino acids and exporters for non-catabolizable amino acids. In *C. glutamicum*, the export of BCAAs and methionine is catalysed by the exporter BrnFE (Kennerknecht *et al.*, 2002; Trötschel *et al.*, 2005; Xie *et al.*, 2012), but evidence is available that BrnFE is not involved in KIC export (Radespiel, 2010). The identification of transporters can be beneficial for biotechnological processes to increase productivity since transport of desired products into the medium often represents a bottleneck. For example, export was identified as a limiting factor for L-isoleucine production with *C. glutamicum* (Morbach *et al.*, 1996), and the production of this BCAA was improved by overexpression of the respective transporter encoded by *brnFE* (Kennerknecht *et al.*, 2002; Xie *et al.*, 2012).

In order to test if a KIC exporter is present in *C. glutamicum* that might be useful to improve KIC overproduction, we searched for genes showing increased mRNA levels during KIC production by performing comparative transcriptome analyses using DNA microarrays as described previously (Vogt *et al.*, 2014). KIC producer strain MV-Leu20 Δ*ilvE* was compared with the wild type to determine differentially expressed genes, resulting in a list of eight candidate transporter genes (Table 3). Each of these genes was deleted in the *C. glutamicum* Δ*ilvE* background, and the resulting double deletion mutants were transformed with pAN6-*leuA*_B018. When cultivated in glucose minimal medium with 0.1 mM IPTG, one of the eight strains, which contained a deletion of cg1121 (annotated as putative permease of the major facilitator superfamily), showed a reduced growth rate $(0.23 \pm 0.01\ h^{-1})$ and a reduced maximal KIC titre

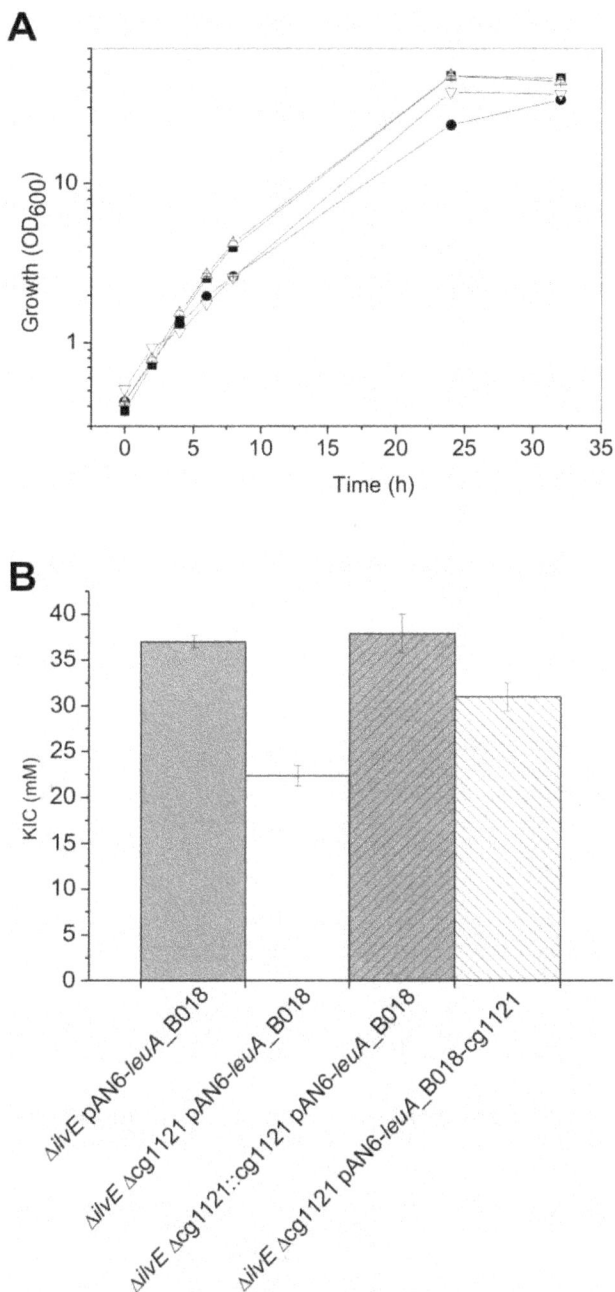

**Fig. 3.** Complementation of the effects on growth and KIC accumulation caused by deletion of cg1121 in strain *C. glutamicum* Δ*ilvE* carrying pAN6-*leuA*_B018.
A. Growth of strains Δ*ilvE* with pAN6-*leuA*_B018 (■), Δ*ilvE* Δcg1121 with pAN6-*leuA*_B018 (●), Δ*ilvE* Δcg1121::cg1121 with pAN6-*leuA*_B018 (△) and Δ*ilvE* Δcg1121 with pAN6-*leuA*_B018-cg1121 (▽) are shown.
B. Maximal KIC concentrations reached after 32 h cultivation in 500 ml baffled shake flasks with 50 ml CGXII minimal medium containing 4% (w/v) glucose and 0.1 mM IPTG at 30°C and 120 rpm on a rotary shaker. The deletion of cg1121 was complemented either by genomic reintegration of cg1121 (Δ*ilvE* Δcg1121::cg1121 pAN6-*leuA*_B018) or by plasmid-borne expression of cg1121 (Δ*ilvE* Δcg1121 pAN6-*leuA*_B018-cg1121). The data represent mean values and standard deviations obtained from three independent cultivations.

(22.4 mM) compared with the reference strain $\Delta ilvE$ pAN6-*leuA*_B018 (0.30 ± 0.01 h$^{-1}$, 37 mM KIC). The phenotype could be complemented by reintegration of cg1121 into the genome of strain $\Delta ilvE$ $\Delta$cg1121 or by plasmid-borne expression of cg1121 (Fig. 3). However, the specific KIC export rates (determined as described by Kennerknecht *et al.*, 2002) of strain $\Delta ilvE$ $\Delta$cg1121 (6.5 ± 1.0 nmol min$^{-1}$ g$_{CDW}^{-1}$) were not significantly reduced compared with that of the reference strain $\Delta ilvE$ (8.0 ± 1.5 nmol min$^{-1}$ g$_{CDW}^{-1}$) and the cytoplasmic KIC concentrations of the two strains at an OD$_{600}$ of 5 were comparable. Therefore, the role of Cg1121 for growth and KIC production remains unclear and needs further investigations.

## Acknowledgements

We thank Dr Nicole Paczia (IBG-1: Biotechnology, Forschungszentrum Jülich, Germany) for her help with the measurements of cytoplasmic 2-ketoisocaproate concentrations.

## Conflict of interest

None declared.

## References

Abe, S., Takayama, K.I., and Kinoshita, S. (1967) Taxonomical studies on glutamic acid-producing bacteria. *J Gen Appl Microbiol* **13:** 279–301.

Aparicio, M., Bellizzi, V., Chauveau, P., Cupisti, A., Ecder, T., Fouque, D., *et al.* (2012) Keto acid therapy in predialysis chronic kidney disease patients: final consensus. *J Ren Nutr* **22:** S22–S24.

Becker, J., and Wittmann, C. (2012) Systems and synthetic metabolic engineering for amino acid production – the heartbeat of industrial strain development. *Curr Opin Biotechnol* **23:** 718–726.

Becker, J., Buschke, N., Bücker, R., and Wittmann, C. (2010) Systems level engineering of *Corynebacterium glutamicum* – reprogramming translational efficiency for superior production. *Eng Life Sci* **10:** 430–438.

Blombach, B., Riester, T., Wieschalka, S., Ziert, C., Youn, J.W., Wendisch, V.F., and Eikmanns, B.J. (2011) *Corynebacterium glutamicum* tailored for efficient isobutanol production. *Appl Environ Microbiol* **77:** 3300–3310.

Burkovski, A. (2008) *Corynebacteria: Genomics and Molecular Biology*. Norfolk, UK: Caister Academic Press.

Bückle-Vallant, V., Krause, F.S., Messerschmidt, S., and Eikmanns, B.J. (2014) Metabolic engineering of *Corynebacterium glutamicum* for 2-ketoisocaproate production. *Appl Microbiol Biotechnol* **98:** 297–311.

Cann, A.F., and Liao, J.C. (2010) Pentanol isomer synthesis in engineered microorganisms. *Appl Microbiol Biotechnol* **85:** 893–899.

Cooper, A.J., Ginos, J.Z., and Meister, A. (1983) Synthesis and properties of the $\alpha$-keto acids. *Chem Rev* **83:** 321–358.

Eggeling, L., and Bott, M. (2005) *Handbook of Corynebacterium glutamicum*. Boca Raton, FL, USA: Taylor & Francis.

Eggeling, I., Cordes, C., Eggeling, L., and Sahm, H. (1987) Regulation of acetohydroxy acid synthase in *Corynebacterium glutamicum* during fermentation of $\alpha$-ketobutyrate to L-isoleucine. *Appl Microbiol Biotechnol* **25:** 346–351.

Escobar, J., Frank, J.W., Suryawan, A., Nguyen, H.V., Van Horn, C.G., Hutson, S.M., and Davis, T.A. (2010) Leucine and $\alpha$-ketoisocaproic acid, but not norleucine, stimulate skeletal muscle protein synthesis in neonatal pigs. *J Nutr* **140:** 1418–1424.

Frunzke, J., Engels, V., Hasenbein, S., Gätgens, C., and Bott, M. (2008) Co-ordinated regulation of gluconate catabolism and glucose uptake in *Corynebacterium glutamicum* by two functionally equivalent transcriptional regulators, GntR1 and GntR2. *Mol Microbiol* **67:** 305–322.

Groeger, U., and Sahm, H. (1987) Microbial production of L-leucine from $\alpha$-ketoisocaproate by *Corynebacterium glutamicum*. *Appl Microbiol Biotechnol* **25:** 352–356.

Hammer, K., Mijakovic, I., and Jensen, P.R. (2006) Synthetic promoter libraries – tuning of gene expression. *Trends Biotechnol* **24:** 53–55.

Heissig, H., Urban, K.A., Hastedt, K., Zunkler, B.J., and Panten, U. (2005) Mechanism of the insulin-releasing action of $\alpha$-ketoisocaproate and related $\alpha$-keto acid anions. *Mol Pharmacol* **68:** 1097–1105.

Ikeda, M., and Nakagawa, S. (2003) The *Corynebacterium glutamicum* genome: features and impacts on biotechnological processes. *Appl Microbiol Biotechnol* **62:** 99–109.

Inui, M., Kawaguchi, H., Murakami, S., Vertes, A.A., and Yukawa, H. (2004) Metabolic engineering of *Corynebacterium glutamicum* for fuel ethanol production under oxygen-deprivation conditions. *J Mol Microbiol Biotechnol* **8:** 243–254.

Kalinowski, J., Bathe, B., Bartels, D., Bischoff, N., Bott, M., Burkovski, A., *et al.* (2003) The complete *Corynebacterium glutamicum* ATCC 13032 genome sequence and its impact on the production of L-aspartate-derived amino acids and vitamins. *J Biotechnol* **104:** 5–25.

Keilhauer, C., Eggeling, L., and Sahm, H. (1993) Isoleucine synthesis in *Corynebacterium glutamicum*: molecular analysis of the *ilvB-ilvN-ilvC* operon. *J Bacteriol* **175:** 5595–5603.

Kennerknecht, N., Sahm, H., Yen, M.R., Patek, M., Saier Jr, M.H., Jr, and Eggeling, L. (2002) Export of L-isoleucine from *Corynebacterium glutamicum*: a two-gene-encoded member of a new translocator family. *J Bacteriol* **184:** 3947–3956.

Kind, S., and Wittmann, C. (2011) Bio-based production of the platform chemical 1,5-diaminopentane. *Appl Microbiol Biotechnol* **91:** 1287–1296.

Kirchner, O., and Tauch, A. (2003) Tools for genetic engineering in the amino acid-producing bacterium *Corynebacterium glutamicum*. *J Biotechnol* **104:** 287–299.

Klaffl, S., Brocker, M., Kalinowski, J., Eikmanns, B.J., and Bott, M. (2013) Complex regulation of the phosphoenolpyruvate carboxykinase gene *pck* and characterization of its GntR-type regulator IolR as a repressor of *myo*-inositol utilization genes in *Corynebacterium glutamicum*. *J Bacteriol* **195:** 4283–4296.

Krause, F.S., Blombach, B., and Eikmanns, B.J. (2010) Metabolic engineering of *Corynebacterium glutamicum* for 2-ketoisovalerate production. *Appl Environ Microbiol* **76:** 8053–8061.

Leclercq-Meyer, V., Marchand, J., Leclercq, R., and Malaisse, W.J. (1979) Interactions of α-ketoisocaproate, glucose and arginine in the secretion of glucagon and insulin from the perfused rat pancreas. *Diabetologia* **17:** 121–126.

Litsanov, B., Brocker, M., and Bott, M. (2012a) Toward homosuccinate fermentation: metabolic engineering of *Corynebacterium glutamicum* for anaerobic production of succinate from glucose and formate. *Appl Environ Microbiol* **78:** 3325–3337.

Litsanov, B., Kabus, A., Brocker, M., and Bott, M. (2012b) Efficient aerobic succinate production from glucose in minimal medium with *Corynebacterium glutamicum*. *Microb Biotechnol* **5:** 116–128.

Marienhagen, J., Kennerknecht, N., Sahm, H., and Eggeling, L. (2005) Functional analysis of all aminotransferase proteins inferred from the genome sequence of *Corynebacterium glutamicum*. *J Bacteriol* **187:** 7639–7646.

Mimitsuka, T., Sawai, H., Hatsu, M., and Yamada, K. (2007) Metabolic engineering of *Corynebacterium glutamicum* for cadaverine fermentation. *Biosci Biotechnol Biochem* **71:** 2130–2135.

Morbach, S., Sahm, H., and Eggeling, L. (1996) L-Isoleucine production with *Corynebacterium glutamicum*: further flux increase and limitation of export. *Appl Environ Microbiol* **62:** 4345–4351.

Möckel, B., Eggeling, L., and Sahm, H. (1992) Functional and structural analyses of threonine dehydratase from *Corynebacterium glutamicum*. *J Bacteriol* **174:** 8065–8072.

Niebisch, A., and Bott, M. (2001) Molecular analysis of the cytochrome $bc_1$-$aa_3$ branch of the *Corynebacterium glutamicum* respiratory chain containing an unusual diheme cytochrome $c_1$. *Arch Microbiol* **175:** 282–294.

Okino, S., Noburyu, R., Suda, M., Jojima, T., Inui, M., and Yukawa, H. (2008) An efficient succinic acid production process in a metabolically engineered *Corynebacterium glutamicum* strain. *Appl Microbiol Biotechnol* **81:** 459–464.

van Ooyen, J., Emer, D., Bussmann, M., Bott, M., Eikmanns, B.J., and Eggeling, L. (2011) Citrate synthase in *Corynebacterium glutamicum* is encoded by two *gltA* transcripts which are controlled by RamA, RamB, and GlxR. *J Biotechnol* **154:** 140–148.

van Ooyen, J., Noack, S., Bott, M., Reth, A., and Eggeling, L. (2012) Improved L-lysine production with *Corynebacterium glutamicum* and systemic insight into citrate synthase flux and activity. *Biotechnol Bioeng* **109:** 2070–2081.

Paczia, N., Nilgen, A., Lehmann, T., Gätgens, J., Wiechert, W., and Noack, S. (2012) Extensive exometabolome analysis reveals extended overflow metabolism in various microorganisms. *Microb Cell Fact* **11:** 122.

Park, J.H., and Lee, S.Y. (2010) Fermentative production of branched chain amino acids: a focus on metabolic engineering. *Appl Microbiol Biotechnol* **85:** 491–506.

Pátek, M. (2007) Branched-chain amino acids. In *Amino Acid Biosynthesis – Pathways, Regulation and Metabolic Engineering*. Wendisch, V.F. (ed.). Berlin, Germany: Springer, pp. 129–162.

Pátek, M., Krumbach, K., Eggeling, L., and Sahm, H. (1994) Leucine synthesis in *Corynebacterium glutamicum*: enzyme activities, structure of *leuA*, and effect of *leuA* inactivation on lysine synthesis. *Appl Environ Microbiol* **60:** 133–140.

Radespiel, T. (2010) Dissertation abstract. URL http://kups.ub.uni-koeln.de/id/eprint/3156.

Radmacher, E., Vaitsikova, A., Burger, U., Krumbach, K., Sahm, H., and Eggeling, L. (2002) Linking central metabolism with increased pathway flux: L-valine accumulation by *Corynebacterium glutamicum*. *Appl Environ Microbiol* **68:** 2246–2250.

Schäfer, A., Tauch, A., Jäger, W., Kalinowski, J., Thierbach, G., and Pühler, A. (1994) Small mobilizable multi-purpose cloning vectors derived from the *Escherichia coli* plasmids pK18 and pK19: selection of defined deletions in the chromosome of *Corynebacterium glutamicum*. *Gene* **145:** 69–73.

Scheele, S., Oertel, D., Bongaerts, J., Evers, S., Hellmuth, H., Maurer, K.H., *et al.* (2013) Secretory production of an FAD cofactor-containing cytosolic enzyme (sorbitol-xylitol oxidase from *Streptomyces coelicolor*) using the twin-arginine translocation (Tat) pathway of *Corynebacterium glutamicum*. *Microb Biotechnol* **6:** 202–206.

Schneider, J., and Wendisch, V.F. (2011) Biotechnological production of polyamines by bacteria: recent achievements and future perspectives. *Appl Microbiol Biotechnol* **91:** 17–30.

Smith, K.M., Cho, K.M., and Liao, J.C. (2010) Engineering *Corynebacterium glutamicum* for isobutanol production. *Appl Microbiol Biotechnol* **87:** 1045–1055.

Tauch, A., Götker, S., Pühler, A., Kalinowski, J., and Thierbach, G. (2002) The alanine racemase gene *alr* is an alternative to antibiotic resistance genes in cloning systems for industrial *Corynebacterium glutamicum* strains. *J Biotechnol* **99:** 79–91.

Trötschel, C., Deutenberg, D., Bathe, B., Burkovski, A., and Krämer, R. (2005) Characterization of methionine export in *Corynebacterium glutamicum*. *J Bacteriol* **187:** 3786–3794.

Vašicová, P., Pátek, M., Nešvera, J., Sahm, H., and Eikmanns, B. (1999) Analysis of the *Corynebacterium glutamicum dapA* promoter. *J Bacteriol* **181:** 6188–6191.

Vogt, M., Haas, S., Klaffl, S., Polen, T., Eggeling, L., van Ooyen, J., and Bott, M. (2014) Pushing product formation to its limit: metabolic engineering of *Corynebacterium glutamicum* for L-leucine overproduction. *Metab Eng* **22:** 40–52.

Wieschalka, S., Blombach, B., Bott, M., and Eikmanns, B.J. (2013) Bio-based production of organic acids with *Corynebacterium glutamicum*. *Microb Biotechnol* **6:** 87–102.

Xie, X., Xu, L., Shi, J., Xu, Q., and Chen, N. (2012) Effect of transport proteins on L-isoleucine production with the L-isoleucine-producing strain *Corynebacterium glutamicum* YILW. *J Ind Microbiol Biotechnol* **39:** 1549–1556.

Yukawa, H., and Inui, M. (2013) *Corynebacterium glutamicum: Biology and Biotechnology*. Heidelberg, Germany: Springer.

Zanchi, N.E., Gerlinger-Romero, F., Guimaraes-Ferreira, L., de Siqueira Filho, M.A., Felitti, V., Lira, F.S., *et al.* (2011) HMB supplementation: clinical and athletic performance-related effects and mechanisms of action. *Amino Acids* **40:** 1015–1025.

Zhu, Y., Li, J., Liu, L., Du, G., and Chen, J. (2011) Production of α-ketoisocaproate via free-whole-cell biotransformation by *Rhodococcus opacus* DSM 43250 with L-leucine as the substrate. *Enzyme Microb Technol* **49:** 321–325.

Zittrich, S., and Krämer, R. (1994) Quantitative discrimination of carrier-mediated excretion of isoleucine from uptake and diffusion in *Corynebacterium glutamicum*. *J Bacteriol* **176:** 6892–6899.

## Supporting information

Additional Supporting Information may be found in the online version of this article at the publisher's web-site:

**Table S1.** DNA oligonucleotides used in this study.

# 18

# Detection and evolutionary analysis of picobirnaviruses in treated wastewater

Shiwei Zhang,[1] Ru Bai,[1] Run Feng,[1] Hongxun Zhang[1] and Lixin Liu[1,2,3]*

[1]College of Life Science, University of Chinese Academy of Sciences, No.19A, Yuquan Road, Shijingshan District, Beijing 100049, China.
[2]Key Laboratory for Polymeric Composite and Functional Materials of Ministry of Education, School of Chemistry and Chemical Engineering, Sun Yat-Sen University, No. 135, Xingang Xi Road, Guangzhou 510275, China.
[3]State Environmental Protection Key Laboratory of Microorganism Application and Risk Control (SMARC), Tsinghua University, Beijing 100084, China.

## Summary

**Wastewater contains numerous viruses. In this study, picobirnaviruses (PBVs) were detected in the stream of a wastewater treatment plant in Changsha, Hunan province, China, and evolutionary analysis of the isolated PBVs was performed. The phylogenetic tree revealed that the PBVs were highly divergent and could be classified into six distinct groups according to their hosts. Among these groups, pairwise comparison of the six groups revealed that the nucleotide distance of group 4 (bootstrap value = 0.92; nucleotide identity = 94%) was the largest. Thus, group 4 might represent a new division of PBVs. Comprehensive analysis of the obtained PBV sequences to investigate their evolutionary history and phylodynamics revealed that group 5 (PBVs from monkey) exhibited maximum polymorphism ($K = 30.582$, $S = 74$, $\eta = 98$, $Pa = 47$) and lowest nucleotide substitutions per site per year (6.54E-3 subs per site per year), except group 4. Maximum clade credibility tree indicated that group 5 appeared earlier than the other groups. In conclusion, this study detected PBVs in treated wastewater in China, and identified a new PBV group. Furthermore, among these PBVs, group 5 was found to survive longer and present a balance between PBVs and their monkey host.**

*For correspondence. E-mail liulixin@mail.sysu.edu.cn

**Funding Information** This project was supported by the Open Fund of State Environmental Protection Key Laboratory of Microorganism Application and Risk Control (No. SMARC 2013D011) and Chinese Academy of Sciences (KAAD-EW-09-1).

## Introduction

Sewage or wastewater may contain numerous viruses disseminated from the hosts and provides an ideal environment for the growth of diverse viruses and their host species (Lodder and de Roda Husman, 2005; Cantalupo et al., 2011). The viruses that exist in the wastewater may come into contact with humans through many ways and, hence, are serious potential threats to human health. Recently, many studies have been carried out to examine the viruses in sewages (van den Berg et al., 2005; Rosario et al., 2009; Symonds et al., 2009). The development of sequencing technique, especially the metagenomic method, has provided a great opportunity to gain a better understanding of viruses (Thurber et al., 2009). However, the vast majority of the sequences of viruses are difficult to read or not related to known viruses, and thus could not be studied (Breitbart et al., 2003).

In the present study, picobirnaviruses (PBVs) were detected in the stream of a wastewater treatment plant in China, and evolutionary analysis of the isolated PBVs was carried out. PBVs widely exist in wastewater and have been classified into *Picobirnaviridae* family by the International Committee of Taxonomy on Viruses. They are small ($\Phi$ 35–40 nm), non-enveloped, dsRNA viruses with a bisegmented genome (Ganesh et al., 2012; Mondal et al., 2013). The large genomic segment or segment 1 is 2.2–2.7 kb and encodes the capsid protein. The small segment or segment 2 is 1.2–1.9 kb and encodes the RNA-dependent RNA polymerase (RdRp). Based on the sequences of this gene, PBVs have been classified into two genogroups represented by the Chinese strain 1-CHN-97 (AF246939, prototype of genogroup I) and US strain 4-GA-91 (AF246940, prototype of genogroup II) (Rosen et al., 2000; Bhattacharya et al., 2007).

PBVs widely exist in the environment and have a wide range of hosts, including humans (Smits et al., 2012), rabbits (Ludert et al., 1995; Green et al., 1999), pigs (Martínez et al., 2010; Smits et al., 2011), cattle (Ghosh et al., 2009), horses (Ganesh et al., 2011), birds (Chandra, 1997), monkeys (Wang et al., 2007; 2012) and foxes (Bodewes et al., 2013). Currently, many studies have been carried out to analyse PBVs at the epidemiologic and phylogenetic levels. Although phylogenetic tree is the most valuable and common technique of computational analysis of viral evolutionary changes, additional methods are required for detailed investigation of the evolution and phylodynamics of PBVs to gain in-depth understanding of how PBVs transmitted among a wide range of hosts (Nates *et al.*, 2011). In the present study, phylogenetic tree was used to analyse the diversity of PBVs, and five comparative techniques, namely phylogenetic analysis, polymorphism analyses, inference of selection pressures acting on PBVs, measurement of evolutionary rate and construction of maximum clade credibility (MCC) tree, were employed to exhaustively study the evolutionary process of PBVs isolated from the stream of a wastewater treatment plant in China.

## Results

### Detection and sequencing of PBVs

The Polymerase Chain Reaction (PCR) method was used to detect PBVs with partial RdRp gene (200 bp) from the six samples of treated wastewater. As shown in Fig. 1, for all the six samples, a distinctive 200 bp band could be detected in agarose gel (2%) with ethidium bromide under UV light. To better understand the diversity of the PBVs in treated wastewater, a total of 300 (~ 50 products per sample) cloned PBV PCR products (~ 200 bp) of the RdRp gene were sequenced by Sangon (Sangon, Beijing, China). After removing the error and similar identity sequences by BLAST analysis, 139 sequences (Table S1) were obtained and deposited in GenBank under accession numbers KJ135791 to KJ135929.

### Phylogenetic analysis

To examine the genetic relatedness between the PBVs from treated wastewater and known PBVs, the top three similar known sequences from the National Center for Biotechnology Information (NCBI) corresponding to the 139 sequences were chosen as reference sequences to construct the phylogenetic tree. A total of 65 known sequences (Table S2) were obtained from the NCBI database by using BLAST analysis. These 65 sequences included monkey strains (16 sequences), avian strains (11 sequences), porcine strains (9 sequences), human strains (7 sequences), fox strain (1 sequence) and uncultured strains (21 sequences). Thus, a total of 204 sequences were used to construct the phylogenetic tree by using MEGA 5.0.

From the constructed phylogenetic tree (Fig. 2), it could be observed that the PBVs from the treated wastewater had a high genetic diversity and that all the sequences had nucleotide identities ranging between 42% and 100%. With regard to the reliability of data analysis, sequences with bootstrap values > 0.8 were categorized as one group. Thus, of the total 139 sequences, 53 sequences (Table S3) were grouped with known host sequences: 26 sequences were grouped with porcine PBVs (average nucleotide identity = 76%; owing to the relatively low nucleotide identity, it was hypothesized that there were two different porcine PBVs); 12 sequences were grouped with avian PBVs (average nucleotide identity = 98%); 11 sequences were grouped with human PBVs (average nucleotide identity = 96%); and four sequences were grouped with monkey PBVs (average nucleotide identity = 81%) (Table S3). It must be noted that four sequences (5–70 strain, 5–75 strain, 5–36 strain and 5–72 strain) were grouped with three different known host strains Mo/CHN-56/2002 strain, Po/E4-14 strain and Hu/1-CHN-01 strain (average nucleotide identity = 85%). Similarly, one sequence (2–24 strain) clustered with Fox/5 strain, Hu/6C2P strain and Un/Florida-12 strain (average nucleotide identity = 90%). This evidence suggests the potential of PBVs for interspecies zoonotic transmission. With regard to the remaining sequences, it was difficult to clearly group them with the unique known host sequences. For example, eight sequences (5–94 strain, 4–52 strain, 5–115 strain, 3-1 strain, 6–7 strain, 1–12 strain, 3–51 strain and 5–77 strain) were clustered into one group with high bootstrap value (1.0) and high average nucleotide identity (95%) and one avian PBV strain Av/AVE 71/2010-3 was in the vicinity of this group of sequences. However, despite a nucleotide identity of 92% and a bootstrap value of 0.24 (< 0.8), this group of strains could not be assigned to the avian PBVs with such weak statistical support. Moreover, some sequences (44 sequences of the 139 sequences) could not be clustered with known sequences; for example, 21 sequences, which were well clustered into one group with a bootstrap value of 0.92 and nucleotide identity of 94%, could not be assigned to any PBV species, because they branched in the deeper part of the phylogenetic tree with no known host strains close to the branches (Fig. 2). Thus, this branch of PBVs might be a new group of PBVs, and more

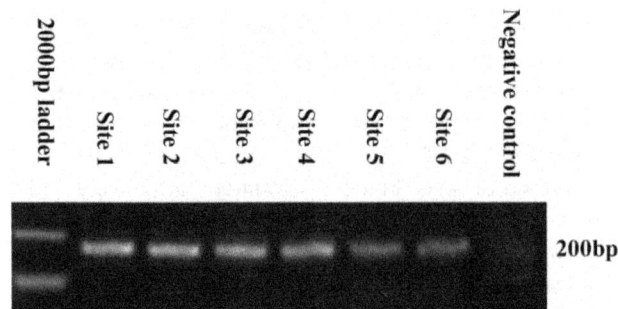

**Fig. 1.** Agarose gel electrophoresis showing amplified products of 200 bases in 40 cycles of amplification by using PBVs primers. M: molecular weight standard marker, site 1–6: PCR product of six sites from treated water plant in Changsha Hunan province (China) with the amplicons length of 200 bp, N: negative control.

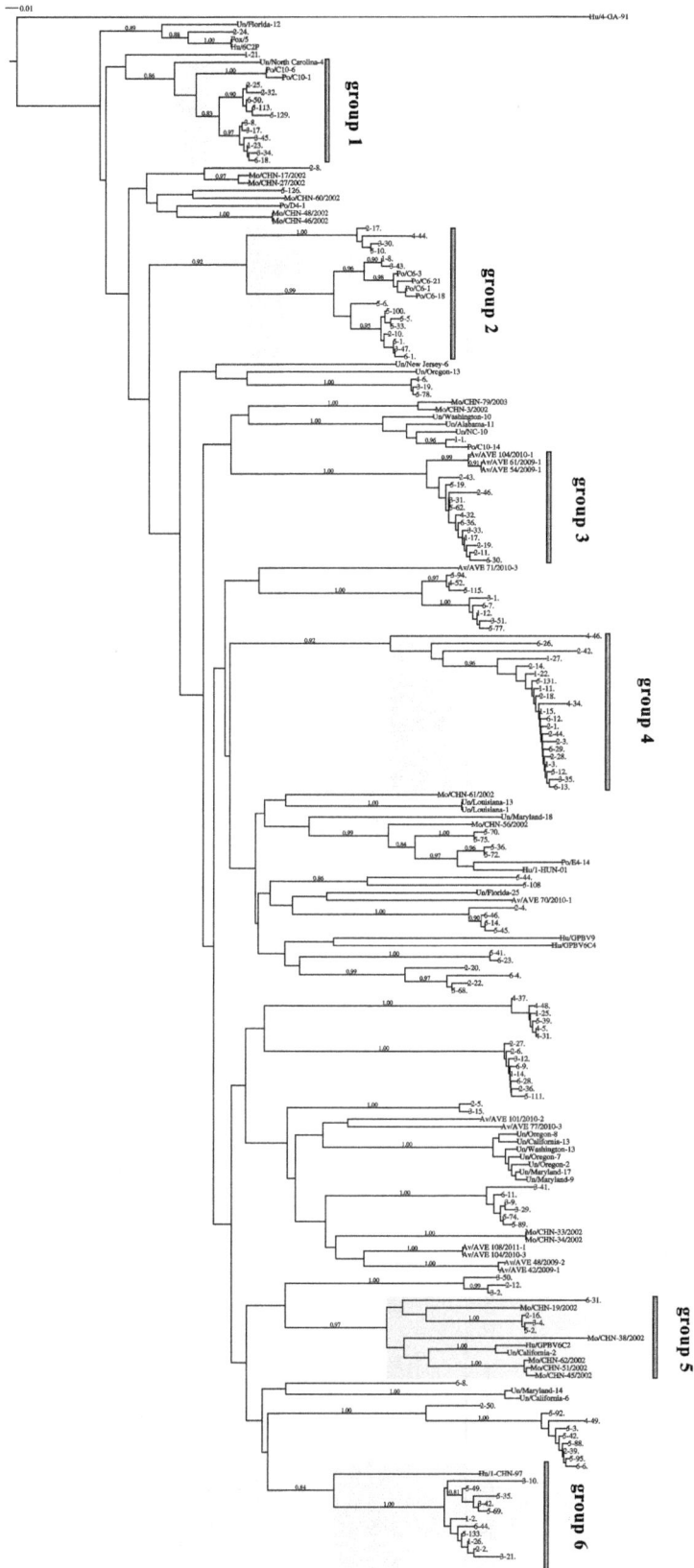

**Fig. 2.** Neighbour-joining (Kimura 2-parameter model) phylogenetic tree of ~ 200 bp segment 2 of the PBVs (genotype I) RNA-dependent RNA polymerase gene. A total of 204 sequences (139 sequences we obtained and 65 referenced sequences) were used to construct phylogenetic tree by MEGA 5.0. Six groups were displayed by different colours. A bootstrap value > 0.8 was indicated for the corresponding nodes, and the tree was statistically supported by bootstrapping with 1000 replicates. Bar, 0.01 substitutions per nucleotide.

backup data were required to confirm this finding. Although the phylogenetic tree revealed that the PBVs showed wide diversity, the viruses of each host species clustered in distinct clades.

A clade with over 10 sequences and bootstrap values > 0.8 was considered as one group (Fig. 2). Thus, a total of six groups (groups 1–6) were obtained. Except group 4 mentioned earlier, which might be a new group of PBVs with no known host, the other five groups were allocated to known hosts. The porcine PBVs were divided into two groups (groups 1 and 2), because the diversity of the two intergroups was noted to be 35%. The hosts of groups 3, 5 and 6 were avian, monkey and human respectively (Fig. 2). The subsequent evolutionary study was based on these six groups, and the corresponding evolutionary models were as follows: Hasegawa, Kishino, and Yano 1985 (HKY) + Gamma distribution (G) for groups 1 and 2, HKY for group 3, Kimura 1980 (K80) + Invariable Sites (I) for group 4, Transversion Model with equal nucleotide frequencies (TVMef) + I for group 5, and Three-Parameter Model (TPM1) for group 6.

### Assessment of polymorphism

Genetic polymorphism analyses and neutrality tests of the six groups were performed. The results of polymorphism analyses (Table 1) for the six groups showed that the K, S, $\eta$ and Pa values were the highest (K = 30.582, S = 74, $\eta$ = 98, Pa = 47) for the monkey strains (group 5) and lowest (K = 6.333, S = 24, $\eta$ = 24, Pa = 10) for the avian strains (group 3), indicating that monkey strains had a relatively high diversity of nucleic acids, whereas the avian strains had low nucleic acid diversity. The rest of the three groups exhibited similar values. Furthermore, the Ka/Ks value of group 5 was the lowest (0.092125), whereas those of group 3 (0.385676) and group 4 (0.370711) were higher. The sequences of group 5 showed lower levels of non-synonymous substitution, suggesting that despite the high diversity of these sequences with comparatively high polymorphism, synonymous substitution was predominant. Tajima's and D tests were not significant, except for group 4, which presented values of −2.17764 and −1.95013 respectively (Table 1). Moreover, the probability of positive selection of group 4 was higher than that of other groups (Table 2). With regard to group 2, three neutral tests were positive, suggesting that there might be a balancing selection. Overall, except group 4, the PBVs obtained from the treated wastewater samples were noted to prefer a neutral choice.

### Measurement of selection pressure

The global $\omega$ value (0.64) was < 1.0 for all the six groups, which indicated that positive selection was not detected

**Table 1.** Polymorphism analyses and neutrality tests of six groups.

| | Seq | Hp | K | S | η | Pa | η(s) | Ka | Ks | Ka/Ks | θ[a] | π[b] | Tajima's | D[d] | F[d] | Host |
|---|---|---|---|---|---|---|---|---|---|---|---|---|---|---|---|---|
| Group 1 | 14 | 13 | 10.604 | 39 | 43 | 18 | 24 | 0.036849 | 0.216865 | 0.178854 | 12.578(3.955) | 0.06321(0.01264) | −0.94103 | −0.918755 | −1.18411 | Porcine |
| Group 2 | 18 | 17 | 18.353 | 52 | 61 | 41 | 17 | 0.045231 | 0.754099 | 0.146385 | 15.677(4.637) | 0.11056(0.01606) | 0.14454 | 0.16258 | 0.18269 | Porcine |
| Group 3 | 12 | 11 | 6.333 | 24 | 24 | 10 | 14 | 0.019783 | 0.11238 | 0.385676 | 7.285(2.412) | 0.03598(0.00552) | −0.90266 | −0.96741 | −1.08234 | Avian |
| Group 4 | 21 | 20 | 10.152 | 64 | 85 | 21 | 64 | 0.05963 | 0.113056 | 0.370711 | 21.680(6.026) | 0.05835(0.01915) | −2.17764[d] | −1.95013[c] | −1.98344 | Uncultured |
| Group 5 | 11 | 11 | 30.582 | 74 | 98 | 47 | 41 | 0.091158 | 0.807 | 0.092125 | 25.265(8.626) | 0.16896(0.01711) | −0.41194 | −0.18922 | −0.28001 | Monkey |
| Group 6 | 12 | 12 | 11.455 | 53 | 56 | 11 | 45 | 0.034495 | 0.275658 | 0.180213 | 17.882(5.921) | 0.06508(0.02254) | −1.7603 | −2.01308 | −2.21957 | Human |

a. Variance of θ (free recombination).
b. Standard deviation of π.
c. Statistical significance P < 0.02 (for Fu and Li's test), P < 0.01 (for Tajima's test).
d. Statistical significance P < 0.05 (for all tests).

Hp, haplotypes; K, average number of pairwise difference; S, number of polymorphic (segregating) sites; η, total number of mutations; Pa, parsimony informative sites; η(s), number of singletons; θ, Watterson's mutation parameter (per sequence calculated from S); π, nucleotide diversity; Ks, rate of synonymous substitutions; Ka, rate of non-synonymous substitutions.

**Table 2.** Positive sites analysis for six groups using SLAC, FEL and REL methods.

|  | SLAC | FEL | REL | Host |
|---|---|---|---|---|
| Group 1 | – | – | – | Porcine |
| Group 2 | – | – | – | Porcine |
| Group 3 | – | – | – | Avian |
| Group 4 | – | – | 2 | Uncultured |
| Group 5 | – | – | 1 | Monkey |
| Group 6 | – | – | – | Human |

on the RdRp gene in the evolutionary processes of PBVs. Further site-by-site analyses of the six groups were conducted by employing single likelihood ancestor counting (SLAC), fixed effects likelihood (FEL) and random effects likelihood (REL) methods with HYPHY package. The results of the SLAC and FEL methods (Table 2) did not reveal any positive selection sites in all the groups. However, the results of the REL method were slightly different, indicating that there were two and one positive selection sites in group 4 (sites 8 and 39) and group 5 (site 36) respectively. Overall, it was concluded that the PBVs preferred neutral selection, which was consistent with the above-mentioned result.

### Substitution rate estimates and MCC tree construct

The mean and 95% lower and upper highest posterior density (HPD) were estimated for the nucleic acid substitution rates for all the groups by using the Markov chain Monte Carlo (MCMC) approach. The results (Table 3) showed that the substitution rates estimated for the PBVs ranged from 4.05E-3 to 1.41E-2 subs per site per year. The substitution rates of PBVs were substantially similar in human strains (1.31E-2 subs per site per year), porcine strains (1.41E-2 and 1.21E-2 subs per site per year), and avian strains (1.33E-2 subs per site per year), whereas that of the monkey PBVs was 6.54E-3 subs per site per year, suggesting that PBVs evolve more slowly in monkey hosts than in human, porcine and avian hosts. This difference in the evolutionary rates of PBVs in different hosts might be owing to the different ecological pressures such as the host immune defence. The MCC phylogenetic trees

**Table 3.** Substitution rate for each group of picobirnaviruses.

| | Mean substitution rate $(10^{-2})$ | Uncorrelated relaxed clock model | Substitution rate HPD $(10^{-2})$ | Host |
|---|---|---|---|---|
| Group 1 | 1.41 | Lognormal | 0.78–2.01 | Porcine |
| Group 2 | 1.21 | Lognormal | 0.91–1.46 | Porcine |
| Group 3 | 1.33 | Exponential | 0.76–2.05 | Avian |
| Group 4 | 0.41 | Exponential | 0.18–0.62 | Uncultured |
| Group 5 | 0.65 | Exponential | 0.31–1.04 | Monkey |
| Group 6 | 1.31 | Exponential | 0.61–1.98 | Human |

for each group are shown in Fig. 3 with different groups marked with different colours. Interestingly, from the MCC tree, it could be deduced that monkey was the earliest host for PBVs, although the nucleic acid substitution rate for monkey PBVs was found to be relatively low. Furthermore, among the four hosts (monkey, pig, avian, human), the nearest host was avian. However, PBVs from monkey and avian hosts were likely to be from the same ancestor. As expected, the porcine host of groups 1 and 2 clustered. Based on the MCC tree construct, the undetermined group 4 was speculated be the human PBVs.

### Discussion

In the present study, the PBV population in wastewater treatment plant in Changsha Hunan province, China, was investigated. The results obtained showed that the PBVs could be detected in 100% of the samples collected (Fig. 1). Furthermore, the findings indicated that PBVs were extremely stable and resistant to treatment, and thus might be a good indicator of faecal pollution. Wastewater is a huge mixture matrix, containing various viruses disseminated from different hosts (Cantalupo et al., 2011). The viruses collected from wastewater often show vast diversity. In the present study, molecular analysis of the PBVs isolated from the collected wastewater samples revealed their high genetic diversity, with all the sequences exhibiting nucleotide identities ranging between 42% and 100%, which is consistent with the findings reported in a previous study of sewage samples from the USA (Symonds et al., 2009). The present study is the first to examine the PBVs from a wastewater treatment plant in China, and a total of 139 sequences of PBVs have been submitted to GenBank (Accession Nos. KJ135791 to KJ135929), thus providing more PBV sequences for further studies. In addition, several important conclusions could be drawn from the phylogenetic analysis of the PBV sequences. First, the PBV population in the wastewater treatment plant in China might have originated from monkey, porcine, avian and human hosts, perhaps owing to the extreme stability and resistance of PBVs to treatment, thus making them ubiquitous with the ability to infect different host species in the same setting. Second, the PBVs exhibited interspecies transmission; for example, four strains (5–70 strain, 5–75 strain, 5–36 strain and 5–72 strain) were allocated to three different known host strains, Mo/CHN-56/2002 strain, Po/E4-14 strain and Hu/1-CHN-01 strain (average nucleotide identity = 85%) (Fig. 2), and one strain (2–24 strain) clustered with Fox/5 strain, Hu/6C2P strain and Un/Florida-12 strain (average nucleotide identity = 90%). These pieces of evidence indicated the potential of PBVs for interspecies zoonotic transmission. Similarly, previous studies on simian genotype I PBVs and PBVs in sewage samples

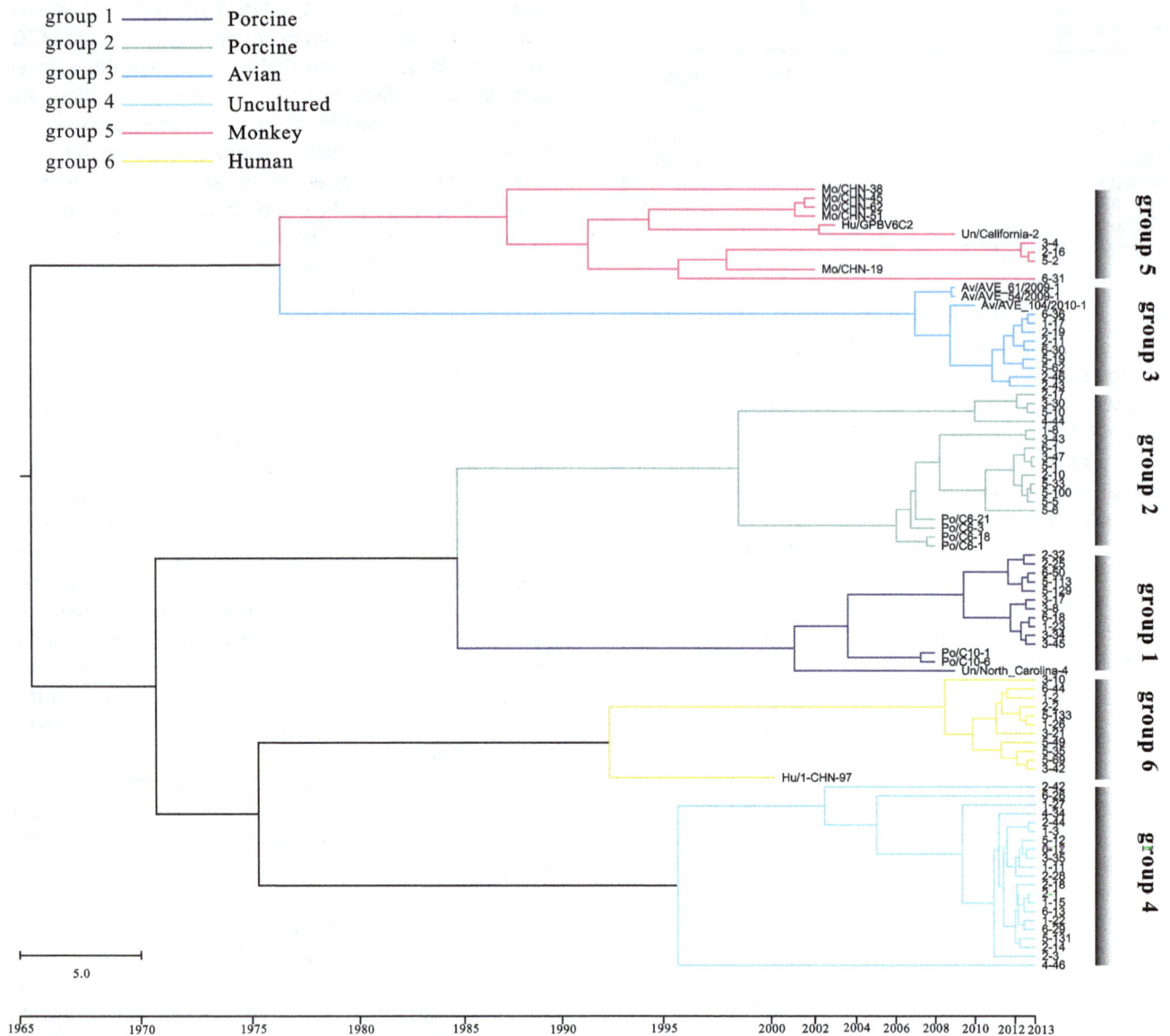

**Fig. 3.** Bayesian maximum credibility (MCC) tree construction based on the sequences of six groups. The branches of different groups were coloured on the basis of different hosts, and at the bottom of the MCC tree represents the years before the last sampling time (2013). Bar, 5 years. The scale on the upper left of the MCC tree showed the different colours of different groups and hosts.

from the USA also suggested that PBVs had the potential for interspecies transmission (Ganesh *et al.*, 2011). Third, the PBVs presented higher sequence diversity than that expected. Only 53 strains of the 139 sequences were assigned to the host, whereas the host of the remaining 86 strains could not be determined, indicating that numerous PBVs are yet to be detected. Fourth, a new group of PBVs, including 21 sequences (group 4), was identified, which clustered well into one group with a bootstrap value of 0.92 and a nucleotide identity of 94%; however, these PBVs could not be assigned to any known PBV species (Fig. 2).

Evolutionary analysis revealed that the two porcine strains (groups 1 and 2), avian strains (group 3) and

human strains (group 6) presented similar results of site-by-site analyses (Table 2) and nucleotide substitution rates (Table 3). Furthermore, it is worth noting that the monkey strains (group 5) exhibited the highest polymorphism ($K = 30.582$, $S = 74$, $\eta = 98$, $Pa = 47$) and comparatively lower nucleotide substitution rate (6.54E-3 subs per site per year). The reason for this seemingly contradictory result may be that the monkey strains appeared earlier than the other hosts in the MCC tree (Fig. 3). Hence, it can be concluded that the relationship between the PBVs and the monkey host is very important and that a balance may exist between them. The new PBV group identified (group 4) showed two positive selection sites (sites 8 and site 39), the lowest evolutionary rate (4.05E-3 subs per site per year) and unique characteristics in the neutrality test. The results of Tajima's and F tests were not significant for all the analysed groups, except for group 4, which showed significant $(P < 0.05)$ and highly significant $(P < 0.02)$ results respectively (Table 1). Similar to most of the recent works, in the present study, a short segment of RdRp gene was used to construct the phylogenetic tree of PBVs and investigate its evolutionary history and phylodynamics. However, the use of one gene as a criterion, particularly, a part of the RdRp gene that is relatively highly conserved, to elucidate the global understanding of one virus is insufficient. Hence, future PBV studies should focus on full-length (segments 1 and 2) sequence analysis to gain a better understanding of the relationship between evolution and pathogenicity of PBVs.

## Experimental procedures

### Collection and concentration of the wastewater samples

Six samples were collected at different points from the stream of a wastewater treatment plant in Changsha, Hunan province, China, in autumn 2012. The recovery of viral particles and nucleic acid extraction was carried out by employing methods previously described (Pina et al., 1998). Briefly, 48 ml of sewage sample were ultracentrifuged at 200 000 × g for 1 h at 4°C. The sediment was dissolved in 4.8 ml of 0.25 N glycine buffer (pH 9.5) on ice for 30 min, and then the mixture was ultracentrifuged at 10 000 × g for 15 min and the suspended solids were removed. The supernatant was ultracentrifuged at 200 000 × g for 1 h at 4°C, and the virus in the pellet was resuspended in 0.1 ml of PBS and stored at −80°C.

### Nucleic acid extraction and reverse transcription

The QIAamp MinElute Virus Spin Kit was used for the extraction of nucleic acids. Briefly, 200 μl of concentrated virus and 200 μl of Buffer AL were added into a 1.5 ml tube at 56°C for 15 s. Then, the tube was centrifuged briefly to remove the drops, and the lysate was carefully applied onto the QIAamp MinElute column and centrifuged at 6000 × g for 1 min. Subsequently, 500 μl of Buffer AW1, Buffer AW2 and ethanol were respectively added to the mixture and centrifuged at 6000 × g for 1 min. Finally, 60 μl of RNase-free water were added to the centre of the membrane. Immediately following nucleic acid extraction, cDNA was synthesized from the extracted RNA by using a First-Strand Synthesis Superscript III Reverse Transcription Kit (Invitrogen, Carlsbad, CA) and stored at −20°C.

### PCR and sequencing of PBVs

A part of the genomic segment 2 of PBVs (genotype I) was amplified by PCR. The specific primers used were PicoB25 TGGTGTGGATGTTTC and PicoB43 ARTGYTGGT CGAACTT (~ 200 bp) (Martínez et al., 2003). The PCR conditions were as follows: initial denaturation at 94°C for 2 min,

followed by 40 cycles at 94°C for 1 min, 49°C for 2 min and 72°C for 3 min, and a final elongation step at 72°C for 5 min and 72°C for 10 min. All the PCR products were visualized by agarose gel (2%) electrophoresis with ethidium bromide under UV light. The EasyPure PCR Purification Kit (TransGen, Beijing, China) was used to purify the PCR products. To understand the diversity of the PBVs isolated from the treated wastewater, the positive PCR products from the six samples were cloned into pMD-19 T-vector (Takara, Dalian, China), and the transformants were screened for inserts by PCR and sequenced by Sangon (Sangon, Beijing, China). All the sequences were trimmed by EditSeq of DNAMAN VERSION 6.0.40. The identities of the positive PCR products were confirmed by comparing the sequences against the GenBank non-redundant database using BLASTN.

### Phylogenetic analysis

To gain a deeper understanding of the diversity of genotype I PBVs detected in the treated wastewater samples, phylogenetic analysis of the cloned PBVs was executed. The top three similar known sequences from NCBI corresponding to every sequence obtained in the present study were chosen as the reference sequences. All the sequences were aligned and a phylogenetic tree was constructed by using MEGA 5.0 (Tamura et al., 2011). Briefly, CLUSTAL of MEGA 5.0 was used for multiple alignments of all the sequences. A neighbour-joining (NJ) phylogenetic tree was constructed using a Kimura 2-parameter model by MEGA 5.0. The Kimura 2-parameter model was chosen because the average pairwise Kimura 2-parameter distance was < 1.0 (0.474), which is appropriate for creating an NJ tree. The phylogenetic tree was rooted with the human PBV strain 4-GA-91 (AF246940), which is the prototype strain for genotype II. Bootstrap analysis was performed with 1000 replicates. The bootstrap values > 80% were depicted at the nodes of the tree and non-significant values were omitted. The best-fit model of nucleotide substitution was selected by using JMODELTEST 0.1.1 based on AIC (Posada, 2008), and the corresponding evolutionary models were generated: HKY + G for groups 1 and 2, HKY for group 3, K80 + I for group 4, TVMef + I for group 5 and TPM1 for group 6.

### Polymorphism analyses

DNA polymorphism analyses are helpful to understand the evolutionary process, which include determination of the number of haplotypes (Hp), average number of pairwise nucleotide differences within the population (K), number of segregating sites (S) (Watterson, 1975), total number of mutations (Eta), parsimony informative sites (Pa), number of singletons [η(s)], and synonymous and non-synonymous substitution rates (Ka/Ks). The neutrality tests employed involve Tajima's test (Tajima, 1989) and Fu and Li's D and F tests (Fu and Li, 1993). The DNASP 5.0 package (Librado and Rozas, 2009) was used to analyse the results of genetic polymorphism and neutrality tests.

### Analyses of selection pressure

Global ω can reflect the direction and strength of selection. In the present study, the global ω was calculated for all the six groups by using HYPHY 2.1.0 (Pond and Muse, 2005). Furthermore, three algorithms, namely the SLAC, FEL and REL methods, were employed to analyse site-specific positive selection pressure for each group by using the same software (Wei *et al.*, 2013). The results of SLAC and FEL were deemed to be significant if $P < 0.05$, whereas for REL, Bayes factor > 20 was considered to be significant.

### Analyses of evolutionary rate and construction of MCC tree

The analyses of evolutionary rate can reflect some ecological pressures to a certain extent. In the present study, Bayesian analysis, using the Bayesian MCMC approach, was used to estimate the rate of molecular evolution by employing BEAST 1.7.2 (Drummond and Rambaut, 2007). For each group, a Bayes factor test was performed using Tracer v1.5 to choose the uncorrelated lognormal (UCLD) or exponential (UCED) distributions of strict and relaxed molecular clock. The strict clock assumes a single evolutionary rate along all branches, and the UCLD and UCED clocks allow evolutionary rates to vary along branches within lognormal and exponential distributions respectively. The UCLD clock was found to be the most appropriate model for groups 1 and 2. With regard to the other groups, the UCED was noted to be the best-fit clock model. The uncertainty was addressed as 95% HPD intervals. Subsequently, MCC tree was constructed by using TREEANNOTATOR 1.7.1 after discarding burn-in of 15%. The MCC tree was visualized with FIGTREE 1.3.1 and CORELDRAW x5.

### Nucleotide sequence accession numbers

All the sequences determined in this study were deposited in GenBank under accession numbers KJ135791 to KJ135929.

### Acknowledgement

We would like to thank Professor Fangqing Zhao in Beijing Institutes of Life Science, Chinese Academy of Sciences for his kind and useful discussion.

### Conflict of interest

The authors declare that they have no conflicts of interests.

### References

van den Berg, H., Lodder, W., van der Poel, W., Vennema, H., and de Roda Husman, A.M. (2005) Genetic diversity of noroviruses in raw and treated sewage water. *Res Microbiol* **156**: 532–540.

Bhattacharya, R., Sahoo, G.C., Nayak, M.K., Rajendran, K., Dutta, P., Mitra, U., *et al.* (2007) Detection of genogroup I and II human picobirnaviruses showing small genomic RNA profile causing acute watery diarrhoea among children in Kolkata, India. *Infect Genet Evol* **7**: 229–238.

Bodewes, R., van der Giessen, J., Haagmans, B.L., Osterhaus, A.D., and Smits, S.L. (2013) Identification of multiple novel viruses, including a parvovirus and a hepevirus, in feces of red foxes. *J Virol* **87**: 7758–7764.

Breitbart, M., Hewson, I., Felts, B., Mahaffy, J.M., Nulton, J., Salamon, P., and Rohwer, F. (2003) Metagenomic analyses of an uncultured viral community from human feces. *J Bacteriol* **185**: 6220–6223.

Cantalupo, P.G., Calgua, B., Zhao, G., Hundesa, A., Wier, A.D., Katz, J.P., *et al.* (2011) Raw sewage harbors diverse viral populations. *MBio* **2**: e00180-11.

Chandra, R. (1997) Picobirnavirus, a novel group of undescribed viruses of mammals and birds: a minireview. *Acta Virol* **41**: 59–62.

Drummond, A.J., and Rambaut, A. (2007) BEAST: Bayesian evolutionary analysis by sampling trees. *BMC Evol Biol* **7**: 214.

Fu, Y.-X., and Li, W.-H. (1993) Statistical tests of neutrality of mutations. *Genetics* **133**: 693–709.

Ganesh, B., Banyai, K., Masachessi, G., Mladenova, Z., Nagashima, S., Ghosh, S., *et al.* (2011) Genogroup I picobirnavirus in diarrhoeic foals: can the horse serve as a natural reservoir for human infection? *Vet Res* **42**: 52.

Ganesh, B., Bányai, K., Kanungo, S., Sur, D., Malik, Y.S., and Kobayashi, N. (2012) Detection and molecular characterization of porcine picobirnavirus in feces of domestic pigs from Kolkata, India. *Indian J Virol* **23**: 387–391.

Ghosh, S., Kobayashi, N., Nagashima, S., and Naik, T.N. (2009) Molecular characterization of full-length genomic segment 2 of a bovine picobirnavirus (PBV) strain: evidence for high genetic diversity with genogroup I PBVs. *J Gen Virol* **90**: 2519–2524.

Green, J., Gallimore, C., Clewley, J., and Brown, D. (1999) Genomic characterisation of the large segment of a rabbit picobirnavirus and comparison with the atypical picobirnavirus of Cryptosporidium parvum. *Arch Virol* **144**: 2457–2465.

Librado, P., and Rozas, J. (2009) DnaSP v5: a software for comprehensive analysis of DNA polymorphism data. *Bioinformatics* **25**: 1451–1452.

Lodder, W., and de Roda Husman, A. (2005) Presence of noroviruses and other enteric viruses in sewage and surface waters in The Netherlands. *Appl Environ Microbiol* **71**: 1453–1461.

Ludert, J., Abdul-Latiff, L., Liprandi, A., and Liprandi, F. (1995) Identification of picobirnavirus, viruses with bisegmented double stranded RNA, in rabbit faeces. *Res Vet Sci* **59**: 222–225.

Martínez, L., Giordano, M.O., Isa, M.B., Alvarado, L., Pavan, J., Rinaldi, D., and Nates, S.V. (2003) Molecular diversity of partial-length genomic segment 2 of human picobirnavirus. *Intervirology* **46**: 207–213.

Martínez, L.C., Masachessi, G., Carruyo, G., Ferreyra, L.J., Barril, P.A., Isa, M.B., *et al.* (2010) Picobirnavirus causes persistent infection in pigs. *Infect Genet Evol* **10**: 984–988.

Mondal, A., Chakravarti, S., Majee, S.B., and Bannalikar, A.S. (2013) Detection of picobirnavirus and rotavirus in diarrhoeic faecal samples of cattle and buffalo calves in Mumbai metropolis, Western India. *Vet Ital* **49**: 357–360.

Nates, S.V., Gatti, M.S.V., and Ludert, J.E. (2011) The picobirnavirus: an integrated view on its biology, epidemiology and pathogenic potential. *Future Virol* **6**: 223–235.

Pina, S., Jofre, J., Emerson, S.U., Purcell, R.H., and Girones, R. (1998) Characterization of a strain of infectious hepatitis E virus isolated from sewage in an area where hepatitis E is not endemic. *Appl Environ Microbiol* **64:** 4485–4488.

Pond, S.L.K., and Muse, S.V. (2005) HyPhy: hypothesis testing using phylogenies. *Bioinformatics* **21:** 676–679.

Posada, D. (2008) jModelTest: phylogenetic model averaging. *Mol Biol Evol* **25:** 1253–1256.

Rosario, K., Nilsson, C., Lim, Y.W., Ruan, Y., and Breitbart, M. (2009) Metagenomic analysis of viruses in reclaimed water. *Environ Microbiol* **11:** 2806–2820.

Rosen, B.I., Fang, Z.-Y., Glass, R.I., and Monroe, S.S. (2000) Cloning of human picobirnavirus genomic segments and development of an RT-PCR detection assay. *Virology* **277:** 316–329.

Smits, S.L., Poon, L.L., van Leeuwen, M., Lau, P.-N., Perera, H.K., Peiris, J.S.M., *et al.* (2011) Genogroup I and II picobirnaviruses in respiratory tracts of pigs. *Emerg Infect Dis* **17:** 2328–2330.

Smits, S.L., van Leeuwen, M., Schapendonk, C.M., Schürch, A.C., Bodewes, R., Haagmans, B.L., and Osterhaus, A.D. (2012) Picobirnaviruses in the human respiratory tract. *Emerg Infect Dis* **18:** 1539–1540.

Symonds, E.M., Griffin, D.W., and Breitbart, M. (2009) Eukaryotic viruses in wastewater samples from the United States. *Appl Environ Microbiol* **75:** 1402–1409.

Tajima, F. (1989) Statistical method for testing the neutral mutation hypothesis by DNA polymorphism. *Genetics* **123:** 585–595.

Tamura, K., Peterson, D., Peterson, N., Stecher, G., Nei, M., and Kumar, S. (2011) MEGA5: molecular evolutionary genetics analysis using maximum likelihood, evolutionary distance, and maximum parsimony methods. *Mol Biol Evol* **28:** 2731–2739.

Thurber, R.V., Haynes, M., Breitbart, M., Wegley, L., and Rohwer, F. (2009) Laboratory procedures to generate viral metagenomes. *Nat Protoc* **4:** 470–483.

Wang, Y., Tu, X., Humphrey, C., McClure, H., Jiang, X., Qin, C., *et al.* (2007) Detection of viral agents in fecal specimens of monkeys with diarrhea. *J Med Primatol* **36:** 101–107.

Wang, Y., Bányai, K., Tu, X., and Jiang, B. (2012) Simian genogroup I picobirnaviruses: prevalence, genetic diversity, and zoonotic potential. *J Clin Microbiol* **50:** 2779–2782.

Watterson, G. (1975) On the number of segregating sites in genetical models without recombination. *Theor Popul Biol* **7:** 256–276.

Wei, K., Chen, Y., and Xie, D. (2013) Genome-scale evolution and phylodynamics of H5N1 influenza virus in China during 1996–2012. *Vet Microbiol* **167:** 383–393.

## Supporting information

Additional Supporting Information may be found in the online version of this article at the publisher's web-site:

**Table S1.** Accession numbers of the 139 sequences we obtained used in this study.
**Table S2.** Accession numbers of the 65 reference sequences from NCBI used in this study.
**Table S3.** The hosts of the sequences we obtained in this study.

# 19

# Environmental factors influencing the distribution of ammonifying and denitrifying bacteria and water qualities in 10 lakes and reservoirs of the Northeast, China

XinYu Zhao,[1,3] Zimin Wei,[2**] Yue Zhao,[2] Beidou Xi,[3*] Xueqin Wang,[2] Taozhi Zhao,[2] Xu Zhang[2] and Yuquan Wei[2]

[1]College of Water Sciences, Beijing Normal University, Beijing 100875, China.

[2]Life Science College, Northeast Agricultural University, Harbin 150030, China.

[3]State Key Laboratory of Environmental Criteria and Risk Assessment, Chinese Research Academy of Environmental Sciences, Beijing 100012, China.

## Summary

This study presents seasonal and spatial variations of the ammonifying bacteria (AB) and denitrifying bacteria (DNB) and physicochemical parameters in 10 lakes and reservoirs in the northeast of China. Water samples were collected in winter (January), spring (March), summer (July) and fall (November) in 2011. The study revealed that physicochemical parameters such as pH, dissolved oxygen (DO), $NH_4^+$-N and nitrate as nitrogen were closely related with the distribution of AB and DNB. Seasonally, the levels of AB presents gradually upward trend from winter to summer, and declines in fall and DNB were higher in spring and fall than summer and lowest in winter. Spatially, the annual average of AB among 10 lakes and reservoirs showed insignificant difference ($P > 0.05$), for DNB, Udalianchi and Lianhuan Lake were lower than others ($P < 0.05$). Regression correlation analysis showed that the levels of AB and DNB had a close relationship with nitrogen nutrition. Three principal components were identified of total variances which are conditionally classified by the 'natural' factor (PC1) and 'nitrogen nutrients' (PC2, PC3). According the principal component scores, cluster analysis detected two distinct groups: (C1) mainly affected by nitrogen nutrients and (C2) natural environmental factors.

For correspondence. *E-mail xibeidou1126@126.com
**E-mail Weizm691120@163.com

Funding Information This study was partially supported by the Scientific Research Foundation for Major Science and Technology Program for Water Pollution Control and Treatment and National Science (No.2009ZX07106-001-001) and Technology Program (No.2012BAJ21B02).

## Introduction

Global nitrogen cycle (N-cycle) has increased attention since nitrogen loading have undoubtedly contributed to an increased occurrence of harmful in freshwaters, estuaries and coastal oceans (Herbert, 1999a). From a human point of view, the eutrophication of aquatic ecosystems by excess nitrogen has led to altered ecosystem function and structure, water quality degradation and economic loss (Bianchi et al., 1994). As a consequence of the high external loading with nitrate as nitrogen ($NO_3^-$-N) and nitrite as nitrogen ($NO_2^-$-N) growth of especially planktonic primary producers spring be enhanced, which can have profound effects on the quality of receiving waters.

On account of most plants and microorganisms in the water system couldn't make use of the nitrogenous organic matter directly, based on this, which should be transformed into absorbable components by microbial degradation. Degradation of organic matter leads to the formation of ammonia as nitrogen ($NH_4^+$-N) via ammoniation, which is either lost to the overlying water or oxidized to $NO_3^-$-N via nitrification at the oxic water interface (LeChevallier, 2003). Denitrification is a key process in the water nitrogen cycle since it decreases the amount of nitrogen available to the primary producers as the gaseous end-products ($N_2O$ and $N_2$) diffuse into the atmosphere. Microbial communities in the aquatic ecosystems play a key role help in the nutrient recycling which involves nitrogen fixation, ammonification, nitrification and denitrification processes carried out by different microorganisms (Altmann, 2003). The ability to ammonify and denitrify is widely distributed among ammonifying and denitrifying bacteria (Payne, 1973; Herbert, 1982). These transformations are not only mediated by a metabolically diverse range of autotrophic and heterotrophic microorganisms but also strongly affected water quality and eutrophication control by the prevailing physicochemical conditions such as proper pH, dissolved oxygen (DO) and concentration of nitrogen nutrition (Davies *et al.*, 1995; Juhna *et al.*, 2007). In order to understand the influence factors of the nitrogen nutrition in water, we need to study the relationship between the environment factors and microorganism. In addition, a number of studies have shown in temperate aquatic ecosystems that microorganism showed distinct seasonal patterns governed principally by seasonal variation (Smith *et al.*, 1985; Yoon and Benner, 1992).

Typical lakes and reservoirs in Northeast of China were important sources of drinking water areas. However, with the development of industry and agriculture in recent years, the water bodies of lakes and reservoirs had experienced water quality deterioration. As far as our knowledge is concerned, there are very few studies examining the dynamics of spatial-temporal variations of ammonifying bacteria (AB) and denitrifying bacteria (DNB), comparing the water quality among the typical

lakes and reservoirs in the northeast of China. The present study's aim was to identify the distribution of the AB and DNB and the relationship with physicochemical characteristics in the water bodies. Then, 10 typical lakes and reservoirs were classified into groups with the similar levels of indicators which would be beneficial for the future management. The main objective was to provide basic information and scientific data for policy makers and for the researchers to deal with similar kinds of water system.

## Results and discussion

### *Physicochemical parameters associated with distribution of ammonifying and denitrifying bacteria*

The seasonal variations of water quality values and correlation coefficients to identity the interrelationships for 10 lakes and reservoirs are given in Fig. 1, pH is a significant indicator for the growth of AB and DNB (Venkatesharaju *et al.*, 2010). Too much acid or alkaline pH would inhibit the growth of AB and DNB (Beversdorf *et al.*, 2013). As

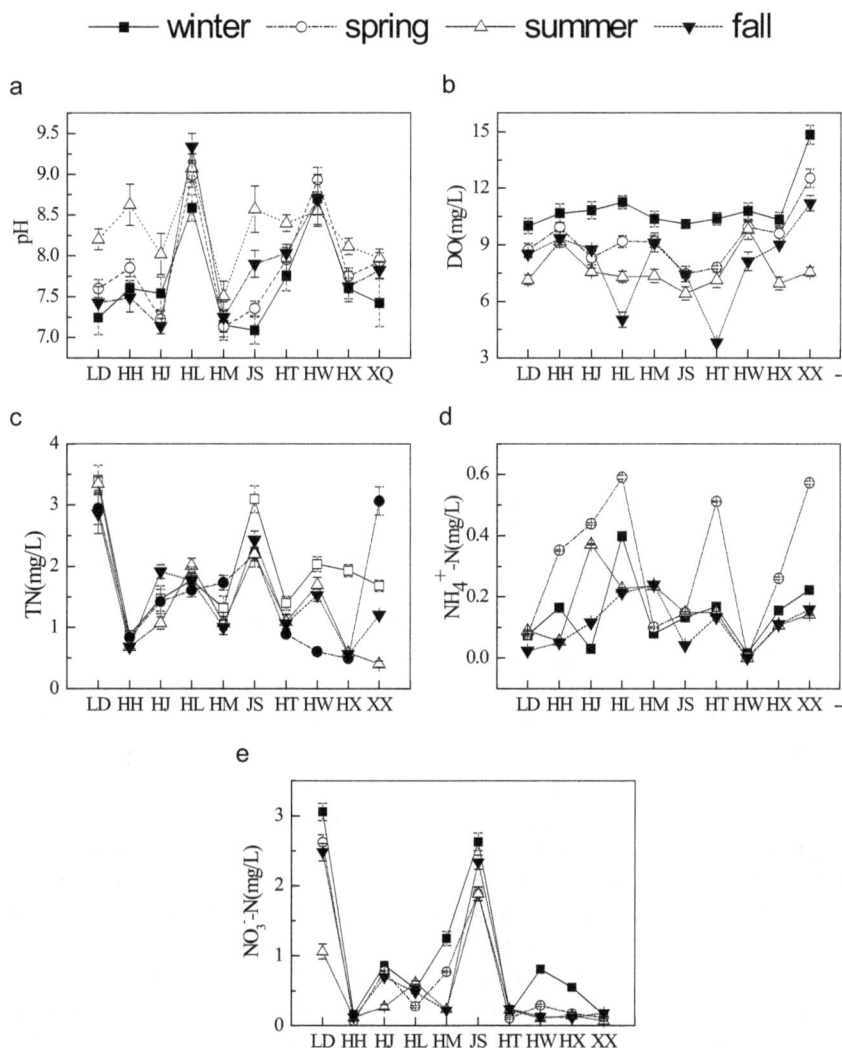

**Fig. 1.** Seasonal changes of variables pH, DO, TN, NH$_4^+$-N and NO$_3^-$-N in 10 lakes and reservoirs.

**Table 1.** Correlation matrix for levels of ammonia and denitrifying bacteria and physicochemical parameters in water samples.

| | TN | $NH_4^+$-N | LgAB | $NO_3^-$-N | $NO_2^-$-N | LgDNB | pH | DO | TSS | EC | TA | $COD_{Mn}$ | $HCO_3^-$ |
|---|---|---|---|---|---|---|---|---|---|---|---|---|---|
| $NH_4^+$-N | 0.034 | | | | | | | | | | | | |
| LgAB | −0.145 | 0.871** | | | | | | | | | | | |
| $NO_3^-$-N | 0.904** | −0.202 | −0.418 | | | | | | | | | | |
| $NO_2^-$-N | 0.789** | −0.287 | −0.236 | 0.735* | | | | | | | | | |
| LgDNB | 0.421 | −0.121 | −0.263 | 0.775* | 0.293 | | | | | | | | |
| pH | −0.178 | 0.616 | 0.639 | −0.368 | −0.255 | −0.090 | | | | | | | |
| DO | −0.239 | 0.807* | 0.304 | −0.785* | −0.340 | −0.744* | −0.214 | | | | | | |
| TSS | 0.460 | −0.141 | −0.189 | 0.488 | 0.666* | 0.233 | −0.424 | −0.161 | | | | | |
| EC | 0.077 | 0.466 | 0.523 | −0.144 | −0.165 | 0.127 | 0.805** | −0.324 | −0.574 | | | | |
| TA | 0.005 | 0.463 | 0.526 | −0.204 | −0.226 | 0.127 | 0.822** | −0.344 | −0.577 | 0.995** | | | |
| $COD_{Mn}$ | −0.458 | 0.799** | 0.512 | −0.646* | −0.629 | −0.210 | 0.675* | 0.247 | −0.517 | 0.523 | 0.546 | | |
| $HCO_3^-$ | −0.002 | 0.472 | 0.536 | −0.217 | −0.231 | 0.100 | 0.831** | −0.324 | −0.583 | 0.995** | 1.000** | 0.556 | |
| $CO_3^{2-}$ | 0.073 | 0.421 | 0.468 | −0.119 | −0.187 | 0.284 | 0.733* | −0.436 | −0.530 | 0.976** | 0.981** | 0.493 | 0.975** |

* means significant difference ($p < 0.05$) ** means significant difference ($p < 0.01$).

shown in Fig. 1A, pH in 10 lakes and reservoirs showed slightly alkaline all year around, especially in summer which might be due to that water was cleaner in summer with lower total suspended solids (TSS; Table 1). There was significant negative correlation between pH and TSS (Table 1). Spatially, pH in Lianhuan Lake (HL) and Udalianchi (HW) were obviously more alkaline than other lakes throughout the year. HW is a volcano dammed lake which has been around lots of peralkaline rocks. For HL, slightly alkaline pH was preferable in water which was due to the high carbonate or bicarbonate (Table 1). Heavy metals could be removed by high carbonate and bicarbonate precipitates (Ahipathy and Puttaiah, 2006).

DO was another important indicator to affect the process of microbial metabolism (Ahipathy and Puttaiah, 2006). Figure 1B also showed seasonal and spatial variation of DO for 10 lakes. Temporally, DO was the highest in winter and the lowest in summer. This is due to the breeding growth of microorganisms during summer, consumed considerable amount of oxygen with the decomposition of organic matter. Spatially, lower DO levels were detected at HL and Taoshan Reservoir (HT) in fall, and the highest annual averages of DO were observed in Xingkai Lake (XQ), which were probably due to the bigger wind and waves.

One of the most relevant factors to affect the distribution of AB and DNB is eutrophication. Total nitrogen (TN), $NH_4^+$-N and $NO_3^-$-N are the main nitrogen nutrients in the water body (Fig. 1C–E). As spring is the cultivation time along the river, a rapid release of ammonium to the water body, $NH_4^+$-N was the highest in March. Variations of $NO_3^-$-N were similar to those of TN, and there was a positive correlation between each other (Table 1). Spatial difference was most obviously reflected in Dahuofang (LD) and Songhua Lake (JS) with significantly high levels of TN and $NO_3^-$-N. The accumulation of $NO_3^-$-N in LD) and Songhua Lake (JS) showed that the nitrogen removal capacity by denitrification was stronger than others (Herbert, 1999a). In contrast, extremely low $NO_3^-$-N stimulated the level of ammonification at the expense of the denitrification in the water (King and Nedwell, 1985).

*Seasonal and changes in distribution of ammonifying and denitrifying bacteria*

The amount of ammonifying bacteria (AB) presents gradually upward trend from winter to summer, and declines in fall were found in almost 10 lakes (Fig. 2). Extremely low temperature in winter goes against the growth of AB, and a marked increase of AB in spring followed a rapid release of ammonium to the water body (Donnelly and Herbert, 1996; Poulin et al., 2007). The upward trend is still preserved in summer, due to the more frequently rainstorms and the rise of temperature in summer, water flow rates speed up which lead to much organic matter contain nitrogen flowed into the surface water as the substrates for AB (George et al., 2004; Djuikom et al., 2006). There were significant positive correlations between $NH_4^+$-N, DO and AB (Table 1). It is indicated that high values of $NH_4^+$-N may have stimulated the growth rate of AB, and ammoniation process also needs oxygen existing in the water body (Yang et al., 2007). Spatially, one-way analysis of variance (ANOVA) showed that the annual average of AB among 10 lakes and reservoirs showed insignificant difference ($P > 0.05$), while lower levels of AB were found in HL and HT in fall. Very low DO is a disadvantage of AB overgrowth of HT and HL (below 6 mg $l^{-1}$) (Fig. 1A).

Seasonal changes have a large impact on the distribution of denitrifying bacteria (Fig. 3). There was significant negative correlation with DO and DNB (Table 1). Lowest numbers of DNB were recorded when DO is at maximum during winter. This coincided with that nitrate can be reduced to $N_2O$ by a number of fermentative anaerobe bacteria, and excessive DO concentrations inhibited the

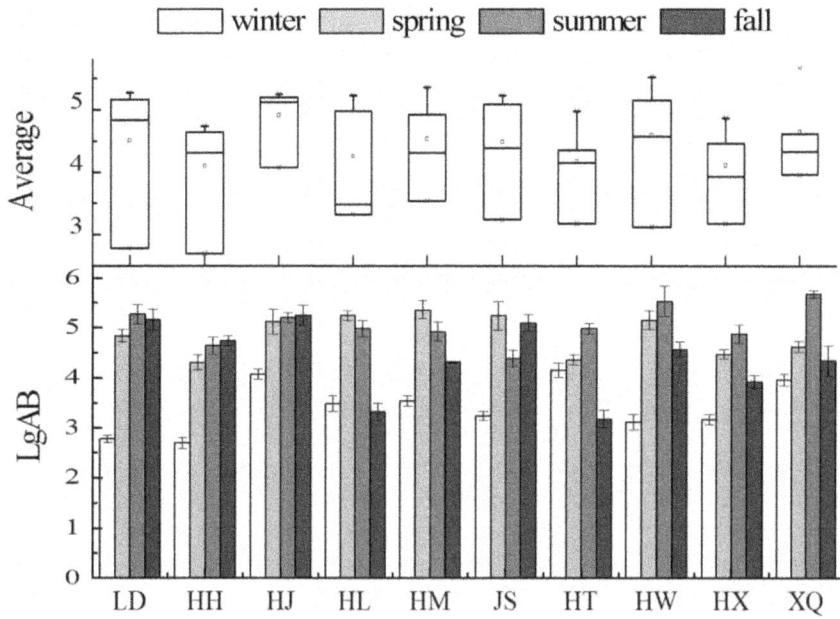

**Fig. 2.** Levels of AB in 10 water reservoirs. Each datum represents the mean ± standard deviation. The average of AB among the 10 reservoirs is not significantly different ($P > 0.05$) according to one-way ANOVA.

growth of DNB (Dunn *et al.*, 1980; MacFarlane and Herbert, 1982; Keith and Herbert, 1983). Generally, the numbers of DNB were higher in spring and fall than in summer. This is in accordance with the study that the capacity for $NO_3^-$-N reduction to $NH_4^+$-N was higher than reduction to $N_2$ which leads to the capacity for denitrification, which has been lower in summer (King and Nedwell, 1985). Spatially, ANOVA showed that the annual averages of denitrifying bacteria in HW and HL are lower than other lakes ($P < 0.05$). For these two lakes, lower levels of DNB were resulted in the alkaline pH value of water which is adverse to DNB growth (Fig. 1). It was

worth mentioning that extremely low DO was also limited to DNB growth in HT (below 4 mg l$^{-1}$) in fall (Fig. 1).

*Nitrogen compounds associated with distribution of ammonifying and denitrifying bacteria*

It is clearly evident from the foregoing section that ammoniation and denitrification in water system were subject to a complex array of regulatory mechanisms involving both physicochemical and biological factors (Herbert, 1999a). Hence, there is a need to understand the relationship between the average microbial biomass

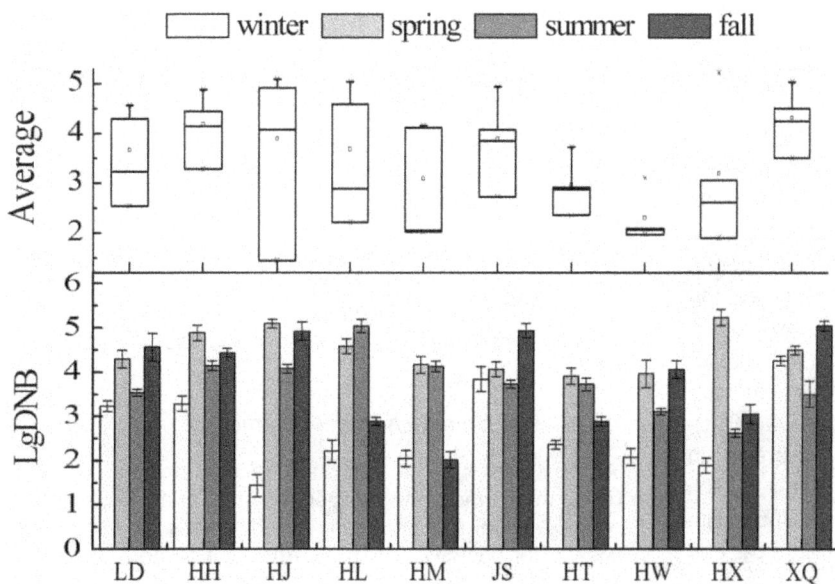

**Fig. 3.** Levels of DNB in 10 water reservoirs. Each datum represents the mean ± standard deviation. The average of DNB in HW, HL, HT is significantly different than other reservoirs ($P < 0.05$) according to one-way ANOVA.

of AB, DNB and nitrogen nutrients for each lake by regression correlation analysis (Fig. 4). Significant correlation with the amount of LgAB and TN or $NH_4^+$-N were found in Hongqipao (HH), Jingbo Lake (HJ), HL, HW, Xingkai Lake (XQ). Strong association were found between LgDNB and $NO_3^-$-N or $NO_2^-$-N in LD, HJ, JS, HT and XQ, while it should be noted that positive correlation only found in LD and JS, indicating that excessive $NO_3^-$-N and $NO_2^-$-N would stimulate the growth of DNB, denitrification would be the dominant process in LD and JS (King and Nedwell, 1985).

### Classification for lakes and reservoirs

In order to explain variance of large dataset of the related indicators with small groups, principal components analysis (PCA) was a conducted pattern recognition technique (Hopke, 1985). Three principal components were investigated with eigen-values > 1 summing almost 85.6% of the total variance in the water dataset (Table 2). The first PC (natural factor; PC1), accounting for 48.618% of the total variance was correlated with total alkalinity (TA), carbonate ($CO_3^{2-}$), electrical conductivity (EC), bicarbonate ($HCO_3^-$) and pH. This 'natural' factor was represented internal environmental characteristic factors of water's natural quality. These factors were mainly influenced from non-point sources such as fields, base erosions, soil erosion and atmosphere deposition (Kannel et al., 2007). Nitrogen nutrients PC2 and PC3, accounting for 22.749% and 14.219% of the total variance, respectively, can be grouped as nutrients. PC2 were correlated primarily with TN, $NO_3^-$-N, $NO_2^-$-N, LgDNB and secondarily with TSS and chemical oxygen demand ($COD_{Mn}$). The sources of these variables were mainly from anthropogenic pollution, such as municipal solid waste. PC3 were primary correlated with LgAB, DO and $NH_4^+$-N. Degradation of organic matter leads to the formation of $NH_4^+$-N based upon the oxic conditions of the water environment (Altmann et al., 2003).

Cluster analysis (CA) was applied to reveal a dendrogram in the 10 lakes and reservoirs (Fig. 5). After determining the number and identity of possible sources affecting surface waters by using PCA, site similarity were calculated next by CA on the principal component scores. It is possible to classify 10 lakes and reservoirs among various source components obtained by PCA which are grouped into distinct pattern of two main clusters: Cluster 1 (C1) composed of nine stations has two sub clusters: C1(a) consists of seven lakes [HH, HT, Xiquanyan (HX), HW, XQ, HJ and Mopanshan (HM)], which are main contributors to PC3. C1(b) consists of JS, LD, which appeared to be related to PC2. Cluster 2 only includes HL, which was mostly related to PC1, second with PC2 and PC3.

Cluster 1(a) showed significant relation with $NH_4^-$-N, DO and AB, accounting for high concentration organic matter, and $NH_4^-$-N was dominant compared with the other dissolved forms of nitrogen, ammonification was stronger in HH, HT, HX, HW, XQ, HJ and HM. Most of these lakes were aquacultures which were adjacent to land or residential areas; the source of $NH_4^-$-N might be associated with nitrogen compounds used in agriculture runoffs or human waste (Wakida and Lerner, 2005). C1 (b) showed that denitrification was a major role in LD and JS. This contributed to the higher level of $NO_3^-$-N compared with other dissolved forms of nitrogen in these areas. Bianchi and colleagues (1994) demonstrated that denitrification occurs in highly turbid estuarine waters with a high nitrate concentration (Bianchi et al., 1994). As the nitrate concentration increased, denitrification became the dominant process (King and Nedwell, 1987). Therefore, typical lake and reservoirs of northeast China are natural highly productive environments which in recent years have been subject to increased anthropogenic inputs of nitrogen arising from such diverse sources such as fertilizer run-off, selvage discharges or aquaculture (Ho et al., 2003). Cluster 2 corresponded to TA, $CO_3^{2-}$, EC, $HCO_3^-$ and pH, composed only HL. Clearly, nitrogen nutrient was not the major pollution in HL, and was associated with the base erosions, soil erosion and atmosphere deposition (Kannel et al., 2007).

### Experimental procedures

#### Field sampling

Water sampling was conducted in winter (January), spring (March), summer (June) and fall (September) in 2011. Ten water storage reservoirs which were located in three provinces of northeast China were selected. They were independent from each other (Fig. 6). These include Heilongjiang Province, Udalianchi, Xingkai Lake, Jingbo Lake, Mopanshan, Hongqipao, Lianhuan Lake, Taoshan Reservoir, Xiquanyan, Jilin province, Songhua Lake, Liaoning province, Dahuofang.

#### Determination of denitrifying and ammonifying bacteria

Ammonifying bacteria was cultivated in peptone ammoniation medium, and DNB was in denitrifying bacteria culture medium (Rodina, 1972). Numbers of AB and DNB in water samples was conducted using most probable number.

#### Analysis of physicochemical parameters

Water samples were collected from approximately 16 m below the surface with an open-mouthed bottle (for analysis of physicochemical parameters) and a sterile 1000 ml glass vessel (for bacteria analyses). For each reservoir, different numbers of sampling points were selected according to the

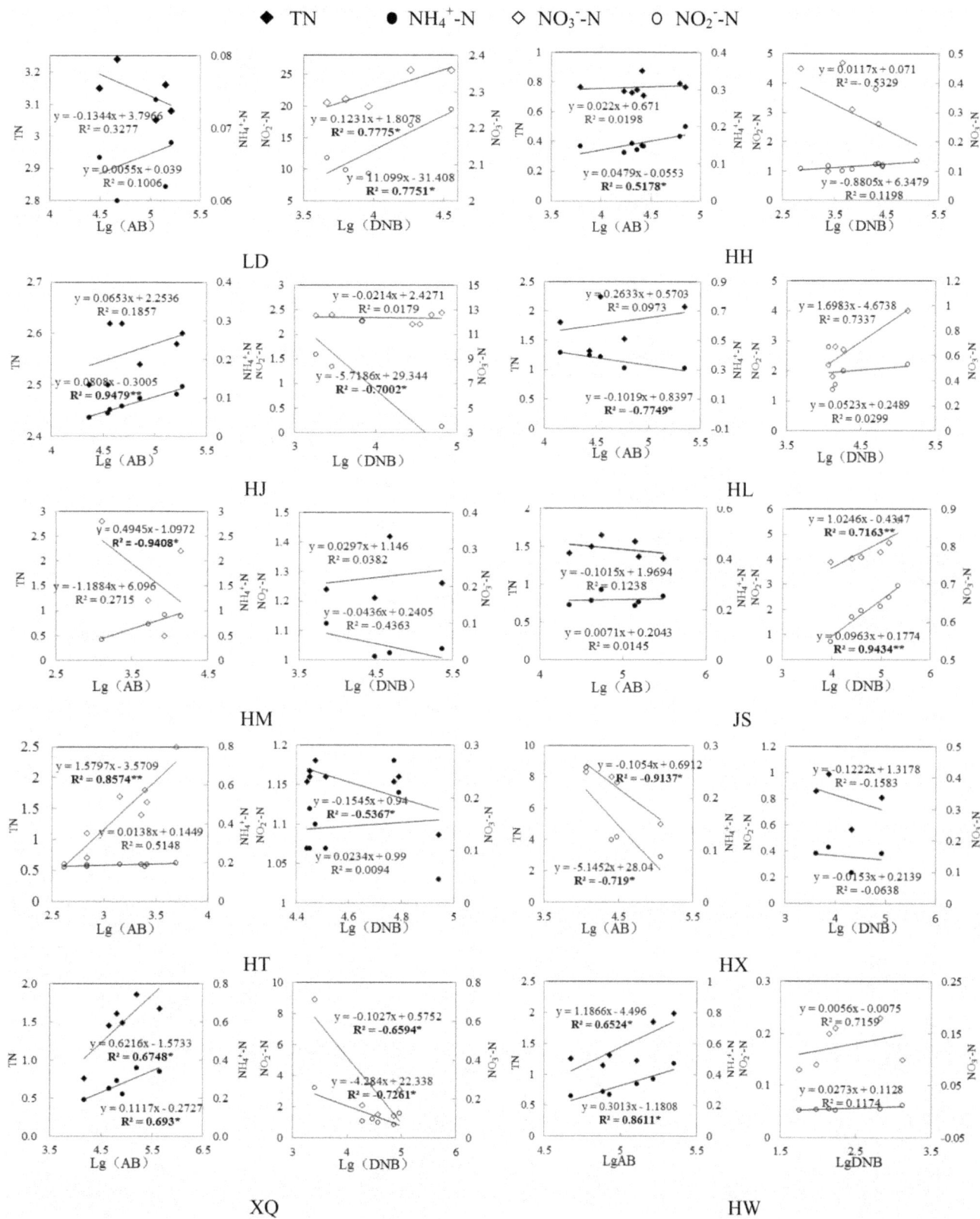

**Fig. 4.** Regression correlation matrix for AB and DNB levels and nitrogen compounds in 10 lakes and reservoirs.

**Table 2.** Rotated (varimax rotation) factor loadings and communalities.

|                    | PC1    | PC2    | PC3    |
|--------------------|--------|--------|--------|
| TA                 | 0.978  | −0.065 | 0.110  |
| $CO_3^{2-}$        | 0.977  | 0.022  | 0.047  |
| EC                 | 0.975  | −0.007 | 0.130  |
| $HCO_3^-$          | 0.974  | −0.078 | 0.123  |
| pH                 | 0.802  | −0.243 | 0.302  |
| TN                 | 0.061  | 0.961  | 0.108  |
| $NO_3^-$-N         | −0.101 | 0.937  | −0.170 |
| $NO_2^-$-N         | −0.202 | 0.851  | 0.031  |
| LgDNB              | 0.140  | 0.689  | −0.386 |
| TSS                | −0.562 | 0.590  | 0.170  |
| $COD_{Mn}$         | 0.520  | −0.572 | 0.505  |
| LgAB               | 0.433  | −0.228 | 0.843  |
| $NH_4^+$-N         | 0.436  | −0.099 | 0.795  |
| DO                 | −0.451 | 0.251  | 0.677  |
| Percent variance (%) | 48.618 | 22.749 | 14.219 |

location or shape, and all of the samplings were conducted in 9:30–11:30 (a.m.). In this study, 12 parameters were detected, which were pH, EC, DO, TN, $NO_3^-$-N, $NO_2^-$-N, $NH_4^+$-N, $COD_{Mn}$, TA, $HCO_3^-$, $CO_3^{2-}$, TSS. The data quality was checked by careful standardization, procedural blank measurements, spiked and duplicate samples. The analysis methods were based on standard methods in water and wastewater monitoring analysis method (4th edition). Part of the data were shown in Fig. 1A–E, and all data processing used ORIGIN (8.0).

### Data analysis

One-way analysis of variance followed by the LSD comparisons test were used to compare AB and DNB levels among the reservoirs in different seasons respectively. Pearson linear correlations and PCA were used to study the relationship between the annual mean of AB, DNB and physicochemical parameters. Regression correlation matrix

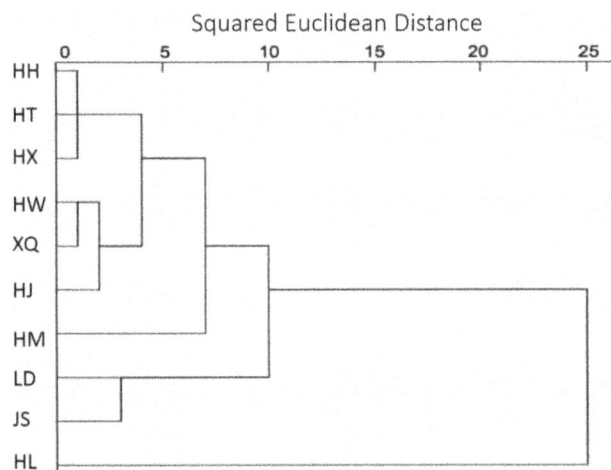

**Fig. 6.** Distribution of 10 water reservoirs.

was used to identify the correlation between levels of AB and DNB and nitrogen compounds in each lake and reservoir. In PCA analysis, factors were denitrified via varimax rotation with eigenvalue > 1. Cluster analysis was calculated by principal component scores. It makes it possible to classify water stations among various source components obtained by PCA. The Baverage's linkage cluster method was applied to the data. All data processing used SPSS (19.0).

### Conclusions

This study indicated that seasonal variation and physicochemical properties of the water would influence the levels of AB and DNB in directly and indirectly way. Statistical analysis demonstrated that AB and DNB were closely related to physicochemical factors such as pH, DO, $NH_4^+$-N and $NO_3^-$-N. Regression correlation analysis showed that AB and DNB were closely related to nitrogen-related indictors and strong positive correlation between DNB and $NO_3^-$-N, $NO_2^-$-N were only found LD and JS. Principle component analysis revealed that the major factors in 10 lakes and reservoirs were: natural factors (PC1: TA, $CO_3^{2-}$, EC, $HCO_3^-$, pH) and nitrogen nutrients (PC2: TN, $NO_3^-$-N, $NO_2^-$, LgDNB; PC3: LgAB, DO, $NH_4^+$-N). The cluster analysis detected two distinct groups: (a) C1 (HH, HT, HX, HW, XQ, HJ and HM) was major affected by $NH_4^+$-N, AB, and ammoniation was stronger; (b) (LD and JS) was mainly affected by the TN, $NO_3^-$-N, $NO_2^-$-N and DNB, denitrification was a major role; C2 (HL) was majorly affected by natural factors.

### Conflict of interest

None declared.

### References

Ahipathy, M.V., and Puttaiah, E.T. (2006) Ecological characteristics of vrishabhavathy River in Bangalore (India). *Environ Geol* **49:** 1217–1222.

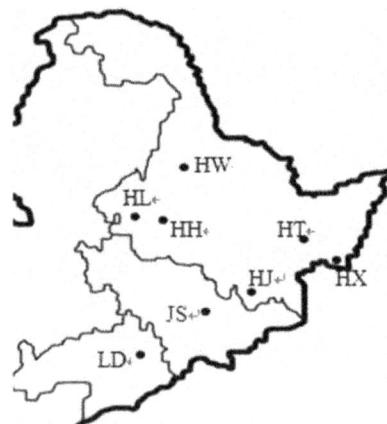

**Fig. 5.** Dendrogram with Baverage's linkage and correlation distance obtained from hierarchical cluster analysis for 10 water reservoirs.

Altmann, D. (2003) *Nitrification in Freshwater Sediments as Studied with Microsensors and Fluorescence In Situ Hybridization.* Ph. D. Dissertation, University of Bremen, Germany.

Beversdorf, L.J., Miller, T.R., and McMahon, K.D. (2013) The role of nitrogen fixation in cyanobacterial bloom toxicity in a temperate, eutrophic lake. *PLoS ONE* **8:** e56103.

Bianchi, M., Bonin, P., and Feliatra, P. (1994) Bacterial nitrification and denitrification rates in the Rhone River plume (northwestern Mediterranean-Sea). *Mar Ecol Prog Ser* **103:** 197–202.

Davies, C.M., Long, J.A., Donald, M., and Ashbolt, N.J. (1995) Survival of fecal microorganisms in marine and freshwater sediments. *Appl Environ Microbiol* **61:** 1888–1896.

Djuikom, E., Njine, T., Nola, M., Sikati, V., and Jugnia, L. (2006) Microbiological water quality of the Mfoundi River watershed at Yaoundé, Cameroon, as inferred from indicator bacteria of fecal contamination. *Environ Monit Assess* **122:** 171–183.

Donnelly, A.P., and Herbert, R.A. (1996) An investigation into the role of bacteria in the remineralisation of organic nitrogen in shallow coastal sediments of the Northern Adriatic Sea. In *Transfer Pathways and Flux of Organic Matter and Related Elements in Water and Sediments of the Northern Adriatic Sea and Their Importance in the Eastern Mediterranean Sea.* Price, N.B. (ed.). EuromargeAS Project, Final Report, European Union, Brussels, pp. 189–197.

Dunn, G.M., Wardell, J.N., Herbert, R.A., and Brown, C.M. (1980) Enrichment, enumeration and characterisation of nitrate-reducing bacteria present in sediments of the River Tay Estuary. *Proc R Soc Edinb Biol Sci* **78:** s47–s56.

George, I., Anzil, A., and Servais, P. (2004) Quantification of fecal coliform inputs to aquatic systems through soil leaching. *Water Res* **38:** 611–618.

Herbert, R.A. (1982) Nitrate dissimilation in marine and estuarine sediments. In *Sediment Microbiology.* Nedwell, D.B., and Brown, C.M. (eds). London, UK: Academic Press, pp. 53–71.

Herbert, R.A. (1999a) Nitrogen cycling in coastal marine ecosystems. *FEMS Microbiol Rev* **23:** 563–590.

Ho, K.C., Chow, Y.L., and Yau, J. (2003) Chemical and microbiological qualities of The East River (Dongjiang) water, with particular reference to drinking water supply in Hong Kong. *Chemosphere* **52:** 1441–1450.

Hopke, P.K. (1985) *Receptor Modeling in Environmental Chemistry.* USA: John Wiley & Sons.

Juhna, T., Birzniece, D., and Rubulis, J. (2007) Effect of phosphorus on survival of *Escherichia coli* in drinking water biofilms. *Appl Environ Microbiol* **73:** 3755–3758.

Kannel, P.R., Lee, S., Kanel, S.R., Khan, S.P., and Lee, Y. (2007) Spatial–temporal variation and comparative assessment of water qualities of urban river system: a case study of the river Bagmati (Nepal). *Environ Monit Assess* **129:** 433–459.

Keith, S.M., and Herbert, R.A. (1983) Dissimilatory nitrate reduction by a strain of Desulfovibrio desulfuricans. *FEMS Microbiol Lett* **18:** 55–59.

King, D., and Nedwell, D.B. (1985) The influence of nitrate concentration upon the end-products of nitrate dissimilation by bacteria in anaerobic salt marsh sediment. *FEMS Microbiol Lett* **31:** 23–28.

King, D., and Nedwell, D.B. (1987) The adaptation of nitrate-reducing bacterial communities in estuarine sediments in response to overlying nitrate load. *FEMS Microbiol Lett* **45:** 15–20.

LeChevallier, M.W. (2003) Conditions favouring coliform and HPC bacterial growth in drinking water and on water contact surfaces. In *Heterotrophic Plate Count Measurement in Drinking Water Safety Management.* IWA Publishing (ed.). Geneva, London, UK: World Health Organization, pp. 177–198.

MacFarlane, G.T., and Herbert, R.A. (1982) Nitrate dissimilation by Vibrio spp. isolated from estuarine sediments. *J Gen Microbiol* **128:** 2463–2468.

Payne, W.J. (1973) Reduction of nitrogenous oxides by microorganisms. *Bacteriol Rev* **37:** 409–452.

Poulin, P., Pelletier, E., and Saint-Louis, R. (2007) Seasonal variability of denitrification efficiency in northern salt marshes: an example from the St. Lawrence Estuary. *Mar Environ Res* **63:** 490–505.

Rodina, A.G. (1972) *Methods in Aquatic Microbiology.* Colwell, R.R., and Zambruski, M.S. (eds). Baltimore. Buterworths, London: University Park Press, p. 461.

Smith, C.J., DeLaune, R.D., and Patrick, W.H. (1985) Fate of riverine nitrate entering an estuary: I. Denitrification and nitrogen burial. *Estuaries* **8:** 15–21.

Venkatesharaju, K., Ravikumar, P., Somashekar, R.K., and Prakash, K.L. (2010) Physico-chemical and bacteriological investigation on the river Cauvery of Kollegal stretch in Karnataka. *J Sci Eng Technol* **6:** 50–59.

Wakida, F.T., and Lerner, D.N. (2005) Non-agricultural sources of groundwater nitrate: a review and case study. *Water Res* **39:** 3–16.

Yang, H., Shen, Z., Zhang, J., and Wang, W. (2007) Water quality characteristics along the course of the Huangpu River (China). *J Environ Sci* **19:** 1193–1198.

Yoon, W.B., and Benner, R. (1992) Denitrification and oxygen consumption in sediments of two south Texas estuaries. *Mar Ecol Prog Ser* **90:** 157–167.

# 20

# Restoration of a Mediterranean forest after a fire: bioremediation and rhizoremediation field-scale trial

Paloma Pizarro-Tobías,[1] Matilde Fernández,[2] José Luis Niqui,[1] Jennifer Solano,[1] Estrella Duque,[2] Juan-Luis Ramos[2]*[†] and Amalia Roca[1]

[1]Bio-Ilíberis R&D, Polígono Industrial Juncaril, Peligros, Granada 18210, Spain.
[2]Estación Experimental del Zaidín-CSIC, Granada, Granada 18008, Spain.

## Summary

Forest fires pose a serious threat to countries in the Mediterranean basin, often razing large areas of land each year. After fires, soils are more likely to erode and resilience is inhibited in part by the toxic aromatic hydrocarbons produced during the combustion of cellulose and lignins. In this study, we explored the use of bioremediation and rhizoremediation techniques for soil restoration in a field-scale trial in a protected Mediterranean ecosystem after a controlled fire. Our bioremediation strategy combined the use of *Pseudomonas putida* strains, indigenous culturable microbes and annual grasses. After 8 months of monitoring soil quality parameters, including the removal of monoaromatic and polycyclic aromatic hydrocarbons as well as vegetation cover, we found that the site had returned to pre-fire status. Microbial population analysis revealed that fires induced changes in the indigenous microbiota and that rhizoremediation favours the recovery of soil microbiota in time. The results obtained in this study indicate that the rhizoremediation strategy could be presented as a viable and cost-effective alternative for the treatment of ecosystems affected by fires.

## Introduction

During high temperatures and lack of rainfall, forest fires represent the most frequent perturbation within

*For correspondence: E-mail juan.ramos@research.abengoa.com

Funding Information Work in our laboratory was supported by Fondo Social Europeo and Fondos FEDER from the European Union, through several projects (BIO2010-17227, Consolider-Ingenio CSD2007-00005, Excelencia 2007 CVI-3010, Excelencia 2011 CVI-7391 and EXPLORA BIO2011-12776-E).

Mediterranean ecosystems (Hernández *et al.*, 1997; Vila-Escalé *et al.*, 2007). Loss of forest mass is an extended concern throughout the Mediterranean basin; in 2012 almost 200 000 hectares (Ha) of forests were affected by fire in Spain. This amount was three times the land area affected in 2011, making it one of the most devastating years for forest biomass in the Iberian Peninsula (https://magrama.gob.es/es/desarrollo-rural/temas/politica-forestal/incendios-forestales/lucha.aspx). While drought and heat are natural causes of wildfires, many occur due to incorrect agricultural management, negligence or as the result of economic interests (Olivella *et al.*, 2006; Vergnoux *et al.*, 2011).

Fire-induced perturbations comprise changes in the physical, mineralogical, chemical and biological properties of soil (Certini, 2005), with levels of severity depending on the intensity and duration of combustion (Campbell *et al.*, 1994; Franklin *et al.*, 1997; DeBano *et al.*, 1998). The immediate effects of fire on soil include: (i) the incineration of associated vegetation cover, which changes nutrient availability and surface organic matter content (Vázquez *et al.*, 1993), (ii) a significant decrease in microbial cell density per gram of soil (DeBano *et al.*, 1998; Certini, 2005), (iii) compositional changes in soil microbial populations (Torres and Honrubia, 1997; Smith *et al.*, 2008), (iv) reduction of water infiltration and rainfall retention, which is required to support plants and, thus, important for resisting erosion (DeBano, 2000; González-Pérez *et al.*, 2004) and (v) the release of several pyrolytic substances as polycyclic aromatic hydrocarbons (PAHs), which are toxic and have a tendency to accumulate in tissues (Vila-Escalé *et al.*, 2007). Soil dynamics depend not only on physicochemical properties, but also on microbiological health because the return of vegetation after a fire is directly impacted by the metabolic activity of microorganisms, which facilitate nutrient cycling (Certini, 2005).

Technology for the remediation of PAH-polluted sites has traditionally been centred on physicochemical treatments (Fernández *et al.*, 2012); however, more recently, the use of microorganisms for *in situ* degradation of pollutants has gained popularity as a bioremediation process (Kuiper *et al.*, 2004; Segura *et al.*, 2009). In these processes, either native degraders or exogenous microorganisms with appropriate metabolic traits are used. When indigenous microbes are used, the process is known as bioaugmentation (Andreoni *et al.*, 2004; Segura *et al.*, 2009).

Due to the slow natural recovery of soil after a perturbation, bioremediation techniques have been developed combining microorganisms with plants to accelerate the recovery of soil properties, increase microbial biomass and accelerate plant recolonization. The general process is referred to as phytoremediation, whereas the process is known as rhizoremediation when plants with root-associated microorganisms are used (Kuiper *et al.*, 2001; 2004; Wood, 2008; Segura *et al.*, 2009; Segura and Ramos, 2012). Hence, in a broad sense, the term bioremediation encompasses rhizoremediation; however, in this article, we will refer to the use of microorganisms alone (without plants) as 'bioremediation' to distinguish it from rhizoremediation (joint plant-microbe processes).

There have been few studies that tackle field-scale bioremediation and/or rhizoremediation assays without using physicochemical treatments (Huang *et al.*, 2004; Bamforth and Singleton, 2005); in addition, these technologies had not been tested in ecosystems affected by fire, emphasizing the importance of results of the current study. The aims of this current field-scale study were to assess bioremediation and rhizoremediation techniques for the recovery of soil health to pre-fire levels. The chosen methods involved the use of rapidly growing pasture seeds to curb erosion; together with the addition of plant growth-promoting rhizobacteria (PGPR) with biodegradative properties to facilitate the degradation of PAHs, to promote seed germination, plant recolonization and vegetation growth. The results suggest that rhizoremediation was an effective and inocuous treatment to be used in the restoration of Mediterranean ecosystems.

## Results

### Bacterial survival

The survival of the introduced microbial consortium (*Pseudomonas putida* strains and indigenous culturable bacteria with biodegradation potential) in burnt bulk soil (Bioremediation treatment), rhizosphere of introduced plants in burnt soil (Rhizoremediation treatment) and in the rhizosphere of plants in pristine soil (Treated control) (Table 1, Fig. 1) was determined by plate counting on selective media; besides, survival of introduced *P. putida* strains was verified by either polymerase chain reaction (PCR) or colony hybridization. The two *P. putida* strains, *P. putida* BIRD-1 and KT2440, that were introduced in soil were able to survive and maintained a population size around $10^6$ colony-forming unit (cfu) $g^{-1}$ soil (from $6.7 \times 10^6 \pm 1.1 \times 10^6$ to $5.7 \times 10^5 \pm 1.7 \times 10^4$ cfu $g^{-1}$ soil) for about six months (26 weeks), and then dropped below detection limits during the aestival season (Fig. 1A and B). No significant differences ($P > 0.05$) were detected in the survival of these strains the rhizosphere of plants in the rhizoremediation treatment compared to the treated pristine treatment in most of the sampling times during the first 24 weeks (6 months). Cell densities of these strains in burnt bulk soil (Bioremediation treatment), were significantly lower ($P < 0.05$) (two orders of magnitude) than in the rhizosphere of plants, whether in burnt or pristine soil. This indicates that survival improved when the strains were associated to plants, especially in the case of *P. putida* BIRD-1 that was detected for seven months in rhizospheric soil.

As expected, the culturable indigenous hydrocarbon-degrading microbial consortium, monitored by plate

**Table 1.** Composition of treatments and strains applied to burnt and pristine soil.

### Burnt soil

| Treatment | Composition | Microorganisms applied | Plant seeds mixture |
| --- | --- | --- | --- |
| Control | Untreated bare soil | None | None |
| Plants control | Non-inoculated plants | None | Peat AVEXIII® Trifolium repens |
| Bioremediation | Microbial consortium | P. putida BIRD-1 (pWW0) P. putida KT2440 (pWW0) Indigenous bacterial consortium | Peat |
| Rhizoremediation | Plants and microbial consortium | P. putida BIRD-1 (pWW0) P. putida KT2440 (pWW0) Indigenous bacterial consortium | Peat AVEXIII® Trifolium repens |

### Pristine soil

| Treatment | Composition | Microorganisms applied | Plant seeds mixture |
| --- | --- | --- | --- |
| Control | Untreated | None | None |
| Treated | Plants and microbial consortium | P. putida BIRD-1 (pWW0) P. putida KT2440 (pWW0) Indigenous bacterial consortium | Peat AVEXIII® Trifolium repens |

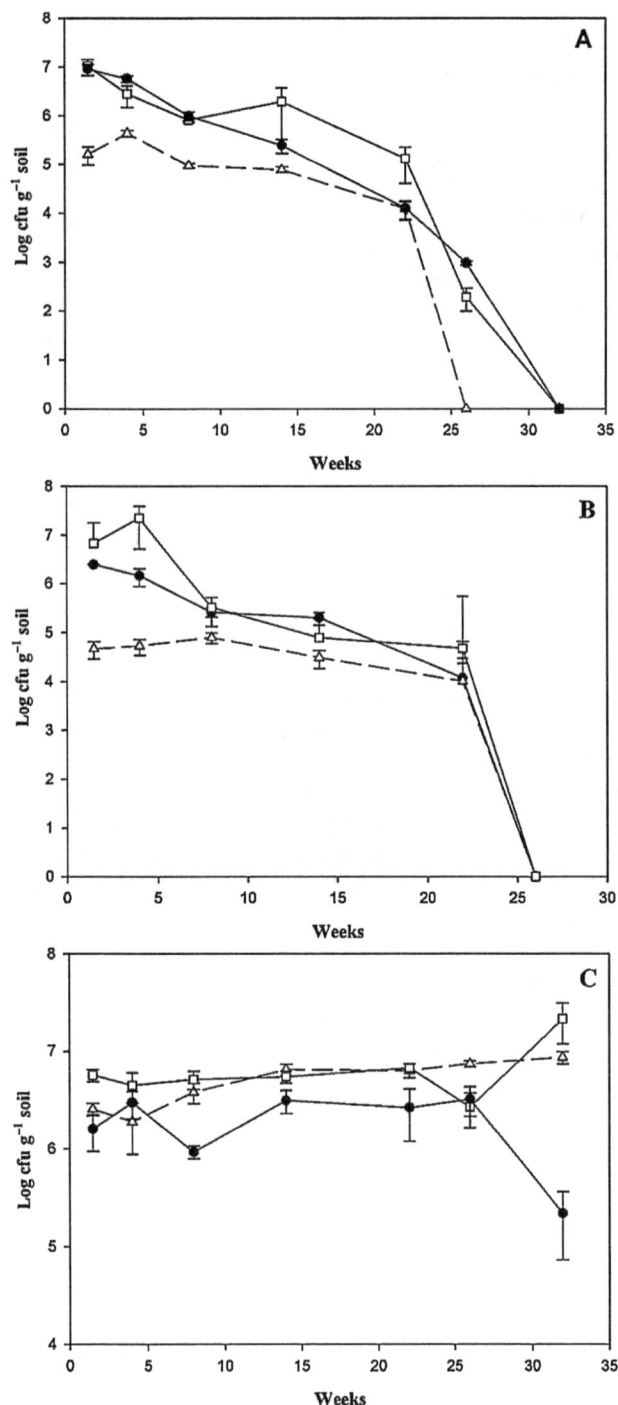

**Fig. 1.** Viable *P. putida* BIRD-1 (A), *P. putida* KT2440 (B) and indigenous microbial consortium (C) in rhizosphere of introduced plants in pristine soil (Treated) (filled circle), in the rhizosphere of introduced plants in burnt soil (Rhizoremediation) (square) and in burnt bulk soil (Bioremediation) (triangle). Data showed as mean (*n* = 3) and error bars refer to standard deviations.

counting with diesel fuel as a sole carbon source (Fig. 1C), revealed a steady population size around $10^6$ cfu g$^{-1}$ soil (from $1.6 \times 10^6 \pm 7.9 \times 10^5$ to $8.6 \times 10^6 \pm 1.6 \times 10^6$ cfu g$^{-1}$ soil). These population levels remained

quite constant until the end of the study, showing no significant differences ($P > 0.05$) regardless of whether the soil was pristine or burnt or whether exogenous *P. putida* strains and/or plants were present at most sampling times (Fig. 1C), except at the end of the study (32 weeks) in which cell densities in pristine soil dropped one order of magnitude.

*Metagenomic analysis of soil microbial population*

In order to determine the consequences of fire on soil microbiota, as well as to study the effect of rhizoremediation treatments over the spectrum of indigenous microbial populations, a metagenomic analysis of 16S RNA for bacterial biodiversity was carried out at month 1 (autumn, November 2008) and 6 months after the beginning of the trial (spring, April 2009) for Control burnt soil (bulk), Control pristine soil (rhizospheric) and soil undergoing Rhizoremediation (rhizospheric).

Rarefaction analyses were performed to compare bacterial richness among pristine, burnt and soil undergoing rhizoremediation. Analyses were based on a minimum of 125 sequences and the number of operational taxonomic units (OTUs) were estimated using a cut-off of 97% for sequence similarity a generally accepted level for comparative analysis of whole and partial 16S rRNA sequences (Konstantinidis *et al.*, 2006). The rarefaction curve (Supporting Information Fig. S2) showed a similar number of OTUs in all soil samples, indicating similar bacterial richness.

The analysis of relative abundance at phylum level (Fig. 2) showed changes in the bacterial community distribution and proportion. *Proteobacteria*, *Acidobacteria* and *Bacteriodetes* were the predominant phyla in all the cases we studied; combined, these three phyla constituted 80% of the total. Specifically, *Acidobacteria*, which was the prevailing phylum in pristine soil and 46% of the total, experienced a remarkable population reduction to 29% of the total in burnt soil. In contrast to *Acidobacteria*, an increase in the proportion of *Proteobacteria* (from 31.3% to 38.6%) and *Bacteriodetes* (from 8.4% to 12.9%) was observed in Control burnt soil.

Less common phyla exhibited significant changes in their relative abundance: *Verrucomicrobia*, which made up 5.3% of the total in pristine soil, dropped to 1.4% in Control burnt soil and *Chloroflexi* and *Gemmatimonadetes* dropped below detection limits; nevertheless, the proportion of these phyla was less than 1% in pristine soil (0.8%). On the other hand, *Actinobacteria* was more abundant in burnt (8.6%) than in pristine samples (3.8%) at the first sampling time point; similarly, *Gemmatimonadetes* was also more predominant in burnt (4.2%) than in pristine (undetectable) soil at the last sampling time point. No bacterial phylum

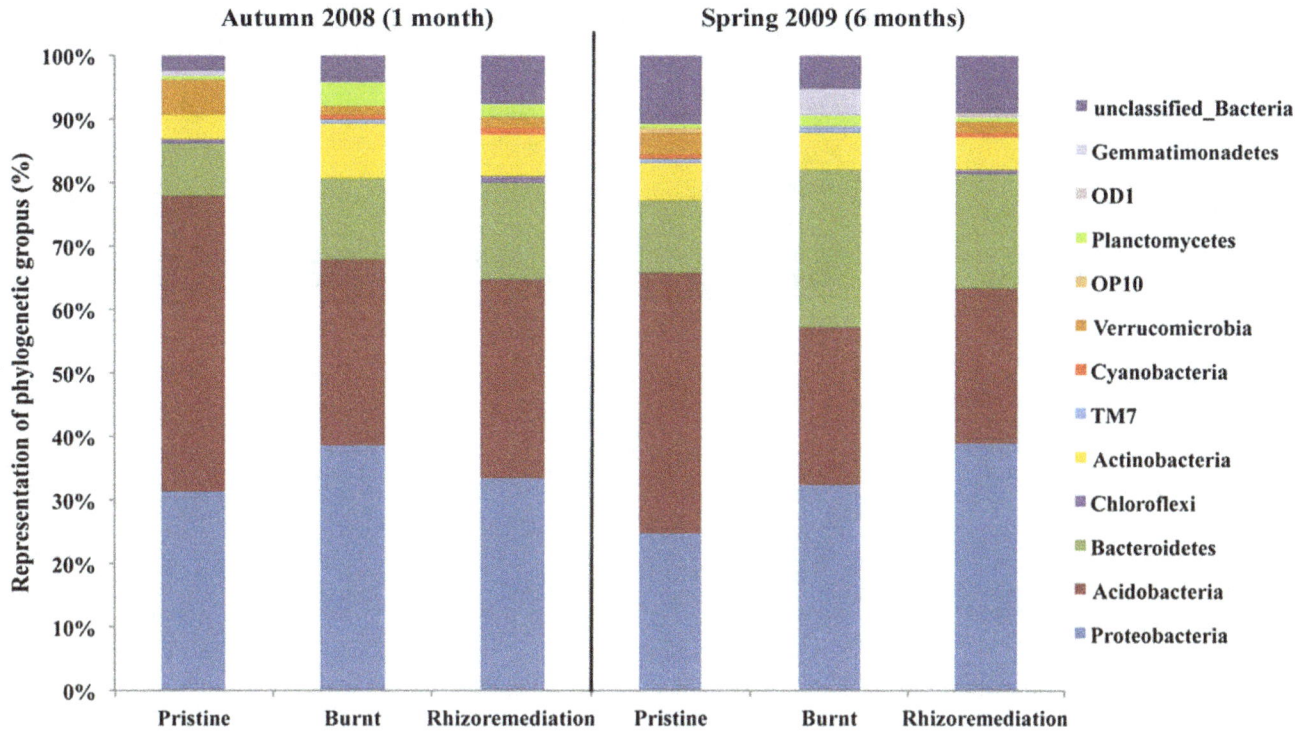

**Fig. 2.** Relative abundance of phylogenetic groups at phylum level based on 16S rRNA genes. Pristine soil, bulk burnt soil and rhizoremediation treatment at two different times of the assay 1 month (autumn) and 6 months (spring).

was reduced exclusively in soil under Rhizoremediation treatment.

A principal coordinate analysis (PCoA) was performed to compare genetic distance matrix between groups (Fig. 3). Statistical differences were observed in the phylogenetic composition of microbial populations between burnt and pristine soil, with intermediate values

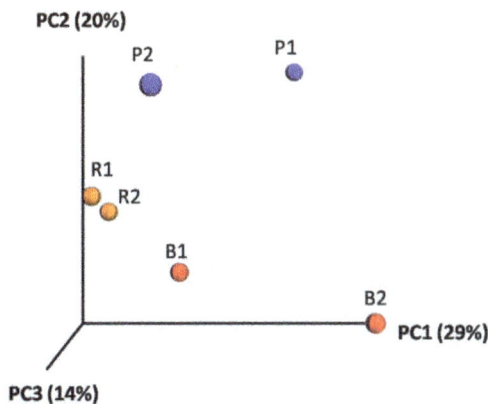

**Fig. 3.** Principal coordinate analysis (PCoA) constructed using FASTUNIFRAC software (Hamady et al., 2010) to compare sample groups from phylogenetic distance-matrix, determining similarity between or differences among pristine soil (P), bulk burnt soil (B) and rhizoremediation treatment (R) at two different times of the assay 1 month (autumn) and 6 months (spring). 3D image obtained through EMPeror software (Vazquez-Baeza et al., 2013).

observed for sites that were treated using Rhizoremediation. Moreover, Control burnt and Control pristine soil microbial populations showed season-induced variations while the Rhizoremediation ones remained quite unaltered through sampling times.

*Soil hydrocarbon monitoring*

At the beginning of the assay, a number of pyrolytic hydrocarbons such as monoaromatic hydrocarbons (BTEX: benzene, toluene, ethylbenzene and xylene) and PAHs that had been generated during the fire were detected (Table 2). An average concentration of $149.7 \pm 17.5$ µg kg$^{-1}$ soil for BTEX and $398.0 \pm 32.4$ µg kg$^{-1}$ soils for PAHs were measured in burnt plots; accordingly, these compounds were below detection limits in Control pristine soil (data not shown).

After 2 months of treatment, BTEX concentrations in the burnt plot dropped below detection limits. Low molecular PAHs, comprising two to four carbon rings, made up the main fraction of the total PAHs ($\approx 95\%$) in burnt plots. The dominant PAH compounds were naphthalene and phenanthrene, which each made up $\approx 30\%$ of the total. Two months after the beginning the assay, the concentration of total PAHs on Plants control and bioremediation treatments was a 40% lower than in Control burnt soil, being up to a 60% lower were soil had been undergoing Rhizoremediation. At this point, most of

**Table 2.** Concentration of pyrolytic hydrocarbons (μg per kg of soil) generated after fire (Initial concentration, $n = 20$), versus the concentration of the same substances measured after 2 months of treatment ($n = 5$) (Control burnt, bioremediation, untreated plants and rhizoremediation treatments). Results are expressed as the mean and standard deviation. ND means below detection limits (10 μg kg$^{-1}$ and 0.5 μg kg$^{-1}$ for BTEX and PAHs respectively).

| Pyrolitic substances | Rings | Compounds | Initial concentration (October 2008) (μg kg$^{-1}$) | Two months after the outset of the study (December 2008) | | | |
| --- | --- | --- | --- | --- | --- | --- | --- |
| | | | | Control burnt (μg kg$^{-1}$) | Plants control (μg kg$^{-1}$) | Bioremediation (μg kg$^{-1}$) | Rhizoremediation (μg kg$^{-1}$) |
| BTEX | 1 | Benzene | 38.1 ± 5.3 | ND | ND | ND | ND |
| | | Toluene | 62.4 ± 7.2 | ND | ND | ND | ND |
| | | Ethylbenzene | 17.0 ± 0.9 | ND | ND | ND | ND |
| | | Xylene | 32.2 ± 4.1 | ND | ND | ND | ND |
| *TOTAL BTEX* | | | 149.7 ± 17.5 | ND | ND | ND | ND |
| PAHs | 2 | Naphthalene | 118.0 ± 8.9 | 31.6 ± 2.3 | 34.0 ± 2.6 | 43.2 ± 2.9 | 26.8 ± 1.3 |
| | 3 | Acenaphthene | ND | ND | ND | ND | ND |
| | | Fluorene | 48.3 ± 3.0 | 1.9 ± 0.7 | 1.7 ± 0.4 | 1.7 ± 0.1 | ND |
| | | Phenanthrene | 121.0 ± 9.9 | 26.9 ± 3.1 | 15 ± 2.0 | 11.5 ± 2.4 | 9.5 ± 0.7 |
| | | Anthracene | 1.6 ± 0.2 | 1.6 ± 0.3 | ND | ND | ND |
| | 4 | Fluoranthene | 26.0 ± 3.1 | 6.4 ± 1.3 | ND | ND | ND |
| | | Pyrene | 29.6 ± 2.0 | 8.4 ± 2.1 | 1.7 ± 0.1 | 1.0 ± 0.3 | 0.5 ± 0.1 |
| | | Benzo(a)Anthracene | 12.5 ± 1.1 | 1.2 ± 0.2 | 1.5 ± 0.3 | ND | ND |
| | | Crysene | 8.6 ± 0.3 | 6.9 ± 2.1 | 0.3 ± 0.1 | ND | ND |
| | 5 | Benzo(b)Fluoranthene | 1.5 ± 0.3 | 2.0 ± 0.3 | ND | ND | ND |
| | | Benzo(k)Fluoranthene | 0.8 ± 0.1 | 0.9 ± 0.1 | ND | ND | ND |
| | | Benzo(a)Pyrene | 17 ± 0.6 | 1.9 ± 0.1 | 0.6 ± 0.1 | ND | ND |
| | 6 | Dibenzo(a,h)Anthracene | 8.1 ± 1.7 | ND | ND | ND | ND |
| | | Benzo(g,h,i)Perylene | 2.0 ± 0.7 | 3.9 ± 0.4 | ND | ND | ND |
| | | Indene | 3.0 ± 0.4 | ND | ND | ND | ND |
| *TOTAL PAHs* | | | 398.0 ± 32.3 | 93.6 ± 10.7 | 54.8 ± 5.6 | 57.4 ± 5.7 | 36.8 ± 2.1 |

the initially measured compounds were still found in the Control burnt soil, whereas in the Rhizoremediation treatment only naphthalene, phenanthrene and pyrene were remaining.

*Soil quality indicators*

Soil quality indicators, such as soil pH and enzyme activity, can be used as an indirect measure of soil quality changes or as indexes of soil disturbance or restoration (Karaca *et al.*, 2011).

*pH.* A significant soil alkalynization ($P < 0.05$) was observed in burnt soil (pH 7.63 ± 0.11) versus control pristine soil (pH 6.53 ± 0.15) (Fig. 4). pH among burnt soil treatments (Control bulk soil, Plants control, Bioremediation and Rhizoremediation) showed no significant differences ($P > 0.05$) in the first 4 months of treatment. Then, all treatments experienced a progressive decrease of pH. Nevertheless, both soils treated with Bioremediation and Rhizoremediation treatments experienced a decrease in pH (6.63 ± 0.05 and 6.73 ± 0.04, respectively), reaching Control pristine soil pH levels (6.70 ± 0.04) after 5 months (22 weeks).

*Enzymatic activity assays.* In this study, the evolution of three different enzymatic activities was monitored (Fig. 5): (i) β–glucosidases, which are involved in the

saccharification of cellulose, (ii) dehydrogenases, which take part in reactions involved in energy transfer in microbial metabolism reactions and (iii) phosphatases, which hydrolyze organic phosphorous compounds to different forms of inorganic phosphorous (Karaca *et al.*, 2011). β–glucosidase and phosphatase are directly involved in C and P cycles respectively (Bandick and Dick, 1999).

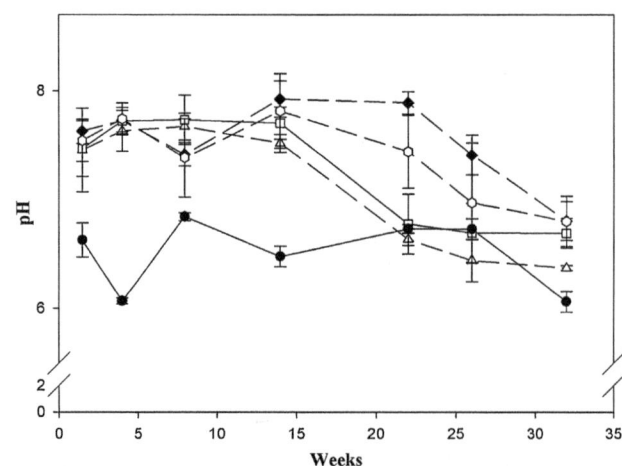

**Fig. 4.** pH measurements performed along the study during 8 months, on pristine soil (P) (filled circle), burnt bulk soil (C) (filled diamond), bioremediation (B) (triangle) and rhizoremediation (R) (square), and plants control (CP) (hexagon). Data are shown as mean ($n = 3$) and error bars refer to standard deviations.

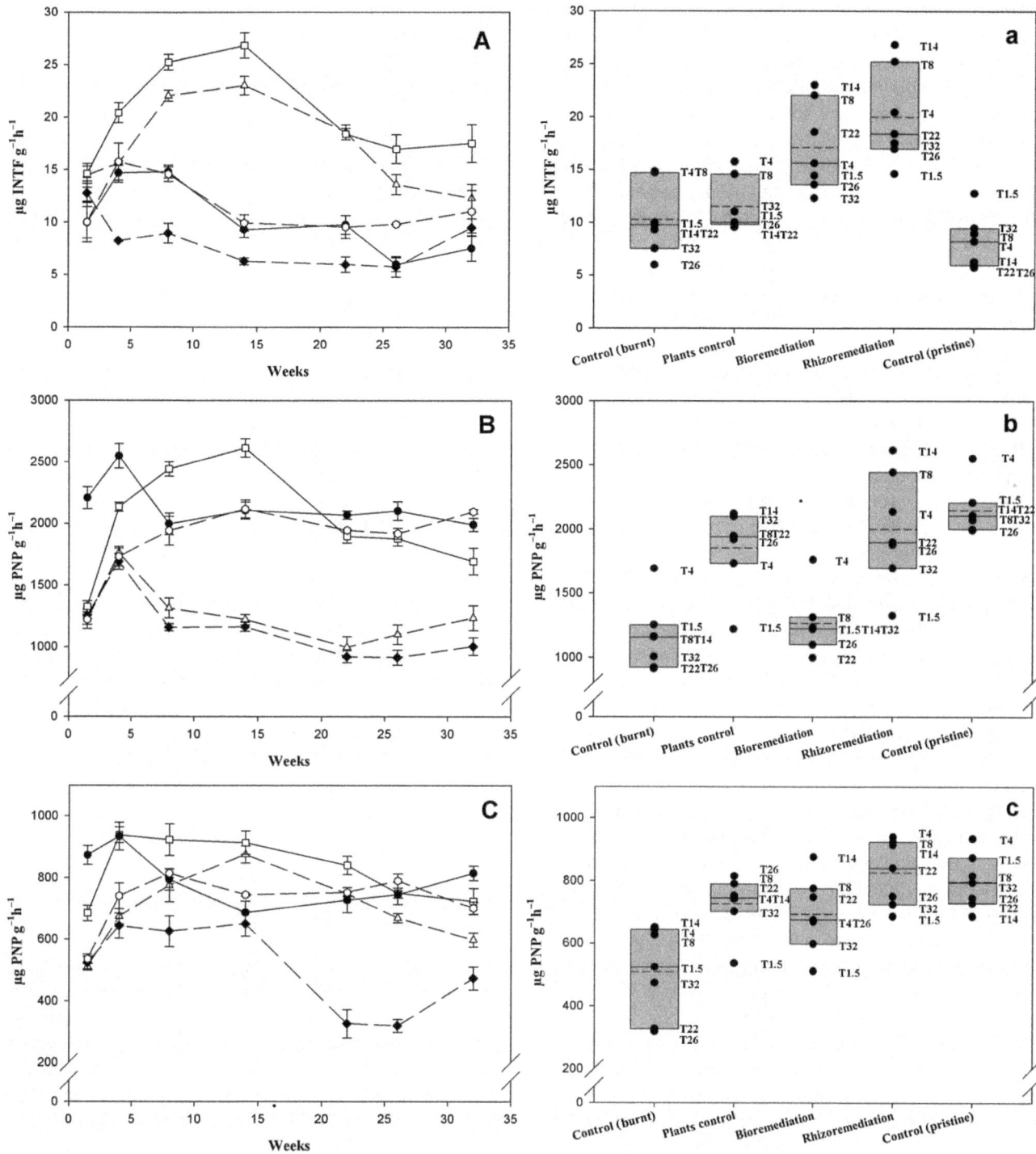

**Fig. 5.** Enzymatic activities measurements performed along the study during 8 months. (Aa) show dehydrogenase activity, (Bb) show phosphatase activity and (Cc) shows β-glucosidase activity assessed on Control pristine soil (filled circle), Control burnt soil (filled diamond), Bioremediation (triangle) and Rhizoremediation (square), and plants control (circle). A–C data are shown as mean ($n = 3$) and error bars refer to standard deviations. a–c data are shown as mean ($n = 3$), vertical boxes show the median (solid line), mean (dash line) and the 5th/95th percentiles.

Dehydrogenase activity (DHA) was measured as the amount of iodonitrotetrazolium violet-formazan (INTF) released from 1 g soil after 20 h incubation in the dark (Fig. 5A). DHA monitoring showed that levels in all treatments varied in the range between $10.0 \pm 1.5$ to $14.6 \pm 0.7$ µg INTF $g^{-1}h^{-1}$, whether burnt or pristine soil at the beginning of the assay. Bioremediation and Rhizoremediation treatments steadily raised DHA levels for about 15 weeks, leading to total increases of 2.5-fold $(23.0 \pm 0.9 \, \mu g \, INTF \, g^{-1}h^{-1})$ and 3-fold $(26.8 \pm 1.2 \, \mu g \, INTF \, g^{-1}h^{-1})$, respectively, over Control burnt soil $(9.3 \pm 0.8 \, \mu g \, INTF \, g^{-1}h^{-1})$. All non-inoculated treatments (Control pristine soil, Control burnt soil and Control plants) remained with lower DHA activity, below $15.0 \, \mu g \, INTFg^{-1}h^{-1}$, along the assay; as can be ascertained from Fig. 5A, treatments grouped in inoculated (Bioremediarion and Rhizoremediation) and non-inoculated (Control pristine and burnt soil and Plants control) along the assay. No differences $(P > 0.05)$ were found between plants control and pristine soil at most sampling times. At the end of the assay (July 2009), DHA in Rhizoremediation treatment remained significantly higher $(P < 0.05)$ $(17.5 \pm 1.8 \, \mu g \, INTFg^{-1}h^{-1})$ when compared with the rest of the treatments $(7.5 \pm 1.2 \, \mu g \, INTFg^{-1}h^{-1}$ in Control burnt soil; $11.3 \pm 2.1 \, \mu g \, INTFg^{-1}h^{-1}$ in Plants control; $12.3 \pm 1.3 \, \mu g \, INTFg^{-1}h^{-1}$ in Bioremediation and $9.5 \pm 0.8 \, \mu g \, INTFg^{-1}h^{-1}$ in Control pristine soil).

Phosphatase activity (Fig. 5B) was found to be $2210 \pm 90 \, \mu g \, p$-Nitrophenol (PNP) $g^{-1}h^{-1}$ in Control pristine soil at the beginning of the assay, whereas activity in the burnt plot was below $1325 \pm 45 \, \mu g \, PNP \, g^{-1}h^{-1}$. Fire led to a significant decrease $(P < 0.05)$ in activity by about 40%. After 8 weeks of treatment, phosphatase activity in the Plants control treatment was found to reach Control pristine soil levels $(2000 \pm 90 \, \mu g \, PNP \, g^{-1}h^{-1})$, and also in Rhizoremediation treatments, where phosphatase activity had increased up to $2440 \pm 60 \, \mu g \, PNP \, g^{-1}h^{-1}$, doubling its initial values. Along the assay, phosphatase activity was higher in treatments with associated plants (Control pristine soil, Plants control and Rhizoremediation) than in those without (Control burnt soil and Bioremediation) (Fig. 5C). Phosphatase activity in Control burnt soil and in Bioremediation treatment remained unaltered between 1000 and 1500 $\mu g \, PNP \, g^{-1}h^{-1}$ along the assay. At the end of the trial, measurements in treatments where plants had been introduced (Rhizoremediation and Plants control) were found to still be higher than those in bulk soil (Control burnt soil and Bioremediation).

At the beginning of the assay, $\beta$–glucosidase activity (Fig. 5C) was also affected by fire, significantly decreasing $(P < 0.05)$ by 20 (Rhizoremediation) and about a 40% (the rest of treatments), when compared with Control pristine soil. These levels were restored after 4 weeks in Rhizoremediation treatment (fluctuating between $750 \pm 40$ and $940 \pm 30 \, \mu g \, PNP \, g^{-1}h^{-1}$), whereas Bioremediation required 3 months to reach Rhizoremediation parameters. Along the assay, $\beta$–glucosidase activity was found to be at Control pristine levels in all treatments, except for Control burnt soil (Fig. 5C). At the end of the trial, $\beta$–glucosidase activity was higher in all treatments when compared with Control burnt soil $(475 \pm 40 \, \mu g \, PNP \, g^{-1}h^{-1})$, especially in Rhizoremediation treatment $(725 \pm 40 \, \mu g \, PNP \, g^{-1}h^{-1})$.

### Plant fitness and effect on the landscape

To evaluate plant fitness, we measured weight, length of roots and aerial parts of plants between the various conditions (Table 3). Introduced plants (Avex III® and Clover) used on this study showed increased size and dry weight when inoculated with bacteria $(P < 0.05)$. Length increases were 25% for clover and 58% for Avex III® plants $(P < 0.05)$; average dry weight increases were 43% for clover and 67% for Avex III® $(P < 0.05)$, after fourteen weeks of treatment (February 2009).

It should be noted that 4 weeks after the beginning of the assay, the plant growth promoting (PGP) effect of the artificial consortium over introduced vegetation development and soil coverage could be perceived in rhizoremediation versus non-inoculated plant subplots (Supporting Information Fig. S3). Finally, the visual impact of bioremediation and rhizoremediation processes on the landscape was documented by a series of photographs (Fig. 6).

### Discussion

#### Bacterial performance in bioremediation and rhizoremediation technologies

The ability to survive in harsh, polluted environments is one of the key requirements of the microorganisms selected for bioremediation processes. For survival, microorganisms need to show evidence of: (i) enhanced adaptation to the particular environment undergoing remediation and (ii) mechanisms to overcome the deleterious effects caused by the pollutant(s).

**Table 3.** Introduced plants measurements after 14 weeks of the beginning of the assay. Results are expressed as the mean $(n = 3)$ and standard deviation. For each introduced plants (Avex III® and Clover) and each parameter, different letters mean statistical differences according to the Tukey test $(P < 0.05)$.

| | Avex III® | | Clover | |
|---|---|---|---|---|
| | Control plants | Rhizoremediation | Control plants | Rhizoremediation |
| Length (cm) | $13.84 \pm 0.99$b | $21.97 \pm 1.65$a | $8.28 \pm 1.04$B | $10.54 \pm 0.68$A |
| Fresh weight (g) | $3.56 \pm 0.69$b | $6.10 \pm 0.95$a | $5.14 \pm 1.62$B | $9.02 \pm 1.29$A |
| Dry weight (g) | $1.19 \pm 0.13$b | $1.98 \pm 0.31$a | $2.04 \pm 0.12$B | $2.92 \pm 0.17$A |

**Fig. 6.** General evolution of the treated burnt parcel along the study (8 months, from October 2008 to June 2009). Clockwise: before treatments setting (October 2008) (A), 4 weeks after beginning of the trial (November 2008) (B), 8 weeks (December 2008) (C), 14 weeks (February 2009) (D), 22 weeks (April 2009) (E) and 32 weeks (July 2009) (F).

*Pseudomonas putida* strains are excellent candidates for soil restoration, especially in rhizoremediation processes because they are good rhizosphere colonizers (Molina *et al.*, 2000; Lugtenberg *et al.*, 2001; Ramos-González *et al.*, 2005; Matilla *et al.*, 2011; Wu *et al.*, 2011; Roca *et al.*, 2013) and metabolically versatile (Palleroni, 1992; 2010). Apart from these important traits, our results revealed that the *Pseudomonas* strains used in this study can overcome the toxic effects of hydrocarbon compounds produced during the combustion of organic matter. Microbial tolerance to hydrocarbons has been linked to the strains' ability to degrade these compounds (Park *et al.*, 2004); nevertheless, tolerance is a relatively complex process involving the activation of extrusion mechanisms, the establishment of oxidative stress responses and an overall fitness programme (Silby *et al.*, 2011; Krell *et al.*, 2012). *Pseudomonas putida* KT2440 and BIRD-1 lack the metabolic potential for hydrocarbon degradation; however, the transfer of the pWW0 plasmid endows them with the ability to degrade some BTEX compounds.

From the multiple parameters presented in this field assay, only drought conditions showed a clear negative effect on introduced *P. putida* strains; in fact, the decrease in added *Pseudomonas* populations concurred with the wilting of plants during the summer season when lack of precipitation and high temperatures converge. Monitoring of native culturable hydrocarbon degraders showed that these populations remained unaltered through fires and

landscape and climate variations, whereas introduced strains failed to survive through climate variations. As Mediterranean ecosystems are constantly affected by fires (Hernández *et al.*, 1997; Vila-Escalé *et al.*, 2007), indigenous microbial adaptation to this environment seems to have occurred (Fonturbel *et al.*, 1995; Choromanska and DeLuca, 2001; Smith *et al.*, 2008).

### Impact of fire and rhizoremediation treatments on soil microbial populations

Fires have been studied for their ability to change soil properties (Vázquez *et al.*, 1993; Certini, 2005) and disrupt indigenous microbial populations (Torres and Honrubia, 1997; DeBano *et al.*, 1998; Certini, 2005; Smith *et al.*, 2008). We carried out biodiversity analysis, which corroborated the resilience of native Mediterranean microbiota because only changes in the bacterial community distribution were observed, with no population loss (Supporting Information Fig. S2). These findings reinforce the hypothesis that indigenous microorganisms in Mediterranean ecosystems exhibit a high level of adaptation to fires (Fonturbel *et al.*, 1995; Hernández *et al.*, 1997; Choromanska and DeLuca, 2001; Vila-Escalé *et al.*, 2007; Smith *et al.*, 2008). Nevertheless, changes in the composition of microbial populations had been previously observed in connection with soil deterioration/ pollution generated by fire: (i) increasing soil pH values because of soil organic matter denaturation lead to a decrease in *Acidobacteria* and to an increase in *Bacteroidetes* populations (Certini, 2005; Smith *et al.*, 2008; Lauber *et al.*, 2009), (ii) the ratio of *Proteobacteria* increased in burnt plots, a phylum where strains with enhanced abilities for PAH metabolism can be found (Mueller *et al.*, 1997; Watanabe *et al.*, 2001), whereas microbial populations that lack this ability do not proliferate in these soils because of the selective pressure these compounds may exert (Martínez *et al.*, 2000) and (iii) the relative abundance of *Gemmatimonadetes* is increased in burnt soils and is also modulated by soil aridness (DeBruyn *et al.*, 2011), pH (Lauber *et al.*, 2009) and the presence of pyrogenic carbon (Khodadad *et al.*, 2011).

The Rhizoremediation treatment may introduce factors that can affect the structure of soil microbial communities, including the introduction of exogenous bacteria, the introduction of exogenous vegetal species and enrichment in indigenous culturable bacteria. Nevertheless, as we mentioned above, the relative abundance of most of the bacterial groups in soils undergoing rhizoremediation was intermediate between burnt and pristine forest soil, which suggests that the main perturbation on indigenous microbial populations was fire. This also suggests that the Rhizoremediation treatment tended to restore the original structure of microbial communities, probably due to the observed restoration of soil characteristics such as pH, vegetal cover and decreases in PAHs.

### Rhizoremediation enhances hydrocarbon degradation

One of the main consequences observed after a fire is the generation of new, toxic and recalcitrant forms of carbon (González-Pérez *et al.*, 2004), such as PAHs (Vila-Escalé *et al.*, 2007), as well as associated volatile hydrocarbons such as BTEX (Bamforth and Singleton, 2005). The PAH profiles observed in the current study comprised 60% of naphthalene and phenanthrene, which is consistent with reported profiles corresponding to wood combustion (Xu *et al.*, 2006; Kim *et al.*, 2011) and, in particular, to pine needles and wood (Conde *et al.*, 2005).

The lower complexity of these monoaromatic chemical structures (Bamforth and Singleton, 2005), as well as the increase in the proportion of bacterial populations with degrading potential, rapidly cleared these compounds from burnt soils, regardless of applied treatment. Furthermore, bio-attenuation mediated decreases in PAHs in untreated soil because of the presence of native degraders.

Nevertheless, Rhizoremediation treatments promoted the almost complete removal of heavier pyrolytic hydrocarbons in a relatively short time, which indicates that the combination of microorganisms, introduced PGPR, native degraders and plants was the most effective method for remediation. Because the introduced *P. putida* strains were not PAHs degraders, native microorganisms played a central role in PAH elimination. Because native populations were not significantly increased, it appears that the rhizosphere exerts a direct effect on stimulating the expression and/or activity of bacterial catabolic pathways, as was proved recently for naphthalene degradation by *P. putida* (Fernández *et al.*, 2012).

### Soil quality parameters indicated best restoration through rhizoremediation

The first, and most noticeable, consequence of a forest fire is the black ash coat (Knicker, 2007) generated by the combustion of the vegetation layer, which leads to the release of cations (Certini, 2005; Smith *et al.*, 2008). This explains the increase in soil pH observed after burning (Fig. 4). Because changes in pH can negatively affect microbial populations and their ability to degrade toxic compounds (Leahy and Colwell, 1990), the stabilization of pH to pre-fire levels is vital in remediation strategies. Our study revealed that Bioremediation and Rhizoremediation treatments were equally effective at restoring pH levels to pre-fire levels. This is also apparent, as non-inoculated plants were less capable of restoring pH levels versus inoculated plants. This soil acidification could be ascribed to the increase of soil microbial activity, as introduced an indigenous microorganisms produce acids and enzymes to solubilize soil nutrients, increasing their availability for plants and microorganisms alike. The same effect takes place for available P (Supporting Information Table S1), which slightly increases after fire and achieved maximal concentration in soil after treatment. *Pseudomonas putida* BIRD-1, one of the microorganisms used in this study, is capable of producing several enzymes and acids in order to solubilize insoluble phosphates (Roca *et al.*, 2013). The use of this bacterium favours the availability of soluble forms of phosphate for plants.

After fire, an increase of organic matter and total N was observed (Supporting Information Table S1), as previously described by Certini (2005). These parameters positively evolve along the treatment, as consequence of introduction of plants in rhizoremediation treatment and the improvement in soil microbial activity. An increase of available K is noticed after fire (Supporting Information Table S1), which is in concordance to parameters described by Khanna and Raison (1986). The proportion of cations ($K^+$) decreases along the study as consequence of lixiviation and/or runoff (Certini, 2005).

An increase in salinity was also noticed after fire (Supporting Information Table S1), as described by several authors (Naidu and Srivasuki, 1994; Hernández *et al.*, 1997), as consequence of the liberation of organic cations from organic matter. Soil conductivity increased after treatment, as consequence of the liberation of ions and cations due to solubilization of soil nutrients mediated by the increase of microbial activity.

In soil ecotoxicology, soil enzyme activities are used as indexes of soil disturbance or restoration because of their sensitivity to natural and anthropogenic induced stresses. These indexes are easy to measure and can detect the impact of microbial activities on nutrient cycling (Nannipieri *et al.*, 2002; Gianfreda *et al.*, 2005; Karaca *et al.*, 2011). In this study, three different enzymatic activities were monitored (Fig. 5): (i) dehydrogenases, which take part in reactions involved in energy transfer in microbial metabolism (Karaca *et al.*, 2011), (ii) phosphatases, which hydrolyze organic phosphorous compounds into different forms of inorganic phosphorous (Karaca *et al.*, 2011) and (iii) β-glucosidases, which are involved in the saccharification of cellulose. β-glucosidases and phosphatases are directly involved in C and P cycles respectively (Bandick and Dick, 1999).

Our results support the previous proposal that soil enzymatic activities are reliable indicators of the health and functionality of microorganisms in response to fire stress (Fioretto *et al.*, 2005) and that hydrocarbon levels can exert negative effects, to varying degrees, on these activities (Kiss *et al.*, 1998). Phosphatase and β-glucosidase activities were, as expected, severely affected by fire (Saa *et al.*, 1993; Eivasi and Bayan, 1996; Boerner *et al.*, 2000; Boerner and Brinkman, 2003). During the study period, phosphatase activity was clearly related to the presence of vegetation and only reached pristine levels with the Rhizoremediation treatment as a result of the secretion of this enzyme by root exudates (Tarafdar and Claassen, 1988). Changes in phosphatase and β-glucosidase activities were similar, remediation hastened recovery to levels found in pristine soils because of the positive effect of the rhizosphere on microbial activity (Valé *et al.*, 2005) and the provision of substrates from rhizodeposition (Morgan and Whipps, 2001) leading to improved enzyme synthesis (Turner *et al.*, 2002). In contrast, DHA measurements showed that soil microbiota was not severely affected by fire, as it is a direct indicator of respiration of viable cells (García *et al.*, 1997). This could be due to the positive 'fertilizing effect' that nutrients from charred necromass provide (Baath and Arnebrant, 1994), as well as to the previously discussed adaptations to fire of native strains. Nevertheless, bioremediation and rhizoremediation treatments showed improvements over untreated plots because of the bioaugmentation of native and added microorganisms, and the development of a vegetation cover.

In all cases, the use of Bioremediation processes enhanced soil quality parameters. Burnt soils reached pristine parameters faster with rhizoremediation providing the most remarkable benefits because of the greater microbial activity provided by the release of enzymes and substrates in root exudates (Badalucco and Kuikman, 2001).

## Landscape restoration

Germination and growth of inoculated plants was rapid in comparison with non-inoculated plants, which were negatively affected by the presence of pyrolytic pollutants. Furthermore, higher degradation rates and microbial activities were observed for inoculated plants. These results emphasize the important link between pollutant removal and the generation of adequate niches for native degraders (Aprill and Sims, 1990; Segura *et al.*, 2009; Segura and Ramos, 2012). Furthermore, the use of PGPR has shown to alleviate pollutant-induced plant stress (Qiu *et al.*, 1994; Kuiper *et al.*, 2001; Zhuang *et al.*, 2007). On this issue, *Pseudomonas putida* BIRD-1 presents itself as an interesting strain in Rhizoremediation due to its robust PGP properties (Matilla *et al.*, 2011; Roca *et al.*, 2013) and proved tolerance to soil hydrocarbons.

The use of PGPR- and native degraders-inoculated pasture plants in this study provided remediation advantages because of (i) the rapid ease of the visual impact through rapidly growing aerial plants, (ii) enhanced soil coverage, (iii) the added substrate and reduced erosion provided by large root surfaces with extensive soil penetration, (iv) the establishment of suitable niches for enhanced degradation processes based on the establishment of microbial consortia and (v) the increase of microbial activity.

## Experimental procedures

### Field experiment site description

The protected Parque Natural de los Montes de Málaga (http://www.juntadeandalucia.es/medioambiente) served as the Mediterranean ecosystem in this study. Located in the south of Spain in the province of Málaga, the park occupies an area close to 5000 Ha (Supporting Information Fig. S1). Average annual temperature is 15°C, and average precipitations are over 600 mm, belonging to the Mediterranean climatic zone. This ecosystem was declared protected in 1989 and comprises *Pinus halepensis*, *Quercus ilex*, *Quercus suber* and *Quercus faginea* forests, accompanied by matching brushwood (*Pistacia lentiscus*, *Rhamnus alaternus*, *Chamaerops humilis*, *Origanum majorana*, *Retama sphaerocarpa*, *Stipa tenacissima*, etc.). The park is also home of endangered species, such as the *Chamaeleo chamaeleon* (chameleon). Other notable vertebrates present in this protected ecosystem comprise the Spanish pond turtle (*Mauremys leprosa*), the endemic iberian worm lizard (*Blanus cinereus*), the short-toed snake eagle (*Circaetus gallicus*), the wild boar (*Sus scrofa*) and an endemic cricket (*Petaloptila malacinata*), among others. Soil geology of this ecosystem consists essentially of sedimentary materials, such as basic sandstones, and slates, having younger deposits of sands, conglomerates, marls and red clays. This mixture of geologic materials composed a high fertile soil that led to a heavy agriculture exploitation, which provoked soil bareness and consequent processes of erosion, leaving some areas of the actual protected ecosystem with undifferentiated lithology, such as the area of our study. The southern Mediterranean region of Spain is heavily affected by fire, the province of Malaga suffered an average of 133 wildfires in the 1992–2002 decade, resulting in an average calcined area of 1631 Ha, concentrated in the aestival season (http://www.magrama.gob.es/es/biodiversidad/temas/incendios-forestales/), so fireguard training and testing of new materials for use in fire extinction is considered to be of great importance (Pausas, 2012). An experimental fire, reaching 450°C on the soil surface [measurement provided by Andalusian Forest Fire Brigade (INFOCA) staff] resulting in the calcination of the brushwood and pine trees (Fig. 6), was induced under the strict supervision of the Andalusian Forest Fire Brigade (INFOCA, http://www.juntadeandalucia.es) in April 2008, allowing us to use it afterwards for this study. Our study was initiated in October 2008.

The burnt plot (N 36° 52.804' - W 004° 21.013') was located on a hill with a high slope that was subdivided into 12 terraces (subplots), each with an area of approximately 100 m². which were cleared of calcined vegetation before setting the assay, including untreated control parcels. A non-burnt plot (pristine soil) was established nearby, which was also divided into terraces (comprising six subplots), for control treatments in pristine soil to be applied in parallel.

## Treatments procedure

Three replicates of the applied treatments (Table 1) were established (3 × 4 in burnt plots and 3 × 2 in pristine plots). Different treatments were tested on different terraces to ease the leaching effect from adjacent conditions; in order to avoid contaminations among inoculated subplots, non-inoculated ones were set up at the top part of the hill.

Bioremediation treatments were designed to analyse *in situ* the role of indigenous microbiota and introduced microbes with or without plants on hydrocarbon removal and soil restoration. The exogenous microbes chosen were two wild-type strains of *Pseudomonas putida* harbouring the catabolic plasmid pWW0 (Ramos *et al.*, 1991), the KT2440 (Bagdasarian *et al.*, 1981) and BIRD-1 strains, being the latter recently reported to promote plant growth (Matilla *et al.*, 2011; Roca *et al.*, 2013). In order to stimulate the natural degradation of the pyrolytic compounds generated after fire, bioaugmentation of the indigenous hydrocarbon-degrading microbes was performed by isolating an indigenous hydrocarbon-degrading microbial consortium from the burnt soil by using diesel fuel as a carbon source, since it has a ≈ 25% of aromatic hydrocarbons (Agency for Substances and Disease Registry, http://www.atsdr.cdc.gov). Rhizoremediation assays were run with a combination of the described microorganisms and two kinds of plants, white clover (*Trifolium repens*) and Avex III® (Fertiprado), which is a commercial pasture seed mixture, composed of annual ryegrass, legumes and vetches and *Avena strigosa*. An organic solid vegetable support was used (commercial peat,

COMPO®) as a carrier for microorganisms. Treatments applied are summarized in Table 1.

(i) *Control:* The control bulk soil subplots remained untreated, providing an insight of the natural environment's recuperation capability; to be compared with any anthropic intervention made though the remediation treatments setting.

(ii) *Control plants:* For each plot, 80 L of peat was homogeneously mixed with Avex III® seeds (10 g m⁻²) and *Trifolium repens* (clover) (5 g m⁻²) were spreaded over the soil surface, then, a slight topsoil work was made to achieve a homogeneous mixture of the peat with the soil and promote seed germination in the dark.

(iii) *Bioremediation:* One litre of each microorganism/consortium (> 10⁹ cfu ml⁻¹) was mixed with 10 L of tap water and then mixed with 80 L of peat. After spreading the mixture over the soil surface, topsoil was slightly worked to achieve a homogeneous mixture of the peat with the soil.

(iv) *Rhizoremediation and treated pristine soil:* For each treated plot, 80 L of peat was homogeneously mixed with Avex III® (10 g m⁻²) and *Trifolium repens* (clover) (5 g m⁻²), then, 1 L of each microorganism/consortium (> 10⁹ cfu ml⁻¹) was mixed with 10 L of tap water and then mixed with the peat-seed mixture. After spreading the mixture over the soil surface, the topsoil was slightly worked to achieve a homogeneous mixture of the peat with the soil and promote seed germination in the dark.

Characteristics that were assayed include bacterial survival, impact on indigenous microbial populations, soil quality parameters, hydrocarbon analysis and visual evolution of the landscape. For soil analyses, five subsamples per plot were collected from the upper 10 cm of topsoil and sieved at < 2 mm.

## Strains and culture media

The bacterial strains and plasmids used in this study are shown in Table 1. *Pseudomonas putida* BIRD-1 was grown in M9 minimal medium supplemented with sodium benzoate (10 mM) as the carbon source (Abril *et al.*, 1989). The catabolic plasmid pWW0 was transferred to *P. putida* BIRD-1 by conjugation, as described by Ramos and colleagues (1991). *Pseudomonas putida* BIRD-1 harbouring the catabolic plasmid pWW0 was grown in M9 minimal medium supplemented with 3-methylbenzoate as carbon source, and supplemented with spectinomycin (100 μg ml⁻¹) and rifampicin (20 μg ml⁻¹). *Pseudomonas putida* KT2442R (pWW0) was grown in M9 minimal medium supplemented with toluene and rifampicin (10 μg ml⁻¹). Cultures were incubated at 30°C and shaken on an orbital platform operating at 200 strokes per minute. Monitoring of survival of each strain in soil was performed by drop-plating dilution series in solid M9 minimal medium supplemented with the required carbon sources and antibiotics, as described previously.

Indigenous culturable microorganisms with the ability to degrade aromatic hydrocarbons were isolated by enrichment from the superficial layer of burnt soil (5 cm depth) using M9 minimal medium supplemented with diesel fuel as a carbon source. Monitoring of survival of soil microorganisms was performed by drop-plating dilution series in solid M9 minimal medium supplemented with diesel fuel in the vapour phase (100 µL per plate), as described above.

To perform bacterial plate count, rhizosphere soil was considered to be soil closely attached to roots sieved through a 2-mm mesh, and bulk soil was obtained from the 5–10 cm of topsoil and then also sieved. One gram of each soil sample was introduced in a Falcon tube with 9 ml of M9 medium and vortexed for 1 min to separate cells form soil particles and resuspend them. Then, dilution series were performed to obtain cfu per gram of soil.

For identity verification of the introduced *P. putida*, BIRD-1 and KT2440 strains, 100 bacterial colonies from each treatment (Treated pristine soil, Bioremediation and Rhizoremediation) were analyzed using REPc fingerprinting, as described by Aranda-Olmedo and colleagues (2002). Colony hybridization was also performed for 100 bacterial colonies per treatment, as described by Sambrook and colleagues (1989) using DNA probes corresponding to the *xylS* gene (to identify the pWW0 catabolic plasmid) or the PP_0314 gene (to identify *P. putida* KT2440); no colony hybridizations were performed for BIRD-1 because its genome was not sequenced at that time.

### Library construction and biodiversity analysis

Total DNA was extracted from approximately 0.5 g of a composed bulk soil sample from Control burnt bulk soil, from 0.5 g of a composed rhizospheric soil sample from Rhizoremediation treatment and from 0.5 g of a composed rhizospheric soil sample from Control pristine soil using the FastDNA kit (Qbiogene, Carlsbad, CA, USA) and purified on agarose gels. The universal Eubacterial primers GM3F (5′-AGAGTTTGATCMTGGC-3′) and GM4R (5′-TACCTTGTTAC GACTT-3′) were used for amplifying the 16S rDNA gene (Muyzer *et al.*, 1993). Each PCR reaction was performed in 50 µL reaction volume containing 5 µL of reaction buffer, 0.2 mM of primers, 2 mM $MgCl_2$, 0.2 mM deoxynucleoside triphosphates, and 2.5 U DNA polymerase. The PCR conditions was as follows: 5 min of denaturation at 95°C, followed by 35 cycles of 1 min at 95°C, 1 min for primer annealing, 2 min at 72°C for primer extension, and a final cycle at 72°C for 10 min. The products of two consecutive PCRs were then pooled and purified through extraction from agarose gels prior to cloning into pGEM-T vectors. The resulting plasmids were transformed in competent *Escherichia coli* DH5α cells and positive transformants were colour-screened on LB plates supplemented with ampicillin (100 µg ml⁻¹), Xgal (80 µg ml⁻¹), and isopropyl-β-D-thiogalactopyranoside (20 mM). Clones with the correct insert were sequenced using the vector primers M13 F (5′-GGAAACAGCTATGA CCATG-3′) and M13 R (5′-GTTGTAAAACGACGGCCAGT-3′).

The quality of the obtained sequences was manually checked using DNA BASER (http://www.dnabaser.com/download/download.html) and verified with BELLEROPHON (Huber *et al.*, 2004) and CHECK_CHIMERA (Maidak *et al.*, 1996), and all chimeric sequences were discarded. These sequences were then compared with those in the GenBank database using the BLASTN tool and the ribosomal database project database with classifier tool and aligned using

CLUSTALW (Thompson *et al.*, 1994). DNA aligned of 16S rDNA gene sequences were used to construct a DNA distance matrix and rarefaction matrices with DOTUR package (Schloss and Handelsman, 2005). FASTUNIFRAC (Hamady *et al.*, 2010) was used to produce PCoA comparing all samples.

### Hydrocarbons measurement

To perform the PAH analysis, soil samples were dried at 40°C and frozen. Defrosted samples were dried completely in a second step using an equivalent weight of mortar-ground anhydrous sodium sulfate. For PAH extraction, approximately 45 g of soil was placed inside a cellulose extraction thimble (Filtros ANOIA, Barcelona, Spain) and extracted with a mixture of dichloromethane:acetone (1:1) for 15 h. Once the extraction was completed, the organic solvents were evaporated and the remaining residue was re-dissolved in a small volume of dichloromethane (4–5 ml). To remove polar compounds, clean-up of the organic extract was performed using Sep-Pak® Plus Florisil cartridges (WATERS Corp., Milford, MA, USA), previously conditioned with 10 ml of dichloromethane. For the next step, dichloromethane was evaporated and the residue was resuspended in 2 ml of acetone. Finally, samples were filtered through a nylon Minisart syringe filter (0,45 µm, 13 mm Ø, Sartorius Stedim Biotech, GmbH, Goettingen, Germany). Analysis of PAH was carried out using an Agilent Technologies high-performance liquid chromatography system 1200 Series (Agilent Technologies, Santa Clara, CA, USA), equipped with a photodiode array detector (DAD, G1315D) and a scanning fluorescence detector (FLD, G1321A). The column used was a ZORBAX Eclipse PAH (Agilent Technologies, 5 µm, 4.6 I.D. × 150 mm). The mobile phase used was an acetonitrile-milli Q water gradient comprising 40% (v/v) acetonitrile from 0 to 1.25 min, programmed 100% (v/v) acetonitrile between 1.25 to 18 min. The initial solvent composition (60% milli Q water, 40% acetonitrile) was then maintained for further 3.5 min (Merck KGaA, Darmstadt, Germany).

Measurement of aromatic volatile organics, such as BTEX was performed by gas chromatography following EPA method 8020 (United States Environmental Protection Agency http://www.epa.gov).

### Measurement of soil quality parameters

*Measurement of pH in soil.* The pH values were measured in air-dried soil, sieved through 2 mm, using a glass combination electrode (soil: water ratio, 1:2.5 w:v), as described by Acosta-Martínez and colleagues (2003).

*Measurement of phosphatase activity in soil.* Phosphatase activity was determined as described by Antolín and colleagues (2005). The amount of 4-nitrophenol (PNP) released from 0.5 g soil, by triplicate, was measured after incubation, in the dark at 37°C for 2 h with 0.115 M 4-nitrophenyl phosphate-disodium (PNPP) as substrate for the enzymatic reaction, in 2 ml of maleate buffer (0.1 M, pH 6.5). Samples were cooled at 2°C for 15 min to stop the enzymatic reaction and 0.5 ml of 0.5 M $CaCl_2$ and 2 ml of 0.5 M NaOH were added and well-mixed. Each sample was centrifuged at $2000 \times g$ for 10 min. A blank experiment by duplicate was performed for each assay, in which the substrate was added to the soil sample after incubation and before stopping the reaction. The amount of PNP per hour released from each soil sample ($\mu$g PNP g$^{-1}$h$^{-1}$) was determined by comparing absorbance measures to a PNP standard curve. Rhizosphere soil was considered to be soil closely attached to roots, sieved through a 2 mm mesh.

*Measurement of β-glucosidase activity in soil.*
β-glucosidase activity was determined as described by García and colleagues (1994). The amount of 4-nitrophenol (PNP) released from 0.5 g of soil, by triplicate, was measured after incubation in the dark, at 37°C for 2 h with 0.5 ml of 50 mM 4-nitrophenyl-β-D-glucopiranoside (PNG) as substrate for the enzymatic reaction, in 2 ml of maleate buffer (0.1 M, pH 6.5). Then, samples were cooled at 2°C for 15 min to stop the enzymatic reaction, and 0.5 ml of 0.5 M CaCl$_2$ and 2 ml of 0.5 M NaOH were added and mixed well. Each sample was centrifuged at $3500 \times g$ for 10 min. A blank experiment, by duplicate, was performed for each assay, in which the substrate was added to the soil sample after incubation and before stopping the reaction. The amount of PNP per hour released from each soil sample ($\mu$g PNP g$^{-1}$h$^{-1}$) was determined by comparing absorbance values to a PNP standard curve.

*Measurement of DHA in soil.* Dehydrogenase activity was determined as described by García and colleagues (1994). The amount of INTF released from 1 g soil, by triplicate, was measured after incubation in the dark, at 37°C for 20 h with 0.2 ml of 0.4% 2-(4-iodophenyl)-3-(4-nitrophenyl)-5-phenylteytrazolium chloride hydrate (INT) as substrate for the enzymatic reaction, and 2 ml of distilled water. Then, 5 ml of an extracting mixture (tetrachloroethylene: acetone (1:1.5 v:v) was added and well-mixed for 2 min. Each sample was centrifuged at $1000 \times g$ for 10 min. A blank experiment, by duplicate, was performed for each assay, without substrate, in which the extracting mixture was added after incubation. The amount of INTF per hour released from each soil sample ($\mu$g INTF g$^{-1}$h$^{-1}$) was determined by comparing absorbance measures to an INTF standard curve.

*Plant biomass monitoring.* Five samples of clover and pasture plants were harvested from each plot at 4, 8 and 14 weeks sampling times. Plants were then manually separated into shoots and roots; fresh weight and length were recorded, and samples were dried in a stove at 90°C for 48 h. Samples were then allowed to cool to room temperature and dry weight was measured. The visual aspect of the area and each plot was photographed at each sampling time.

*Statistical analyses*

A descriptive statistical analysis (the mean and absolute error) was calculated for each parameter. Also, we performed some inferential statistical analyses, such as analysis of variance (two-way analysis of variance) within treatments and soil type (burnt and non-burnt), assuming a normal distribution of the data and homoscedasticity. For post-hoc analysis, we used the Tukey test ($P < 0.05$) to determine changes in the analyzed parameters for each treatment.

## Conclusions

The use of Bioremediation and Rhizoremediation strategies in the current trial did not harm indigenous microbiota, and the release of non-native microbes only remained detectable in the soil for about 6 months after inoculation. Rhizoremediation treatments improved ecosystem resilience, accelerating its natural ability to return to the initial pre-fire state. The strains used in this study have proved their ability to survive in burnt soils, while exhibiting and a strong capacity to promote plant growth and development, making them suitable candidates for future use in the restoration of ecosystems affected by fires.

## Acknowledgement

We thank Ben Pakuts for editing of the manuscript in English and M. Mar Fandila for secretarial assistance.

## Conflict of interest

None declared.

## References

Abril, M.A., Michán, C., Timmis, K.N., and Ramos, J.L. (1989) Regulator and enzyme specificities of TOL plasmid-encoded upper pathway for degradation of aromatic hydrocarbons. *J Bacteriol* **171:** 6782–6790.

Acosta-Martínez, V., Zobeck, T.M., Gill, T.E., and Kennedy, A.C. (2003) Enzyme activities and microbial community structure in semiarid agricultural soils. *Biol Fertil Soils* **38:** 216–227.

Andreoni, V., Cavalca, L., Rao, M.A., Nocerino, G., Bernasconi, S., Dell Amico, E., *et al.* (2004) Bacterial communities and enzyme activities of PAHs polluted soils. *Chemosphere* **57:** 401–412.

Antolín, M.C., Pascual, I., García, C., Polo, A., and Sánchez-Díaz, M. (2005) Growth, yield and solute content of barley in soils treated with sewage sludge under semiarid Mediterranean conditions. *Field Crops Res* **94:** 224–237.

Aprill, W., and Sims, R.C. (1990) Evaluation of the use of prairie grasses for stimulating polycyclic aromatic hydrocarbon treatment in soil. *Chemosphere* **20:** 253–265.

Aranda-Olmedo, I., Tobes, R., Manzanera, M., Ramos, J.L., and Marqués, S. (2002) Specie-specific repetitive extragenic palindromic (REP) sequences in *Pseudomonas putida*. *Nucleic Acids Res* **30:** 1826–1833.

Baath, E., and Arnebrant, K. (1994) Growth rate and response of bacterial communities to pH in limed and ash treated forest soils. *Soil Biol Biochem* **26:** 995–1001.

Badalucco, L., and Kuikman, P.J. (2001) Mineralization and immobilization in the rhizosphere. In *The Rhizosphere.*

*Biochemistry and Organic Substances at the Soil-Plant Interface*. Pinton, R., Varanini, Z., and Nannipieri, P. (eds). New York, NY, USA: Marcel Dekker, pp. 141–196.

Bagdasarian, M., Lurz, R., Rückert, B., Franklin, F.C.H., Bagdasarian, M.M., Frey, J., and Timmis, K.N. (1981) Specific purpose plasmid cloning vectors. II. Broad host range, high copy number, RSF1010-derived vectors, and a host-vector system for gene cloning in *Pseudomonas*. *Gene* **16:** 237–247.

Bamforth, S., and Singleton, I. (2005) Bioremediation of polycyclic aromatic hydrocarbons: current knowledge and future directions. *J Chem Technol Biotechnol* **80:** 723–736.

Bandick, A.K., and Dick, R.P. (1999) Field management effects on soil enzyme activities. *Soil Biol Biochem* **31:** 1471–1479.

Boerner, R.E.J., and Brinkman, J.A. (2003) Fire frequency and soil enzyme activity in southern Ohio oak-hickory forests. *Appl Soil Ecol* **23:** 137–146.

Boerner, R.E.J., Decker, K.L.M., and Sutherland, E.K. (2000) Prescribed burning effects on soil enzyme activity in a Southern Ohio, Harwood forest: a landscape-scale analysis. *Soil Biol Biochem* **32:** 899–908.

Campbell, G.S., Jungbauer, J.D., Jr, Bidlake, W.R., and Hungerford, R.D. (1994) Predicting the effect of temperature on soil thermal conductivity. *Soil Sci* **158:** 307–313.

Certini, G. (2005) Effects of fire on properties of forest soils: a review. *Oecologia* **143:** 1–10.

Choromanska, U., and DeLuca, T.H. (2001) Prescribed fire alters the impact of wildfire on soil biochemical properties in a ponderosa pine forest. *Soil Sci Soc Am J* **65:** 232–238.

Conde, J.C., Ayala, J.H., Afonso, A.M., and Gonzalez, V. (2005) Emissions of polycyclic aromatic hydrocarbons from combustion of agricultural and sylvicultural debris. *Atmos Environ* **39:** 6654–6663.

DeBano, L.F. (2000) The role of fire and soil heating on water repellency in wildland environments: a review. *J Hydrol* **231–232:** 195–206.

DeBano, L.F., Neary, D.G., and Ffolliott, P.F. (1998) *Fire Effects on Ecosystems*. New York, NY, USA: Wiley.

DeBruyn, J.M., Nixon, L.T., Fawaz, M.N., Johnson, A.M., and Radosevich, M. (2011) Global biogeography and quantitative seasonal dynamics of *Gemmatimonadetes* in soil. *Appl Environ Microbiol* **77:** 6295–6300.

Eivasi, F., and Bayan, M.R. (1996) Effects of long-term prescribed burning on the activity of selected soil enzymes in an oak-hickory forest. *Can J Forest Res* **26:** 1799–1804.

Fernández, M., Niqui-Arroyo, J.L., Conde, S., Ramos, J.L., and Duque, E. (2012) Enhanced tolerance to naphthalene and enhanced rhizoremediation performance for *Pseudomonas putida* KT2440 via the NAH$_7$ catabolic plasmid. *Appl Environ Microbiol* **78:** 5104–5110.

Fioretto, A., Papa, S., and Pellegrino, A. (2005) Effects of fire on soil respiration, ATP content and enzymeactivities in Mediterranean maquis. *Appl Veg Sci* **8:** 13–20.

Fonturbel, M.T., Vega, J.A., Bara, S., and Bernardez, I. (1995) Influence of prescribed burning of pine stands in NW Spain on soil microorganisms. *Eur J Soil Biol* **31:** 13–20.

Franklin, S.B., Robertson, P.A., and Fralish, J.S. (1997) Small-scale fire temperature patterns in upland *Quercus* communities. *J Appl Ecol* **34:** 613–630.

García, C., Hernández, T., and Costa, F. (1994) Microbial activity in soils under Mediterranean environmental conditions. *Soil Biol Biochem* **26:** 1185–1191.

García, C., Hernández, T., and Costa, F. (1997) Potential use of dehydrogenase activity as an index of microbial activity in degraded soils. *Commun Soil Sci Plant Anal* **28:** 123–134.

Gianfreda, L., Rao, M.A., Piotrowska, A., Palumbo, G., and Colombo, C. (2005) Soil enzyme activities as affected by anthropogenic alterations: intensive agricultural practices and organic pollution. *Sci Total Environ* **341:** 265–279.

González-Pérez, J.A., González-Vila, F.J., Almendros, G., and Knicker, H. (2004) The effect of fire on soil organic matter – a review. *Environ Int* **30:** 855–870.

Hamady, M., Lozupone, C., and Knight, R. (2010) Fast UniFrac: facilitating high-throughput phylogenetic analyses of microbial communities including analysis of pyrosequencing and PhyloChip data. *ISME J* **4:** 17–27.

Hernández, T., Garcia, C., and Reinhardt, I. (1997) Short-term effect of wildfire on the chemical, biochemical and microbiological properties of Mediterranean pine forest soils. *Biol Fertil Soils* **25:** 109–116.

Huang, X., El-Alawi, Y., Penrose, D., Glick, B., and Greenberg, B. (2004) A multi-process phytoremediation system for removal of polycyclic aromatic hydrocarbons from contaminated soils. *Environ Pollut* **130:** 465–476.

Huber, T., Faulkner, G., Hugenholtz, P., and Bellerophon, A. (2004) A program to detect chimeric sequences in multiple sequence alignments. *Bioinformatics* **20:** 2317–2319.

Karaca, A., Camci Cetin, S., Can Turgay, O., and Kizilkaya, R. (2011) Soil enzymes as indication of soil quality. In *Soil Enzymology, Soil Biology*, **Vol. 22**. Shukla, G., and Varma, A. (eds). Berlin, Germany: Springer, pp. 119–148. 2011.

Khanna, P.K., and Raison, R.J. (1986) Effects of fire intensity on solution chemistry of surface soil under a Eucalyptus pauciflora forest. *Aust J Soil Res* **24:** 423–434.

Khodadad, C.L.M., Zimmerman, A.R., Green, S.J., Uthandi, S., and Foster, J.S. (2011) Taxa-specific changes in soil microbial community composition induced by pyrogenic carbon amendments. *Soil Biol Biochem* **43:** 385–392.

Kim, E.J., Choi, S.D., and Chang, Y.S. (2011) Levels and patterns of polycyclic aromatic hydrocarbons (PAHs) in soils after forest fires in South Korea. *Environ Sci Pollut Res Int* **18:** 1508–1517.

Kiss, S., Pasca, D., and Dragan-Bularda, M. (1998) *Development in Soil Science 26: Enzymology of Disturbed Soils*. Amsterdam, The Netherlands: Elsevier.

Knicker, H. (2007) How does fire affect the nature and stability of soil organic nitrogen and carbon? A review. *Biogeochemistry* **85:** 91–118.

Konstantinidis, K.T., Ramette, A., and Tiedje, J.M. (2006) The bacterial species definition in the genomic era. *Phil Trans R Soc Lond B Biol Sci* **361:** 1929–1940.

Krell, T., Lacal, J., Guazzaroni, M.E., Busch, A., Silva-Jiménez, H., Fillet, S., *et al.* (2012) Responses of *Pseudomonas putida* to toxic aromatic carbon sources. *J Biotechnol* **160:** 25–32.

Kuiper, I., Bloemberg, G.V., and Lugtenberg, B.J.J. (2001) Selection of a plant-bacterium pair as a novel tool for rhizostimulation of polycyclic aromatic hydrocarbon-degrading bacteria. *Mol Plant Microbe Interact* **14**: 1197–1205.

Kuiper, I., Lagendijk, E.L., Bloemberg, G.V., and Lugtenberg, B.J. (2004) Rhizoremediation: a beneficial plant-microbe interaction. *Mol Plant Microbe Interact* **17**: 6–15.

Lauber, C.L., Hamady, M., Knight, R., and Flerer, N. (2009) Pyrosequencing-based assessment of soil pH as a predictor of soil bacterial community structure at the continental scale. *Appl Environ Microbiol* **75**: 5111–5120.

Leahy, J., and Colwell, R. (1990) Microbial degradation of hydrocarbons in the environment. *Microbiol Rev* **54**: 305–315.

Lugtenberg, B.J., Dekkers, L., and Bloemberg, G.V. (2001) Molecular determinants of rhizosphere colonization by *Pseudomonas*. *Annu Rev Phytopathol* **39**: 461–490.

Maidak, B.L., Olsen, G.J., Larsen, N., Overbeek, R., McCaughey, M.J., and Woese, C.R. (1996) The ribosomal database project (RDP). *Nucleic Acids Res* **24**: 82–85.

Martínez, M., Díaz-Ferrero, J., Martí, R., Broto-Puig, F., Comellas, L., and Rodríguez-Larena, M. (2000) Analysis of dioxin-like compounds in vegetation and soil samples burned in Catalan forest fire. Comparison with the corresponding unburned material. *Chemosphere* **41**: 1927–1935.

Matilla, M.A., Pizarro-Tobías, P., Roca, A., Fernández, M., Duque, E., Molina, L., *et al.* (2011) Complete genome of the plant growth-promoting rhizobacterium *Pseudomonas putida* BIRD-1. *J Bacteriol* **193**: 1290.

Molina, L., Ramos, C., Duque, E., Ronchel, M.C., García, J.M., Wyke, L., and Ramos, J.L. (2000) Survival of *Pseudomonas putida* KT2440 in soil and in the rhizosphere of plants under greenhouse and environmental conditions. *Soil Biol Biochem* **32**: 315–321.

Morgan, J.A.W., and Whipps, J.M. (2001) Methodological approaches to the study of rhizosphere carbon flow and microbial population dynamics. In *The Rhizosphere. Biochemistry and Organic Substances at the Soil-Plant Interface*. Pinton, R., Varanini, Z., and Nannipieri, P. (eds). New York, NY, USA: Marcel Dekker, pp. 373–410.

Mueller, J.G., Devereux, R., Santavy, D.L., Lantz, S.E., Willis, S.G., and Pritchard, P.H. (1997) Phylogenetic and physiological comparisons of PAH-degrading bacteria from geographically diverse soils. *Antonie Van Leeuwenhoek* **71**: 329–343.

Muyzer, G., de Waal, E.C., and Uitterlinden, A.G. (1993) Profiling of complex microbial populations by denaturing gradient gel electrophoresis analysis of polymerase chain reaction-amplified genes coding for 16S rRNA. *Appl Environ Microbiol* **59**: 695–700.

Naidu, C.V., and Srivasuki, K.P. (1994) Effect of forest fire on soil characteristics in different areas of Seshachalam hills. *Annals of Forestry* **2**: 166–173.

Nannipieri, P., Kandeler, E., and Ruggiero, P. (2002) Enzyme activities and microbiological and biochemical processes in soil. In *Enzymes in the Environment. Activity, Ecology and Applications*. Burns, R.G., and Dick, R.P. (eds). New York, NY, USA: Marcel Dekker, pp. 1–33.

Olivella, M.A., Ribalta, T.G., de Febrer, A.R., Mollet, J.M., and de las Heras, F.X.C. (2006) Distribution of polycyc-lic aromatic hydrocarbons in riverine waters during Mediterranean forest fires. *Sci Total Environ* **355**: 156–166.

Palleroni, N.J. (1992) *Introduction to the Pseudomonadaceae. The Prokaryotes, A Handbook on the Biology of Bacteria, Ecophysiology, Isolation, Identification and Applications*, **Vol. III**, 2nd edn. Balows, A., Truper, H.G., Dworkin, M., Harder, W., and Schleifer, K.H. (eds). New York, NY, USA: Springer, pp. 3071–3085.

Palleroni, N.J. (2010) The *Pseudomonas* story. *Environ Microbiol* **12**: 1377–1383.

Park, W., Jeon, C.O., Cadillo, H., DeRito, C., and Madsen, E.L. (2004) Survival of naphthalene-degrading *Pseudomonas putida* NCIB 9816-4 in naphthalene-amended soils: toxicity of naphthalene and its metabolites. *Appl Microbiol Biotechnol* **64**: 429–435.

Pausas, J.G. (2012) *Incendios forestales: una visión desde la ecología*. Madrid, Spain: CSIC-Catarata, pp. 1–119.

Qiu, X., Shah, S.I., Kendall, E.W., Sorensen, D.L., Sims, R.C., and Engelke, M.C. (1994) Grass-enhanced bioremediation for clay soils contaminated with polynuclear aromatic hydrocarbons. In *Bioremediation through Rhizosphere Technology*. Anderson, T.A., and Coats, J.R. (eds). Washington, DC, USA: American Chemical Society, pp. 142–157.

Ramos, J.L., Duque, E., and Ramos-González, M.I. (1991) Survival in soils of an herbicide-resistant *Pseudomonas putida* strain bearing a recombinant TOL plasmid. *Appl Environ Microbiol* **54**: 260–266.

Ramos-González, M.I., Campos, M.J., and Ramos, J.L. (2005) Analysis of *Pseudomonas putida* KT2440 gene expression in the maize rhizosphere: in vivo expression technology capture and identification of root-activated promoters. *J Bacteriol* **187**: 4033–4041.

Roca, A., Pizarro-Tobías, P., Udaondo, Z., Fernández, M., Matilla, M.A., Molina-Henares, M.A., *et al.* (2013) Analysis of the plant growth-promoting properties encoded by the genome of the rhizobacterium *Pseudomonas putida* BIRD-1. *Environ Microbiol* **15**: 780–794.

Saa, A., Trasar-Cepeda, M.C., Gill-Sotres, F., and Carballas, T. (1993) Changes in soil phosphorus and acid phosphatase activity immediately following forest fires. *Soil Biol Biochem* **25**: 1223–1230.

Sambrook, J., Fritsch, E.F., and Maniatis, T. (1989) *Molecular Cloning: A Laboratory Manual*, 2nd edn. New York, NY, USA: Cold Spring Harbor Laboratory Press.

Schloss, P.D., and Handelsman, J. (2005) Introducing DOTUR, a computer program for defining operational taxonomic units and estimating species richness. *Appl Environ Microbiol* **71**: 1501–1506.

Segura, A., and Ramos, J.L. (2012) Plant-bacteria interactions in the removal of pollutants. *Curr Opin Biotechnol* **24**: 1–7.

Segura, A., Rodríguez-Conde, S., Ramos, C., and Ramos, J.L. (2009) Bacterial responses and interactions with plants during rhizoremediation. *Microb Biotechnol* **4**: 452–464.

Silby, M.W., Winstanley, C., Godfrey, S.A., Levy, S.B., and Jackson, R.W. (2011) *Pseudomonas* genomes: diverse and adaptable. *FEMS Microbiol Rev* **35**: 652–680.

Smith, N.R., Kishchuk, B.E., and Mohn, W.W. (2008) Effects of wildfire and harvest disturbances on forest soil bacterial communities. *Appl Environ Microbiol* **74**: 216–224.

Tarafdar, J.C., and Claassen, N. (1988) Organic phosphorus compounds as a phosphorus source for higher plants through the activity of phosphatases produced by plant roots and microorganisms. *Biol Fertil Soils* **5**: 308–312.

Thompson, J.D., Higgins, D.G., and Gibson, T.J. (1994) CLUSTAL W: improving the sensitivity of progressive multiple sequence alignment through sequence weighting, position-specific gap penalties and weight matrix choice. *Nucleic Acids Res* **22**: 4673–4680.

Torres, P., and Honrubia, M. (1997) Changes and effects of a natural fire on ectomycorrhizal inoculum potential of soil in a *Pinus halepensis* forest. *For Ecol Manag* **96**: 189–196.

Turner, B., Hopkins, D.W., Haygarth, P.M., and Ostle, N. (2002) β-Glucosidase activity in pasture soils. *Appl Soil Ecol* **20**: 157–162.

Vazquez-Baeza, Y., Pirrung, M, Gonzalez, A., Knight, R. (2013) EMPeror: a tool for visualizing high-throughput microbial community data. *GigaScience 2013* **2**: 16

Vázquez, F.J., Acea, M.J., and Carballas, T. (1993) Soil microbial populations after wildfire. *FEMS Microbiol Ecol* **13**: 93–104.

Valé, M., Nguyen, C., Dambrine, E., and Dupouey, J.L. (2005) Microbial activity in the rhizosphere soil of six herbaceous species cultivated in a greenhouse is correlated with shoot biomass and root C concentrations. *Soil Biol Biochem* **37**: 2329–2333.

Vergnoux, A., Malleret, L., Laurence, A., Doumenq, P., and Theraulaz, F. (2011) Impact of forest fires on PAH level and distribution in soils. *Environ Res* **111**: 193–198.

Vila-Escalé, M., Vegas-Vilarrúbia, T., and Prat, N. (2007) Release of polycyclic aromatic compounds into a Mediterranean creek (Catalonia, NE Spain) after a forest fire. *Water Res* **41**: 2171–2179.

Watanabe, K., Kodama, Y., and Harayama, S. (2001) Design and evaluation of PCR primers to amplify bacterial 16S ribosomal DNA fragments used for community fingerprinting. *J Microbiol Methods* **44**: 253–262.

Wood, T.K. (2008) Molecular approaches in bioremediation. *Curr Opin Biotechnol* **19**: 572–578.

Wu, X., Monchy, S., Taghavi, S., Zhu, W., Ramos, J.L., and van der Lelie, D. (2011) Comparative genomics and functional analysis of niche specific adaptation in *Pseudomonas putida*. *FEMS Microbiol Rev* **35**: 299–323.

Xu, S., Liu, W., and Tao, S. (2006) Emission of polycyclic aromatic hydrocarbons in China. *Environ Sci Technol* **40**: 702–708.

Zhuang, X., Chen, J., Shim, H., and Bai, Z. (2007) New advances in plant growth-promoting rhizobacteria for bioremediation. *Environ Int* **33**: 406–413.

## Supporting information

Additional Supporting Information may be found in the online version of this article at the publisher's web-site:

**Fig. S1.** Location of *Parque Natural de los Montes de Málaga* (green region) in the South of Spain. The red spot in the map corresponds to the location of the burnt site object of this study.

**Fig. S2.** Rarefaction analysis for pristine soil (P), bulk burnt soil (B) and rhizoremediation treatment (R) at two different times of the assay 1 month (autumn) and 6 months (spring). Rarefaction curves were constructed with DOTUR software.

**Fig. S3.** Introduced plants growth and soil coverage after 4 weeks of treatment on burnt soil. (A) shows non-inoculated plants and (B) shows plants on rhizoremediation treatment.

**Table S1.** Soil physic-chemical parameters for assessment of soil recovery, performed according to ORDEN 5/12/1975.

# 21

# Some (bacilli) like it hot: genomics of *Geobacillus* species

David J. Studholme*

*Biosciences, University of Exeter, Geoffrey Pope Building, Stocker Road, Exeter EX4 4QD, UK.*

## What are *Geobacillus*?

The genus *Geobacillus* includes thermophilic Gram-positive spore-forming bacteria that form a phylo-genetically coherent clade within the family *Bacillaceae*. They are of great interest for biotechnology (as discussed below). These thermophiles seem to be ubiquitous; viable *Geobacillus* spores can be isolated in large quantities not only from hot environments such as hydrothermal vents, but also, paradoxically, from cool soils and cold ocean sediments (Zeigler, 2014).

These bacteria were previously categorized as 'Group 5' within the genus *Bacillus* but were subsequently split into the new genus *Geobacillus* (Nazina *et al.*, 2001). Many *Geobacillus* strains were previously described as belonging to a single species *Bacillus stearothermophilus*, but it was clear that there was great heterogeneity in physiology, preferred temperature range and other phenotypic characteristics among these strains. For example, see Fig. 1 showing three distinct colony morphologies among three strains described as '*B. stearothermophilus*'. It is now absolutely clear that there are several distinct species within *Geobacillus* and these can be distinguished by both genotype and phenotype (Nazina *et al.*, 2001; Banat *et al.*, 2004; Zeigler, 2005; Dinsdale *et al.*, 2011; Coorevits *et al.*, 2012).

## Why are *Geobacillus* species of interest for biotechnology?

*Geobacillus* spp. are of interest for biotechnology as source of thermostable enzymes and natural products,

*For correspondence. E-mail d.j.studholme@exeter.ac.uk

**Funding Information** Work in my laboratory is currently supported by Biotechnology and Biological Sciences Research Council (BBSRC) grants BB/H016120/1, BB/I024631/1, BB/I025956/1, BB/K003240/2 and BB/L012499/1.

digesters of lignocellulose, bioremediators of hydrocarbons, producers of bio-fuel, cellular factories for heterologous expression of enzymes and as hosts for directed evolution (Wiegel *et al.*, 1985; Niehaus *et al.*, 1999; Couñago and Shamoo, 2005; Marchant *et al.*, 2006; Cripps *et al.*, 2009; Taylor *et al.*, 2009; Tabachnikov and Shoham, 2013). Industrially important enzymes originating from *Geobacillus* spp. include lipases (Schmidt-Dannert *et al.*, 1998), glycoside hydrolases (Fridjonsson *et al.*, 1999; Bartosiak-Jentys *et al.*, 2013; Suzuki *et al.*, 2013), N-acylhomoserine lactonase (Seo *et al.*, 2011) and DNA polymerase I (Sandalli *et al.*, 2009) and protease (Chen *et al.*, 2004) among others. The advantages of using thermophilic bacteria as whole-cell biocatalysts were recently discussed in this journal (Taylor *et al.*, 2011) and include reduced risk of contamination, acceleration of biochemical processes and easier maintenance of anaerobic conditions. These bacteria also tend to ferment a wide range of substrates, utilizing both cellobiose and pentose sugars. In the context of bioethanol production, there is the additional advantage of reduced cooling costs and easier removal and recovery of the volatile product by sparging or partial vacuum thus also avoiding ethanol poisoning of the bacteria (Taylor *et al.*, 2009). Less positively, *Geobacillus* spp. are common contaminants in the dairy and food industries (Burgess *et al.*, 2010).

## Which genomes have been sequenced?

At the time of writing (28 July 2014), 29 *Geobacillus* genome sequences are available (Table 1). These include representatives of all the major phylogenetic groups within the genus and include representatives of the species *G. thermoleovorans*, *G. kaustophilus*, *G. thermocatenulatus*, *G. thermodenitrificans*, *G. stearothermophilus*, *G. caloxylosilyticus* and *G. thermoglucosidans* (formerly *G. thermoglucosidasius*) as well as several strains that have not been assigned to named species (Fig. 2). Genome sequences are also available for some other thermophilic members of the *Bacillaceae*, such as *Paenibacillus lautus* (Mead *et al.*, 2012) and *Bacillus coagulans* (Xu *et al.*, 2013) and for *Geobacillus*-infecting

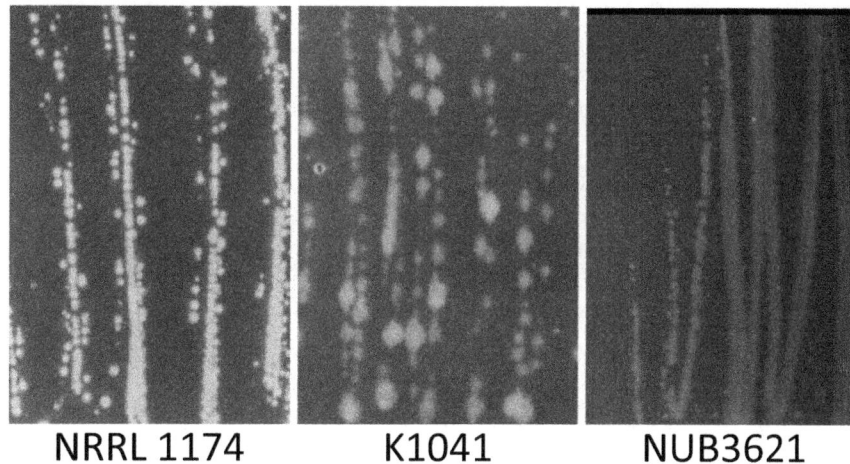

**Fig. 1.** Diverse colony morphologies of strains classified as '*G. stearothermophilus*'. Strains NRRL 1174, K1041 and NUB3621 were streaked-out on tryptic soy broth plates and incubated overnight at 50°C. Plates were photographed under identical conditions.

**Table 1.** *Geobacillus* strains whose genomes have been sequenced as of 26 July 2014.

| Species and strain | Motivation for sequencing | Accession number | References |
|---|---|---|---|
| *G. caldoxylosilyticus* CIC9 | Not known | NZ_AMRO01000000.1 | n. a. |
| *G. caldoxylosilyticus* NBRC 107762 | Not known | BAWO01000000.1 | n. a. |
| *G. kaustophilus* GBlys | Lysogenic, containing an integrated prophage | NZ_BASG01000001.1 | (Doi *et al.*, 2013) |
| *G. kaustophilus* HTA426 | Source of novel glycoside hydrolases (6-phospho-β-glycosidase and β-fucosidase) | NC_006510.1 | (Takami *et al.*, 2004) |
| *G.* sp. A8 | Not known | NZ_AUXP01000001.1 | n. a. |
| *G.* sp. C56-T3 | Not known | NC_014206.1 | n. a. |
| *G.* sp. CAMR12739 | Hemicellulose degradation | JHUR01000001.1 | (De Maayer *et al.*, 2014) |
| *G.* sp. CAMR5420 | Hemicellulose degradation | JHUS01000001.1 | (De Maayer *et al.*, 2014) |
| *G.* sp. FW23 | Potential for degradation and utilization of oil (bioremediation of oil spills) | JGCJ01000001.1 | (Pore *et al.*, 2014) |
| *G.* sp. G11MC16 | Not known | NZ_ABVH01000001.1 | n. a. |
| *G.* sp. GHH01 | Source if thermostable and thermo-active secreted lipase | NC_020210.1 | (Wiegand *et al.*, 2013) |
| *G.* sp. JF8 | Degrades biphenyl and polychlorinated biphenyls (PCB) | NC_022080.4 | (Shintani *et al.*, 2014) |
| *G.* sp. MAS1 | Potential source of useful enzyme-encoding genes | NZ_AYSF01000001.1 | (Siddiqui *et al.*, 2014) |
| *G.* sp. WCH70 | Not known | NC_012793.1 | n. a. |
| *G.* sp. WSUCF1 | Abel to grow on lignocellulosic substrates | NZ_ATCO01000001.1 | (Bhalla *et al.*, 2013) |
| *G.* sp. Y4.1MC1 | Not known | NC_014650.1 | n. a. |
| *G.* sp. Y412MC52 | Not known | NC_014915.1 | n. a. |
| *G.* sp. Y412MC61 | Not known | NC_013411.1 | n. a. |
| *G. stearothermophilus* ATCC 7953 | Not known | JALS01000001.1 | n. a. |
| *G. stearothermophilus* NUB3621 | Genetically amenable host strain for metabolic engineering | AOTZ01000001.1 | (Blanchard *et al.*, 2014) |
| *G. thermocatenulatus* GS-1 | Not known | JFHZ01000001.1 | n. a. |
| *G. thermodenitrificans* NG80-2 | Denitrification and degradation of long-chain alkanes, facilitating oil recovery in oil reservoirs | NC_009328.1 | (Feng *et al.*, 2007) |
| *G. thermodenitrificans* subsp. *thermodenitrificans* DSM 465 | Comparative genomics between the alkane-utilizing NG80-2 and this strain which is unable to utilize alkanes | NZ_AYKT01000001.1 | (Yao *et al.*, 2013) |
| *G. thermoglucosidans* TNO-09.020 | Contaminant in dairy-processing environment | NZ_CM001483.1 | (Zhao *et al.*, 2012) |
| *G. thermoglucosidasius* C56-YS93 | Not known | NC_015660.1 | n. a. |
| *G. thermoglucosidasius* NBRC 107763 | Not known | BAWP01000001.1 | n. a. |
| *G. thermoleovorans* B23 DNA | Alkane degrader with unidentified alkane monooxygenase | BATY01000001.1 | (Boonmak *et al.*, 2013) |
| *G. thermoleovorans* CCB_US3_UF5 | Not known | NC_016593.1 | (Muhd Sakaff *et al.*, 2012) |

Names are given as found in the GenBank sequence database. n.a., not available.

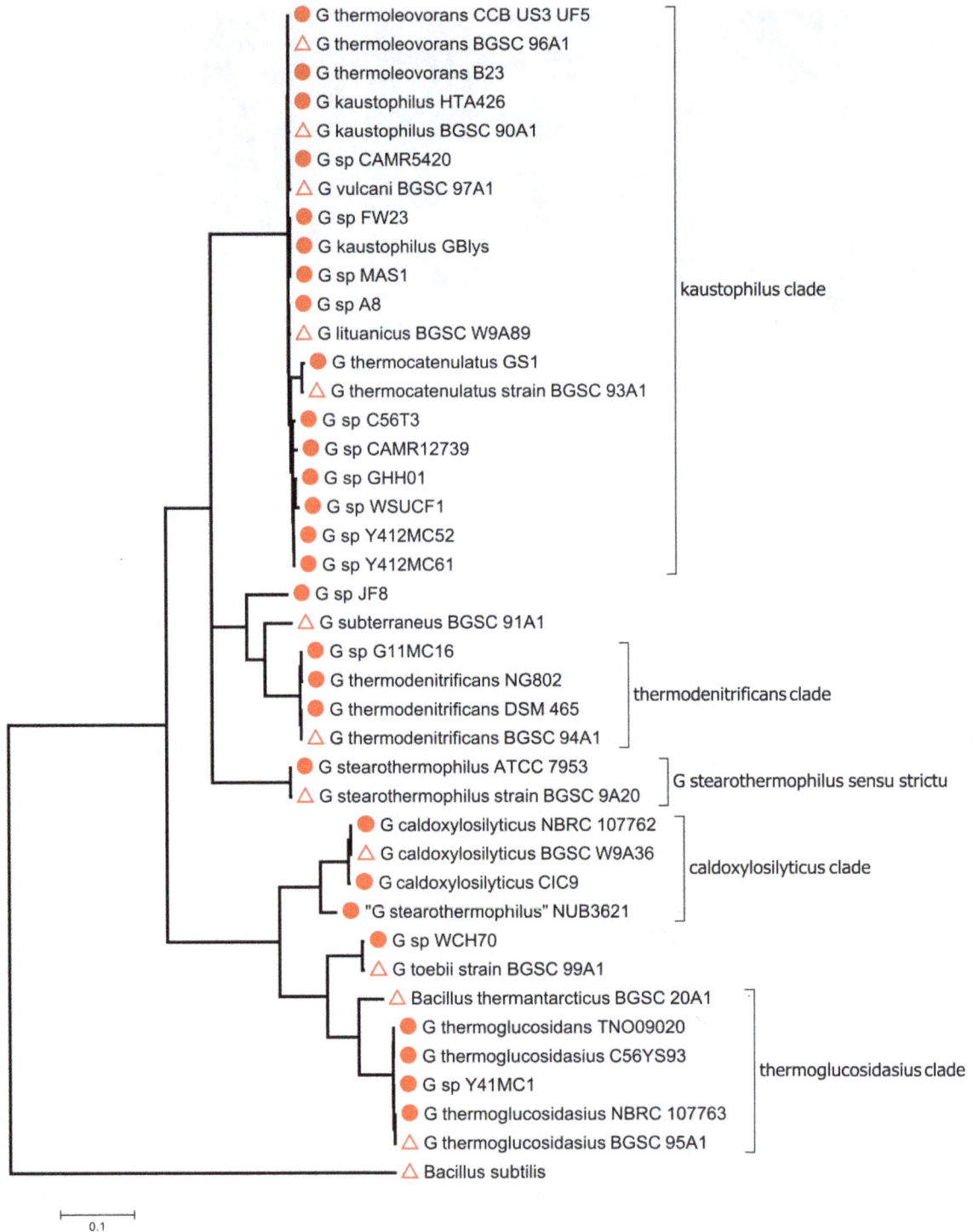

**Fig. 2.** Phylogenetic relationships among sequenced strains of *Geobacillus* inferred from a multiple sequence alignment of *recN* sequences. The circles indicate strains whose genomes have been sequenced, as listed in Table 1. The triangles indicate type strains of the various *Geobacillus* species; *recN* sequences from these are taken from a previous phylogenetic analysis by Zeigler (2005). The maximum-likelihood tree was generated using MEGA6 (Tamura *et al.*, 2013).

bacteriophage (Marks and Hamilton, 2014), but these will not be discussed here. The team who sequenced the genome of *Geobacillus* sp. MAS1 described this strain as '*G. thermopakistaniensis*', but this is not a validly named species and no justification was provided for its proposal as a new species (Siddiqui *et al.*, 2014). On the basis of its *recN* sequence, a useful phylogenetic marker for *Geobacillus* spp. (Zeigler, 2005), strain MAS1 is closely related to the type strains of *G. kaustophilus* and *G. thermoleovorans* (Fig. 2). Strain NUB3621 was described as '*G. stearothermophilus*' but as has been previously noted (Studholme *et al.*, 1999; Zeigler, 2005; Blanchard *et al.*, 2014), this strain is phylogenetically distinct from *B. stearothermophilus sensu strictu* and is more closely related to *G. caldoxylsilyticus* and, to a lesser extent, *G. thermoglucosidans* (Fig. 2). For more than half of the sequenced genomes, papers have been published describing and/or announcing the sequence data and usually indicating the particular features of the strain that motivated its sequencing. An insightful discussion of the biological lessons from *Geobacillus* genomes was previously published earlier this year, including surveys of genes involved in breakdown of plant-derived lignocellulose (Zeigler, 2014); but at that time, only 10 genome sequences were available.

The phylogenetic group within *Geobacillus* most richly represented by genome sequences is the clade containing *G. thermoleovorans*, *G. kaustophilus* and *G. thermocatenulatus* (see the 'kaustophilus clade' in Fig. 2). Based solely of sequences of the *recN* phylogenetic marker, it is not possible to precisely resolve relationships among sequenced strains within this group (Fig. 2). However, the availability of complete genome sequence data enables phylogenetic analysis based on single-nucleotide variants over the entire core genome, offering much greater resolution (Fig. 3A). According to the core-genome-wide phylogenetic analysis, the two strains assigned as *G. kaustophilus* do not form a phylogenetically coherent monophyletic clade. On the other hand, the two strains of *G. thermoleovorans* are closely related and share 99.4% nucleotide sequence identity [based on MUMMER2 alignments (Delcher *et al.*, 2002)]. Strain FW23 also appears to fall within this clade and, subject to phenotypic characterization, can probably be considered a member of this species too. *Geobacillus thermocatenulatus* GS-1 is much more divergent, sharing only 94% to 95% identity with the other strains in the clade, which is consistent with the *recN*-based analysis (Fig. 2). Strains Y412MC52 and YP412MC61 appear to be extremely closely related to each other, sharing 99.8% sequence identity and showing no detectable differences in gene content. Nucleotide sequence identities between clades are much lower; between *G. kaustophilus* and *G. thermoglucosidans*, there is approximately 84% identity.

The considerable amount of reticulation in the phylogenetic network (Fig. 3A) suggests significant horizontal genetic transfer within and among these species. This is further illustrated by the extent of variation in the variable component of the genome (Fig. 3B). Out of 3887 genes on the chromosome of *G. thermoleovorans* CCB US3 UF5, a total of 931 (approximately 24%) are variable

(that is, they are absent from at least one of the other sequenced genomes). The global pattern of gene content (Fig. 3B) broadly reflects the phylogenetic relationships (Fig. 3A): according to gene content, the genomes fall into four main clusters, indicated by four different colours of shading in Fig. 3B, which correspond to four zones of the phylogenetic network, shaded with the same colours in Fig. 3A. However, there are numerous genes whose distribution across the genomes is incongruent with coregenome phylogeny, again suggesting extensive horizontal transfer.

## What benefits has the sequencing of *Geobacillus* genomes brought?

The availability of complete *Geobacillus* genome sequences has enabled or accelerated the discovery, cloning and exploitation of natural products. For example, the availability of the NG80-2 genome sequence (Feng *et al.*, 2007) enabled the discovery of thermostable homologues of the lantibiotic nisin in *G. thermodenitrificans* (Begley *et al.*, 2009; Garg *et al.*, 2012), opening the possibility of replacing nisin as a food preservative and veterinary antibiotic with more-stable alternatives. Lantibiotics appear to be widely distributed among sequenced *Geobacillus* species. For example, the genome of *G. kaustophilus* HTA426 contains two lantibiotic-biosynthesis gene clusters (centred on the genes for YP_146139 and YP_146147) that are both conserved in the recently sequenced *Geobacillus* sp. CAMR12739. The NG80-2 genome sequence also enabled discovery of the first nitrous oxide reductase gene from a Gram-positive, and a novel thermophilic long-chain alkane monooxygenase (Feng *et al.*, 2007). Furthermore, the genome sequence enabled proteomics-level confirmation of pathways for catabolism of long-chain alkanes (Feng *et al.*, 2007) and aromatics (Li *et al.*, 2012).

Many of the *Geobacillus* genome sequencing projects reported genes potentially encoding thermostable homologues of useful enzymes. In some cases, the genome sequences have been used to clone and express the genes of interest and characterize the enzyme for biotechnological potential. For example, the genome of *G. kaustophilus* HTA426 was recently mined for members of the glycoside hydrolase family 1, which have potential uses in synthesizing therapeutic oligosaccharides (Suzuki *et al.*, 2013). The genome sequence of the alkane-utilizing *G. thermoleovorans* B23 (Boonmak *et al.*, 2013) revealed a cluster of three long-chain alkane monooxygenase genes with homology to that of NG80-2 that showed activity *in vivo* when heterologously expressed in *Pseudomonas fluorescens* (Boonmak *et al.*, 2014). Recently, a novel thermostable endo-xylanase was cloned and expressed from *Geobacillus* sp. WSUCF1

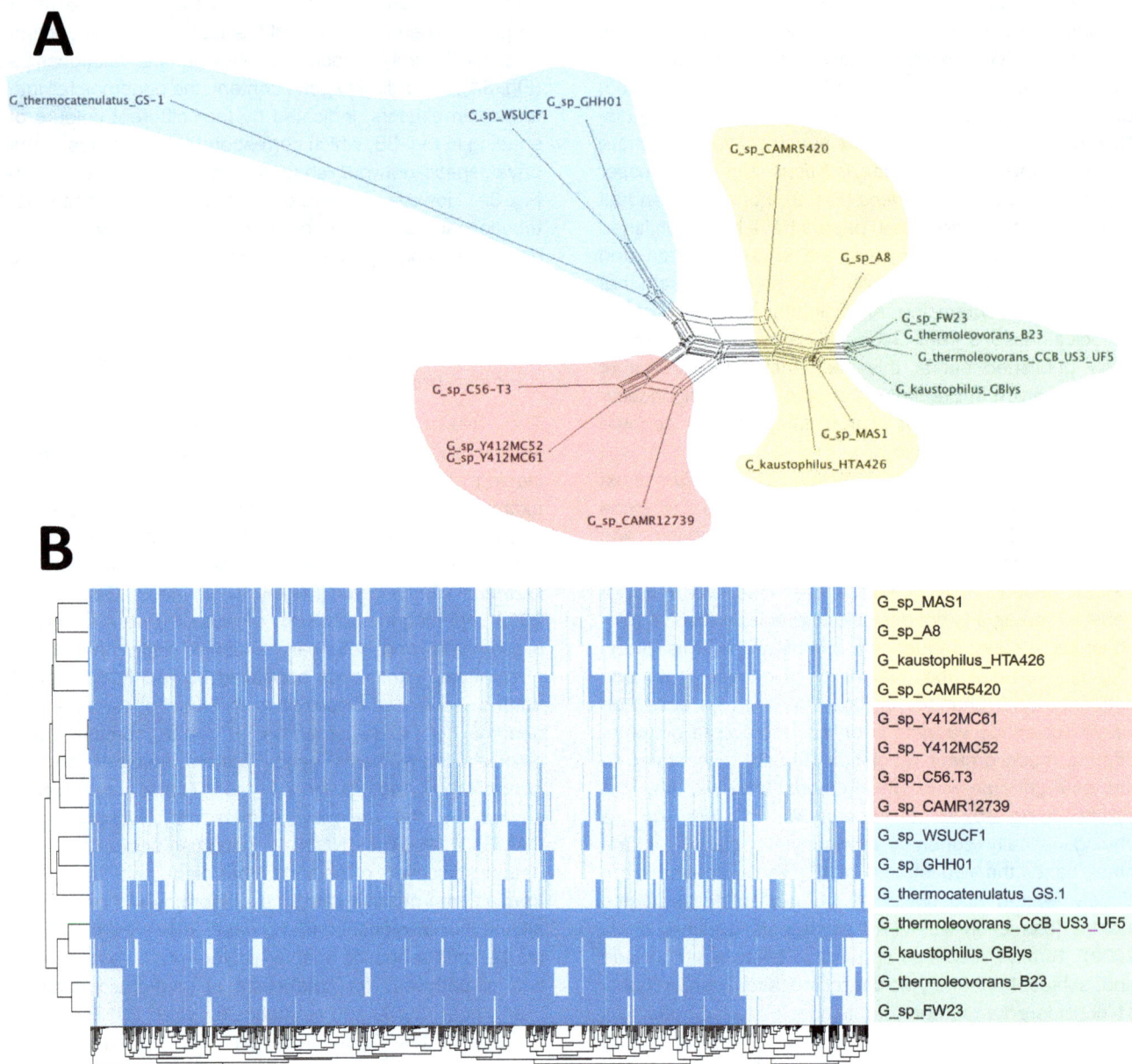

**Fig. 3.** Relationships among sequenced genomes within the *G. kaustophilus* clade resolved using whole-genome sequence data. The phylogenetic network in panel A was based on a concatenation of 1722 variant single-nucleotide sites in 1 874 967 nucleotides of the core genome present in all 15 genomes. The network was generated using the NEIGHBORNET algorithm (Bryant and Moulton, 2004) implemented in the SPLITSTREE software package (Huson, 1998). The heat-map in B indicates the presence (dark blue) and absence (light blue) of each of 931 non-core genes from the genome of *G. thermoleovorans* CCB US3 UF6 across the same 15 genomes appearing in A. The gene-content clusters are shaded in the same colours in both panels. The heat-map was rendered using Raivo Kolde's pheatmap package in R (R Development Core Team, R, 2013).

(Bhalla *et al.*, 2014) following the sequencing of its genome (Bhalla *et al.*, 2013).

Genome sequencing has revealed that interesting traits are often encoded on chromosomes rather than on the chromosome. For example, the biphenyl-degrading pathway of *Geobacillus* sp. JF8 (Mukerjee-Dhar *et al.*, 2005; Shintani *et al.*, 2014) and the long-chain alkane monooxygenase of *G. thermodenitrificans* NG80-2 (Feng

*et al.*, 2007) are both located on plasmids. The dynamic loss and gain of such mobile elements presumably explains, in part, the physiological differences between natural isolates of *Geobacillus* spp. and it also suggests that these bacteria might be engineered to express new traits by introduction of recombinant plasmids. Indeed, progress has been made in developing plasmid shuttle vectors for heterologous expression in *Geobacillus* spp.

(Thompson *et al.*, 2008; Bartosiak-Jentys *et al.*, 2013).

The value of genome sequencing goes beyond cataloguing potentially useful enzymes, as exemplified by the recently published genomic study of strain NUB3621 (Blanchard *et al.*, 2014). Some previous attempts to fully exploit the potential of *Geobacillus* strains as whole-cell catalysts have been frustrated by the paucity of genetic and genomic resources (my own PhD research project in the mid-1990s being a case in point; Studholme, 1998). However, strain NUB3621 is a promising laboratory workhorse strain. It is one of the few *Geobacillus* strains that has been shown to be readily transformable with plasmid DNA (Wu and Welker, 1989); protocols have been developed for genetic analysis (Chen *et al.*, 1986) and a genetic map has been available for more than two decades (Vallier and Welker, 1990). Strain NUB3621 is a mutant derived from wild-type strain NUB36 that lacks its parent strain's restriction-modification system and this probably contributes to transformation efficiency. Incidentally, and consistent with this, we observed that transformation efficiency was significantly affected by the methylation status of the plasmid DNA (Thompson *et al.*, 2008).

Being one of the most genetically amenable *Geobacillus* strains, NUB3621 was obviously a high priority for genome sequencing. But rather than simply announcing and describing its genome sequence, the authors went on to show how the genome sequence could be exploited to further develop the strain as a host for heterologous expression and metabolic engineering (Blanchard *et al.*, 2014). Specifically, they used the genome sequence to clone two promoters and incorporated them into plasmid vectors: one for inducible gene expression and one constitutive. The authors also mention that they tried other promoters that did not work so well; presumably, the availability of the genome sequence allowed them to relatively quickly screen a number of candidates until they found the best ones. The combination of a genome sequence, allowing relatively facile construction of expression and/or knock-out constructs and a global view of metabolism, along with transformability and a wide range of growth temperatures [between 39 and 75°C (Wu and Welker, 1991)] make NUB3621 a strong candidate as the preferred thermophilic host for rationally designed metabolic engineering.

## What's next?

The availability of complete (or nearly complete) genome sequences for nearly 30 *Geobacillus* strains (Table 1) as well as large-scale proteomic data for at least one (Feng *et al.*, 2007; Li *et al.*, 2012) should certainly accelerate cloning, expression and characterization of novel thermostable and thermo-active enzymes, at least in an academic research context. However, there has been relatively little industrial uptake of enzymes from thermophiles, with much greater use of proteins originating from mesophiles but engineered for thermo-stability (Haki and Rakshit, 2003; Taylor *et al.*, 2011). The convergence of genomic data and transformability, at least for strain NUB3621, should help to remove the barriers to greater exploitation of thermophiles. However, genome sequences are not yet publicly available for the handful of other readily transformable *Geobacillus* strains such as *G. thermodenitrificans* K1041 (Narumi *et al.*, 1992), *G. stearothermophilus* IFO 12550 (Imanaka *et al.*, 1982), NRRL 1174 (Liao *et al.*, 1986) and *G. thermoglucosidasius* TN (Thompson *et al.*, 2008). Furthermore, although it is possible to predict the metabolic networks of bacteria from complete genome sequence, there is a need for comprehensive testing of these predictions through metabolomics. Only then can we rationally design genetic interventions to predictably manipulate metabolism. And finally, palaeo-genomics of ancient *Geobacillus* spores, which may be viable after billions of years of dormancy, might shed light on population-genetics and evolutionary processes over timescales that we previously assumed to be intractable (Nicholson, 2003; Zeigler, 2014).

## Acknowledgements

The strains K1041, NUB3621 and NRRL1174 shown in Fig. 1 were kindly given by I. Narumi (Japan Atomic Energy Research Institute, Takasaki, Japan), N. Welker (Northwestern University, Evanston, USA) and H. Liao (Cangene, Ontario, Canada) respectively.

## Conflict of interest

None declared.

## References

Banat, I.M., Marchant, R., and Rahman, T.J. (2004) *Geobacillus debilis* sp. nov., a novel obligately thermophilic bacterium isolated from a cool soil environment, and reassignment of *Bacillus pallidus* to *Geobacillus pallidus* comb. nov. *Int J Syst Evol Microbiol* **54:** 2197–2201.

Bartosiak-Jentys, J., Hussein, A.H., Lewis, C.J., and Leak, D.J. (2013) Modular system for assessment of glycosyl hydrolase secretion in *Geobacillus thermoglucosidasius*. *Microbiology* **159:** 1267–1275.

Begley, M., Cotter, P.D., Hill, C., and Ross, R.P. (2009) Identification of a novel two-peptide lantibiotic, lichenicidin, following rational genome mining for LanM proteins. *Appl Environ Microbiol* **75:** 5451–5460.

Bhalla, A., Kainth, A.S., and Sani, R.K. (2013) Draft genome sequence of lignocellulose-degrading thermophilic bacterium *Geobacillus* sp. strain WSUCF1. *Genome Announc* **1:** pii: e00595-13. doi:10.1128/genomeA.00595-13.

Bhalla, A., Bischoff, K.M., Uppugundla, N., Balan, V., and Sani, R.K. (2014) Novel thermostable endo-xylanase cloned and expressed from bacterium *Geobacillus* sp. WSUCF1. *Bioresour Technol* **165:** 314–318.

Blanchard, K., Robic, S., and Matsumura, I. (2014) Transformable facultative thermophile *Geobacillus stearothermophilus* NUB3621 as a host strain for metabolic engineering. *Appl Microbiol Biotechnol* **98:** 6715–6723.

Boonmak, C., Takahasi, Y., and Morikawa, M. (2013) Draft genome sequence of *Geobacillus thermoleovorans* strain B23. *Genome Announc* **1:** pii: e00944-13. doi:10.1128/genomeA.00944-13.

Boonmak, C., Takahashi, Y., and Morikawa, M. (2014) Cloning and expression of three *ladA*-type alkane monooxygenase genes from an extremely thermophilic alkane-degrading bacterium *Geobacillus thermoleovorans* B23. *Extremophiles* **18:** 515–523.

Bryant, D., and Moulton, V. (2004) Neighbor-net: an agglomerative method for the construction of phylogenetic networks. *Mol Biol Evol* **21:** 255–265.

Burgess, S.A., Lindsay, D., and Flint, S.H. (2010) Thermophilic bacilli and their importance in dairy processing. *Int J Food Microbiol* **144:** 215–225.

Chen, X.-G., Stabnikova, O., Tay, J.-H., Wang, J.-Y., and Tay, S.T.-L. (2004) Thermoactive extracellular proteases of *Geobacillus caldoproteolyticus*, sp. nov., from sewage sludge. *Extremophiles* **8:** 489–498.

Chen, Z.F., Wojcik, S.F., and Welker, N.E. (1986) Genetic analysis of *Bacillus stearothermophilus* by protoplast fusion. *J Bacteriol* **165:** 994–1001.

Coorevits, A., Dinsdale, A.E., Halket, G., Lebbe, L., De Vos, P., Van Landschoot, A., and Logan, N.A. (2012) Taxonomic revision of the genus *Geobacillus*: emendation of *Geobacillus*, *G. stearothermophilus*, *G. jurassicus*, *G. toebii*, *G. thermodenitrificans* and *G. thermoglucosidans* (*nom. corrig.*, formerly '*thermoglucosidasius*'); transfer of *Bacillus thermantarcticus*. *Int J Syst Evol Microbiol* **62:** 1470–1485.

Couñago, R., and Shamoo, Y. (2005) Gene replacement of adenylate kinase in the gram-positive thermophile *Geobacillus stearothermophilus* disrupts adenine nucleotide homeostasis and reduces cell viability. *Extremophiles* **9:** 135–144.

Cripps, R.E., Eley, K., Leak, D.J., Rudd, B., Taylor, M., Todd, M., *et al.* (2009) Metabolic engineering of *Geobacillus thermoglucosidasius* for high yield ethanol production. *Metab Eng* **11:** 398–408.

De Maayer, P., Williamson, C.E., Vennard, C.T., Danson, M.J., and Cowan, D.A. (2014) Draft genome sequences of *Geobacillus* sp. strains CAMR5420 and CAMR12739. *Genome Announc* **2:** pii: e00567-14. doi:10.1128/genomeA.00567-14.

Delcher, A.L., Phillippy, A., Carlton, J., and Salzberg, S.L. (2002) Fast algorithms for large-scale genome alignment and comparison. *Nucleic Acids Res* **30:** 2478–2483.

Dinsdale, A.E., Halket, G., Coorevits, A., Van Landschoot, A., Busse, H.-J., De Vos, P., and Logan, N.A. (2011) Emended descriptions of *Geobacillus thermoleovorans* and *Geobacillus thermocatenulatus*. *Int J Syst Evol Microbiol* **61:** 1802–1810.

Doi, K., Mori, K., Martono, H., Nagayoshi, Y., Fujino, Y., Tashiro, K., *et al.* (2013) Draft genome sequence of *Geobacillus kaustophilus* GBlys, a lysogenic strain with bacteriophage OH2. *Genome Announc* **1:** pii: e00634-13. doi:10.1128/genomeA.00634-13.

Feng, L., Wang, W., Cheng, J., Ren, Y., Zhao, G., Gao, C., *et al.* (2007) Genome and proteome of long-chain alkane degrading *Geobacillus thermodenitrificans* NG80-2 isolated from a deep-subsurface oil reservoir. *Proc Natl Acad Sci USA* **104:** 5602–5607.

Fridjonsson, O., Watzlawick, H., Gehweiler, A., and Mattes, R. (1999) Thermostable alpha-galactosidase from *Bacillus stearothermophilus* NUB3621: cloning, sequencing and characterization. *FEMS Microbiol Lett* **176:** 147–153.

Garg, N., Tang, W., Goto, Y., Nair, S.K., and van der Donk, W.A. (2012) Lantibiotics from *Geobacillus thermodenitrificans*. *Proc Natl Acad Sci USA* **109:** 5241–5246.

Haki, G.D., and Rakshit, S.K. (2003) Developments in industrially important thermostable enzymes: a review. *Bioresour Technol* **89:** 17–34.

Huson, D.H. (1998) SplitsTree: analyzing and visualizing evolutionary data. *Bioinformatics* **14:** 68–73.

Imanaka, T., Fujii, M., Aramori, I., and Aiba, S. (1982) Transformation of *Bacillus stearothermophilus* with plasmid DNA and characterization of shuttle vector plasmids between *Bacillus stearothermophilus* and *Bacillus subtilis*. *J Bacteriol* **149:** 824–830.

Li, Y., Wu, J., Wang, W., Ding, P., and Feng, L. (2012) Proteomics analysis of aromatic catabolic pathways in thermophilic *Geobacillus thermodenitrificans* NG80-2. *J Proteomics* **75:** 1201–1210.

Liao, H., McKenzie, T., and Hageman, R. (1986) Isolation of a thermostable enzyme variant by cloning and selection in a thermophile. *Proc Natl Acad Sci USA* **83:** 576–580.

Marchant, R., Sharkey, F.H., Banat, I.M., Rahman, T.J., and Perfumo, A. (2006) The degradation of n-hexadecane in soil by thermophilic geobacilli. *FEMS Microbiol Ecol* **56:** 44–54.

Marks, T.J., and Hamilton, P.T. (2014) Characterization of a thermophilic bacteriophage of *Geobacillus kaustophilus*. *Arch Virol*. doi:10.1007/s00705-014-2101-8.

Mead, D.A., Lucas, S., Copeland, A., Lapidus, A., Cheng, J.-F., Bruce, D.C., *et al.* (2012) Complete genome sequence of *Paenibacillus* strain Y4.12MC10, a novel *Paenibacillus lautus* strain isolated from Obsidian hot spring in Yellowstone National Park. *Stand Genomic Sci* **6:** 381–400.

Muhd Sakaff, M.K.L., Abdul Rahman, A.Y., Saito, J.A., Hou, S., and Alam, M. (2012) Complete genome sequence of the thermophilic bacterium *Geobacillus thermoleovorans* CCB_US3_UF5. *J Bacteriol* **194:** 1239.

Mukerjee-Dhar, G., Shimura, M., Miyazawa, D., Kimbara, K., and Hatta, T. (2005) *bph* genes of the thermophilic PCB degrader, *Bacillus* sp. JF8: characterization of the divergent ring-hydroxylating dioxygenase and hydrolase genes upstream of the Mn-dependent BphC. *Microbiology* **151:** 4139–4151.

Narumi, I., Sawakami, K., Nakamoto, S., Nakayama, N., Yanagisawa, T., Takahashi, N., and Kihara, H. (1992) A newly isolated *Bacillus stearotheromophilus K1041* and its transformation by electroporation. *Biotechnol Tech* **6:** 83–86.

Nazina, T.N., Tourova, T.P., Poltaraus, A.B., Novikova, E., V, Grigoryan, A.A., Ivanova, A.E., *et al.* (2001) Taxonomic study of aerobic thermophilic bacilli: descriptions of *Geobacillus subterraneus* gen. nov., sp. nov. and *Geobacillus uzenensis* sp. nov. from petroleum reservoirs and transfer of *Bacillus stearothermophilus*, *Bacillus thermocatenulatus*, *Bacillus thermoleovorans*, *Bacillus kaustophilus*, *Bacillus thermodenitrificans* to *Geobacillus* as the new combinations *G. stearothermophilus*, *G. th. Int J Syst Evol Microbiol* **51:** 433–446.

Nicholson, W.L. (2003) Using thermal inactivation kinetics to calculate the probability of extreme spore longevity: implications for paleomicrobiology and lithopanspermia. *Orig Life Evol Biosph* **33:** 621–631.

Niehaus, F., Bertoldo, C., Kähler, M., and Antranikian, G. (1999) Extremophiles as a source of novel enzymes for industrial application. *Appl Microbiol Biotechnol* **51:** 711–729.

Pore, S.D., Arora, P., and Dhakephalkar, P.K. (2014) Draft genome sequence of *Geobacillus* sp. strain FW23, isolated from a formation water sample. *Genome Announc* **2:** pii: e00352-14. doi:10.1128/genomeA.00352-14.

R Development Core Team, R. (2013) R: a language and environment for statistical computing. *R Found Stat Comput* **1:** 409.

Sandalli, C., Singh, K., Modak, M.J., Ketkar, A., Canakci, S., Demir, I., and Belduz, A.O. (2009) A new DNA polymerase I from *Geobacillus caldoxylosilyticus* TK4: cloning, characterization, and mutational analysis of two aromatic residues. *Appl Microbiol Biotechnol* **84:** 105–117.

Schmidt-Dannert, C., Pleiss, J., and Schmid, R.D. (1998) A toolbox of recombinant lipases for industrial applications. *Ann N Y Acad Sci* **864:** 14–22.

Seo, M.-J., Lee, B.-S., Pyun, Y.-R., and Park, H. (2011) Isolation and characterization of N-acylhomoserine lactonase from the thermophilic bacterium, *Geobacillus caldoxylosilyticus* YS-8. *Biosci Biotechnol Biochem* **75:** 1789–1795.

Shintani, M., Ohtsubo, Y., Fukuda, K., Hosoyama, A., Ohji, S., Yamazoe, A., *et al.* (2014) Complete genome sequence of the thermophilic polychlorinated biphenyl degrader *Geobacillus* sp. strain JF8 (NBRC 109937). *Genome Announc* **2:** e01213–13.

Siddiqui, M.A., Rashid, N., Ayyampalayam, S., and Whitman, W.B. (2014) Draft genome sequence of *Geobacillus thermopakistaniensis* strain MAS1. *Genome Announc* **2:** pii: e00559-14. doi:10.1128/genomeA.00559-14.

Studholme, D.J. (1998) Metabolic engineering of thermophilic Bacillus species for ethanol production. PhD Thesis. London: Deparment of Biochemistry, Imperial College.

Studholme, D.J., Jackson, R.A., and Leak, D.J. (1999) Phylogenetic analysis of transformable strains of thermophilic *Bacillus* species. *FEMS Microbiol Lett* **172:** 85–90.

Suzuki, H., Okazaki, F., Kondo, A., and Yoshida, K. (2013) Genome mining and motif modifications of glycoside hydrolase family 1 members encoded by *Geobacillus kaustophilus* HTA426 provide thermostable 6-phospho-β-glycosidase and β-fucosidase. *Appl Microbiol Biotechnol* **97:** 2929–2938.

Tabachnikov, O., and Shoham, Y. (2013) Functional characterization of the galactan utilization system of *Geobacillus stearothermophilus*. *FEBS J* **280:** 950–964.

Takami, H., Takaki, Y., Chee, G.-J., Nishi, S., Shimamura, S., Suzuki, H., *et al.* (2004) Thermoadaptation trait revealed by the genome sequence of thermophilic *Geobacillus kaustophilus*. *Nucleic Acids Res* **32:** 6292–6303.

Tamura, K., Stecher, G., Peterson, D., Filipski, A., and Kumar, S. (2013) MEGA6: Molecular Evolutionary Genetics Analysis version 6.0. *Mol Biol Evol* **30:** 2725–2729.

Taylor, M.P., Eley, K.L., Martin, S., Tuffin, M.I., Burton, S.G., and Cowan, D.A. (2009) Thermophilic ethanologenesis: future prospects for second-generation bioethanol production. *Trends Biotechnol* **27:** 398–405.

Taylor, M.P., Zyl, L., van Tuffin, I.M., Leak, D.J., and Cowan, D.A. (2011) Genetic tool development underpins recent advances in thermophilic whole-cell biocatalysts. *Microb Biotechnol* **4:** 438–448.

Thompson, A.H., Studholme, D.J., Green, E.M., and Leak, D.J. (2008) Heterologous expression of pyruvate decarboxylase in *Geobacillus thermoglucosidasius*. *Biotechnol Lett* **30:** 1359–1365.

Vallier, H., and Welker, N.E. (1990) Genetic map of the *Bacillus stearothermophilus* NUB36 chromosome. *J Bacteriol* **172:** 793–801.

Wiegand, S., Rabausch, U., Chow, J., Daniel, R., Streit, W.R., and Liesegang, H. (2013) Complete genome sequence of *Geobacillus* sp. strain GHH01, a thermophilic lipase-secreting bacterium. *Genome Announc* **1:** e0009213. doi:10.1128/genomeA.00092-13.

Wiegel, J., Ljungdahl, L.G., and Demain, A.L. (1985) The importance of thermophilic bacteria in biotechnology. *Crit Rev Biotechnol* **3:** 39–108.

Wu, L., and Welker, N.E. (1991) Temperature-induced prote in synthesis in *Bacillus stearothermophilus* NUB36. *J Bacteriol* **173:** 4889–4892.

Wu, L.J., and Welker, N.E. (1989) Protoplast transformation of *Bacillus stearothermophilus* NUB36 by plasmid DNA. *J Gen Microbiol* **135:** 1315–1324.

Xu, K., Su, F., Tao, F., Li, C., Ni, J., and Xu, P. (2013) Genome sequences of two morphologically distinct and thermophilic *Bacillus coagulans* strains, H-1 and XZL9. *Genome Announc* **1:** 4563–4564.

Yao, N., Ren, Y., and Wang, W. (2013) Genome sequence of a thermophilic *Bacillus*, *Geobacillus thermodenitrificans* DSM465. *Genome Announc* **1:** pii: e01046-13. doi:10.1128/genomeA.01046-13.

Zeigler, D.R. (2005) Application of a recN sequence similarity analysis to the identification of species within the bacterial genus *Geobacillus*. *Int J Syst Evol Microbiol* **55:** 1171–1179.

Zeigler, D.R. (2014) The *Geobacillus* paradox: why is a thermophilic bacterial genus so prevalent on a mesophilic planet? *Microbiology* **160:** 1–11.

Zhao, Y., Caspers, M.P., Abee, T., Siezen, R.J., and Kort, R. (2012) Complete genome sequence of *Geobacillus* *thermoglucosidans* TNO-09.020, a thermophilic spore-former associated with a dairy-processing environment. *J Bacteriol* **194:** 4118.

# Permissions

# List of Contributors

**Shlomo Sela (Saldinger)**
Department of Food Quality and Safety, Institute for Postharvest and Food Sciences, Bet Dagan, Israel

**Shulamit Manulis-Sasson**
Department of Plant Pathology and Weed Research, Agricultural Research Organization (ARO), The Volcani Center, Bet-Dagan, Israel

**Ingemar Nærdal**
Sector for Biotechnology and Nanomedicine, Department of Molecular Biology, SINTEF Materials and Chemistry, Trondheim, Norway

**Johannes Pfeifenschneider**
Genetics of Prokaryotes, Faculty of Biology & CeBiTec, Bielefeld University, Bielefeld, Germany

**Trygve Brautaset**
Sector for Biotechnology and Nanomedicine, Department of Molecular Biology, SINTEF Materials and Chemistry, Trondheim, Norway
Department of Biotechnology, Norwegian University of Science and Technology, Trondheim, Norway

**Volker F. Wendisch**
Genetics of Prokaryotes, Faculty of Biology & CeBiTec, Bielefeld University, Bielefeld, Germany

**Gabriele Berg**
Institute of Environmental Biotechnology, Graz University of Technology & ACIB Austrian Centre of Industrial Biotechnology, Graz, Austria

**Alexis Broquet**
Faculté de Medicine, Laboratoire UPRES EA 3826, Université de Nantes, Nantes, France

**Karim Asehnoune**
Faculté de Medicine, Laboratoire UPRES EA 3826, Université de Nantes, Nantes, France
Pôle Anesthésie Réanimations, Service d'Anesthésie Réanimation Chirurgicale, Hôtel Dieu, CHU Nantes, Nantes, France

**Maike Kortmann**
Institute of Bio- and Geosciences, IBG-1: Biotechnology, Forschungszentrum Jülich, Jülich D-52425, Germany

**Vanessa Kuhl**
Institute of Bio- and Geosciences, IBG-1: Biotechnology, Forschungszentrum Jülich, Jülich D-52425, Germany

**Simon Klaffl**
Institute of Bio- and Geosciences, IBG-1: Biotechnology, Forschungszentrum Jülich, Jülich D-52425, Germany

**Michael Bott**
Institute of Bio- and Geosciences, IBG-1: Biotechnology, Forschungszentrum Jülich, Jülich D-52425, Germany

**Panagiota-Myrsini Chronopoulou**
School of Biological Sciences, University of Essex, Wivenhoe Park, Colchester CO4 3SQ, UK

**Gbemisola O.Sanni**
School of Biological Sciences, University of Essex, Wivenhoe Park, Colchester CO4 3SQ, UK

**Daniel I. Silas-Olu**
School of Biological Sciences, University of Essex, Wivenhoe Park, Colchester CO4 3SQ, UK

**Jan Roelof van derMeer**
Department of Fundamental Microbiology, University of Lausanne, Lausanne, Switzerland

**Kenneth N. Timmis**
Institute of Microbiology, Technical University Braunschweig, Braunschweig, Germany

**Corina P. D. Brussaard**
Department of Biological Oceanography, Royal Netherlands Institute for Sea Research (NIOZ), Den Burg, The Netherlands

**Terry J. McGenity**
School of Biological Sciences, University of Essex, Wivenhoe Park, Colchester CO4 3SQ, UK

**Beatriz Mesa-Pereira**
Centro Andaluz de Biología del Desarrollo, CSIC, Junta de Andalucía, Universidad Pablo de Olavide, Carretera de Utrera, Km. 1, Seville, 41013, Spain

**Carlos Medina**
Centro Andaluz de Biología del Desarrollo, CSIC, Junta de Andalucía, Universidad Pablo de Olavide, Carretera de Utrera, Km. 1, Seville, 41013, Spain

**Eva MaríaCamacho**
Centro Andaluz de Biología del Desarrollo, CSIC, Junta de Andalucía, Universidad Pablo de Olavide, Carretera de Utrera, Km. 1, Seville, 41013, Spain

**Amando Flores**
Centro Andaluz de Biología del Desarrollo, CSIC, Junta de Andalucía, Universidad Pablo de Olavide, Carretera de Utrera, Km. 1, Seville, 41013, Spain

**Eduardo Santero**
Centro Andaluz de Biología del Desarrollo, CSIC, Junta de Andalucía, Universidad Pablo de Olavide, Carretera de Utrera, Km. 1, Seville, 41013, Spain

**Susanne Melzer**
LIFE – Leipzig Research Center for Civilization Diseases Martin-Luther University Halle-Wittenberg, Halle-Wittenberg, Germany
Department of Pediatric Cardiology, Heart Center Leipzig Martin-Luther University Halle-Wittenberg, Halle-Wittenberg, Germany

**Gudrun Winter**
Department of Biology: Physiology and Biochemistry of Plants, University of Konstanz, Konstanz Martin-Luther University Halle-Wittenberg, Halle-Wittenberg, Germany

**Kathrin Jäger**
Interdisciplinary Center for Clinical Research, Core Unit Fluorescence-Technology Martin-Luther University Halle-Wittenberg, Halle-Wittenberg, Germany

**Thomas Hübschmann**
Department of Environmental Microbiology, Helmholtz Center for Environmental Research, Leipzig Martin-Luther University Halle-Wittenberg, Halle-Wittenberg, Germany

**Gerd Hause**
Biocenter Martin-Luther University Halle-Wittenberg, Halle-Wittenberg, Germany

**FrankSyrowatka**
Interdisciplinary Center for Material Sciences, Martin-Luther University Halle-Wittenberg, Halle-Wittenberg, Germany

**Hauke Harms**
Department of Environmental Microbiology, Helmholtz Center for Environmental Research, Leipzig Martin-Luther University Halle-Wittenberg, Halle-Wittenberg, Germany

**Attila Tárnok**
Department of Pediatric Cardiology, Heart Center Leipzig Martin-Luther University Halle-Wittenberg, Halle-Wittenberg, Germany
Translational Center for Regenerative Medicine, University of Leipzig Martin-Luther University Halle-Wittenberg, Halle-Wittenberg, Germany

**Susann Müller**
Department of Environmental Microbiology, Helmholtz Center for Environmental Research, Leipzig Martin-Luther University Halle-Wittenberg, Halle-Wittenberg, Germany

**A. Oslizlo**
Department of Food Science and Technology, Biotechnical Faculty, University of Ljubljana, Ljubljana, Slovenia

**P. Stefanic**
Department of Food Science and Technology, Biotechnical Faculty, University of Ljubljana, Ljubljana, Slovenia

**S. Vatovec**
Department of Food Science and Technology, Biotechnical Faculty, University of Ljubljana, Ljubljana, Slovenia

**S.Beigot Glaser**
National Laboratory for Health, Environment and Food, Maribor, Slovenia

**M. Rupnik**
National Laboratory for Health, Environment and Food, Maribor, Slovenia
Faculty of Medicine, University of Maribor, Maribor, Slovenia
Centre of Excellence for Integrated Approaches in Chemistry and Biology of Proteins, Ljubljana, Slovenia

**I. Mandic-Mulec**
Department of Food Science and Technology, Biotechnical Faculty, University of Ljubljana, Ljubljana, Slovenia

**Shaman Narayanasamy**
Luxembourg Centre for Systems Biomedicine, University of Luxembourg, 7 avenue des Hauts-Fourneaux, Esch-Sur-Alzette L-4362, Luxembourg

**Emilie E. L. Muller**
Luxembourg Centre for Systems Biomedicine, University of Luxembourg, 7 avenue des Hauts-Fourneaux, Esch-Sur-Alzette L-4362, Luxembourg

**Abdul R. Sheik**
Luxembourg Centre for Systems Biomedicine, University of Luxembourg, 7 avenue des Hauts-Fourneaux, Esch-Sur-Alzette L-4362, Luxembourg

**Paul Wilmes**
Luxembourg Centre for Systems Biomedicine, University of Luxembourg, 7 avenue des Hauts-Fourneaux, Esch-Sur-Alzette L-4362, Luxembourg

**Thomas Schwartz**
Institute of Functional Interfaces (IFG) and Institute of Toxicology and Genetics (ITG), Campus

North, Karlsruhe Institute of Technology (KIT), Hermann von Helmholtz Platz 1, Eggenstein-Leopoldshafen D-76344, Germany

**Olivier Armant**
Institute of Toxicology and Genetics (ITG), Campus North, Karlsruhe Institute of Technology (KIT), Hermann von Helmholtz Platz 1, Eggenstein-Leopoldshafen D-76344, Germany

**NancyBretschneider**
Genomatix GmbH, Bayerstr. 85a, Munich D-80335, Germany

**Alexander Hahn**
Genomatix GmbH, Bayerstr. 85a, Munich D-80335, Germany

**Silke Kirchen**
Institute of Functional Interfaces (IFG) and Institute of Toxicology and Genetics (ITG), Campus
North, Karlsruhe Institute of Technology (KIT), Hermann von Helmholtz Platz 1, Eggenstein-Leopoldshafen D-76344, Germany

**Martin Seifert**
Genomatix GmbH, Bayerstr. 85a, Munich D-80335, Germany

**Andreas Dötsch**
Institute of Functional Interfaces (IFG) and Institute of Toxicology and Genetics (ITG), Campus
North, Karlsruhe Institute of Technology (KIT), Hermann von Helmholtz Platz 1, Eggenstein-Leopoldshafen D-76344, Germany

**Karen Roos**
Bacterial Vaccines and Immune Sera, Department of Veterinary Medicine, Paul Ehrlich Institute, Langen 63225, Germany

**Esther Werner**
Bacterial Vaccines and Immune Sera, Department of Veterinary Medicine, Paul Ehrlich Institute, Langen 63225, Germany

**Holger Loessner**
Bacterial Vaccines and Immune Sera, Department of Veterinary Medicine, Paul Ehrlich Institute, Langen 63225, Germany

**Ryan M. Summers**
Department of Chemical and Biological Engineering, The University of Alabama, Tuscaloosa, AL 35487, USA

**Sujit K. Mohanty**
Department of Chemical and Biochemical Engineering The University of Iowa, Coralville, IA 52241, USA

**SridharGopishetty**
Center for Biocatalysis and Bioprocessing, The University of Iowa, Coralville, IA 52241, USA

**Mani Subramanian**
Department of Chemical and Biochemical Engineering and Center for Biocatalysis and Bioprocessing, The University of Iowa, Coralville, IA 52241, USA
Center for Biocatalysis and Bioprocessing, The University of Iowa, Coralville, IA 52241, USA

**Sebastien Terrat**
INRA, UMR1347 Agroécologie, Plateforme GenoSol, Dijon, France

**Pierre Plassart**
INRA, UMR1347 Agroécologie, Plateforme GenoSol, Dijon, France

**EmilieBourgeois**
INRA, UMR1347 Agroécologie, Dijon, France

**Stéphanie Ferreira**
Equipe R&D Santé-Environnement, Campus Pasteur, Genoscreen, Lille, France

**Samuel Dequiedt**
INRA, UMR1347 Agroécologie, Plateforme GenoSol, Dijon, France

**Nathalie Adele-Dit-De-Renseville**
Equipe R&D Santé-Environnement, Campus Pasteur, Genoscreen, Lille, France

**PhilippeLemanceau**
INRA, UMR1347 Agroécologie, Dijon, France

**Antonio Bispo**
Service Agriculture et Forêt, ADEME, Angers Cedex 01, France

**Abad Chabbi**
INRA-URP3F, Lusignan, France

**Pierre-Alain Maron**
INRA, UMR1347 Agroécologie, Plateforme GenoSol, Dijon, France
INRA, UMR1347 Agroécologie, Dijon, France

**Lionel Ranjard**
INRA, UMR1347 Agroécologie, Plateforme GenoSol, Dijon, France
INRA, UMR1347 Agroécologie, Dijon, France

**Lv-Hui Sun**
Department of Animal Nutrition and Feed Science, College of Animal Science and Technology, Huazhong Agricultural University, Wuhan, Hubei 430070, China

**Ni-Ya Zhang**
Department of Animal Nutrition and Feed Science, College of Animal Science and Technology, Huazhong Agricultural University, Wuhan, Hubei 430070, China

**Ran-Ran Sun**
Department of Animal Nutrition and Feed Science, College of Animal Science and Technology, Huazhong Agricultural University, Wuhan, Hubei 430070, China

**XinGao**
Department of Animal Nutrition and Feed Science, College of Animal Science and Technology, Huazhong Agricultural University, Wuhan, Hubei 430070, China

**Changqin Gu**
Department of Animal Nutrition and Feed Science, College of Animal Science and Technology, Huazhong Agricultural University, Wuhan, Hubei 430070, China

**Christopher Steven Krumm**
Department of Animal Science, Cornell University, Ithaca, NY 14853, USA

**De-Sheng Qi**
Department of Animal Nutrition and Feed Science, College of Animal Science and Technology, Huazhong Agricultural University, Wuhan, Hubei 430070, China

**Gaoge Xu**
College of Plant Protection, Nanjing Agricultural University, China/Key Laboratory of Integrated Management of Crop Diseases and Pests (Nanjing Agricultural University), Ministry of education, Nanjing 210095, China

**Yuxin Zhao**
College of Plant Protection, Nanjing Agricultural University, China/Key Laboratory of Integrated Management of Crop Diseases and Pests (Nanjing Agricultural University), Ministry of education, Nanjing 210095, China

**Liangcheng Du**
Department of Chemistry, University of Nebraska-Lincoln, Lincoln, NE 68588, USA

**Guoliang Qian**
College of Plant Protection, Nanjing Agricultural University, China/Key Laboratory of Integrated Management of Crop Diseases and Pests (Nanjing Agricultural University), Ministry of education, Nanjing 210095, China

**Fengquan Liu**
College of Plant Protection, Nanjing Agricultural University, China/Key Laboratory of Integrated Management of Crop Diseases and Pests (Nanjing Agricultural University), Ministry of education, Nanjing 210095, China
Institute of Plant Protection, Jiangsu Academy of Agricultural Science, Nanjing 210014, China

**Michael Vogt**
Institute of Bio- and Geosciences, IBG-1: Biotechnology, Forschungszentrum Jülich, D-52425 Jülich, Germany

**Sabine Haas**
Institute of Bio- and Geosciences, IBG-1: Biotechnology, Forschungszentrum Jülich, D-52425 Jülich, Germany

**Tino Polen**
Institute of Bio- and Geosciences, IBG-1: Biotechnology, Forschungszentrum Jülich, D-52425 Jülich, Germany

**Jan van Ooyen**
Institute of Bio- and Geosciences, IBG-1: Biotechnology, Forschungszentrum Jülich, D-52425 Jülich, Germany

**Michael Bott**
Institute of Bio- and Geosciences, IBG-1: Biotechnology, Forschungszentrum Jülich, D-52425 Jülich, Germany

**Shiwei Zhang**
College of Life Science, University of Chinese Academy of Sciences, No.19A, Yuquan Road, Shijingshan District, Beijing 100049, China

**Ru Bai**
College of Life Science, University of Chinese Academy of Sciences, No.19A, Yuquan Road, Shijingshan District, Beijing 100049, China

**Run Feng**
College of Life Science, University of Chinese Academy of Sciences, No.19A, Yuquan Road, Shijingshan District, Beijing 100049, China

**Hongxun Zhang**
College of Life Science, University of Chinese Academy of Sciences, No.19A, Yuquan Road, Shijingshan District, Beijing 100049, China

**Lixin Liu**
College of Life Science, University of Chinese Academy of Sciences, No.19A, Yuquan Road, Shijingshan District, Beijing 100049, China
Key Laboratory for Polymeric Composite and Functional Materials of Ministry of Education, School of Chemistry and Chemical Engineering, Sun Yat-Sen University, No. 135, Xingang Xi Road, Guangzhou 510275, China

State Environmental Protection Key Laboratory of Microorganism Application and Risk Control (SMARC), Tsinghua University, Beijing 100084, China

**XinYu Zhao**
College of Water Sciences, Beijing Normal University, Beijing 100875, China

**Zimin Wei**
Life Science College, Northeast Agricultural University, Harbin 150030, China

**Yue Zhao**
Life Science College, Northeast Agricultural University, Harbin 150030, China

**Beidou Xi**
State Key Laboratory of Environmental Criteria and Risk Assessment, Chinese Research Academy of Environmental Sciences, Beijing 100012, China

**Xueqin Wang**
Life Science College, Northeast Agricultural University, Harbin 150030, China

**Taozhi Zhao**
Life Science College, Northeast Agricultural University, Harbin 150030, China

**Xu Zhang**
Life Science College, Northeast Agricultural University, Harbin 150030, China

**Yuquan Wei**
Life Science College, Northeast Agricultural University, Harbin 150030, China

**Paloma Pizarro-Tobías**
Bio-Ilíberis R&D, Polígono Industrial Juncaril, Peligros, Granada 18210, Spain

**Matilde Fernández**
Estación Experimental del Zaidín-CSIC, Granada, Granada 18008, Spain

**JoséLuis Niqui**
Bio-Ilíberis R&D, Polígono Industrial Juncaril, Peligros, Granada 18210, Spain

**Jennifer Solano**
Bio-Ilíberis R&D, Polígono Industrial Juncaril, Peligros, Granada 18210, Spain

**Estrella Duque**
Estación Experimental del Zaidín-CSIC, Granada, Granada 18008, Spain

**Juan-Luis Ramos**
Estación Experimental del Zaidín-CSIC, Granada, Granada 18008, Spain

**Amalia Roca**
Bio-Ilíberis R&D, Polígono Industrial Juncaril, Peligros, Granada 18210, Spain

**David J. Studholme**
Biosciences, University of Exeter, Geoffrey Pope Building, Stocker Road, Exeter EX4 4QD, UK